Energy Systems, Devices and Applications

Energy Systems, Devices and Applications

Edited by John Mosley

SYRAWOOD
PUBLISHING HOUSE

New York

Published by Syrawood Publishing House,
750 Third Avenue, 9th Floor,
New York, NY 10017, USA
www.syrawoodpublishinghouse.com

Energy Systems, Devices and Applications
Edited by John Mosley

© 2018 Syrawood Publishing House

International Standard Book Number: 978-1-68286-611-5 (Hardback)

Cataloging-in-Publication Data

Energy systems, devices and applications / edited by John Mosley.
    p. cm.
Includes bibliographical references and index.
ISBN 978-1-68286-611-5
1. Power resources. 2. Energy facilities. 3. Energy conservation--Equipment and supplies. I. Mosley, John.
TJ163.24 .E53 2018
333.79--dc23

TABLE OF CONTENTS

PREFACE

Every book is initially just a concept; it takes months of research and hard work to give it the final shape in which the readers receive it. In its early stages, this book also went through rigorous reviewing. The notable contributions made by experts from across the globe were first molded into patterned chapters and then arranged in a sensibly sequential manner to bring out the best results.

This book elucidates the concepts and innovative models around prospective developments with respect to energy. It presents the importance and use of energy in the present environment. Energy in all its forms helps life, be it in kinetic form or thermal, electrical, potential, gravitational or even magnetic form. The aim of this book is to provide readers with a thorough understanding of this vast subject. It is a compilation of chapters that discuss the most vital concepts and emerging trends in the field of energy. Those with an interest in this field would find the book helpful.

It has been my immense pleasure to be a part of this project and to contribute my years of learning in such a meaningful form. I would like to take this opportunity to thank all the people who have been associated with the completion of this book at any step.

Editor

The demand side in economic models of energy markets: the challenge of representing consumer behavior

Frank C. Krysiak and Hannes Weigt*

Department of Business and Economics, University of Basel, Basel, Switzerland

Energy models play an increasing role in the ongoing energy transition processes either as tools for forecasting potential developments or for assessments of policy and market design options. In recent years, these models have increased in scope and scale and provide a reasonable representation of the energy supply side, technological aspects and general macroeconomic interactions. However, the representation of the demand side and consumer behavior has remained rather simplistic. The objective of this paper is twofold. First, we review existing large-scale energy model approaches, namely bottom-up and top-down models, with respect to their demand-side representation. Second, we identify gaps in existing approaches and draft potential pathways to account for a more detailed demand-side and behavior representation in energy modeling.

Keywords: energy modeling, bottom-up, top-down, demand side, consumer behavior

Edited by:
Tobias Brosch,
University of Geneva, Switzerland

Reviewed by:
Jay Zarnikau,
The University of Texas, USA
Sonia Yeh,
University of California Davis, USA

*Correspondence:
Frank C. Krysiak,
Department of Business and
Economics, University of Basel, Peter
Merian-Weg 6, Basel CH-4002,
Switzerland
frank.krysiak@unibas.ch

Specialty section:
This article was submitted to Energy
Systems and Policy, a section of the
journal Frontiers in Energy Research

Citation:
Krysiak FC and Weigt H (2015) The
demand side in economic models of
energy markets: the challenge of
representing consumer behavior.
Front. Energy Res. 3:24.

Introduction

Reducing energy demand, or at least its growth, is one of the central objectives in the transition processes in many national and international energy markets. For example, the European vision of a low-carbon economy identifies energy efficiency as a key driver of the transition (European Commission, 2014), the Swiss Energy Strategy 2050 aims for a significant reduction of *per capita* energy consumption of 54% by 2050 (SFOE, 2012), and the IEA's World Energy Outlook considers a reduction in energy consumption as one of the main measures to achieve a significant reduction in CO2 emissions (IEA, 2014a).

Despite this importance of the energy demand side, there still exist significant knowledge gaps as to what factors determine energy demand and how it can be influenced. Besides descriptive statistics on specific energy consumption patterns and profiles and the technological linkage between service demand and energy needs (for example, different options to satisfy transport or heating demand), little is known about the underlying decision and behavioral processes. The fact that consumers seldom demand energy in itself but services and products which require energy for their provision links this challenge to a general understanding of consumption decision processes.

The energy demand aspects extend into the modeling dimension. In the last decades, energy system and market modeling has gained an increasingly important role within the policy process; i.e., forecasts based on models like the IEA World Energy Outlook using the World Energy Model (IEA, 2014b) or the Energy Trends of the European Commission based on the PRIMES model (European Commission, 2013) are important resources for economic and political decision makers. Model-based scenarios also form the basis of energy market processes like the network development planning in Germany and Switzerland (SFOE, 2013; NEP, 2014). Finally, models are used for *ex post* policy evaluation and are, apart from field experiments, the only way to gain knowledge about the necessary intensity of policy interventions. The existing energy system and market models were designed with a focus on the different technology options on the supply and transport side whereas demand was often assumed to be derived from external drivers like GDP or following classic price and substitution elasticities. Thus, they are limited in their capability to capture important psychological or social elements and aspects beyond the technology or price dimension.

The objective of this paper is to assess the role of the demand side and consumer behavior within economic energy market modeling, identify gaps in existing approaches, and design potential pathways to account for a more detailed demand side and behavior representation in energy modeling. In Section "Review on the Demand-Side Representation in Energy Market Models", we review existing model approaches for energy markets used for policy design and evaluation with special focus on their demand-side representation. Section "Energy Demand: Toward Richer Models" provides concepts to extend the existing models to facilitate a more detailed description of energy demand. In Section "Transferring New Approaches into Numerical Modeling", we discuss how these concepts can be used in numerical modeling and Section "Conclusion" concludes.

Review on the Demand-Side Representation in Energy Market Models

There exists a multitude of modeling approaches for energy-related questions. Within this section, we focus on large-scale models covering markets, sectors, or the whole energy system and economy.[1] Generally, those types of models can roughly be

[1] Small-scale analyses focusing on single processes or regional aspects, e.g., like micro-grids or single building optimizations, can address detailed demand aspects but naturally cannot capture general policy or market interactions.

clustered in two streams: bottom-up (BU) and top-down (TD) models that are both able to address a specific range of relevant drivers (**Figure 1**). The former cover techno-economic models that provide a detailed representation of technical aspects of a market or energy sectors, like conversion or transport specifications, as well as microeconomic market representations that address the interaction of different market participants, like producers, traders, and consumers within wholesale markets. The latter cover macroeconomic models that are able to capture the interaction among several sectors and overall welfare effects. A related differentiation between the two clusters would be the terms "disaggregated" for BU models and "aggregated" for TD models (Böhringer and Rutherford, 2009).

Usually, the strengths of one model cluster are the weaknesses of the other. BU models allow a detailed representation of specific market characteristics, the impact of policies on a sector, and the costs and challenges of technological change. However, their focus on a single sector or a set of (energy) sectors limits the possibility to capture further cross-sectoral effects, the price driven influences are often limited to cost optimization, and they omit overall economy impacts like employment, trade, and income effects. Consequently for TD models the reverse is true (Herbst et al., 2012).

Following Hourcade et al. (2006) energy models can be structured along three dimensions: the technological, microeconomic, and macroeconomic detail. Generally, applied energy models are tailored to capture a specific dimension and have to omit other aspects. The first dimension represents the technological explicitness of a model including their ability to capture technological restrictions and how policies affect technological developments. The second refers to behavioral realism of the model including the representation of consumer choice and the impact of market structures on policy effectiveness. The third refers to macroeconomic feedbacks linking energy supply and demand to the general economic structure and development.

Bottom-up models typically rank high with respect to technological details and allow the modeling of different

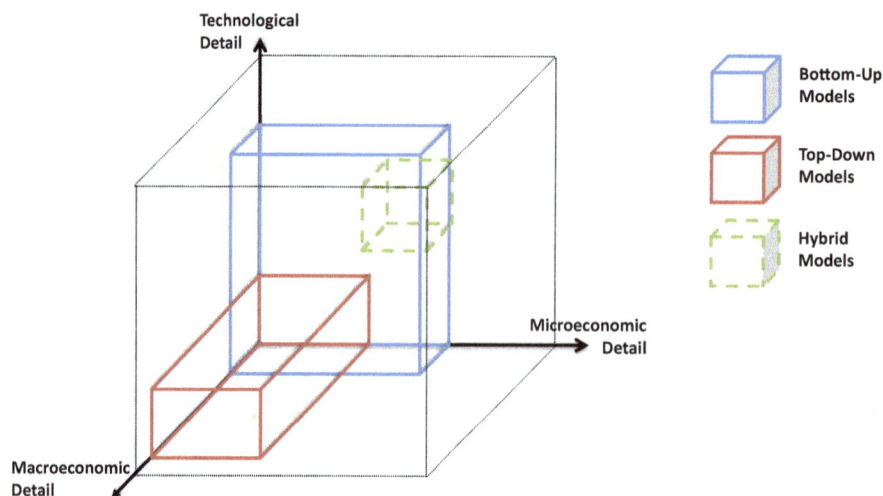

FIGURE 1 | Top-down and bottom-up model dimensions.

technology developments. Similarly, they allow the inclusion of microeconomic aspects like strategic company behavior or game theoretic approaches. However, typically BU model approaches that rank high on the microeconomic dimension need to rely on mixed complementarity formulations. This in turn limits the representation of technological details that require mixed-integer formulations like unit commitment of power plants. The opposite also holds. In contrast, TD models capture the macroeconomic interactions of economies and are based on the basic microeconomic rational of utility maximizing agents but lack technological detail and typically also fall short in addressing more detailed microeconomic behavior beyond perfect competitive market interaction.

Theoretically, a hybrid model capturing all three dimensions would provide the most structure for evaluating energy markets (Hourcade et al., 2006), but potentially at the expense of model focus and transparency of results. In recent years, research on models breaching the gaps between TD and BU has increased with several approaches of hybrid modeling emerging. In the following sections, we briefly present selected BU and TD models and methods as well as hybrid and other model approaches focusing on their representation of the energy demand side.

Bottom-Up Techno-Economic Models

Bottom-up models are characterized by their high degree of detail on the technology side and their representation of market structure and market architecture aspects. They are basically disaggregated representations of specific sectors or markets and therefore have to omit the more general economic interactions. They can be used both for short-term evaluations, like electricity market dispatch analysis, and long-term simulations, like investment scenarios. However, BU models rely on a set of externally defined parameters, which capture those economic aspects that are not covered by the model-like economic growth and demand or fuel prices of energy sectors. Defining these parameters is challenging, in particular for long-term evaluations, but BU models are one of the few options to simulate the impact of future conditions that deviate considerably from historic or current market conditions.

Bottom-up models are typically formulated as optimization problem or complementarity problem.[2] Especially techno-economic models that focus on supply and transport restrictions or operational details often rely on linear optimization techniques.

The large-scale energy system models are the IEA World Energy Model (IEA, 2014b), the PRIMES model (E3MLab/ICCS, 2014), the POLES model (Enerdata, 2014), and the MARKAL/TIMES family of models (ETSAP, 2014), which are covering several regions and sectors in a partial equilibrium setup. Typically, such large-scale energy system models consist of several modules or sub-models covering specific regions, sectors, or value-chain elements, which are linked via iterative simulations. Due to their long-term perspective, they capture investment

decisions but neglect short-term dynamics. In recent years, those model families already started to breach the gaps between BU and TD modeling by integrating more macroeconomic aspects into their models.

Sectoral models focus on one specific fuel and the underlying markets. Consequently, there exist a large number of different models for the specific energy markets; for example, for electricity markets, the ELMOD model (Leuthold et al., 2012) or the DIMENSION model (Richter, 2011) for Europe and RFF's Haiku model for the US (Paul et al., 2009), the World Gas Model (Egging et al., 2010) or the COLUMBUS model (Hecking and Panke, 2012) for the global gas markets, the TREMOVE model for the transport sector (Capros and Siskos, 2012), or residential stock models for the building sector [see Kavgic et al. (2010) for a review], to name a few. The focus on a specific market or sector enables those models to capture much more details, such as network restrictions (e.g., pipeline or transmission line capacities), specific technology restrictions (e.g., power plant start-up times), and detailed regional and temporal resolutions (e.g., daily or seasonal demand patterns), than the large-scale energy system models mentioned above. Their specific model setup varies strongly and is often tailored to the specific research question at hand; that is, short-term technical questions, long-term investment aspects, or market design and strategic interaction, and consequently includes linear, non-linear, and equilibrium approaches.

In general, BU models are well suited to evaluate changes and impacts on the supply and transport side of energy markets. They can capture a wide range of different production restrictions and facilitate a corresponding detailed evaluation of policies. But their mathematical structure limits the representation of demand-side behavior. There are roughly speaking two main types of BU models with respect to price and demand-side representation, both being widely applied:

First, techno-economic BU models designed as *linear or linear mixed-integer optimization* problems. They are required to take the demand side as a fixed input and thus cannot capture price or budget feedbacks. Changes in the demand side can only be incorporated as shifts of the load level, for example, via a new hourly demand profile, due to demand-side management technologies; an increasing demand level, due to economic growth; or different demand scenarios based on energy efficiency assumptions. Furthermore, the linear structure leads to a classical cost optimal result that corresponds to a perfect competitive market framework, whereas imperfect competition cannot easily be captured within this model framework.

Second, BU models designed as *complementarity problems or non-linear optimization problems* incorporate demand-side functionalities, typically a relation between demand and price. Non-linear optimization problems can include welfare maximization instead of a pure cost minimization as the objective. This captures the price interaction but still keeps the models limited to perfect competitive benchmark outcomes. In addition, BU models using the equilibrium framework allow the representation of multiple agents with individual optimization rationales and thereby facilitate the simulation of strategic firm behavior, imperfect competition, or the impact of structural changes. Similar to the linear type BU models, the demand functionalities need to be

---

[2]Note that also TD models are formulated as complementarity problems. The differentiation between the BU and TD complementarity problems is in their coverage: while TD models are formulated as general equilibrium covering the whole economy with capital and labor effects, BU models only capture a subset of sectors or only a single sector and therefore are also termed partial equilibrium models.

externally defined, especially regarding the price elasticities. Consequently, general economic interrelations, such as budget effects or substitution-effects across markets, cannot be captured directly. However, the endogenous price formation makes it possible to cover direct price-quantity effects within the respective sector.

Top-Down CGE Models

Top-down models aim at representing the whole economy instead of only energy sectors and thereby capture the feedback effects across the economy. This modeling approach requires a high degree of aggregation and cannot represent the same technological detail as BU models. The most prominent macroeconomic model approach in energy economics are computable general equilibrium (CGE) models. Those models have a highly aggregated representation of the energy system and the other sectors of an economy. The equilibrium concept ensures that all modeled markets clear (supply equals demand on each market), given supply and demand characteristics. This equilibrium is obtained by endogenous price adjustments following the microeconomic rational of utility maximizing agents and profit maximizing firms. However, the agents in CGE models are highly aggregated; most often, a representative household is used. Due to their aggregation level and equilibrium concept, CGE models are well suited for long-term evaluations of changes in the policy or market frame and not for short-term operational simulations.

Due to the high abstraction level of CGE models, the production technology process is transferred into production functions with constant elasticities of substitution (ESUB). The different inputs and outputs are linked via nesting structures; that is, the energy input into a production function is itself an aggregate of different energy types, like electricity and fossil fuels that can be substituted for each other. As these elasticities determine the degree of substitution between inputs, they are thereby an important driver of the effects of policy changes. To capture the effect of technological change in the energy sector – basically a shift of the production functions – exogenous shift parameters are often used, like the autonomous energy efficiency index (AEEI). The AEEI represents a price-independent energy efficiency increase, which is sometimes used to carry out sensitivity analyses.

The same logic is applied to the demand side of CGE models. **Figure 2** shows an exemplary demand-side structure for

a CGE model with detailed energy specifications. Demand is derived from maximizing the utility function of a representative household, given a budget restriction. Consumption and other "goods," such as leisure time, form the aggregated utility good. Consumption itself is split into direct energy use and consumption of other goods. The energy use in turn can be satisfied by different fuels which can be substituted for each other given the elasticity σ^E; that is, switching from oil to electricity for heating. The energy needed for the other goods, the embedded energy, is obtained by a similar structure on the production side. This allows CGE models to capture indirect energy effects due to changes in consumption.

There exist a large number of CGE models that address different economic aspects. Bergman (2005) provides a general introduction to CGE models and a review on different environmental- and resource-related CGE models. Those models can be (broadly) clustered into global, multi-regional, and single-country CGEs. Within the first group, examples are given by the MIT-EPPA model (Paltsev et al., 2005) and the DICE and RICE model family (Nordhaus, 2012). The GEM-E3 model of the European Commission (Capros et al., 2013), the related GEMINI-E3 model (Bernard and Vielle, 2008), and the PACE modeling framework (Böhringer et al., 2009) are examples for energy-related multi-regional models. Finally, the GENESwIS for Switzerland (Vöhringer, 2012) and the MIT U.S. Regional Energy Policy Model (Lanz and Rausch, 2011) are examples for single-country CGEs.

Top-down computable general equilibrium models are well suited to capture price-based demand side effects across different sectors via budget effects. This is particularly important for the estimation of rebound effects that result from such indirect effects. They also are well suited for public finance evaluations of taxes and other instruments. However, the underlying parameters for the different ESUBs and the AEEI are typically based on estimates and expert judgments (Bataille et al., 2006). This poses two challenges: first, data and estimations on both ESUBs, and particular AEEI, are incomplete, and second, estimates based on past and present data do not necessarily have to be an accurate description of future behavior making TD models less suited for the analysis of extensive system shifts in comparison to BU models.

Hybrid Models and Other Model Approaches

Due to the limitations of both BU and TD approaches, researchers are developing methods to merge both lines of models for policy analyses. The resulting hybrid models can be clustered into three categories (Böhringer and Rutherford, 2008): First, soft-linked models, in which independent BU and TD models are linked by passing data between the models or via direct convergence mechanisms, as, for example, in Schäefer and Jacoby (2006), who link the MIT-EPPA CGE model with the MARKAL model. This approach faces the challenge of consistency of the disaggregated and aggregated results; that is, the electricity generation of different power plant types of a BU electricity model run need to match the aggregated fuel consumption of the electricity sector in the TD model. Second, a reduced form version of one model is incorporated into another model, as, for example, in Bosetti et al. (2006) or Leimbach et al. (2009). Third, integrating technological details directly via the mixed complementarity problem formulation of

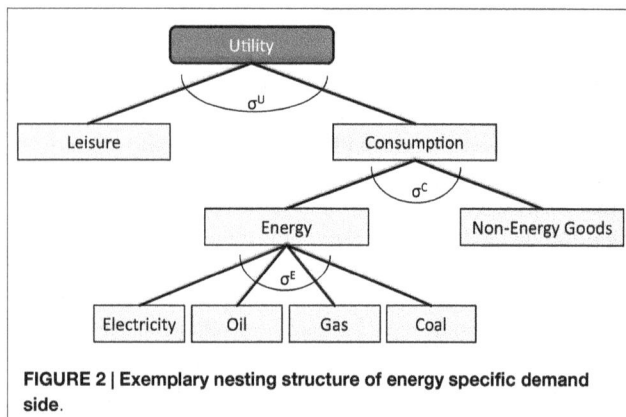

FIGURE 2 | Exemplary nesting structure of energy specific demand side.

a CGE model, see, for example, Böhringer and Rutherford (2008, 2009).

From a demand perspective, hybrid approaches facilitate the combination of detailed sectoral effects, like shifts in demand profiles, with general macroeconomic feedbacks, such as indirect rebound effects. This is of particular relevance for energy efficiency evaluations. Furthermore, the potential to model higher temporal resolutions in BU approaches makes it possible to combine short- and long-term economic feedbacks. Nevertheless, the demand representation is still limited to the above presented characteristics and focused on quantity-price relations and/or externally defined levels and trends.

In addition to CGE models and optimization and partial equilibrium BU models, there are a number of additional model approaches in energy economics [see Catenazzi (2009) and Herbst et al. (2012)]. These include input–output models, system dynamics approaches, and econometric models. The latter often include multiple consumer groups [i.e., the E3ME model has 13 types of household, Camecon (2014)]. But due to their reliance on historic data, they are not well suited to analyze significant system shifts. On the BU side, there are furthermore simulation models and agent-based models. The former are often more technology driven and can represent whole energy systems with great detail; see, for example, the LEAP model (Heaps, 2012). The latter results from a relatively new model approach in energy economics [e.g., see Weidlich and Veit (2008) for a review of electricity market related agent-based models]. Instead of a closed mathematical market formulation, individual market participants are modeled as agents with autonomous behavior that interact with each other. This makes it possible to model different behavior of the market participants and thereby capture choice related aspects.

Summarizing the different existing energy model approaches, we see that they are typically designed to capture supply side related market aspects while demand-side aspects are much less detailed. This is partly a result of the underlying computational structure but also a result of the historic market development; for a long time, electricity and natural gas systems were regulated markets in which cost optimal energy supply was the main focus. Furthermore, most of the recent energy-related developments took place on the supply side, such as, the emergence of renewable energy technologies.

It is thus not surprising that existing models typically lack endogenous demand-side influences beside price-quantity relations. Furthermore, most models treat the demand side as an aggregate with little detail on specific consumer aspects and differentiated consumers.[3]

Despite these problems, existing models are well suited to analyze small, price-induced changes on the demand side as well as the effects of pre-defined (scenario-based) changes to energy consumption on markets and energy supply. However, with an increasing focus on energy efficiency and the liberalization of former monopolistic markets, the demand side will become increasingly important: policies directly aimed at end users will increase, companies will need to compete for consumers with better products or services, and finally consumer will also become active market participants providing their own energy supply and storage potential as 'prosumers'. In particular, it will be necessary to develop models that capture consumer choices with respect to energy provision and that can describe the relation between changes in individual behavior and demand-side policies.

Energy Demand: Toward Richer Models

As discussed above, most applied economic models describe energy demand as being a function of prices and income only. From a theoretical perspective, this is warranted by the basic microeconomic model of consumer choice, where an individual maximizes her utility $U(e, x)$ over a bundle of energy goods e and other goods x subject to the condition that total expenditure does not exceed income y for a given vector of energy prices z and other prices p:

$$\max_{e,x \geq 0} U(e, x),$$
$$s.t.\ z\,e + p\,x \leq y.$$

This results in a demand function for energy $e = f(z, p, y)$. Under conventional assumptions, demand for each energy good is a decreasing function of this good's price and an increasing or a decreasing function of the prices of other goods, depending on substitution possibilities. Typically, the above consumer is used in the sense of a (descriptive) representative consumer, that is, the consumer is used as an "average" of all consumers, so that the characteristics of aggregate demand (over all consumers) are identical to the characteristics of this consumer's energy demand. This approach forms the basis of most CGE models, whereas BU models rely on further simplifications.

The above basic setup is useful to describe the response of energy demand to price changes, in particular, the effects of changes in energy markets or of some policy instruments, such as energy taxation. In fact, numerous studies have assessed the price responsiveness of demand for different energy goods, see, for example, Filippini (2011) or Krishnamurthy and Kriström (2015). Furthermore, it can be used to examine simple indirect phenomena, like the above discussed rebound effects.

However, to assess other types of demand-side policies or more general effects, the model lacks structure. A simple but powerful extension is to consider heterogeneous consumers, for example, groups of consumers that differ regarding their income or preferences. Such an extension makes it possible to assess the distributive impacts of energy policies. Furthermore, such a model can be used to assess potential benefits of group-specific interventions.

But even with this extension, the model does not capture many effects that have been found to be relevant in field studies.[4]

---

[3]Note that especially energy system models often rely on different demand modules (e.g., one for transport demand, one for heating demand etc.) and combine/aggregate detailed consumer information to derive those modules. Many demand aspects are therefore part of the parameterization and not endogenous model aspects.

[4]For a review of energy-related intervention studies, see, e.g., Abrahamse et al. (2005).

Most importantly, the above model assumes that consumers are perfectly aware of all actions to reduce their energy consumption, so that information-based policies are ineffective by definition, and that there are no interactions among consumers, apart from market interactions. Furthermore, preferences are considered as being given and constant, so that there is no leeway for changes to individual lifestyles that are not "forced" by changing conditions (such as, prices or income).

In the next subsections, we will discuss how the above model can be adjusted in simple ways to capture the potential relevance of information, social interactions, and changing preferences.

Modeling the Influence of Information on Energy Demand

To make room for potential effects of information-based approaches to steer energy demand, a necessary assumption is that consumers are not perfectly aware of all options for changing their energy demand. For example, they might not know which energy-efficient appliances exist, what quality and prices they have, and where they can be bought. Thus, if they want to change their behavior, they need to *search* for new solutions. There is a long tradition of search models in economics, with applications mostly to labor markets, explaining price dispersion, and innovation. Chandra and Tappata (2011) use such a model to explain differences in gasoline prices among stations; Kortum (1997) as well as Makri and Lane (2007) use a search model to explain how firms find new technological solutions.

To transfer the main insights of these models to individual energy demand, it is useful to assume that consumers need to invest in appliances (some goods x, in our above notation) to alter their ability of adjusting energy use e. However, they are not aware of the properties of the relevant goods x and thus need to spent time or money searching for an appliance that meets their requirements. From a modeling perspective, we could assume that consumers know a distribution of possible characteristics of appliances, that is, they know which qualities, costs, usage characteristics, and energy reductions are technically feasible. However, without gathering information, they do not know which appliance has which properties.

Thus consumers can either buy an appliance without this information or invest time (modeled via fixed opportunity costs S) to ascertain the characteristics of one appliance (they randomly draw an appliance from the overall distribution and learn its properties). If they invest in this search, they can afterward decide to buy this good or to research another one. This decision will be made based on the overall distribution of possible characteristics, that is, on their knowledge what is feasible; whenever the good comes sufficiently close to having the preferred characteristics among all feasible goods, a consumer will not invest in a new search (the probability of finding a better solution is too small) and rather buy this good.

Such a model is able to describe some interesting effects. First, changing energy consumption induces one-time costs (search costs). Thus potential gains in energy efficiency will only be reaped, if these gains compensate for the search costs, in other words, small changes to energy prices will have little, but somewhat larger changes might have substantial effects. Furthermore,

the model explains why different consumers will resort to different solutions in the short run (and thus explain technological variety, e.g., different alternatives to conventional light bulbs) but might converge to similar solutions later on, when they observe the choices of others. Finally, and most importantly, the model can describe an impact of information-based policies. Such policies would lead to a reduction of search costs, implying an earlier start of the search process and thus making it easier to reap small gains in energy efficiency.

However, as preferences remain unchanged, the model also highlights an important constraint of information-based policies: Such policies only reduce frictions, and they do not alter a consumer's overall assessment of whether it is beneficial to reduce his energy consumption. Thus in the context of this framework, information-based policies will be ineffective; if consumers do not reduce their energy consumption, because the individual gains (savings from using less energy) do not cover the individual costs (in terms of expenses or reduced quality of life).

Social Interactions and Social Norms

A different way of influencing individual behavior is to provide information about the behavior of others or (implicit) information about social norms regarding energy consumption. This approach has been found to be effective in a number of studies. For example, Allcott (2011) shows in a large-scale field study with 600,000 households that using such non-price instruments can have similar short-run effects on total energy consumption as an 11–20% increase in energy prices.

Again, there is some tradition in other fields of economics of modeling social norms. A convenient approach is to include a "disutility" of not meeting a social norm in the description of individual behavior, see, for example, Lindbeck et al. (1999), where such an approach is used in the context of social security. Other contexts where this modeling approach is used are the explanation of tipping behavior, see, for example, Azar (2004), and green consumption, as in Nyborg et al. (2006).

In a general framework, this can be modeled by a slight extension of the above basic model. To this end, assume that the utility of individual i (out of n individuals) depends not only on her consumption (e_i, x_i) but also on a social norm N:

$$\max_{e,x \geq 0} U_i(e_i, x_i, N),$$
$$s.t. \ z \, e_i + p \, x_i \leq y_i.$$

The social norm is in turn a result of the behavior of all individuals in the society (which might, however, have different influence on norm formation):

$$N = g(e_1, x_1, e_2, x_2, \ldots, e_n, x_n).$$

In such a model, changes in individual behavior can result in adjustments of social norms, which in turn will lead to further changes in individual behavior.[5] This approach thus introduces

---

[5]To ensure that this process converges, it can be useful to assume that the second (norm-induced) effect on individual behavior is always smaller than the original change in behavior.

a feedback effect in the basic model. Furthermore, it is possible that there are several equilibria, for example, an equilibrium with low and one with high energy consumption, which are each stabilized via the endogenously formed social norm (Lindbeck et al., 1999).

An information-based policy could be described either as manipulating the social norm or as making people more aware of an existing norm. In the first case, it might be possible to suggest that the norm is low energy consumption, which could move the system to an equilibrium with lower energy consumption (if multiple equilibria exist). In the second case, the policy could increase the disutility from not being close to the norm, which would induce both a direct change in behavior and an according adjustment of the norm. Such increases in disutility could be achieved by providing information about the behavior of other consumers. An example is given in Traxler (2010), who shows how changing the beliefs of tax payers regarding the incidence of tax evasion (and thus their disutility from not meeting a social norm) can change overall outcomes rather drastically.

A slightly more elaborate version of the above model would not use a single social norm but rather a set of group-specific norms, whose formation may be interrelated. This would facilitate the modeling of social interactions or peer pressure within groups.

However, a major problem is the quantification of the effects that social norms have on individual decisions. Some authors argue [see, e.g., Camerer and Fehr (2004) or Krupka and Weber (2013)] that laboratory experiments can be used to gain at least an approximate quantification. Others, such as Levitt and List (2007), are more critical and point out that questions regarding a limited transferability of experimental situations to every-day-behavior have particular relevance for the case of adherence to social norms. Field experiments provide another option, see, for example, Shang and Croson (2009), who study the influence of social information on public good provision. However, as field experiments are rather costly, this option usually implies a transfer across contexts and countries, as it is not possible to implement a field experiment in every situation where the influence of social norms on energy use needs to be assessed.

Modeling Changing Preferences and Sufficiency

Another, much discussed, approach toward reducing energy consumption is *sufficiency*. This term is used in the literature in different ways [see, Oikonomou et al. (2009) or Alcott (2008) for an overview]. Most importantly, sufficiency needs to be disentangled from efficiency, which is not trivial, as the economic concept of efficiency covers both changes in technology and changes in behavior.

Often, sufficiency is considered to be an enforced or voluntary frugal way of living (Oikonomou et al., 2009). In case of enforced frugality, this might imply reduced individual well-being. In contrast, if sufficiency is to be chosen voluntarily, an individual has to get a sufficient recompense for the reduced consumption, which might take the form of an increased self-esteem, utility from contributing to a socially desirable outcome, or an increase in leisure time (due to be able to cope with less income).

However, a salient question is if sufficiency gains exist, why have they not yet been fully reaped? Building on the concept of social norms discussed in the preceding subsection, one argument might be that different societal equilibria exist and individuals are "trapped" in a situation where the benefits of sufficiency cannot be reaped, because they depend on similar behavior by others. This would reduce the problem of modeling sufficiency to the cases discussed in the preceding subsection and interventions toward sufficiency would need to address social norms.

A different approach to sufficiency would be to remain on the individual level and to assume that individuals can only assess the quality of life in situations that they have already experienced. Thus they know how to live in the way they are currently living and how to react to small shocks. However, there might be different ways of living that reduce energy consumption without sacrificing well-being that the individual has not yet experienced and thus does not know.

In terms of modeling, we could assume that preferences consist of a set of local preferences (each defined in a neighborhood of a given consumption bundle) out of which an individual knows only one (her current) local preferences. The other preferences (i.e., ways of living) are not known to exist but their properties (how much utility can be gained, how goods can be substituted) are uncertain until this way of living has been tried. In such a setting, a risk-averse consumer would not alter his way of living until "forced" to do so (either by changing energy prices or by other interventions). Once a new way of living has been tried, the respective local preferences become known. If the driving force of the change vanishes (energy prices come down again), the consumer might either maintain this way of living or switch back to her original consumption pattern.

The benefit of this approach is that it captures much of the essence of the sufficiency concept and introduces an effect into the energy economic modeling that is not present so far: a one-time intervention can have lasting effects for some but not all parts of the population. For example, an oil price shock might initially increase the number of people not using cars. However, once oil price go down again, some consumers might switch back to their original way of living, whereas others have experienced a new and preferred lifestyle, which they voluntarily maintain.

However, it should be noted that if sufficiency gains are to be depicted in a model, this model cannot use *per capita* consumption, GDP, or total costs to assess demand-side policies. Rather, a measure of welfare has to be used that is based on individual utility and that captures either utility derived from adhering to social norms or the above mentioned uncertainty. Whereas this is common in theory, it is hard to implement in numerical models, as the necessary data is lacking.

Transferring New Approaches into Numerical Modeling

Obviously, existing numerical energy models will need adjustments and extensions to address the challenges in relation to increased energy efficiency and demand-side policies. For all changes, a necessary first step is the inclusion of heterogeneous consumers into the existing model structures. For CGE models, this basically refers to a more disaggregated structure on the demand side of the market transferring the oftentimes single representative household into several household types; for example, households representing different income classes that differ in their demand elasticities for specific goods.[6] This is less a modeling challenge, as the basic computational model structure remains unaltered, but more a question of data availability. Detailed data on different household types, their income, the split of income across sources, and consumption choices would be needed. For BU models, such a disaggregation is possible but will only result in a differently shaped aggregate demand function without much impact on overall computational model structure. Again, the main bottleneck for such a development is data availability like sufficient spatial or temporal resolution.

Building on this, it might be feasible to include richer models, such as those presented in Sections "Modeling the Influence of Information on Energy Demand," "Social Interactions and Social Norms," and "Modeling Changing Preferences and Sufficiency." Some extensions might be fairly easy to achieve, for example, the basic structure of norm-based interactions does not differ much conceptually from the inclusion of public knowledge on the production side in endogenous growth [see, for example, Bretschger and Suphaphiphat (2014)] and should thus be transferable to numerical CGE modeling. Including sufficiency or search processes would be much more difficult, as this requires the inclusion of uncertainty, which is hard to achieve in large-scale numerical models.

For BU models, a stepwise or time-dependent model structure as used in dynamic investment models, unit commitment models, or rolling planning models can be used as starting point. Within a period t the consumption decision is derived from externally defined parameters including, for example, norm driven aspects. The resulting consumption will then have an influence on the impact of norms in the following period $t+1$. Whether this influence is handled outside the model, that is, by adjusting the demand function accordingly, or within the model depends on the scope and structure of the model. The former should easily be accommodated by most BU model approaches, including linear optimization problem following a myopic logic. The latter introduces dynamic elements similar to path dependent investment aspects which increases the model complexity.

However, the proposed concepts require a quantification of their effects before they can be included into numerical models. Given our current knowledge on energy demand and particular on non-price driven influences this represents a significant non-modeling challenge. Consequently, to properly address those aspects in economic models we will first need a better understanding of the fundamental drivers of consumers energy demand.

Conclusion

Overall, this paper has two main messages. First, most of the currently available applied energy models do not use sophisticated approaches to describe the demand side. In fact, most models cannot describe or assess demand-side interventions apart from price changes. However, the second part shows that this is not a restriction imposed by the general economic approach to modeling consumer behavior. Much richer models are feasible and are used in other fields of economics. In particular, it is feasible to model many effects, such as social norm or social interactions that have been found to be relevant in field studies.

In our view, there are two reasons why these approaches are currently not used in energy modeling. First, there is a lack of demand. For decades, energy policy has focused on the supply side; whereas billions have been spent to enact changes in energy supply, demand-side policies have typically a small budget.[7] Accordingly, demand for policy assessments is biased toward supply side policies and thus most applied energy models have a highly detailed supply and a fairly simple demand structure.

Second, applied modeling requires not only concepts but also data. Whereas data on energy supply is abundant, there is a lack of data regarding the structure of energy consumption and its main determinants apart from prices and technologies. Few countries have a micro census that includes more than some elementary energy-related items, so that projects aiming for a better description of the demand side have to collect their own data. Given the different foci of such projects, there is little chance of combining their data to a sufficiently broad database.

As energy strategies in many countries are based on a strong reduction in *per capita* energy consumption, the first reason will vanish rather rapidly; the need for more qualified assessments of demand-side policies will strongly increase within the next years. However, the second bottleneck (missing data) will not dissolve in a likewise manner. Thus if better models of energy consumption are desirable, generating the necessary data should be the main priority.

The need for detailed data also extends to a more general lack of understanding the fundamental drivers and mechanisms of energy demand beyond the technological layer. Overcoming this knowledge gap will require fundamental research in social and political science as well as psychological and consumer behavior research and the transfer of those insights into the economic model community. How such an integrated interdisciplinary framework could be achieved is addressed in Burger et al. (2015) in the same issue of Frontiers in Energy Research.

Acknowledgment

We would like to thank Jan Abrell and Sebastian Rausch for helpful input. This research is part of the activities of SCCER CREST (Swiss Competence Center for Energy Research), which is financially supported by the Swiss Commission for Technology and Innovation (CTI).

---

[6] There are numerous examples of this approach in the context of social security evaluation and climate policy. For example, Nijkamp et al. (2005) use this approach to study inequality across countries under different international climate policy regimes and Yang (2010) uses different local households in a CGE model.

[7] A notable exception is policies targeting energy efficiency in residential buildings.

References

Abrahamse, W., Steg, L., Vlek, C., and Rothengatter, T. (2005). A review of intervention studies aimed at household energy conservation. *J. Environ. Psychol.* 25, 273–291. doi:10.1016/j.jenvp.2005.08.002

Alcott, B. (2008). The sufficiency strategy: would rich-world frugality lower environmental impact? *Ecol. Econ.* 64, 770–786. doi:10.1016/j.ecolecon.2007.04.015

Allcott, H. (2011). Social norms and energy conservation. *J. Public Econ.* 95, 1082–1095. doi:10.1016/j.jpubeco.2011.03.003

Azar, O. (2004). What sustains social norms and how they evolve? The case of tipping. *J. Econ. Behav. Organ.* 54, 49–64. doi:10.1016/j.jebo.2003.06.001

Bataille, C., Jaccard, M., Nyboer, J., and Rivers, N. (2006). Towards general equilibrium in a technology-rich model with empirically estimated behavioral parameters. *Energy J.* 27, 93–112.

Bergman, L. (2005). CGE modeling of environmental policy and resource management. *Handbook Environ. Econ.* 3, 1273–1306. doi:10.1016/S1574-0099(05)03024-X

Bernard, A., and Vielle, M. (2008). GEMINI-E3, a general equilibrium model of international–national interactions between economy, energy and the environment. *Comput. Manag. Sci.* 5, 173–206. doi:10.1007/s10287-007-0047-y

Böhringer, C., Löschel, A., and Rutherford, T. (2009). "Policy analysis based on computable equilibrium (pace)," in *Modelling Sustainable Development: Transitions to a Sustainable Future*, eds V. Bosetti, R. Gerlagh, and S. P. Schleicher (Cheltenham: Edward Elgar), 1648–1661.

Böhringer, C., and Rutherford, T. F. (2008). Combining bottom-up and top-down. *Energy Econ.* 30, 574–596. doi:10.1016/j.eneco.2007.03.004

Böhringer, C., and Rutherford, T. F. (2009). Integrated assessment of energy policies: decomposing top-down and bottom-up. *J. Econ. Dyn. Control* 33, 1648–1661. doi:10.1016/j.jedc.2008.12.007

Bosetti, V., Carraro, C., Galeotti, M., Massetti, E., and Tavoni, M. (2006). WITCH: a world induced technical change hybrid model. *Energy J.* 27, 13–37.

Bretschger, L., and Suphaphiphat, N. (2014). Effective climate policies in a dynamic North-South model. *Eur. Econ. Rev.* 69, 59–77. doi:10.1016/j.euroecorev.2013.08.002

Burger, P., Bezençon, V., Bornemann, B., Brosch, T., Carabias-Hütter, V., Farsi, M., et al. (2015). Advances in understanding energy consumption behavior and the governance of its change – outline of an integrated framework. *Front. Energy Res.*

Camecon. (2014). Available at http://www.camecon.com/EnergyEnvironment/EnergyEnvironmentEurope/ModellingCapability/E3ME.aspx

Camerer, C., and Fehr, E. (2004). "Measuring social norms and preferences using experimental games," in *A Guide for Social Sciences in Foundations of Human Society: Economic Experiments and Ethnographic Evidence from Fifteen Small-Scale Societies*, ed. J. Henrich (Oxford: Oxford University Press), 55–95.

Capros, P., and Siskos, P. (2012). *PRIMES-TREMOVE Transport Model v3 Model Description*. Athens: Energy-Economy-Environment Modeling Laboratory.

Capros, P., Van Regemorter, D., Paroussos, L., Karkatsoulis, P., Fragkiadakis, C., Tsani, S., et al. (2013). *GEM-E3 Model Documentation* (No. JRC83177). Seville: Institute for Prospective and Technological Studies, Joint Research Centre.

Catenazzi, G. (2009). *Advances in Techno-Economic Energy Modeling* (Doctoral Dissertation, ETH Zurich). Zurich: Swiss Federal Institute of Technology (ETH).

Chandra, A., and Tappata, M. (2011). Consumer search and dynamic price dispersion: an application to gasoline markets. *Rand J. Econ.* 42, 681–704. doi:10.1111/j.1756-2171.2011.00150.x

E3MLab/ICCS. (2014). *PRIMES MODEL 2013-2014 Detailed Model Description*. Athens: E3MLab/ICCS at National Technical University of Athens.

Egging, R., Holz, F., and Gabriel, S. A. (2010). The world gas model: a multi-period mixed complementary model for the global natural gas market. *Energy* 35, 4016–4029. doi:10.1016/j.energy.2010.03.053

Enerdata. (2014). Available at: http://www.enerdata.net/enerdatauk/solutions/energy-models/poles-model.php

ETSAP. (2014). Available at: http://www.iea-etsap.org/web/Times.asp

European Commission. (2013). *Trends to 2050: EU Energy, Transport and GHG Emissions Reference Scenario 2013*. Luxemburg: International Energy Agency.

European Commission. (2014). *Energy Efficiency and its Contribution to Energy Security and the 2030 Framework for Climate and Energy Policy*. Brussels: International Energy Agency.

Filippini, M. (2011). Short- and long-run time-of-use price elasticities in Swiss residential electricity demand. *Energy Policy* 39, 5811–5817. doi:10.1016/j.enpol.2011.06.002

Heaps, C. G. (2012). *Long-range Energy Alternatives Planning (LEAP) System. [Software version 2014.0.1.20]*. Somerville, MA: Stockholm Environment Institute. Available at: www.energycommunity.org

Hecking, H., and Panke, T. (2012). *COLUMBUS-A Global Gas Market Model* (No. 12/06). Cologne: EWI Working Paper.

Herbst, A., Toro, F., Reitze, F., and Jochem, E. (2012). Introduction to energy systems modelling. *Swiss J. Econ. Stat.* 148, 111–135.

Hourcade, J. C., Jaccard, M., Bataille, C., and Ghersi, F. (2006). Hybrid modeling: new answers to old challenges introduction to the special issue of the energy journal. *Energy J.* 27, 1–11.

IEA. (2014a). *World Energy Outlook 2014*. Paris: International Energy Agency.

IEA. (2014b). *World Energy Model Documentation 2014 Version*. Paris: International Energy Agency (IEA).

Kavgic, M., Mavrogianni, A., Mumovic, D., Summerfield, A., Stevanovic, Z., and Djurovic-Petrovic, M. (2010). A review of bottom-up building stock models for energy consumption in the residential sector. *Build. Environ.* 45, 1683–1697. doi:10.1016/j.buildenv.2010.01.021

Kortum, S. (1997). Research, patenting and technological change. *Econometrica* 65, 1389–1419. doi:10.2307/2171741

Krishnamurthy, B., and Kriström, B. (2015). A cross-country analysis of residential electricity demand in 11 OECD-countries. *Resource Energy Econ.* 39, 68–88. doi:10.1016/j.reseneeco.2014.12.002

Krupka, E., and Weber, R. (2013). Identifying social norms using coordination games: why does dictator game sharing vary? *J. Eur. Econ. Assoc.* 11, 495–524. doi:10.1111/jeea.12006

Lanz, B., and Rausch, S. (2011). General equilibrium, electricity generation technologies and the cost of carbon abatement: a structural sensitivity analysis. *Energy Econ.* 33, 1035–1047. doi:10.1016/j.eneco.2011.06.003

Leimbach, M., Bauer, N., Baumstark, L., and Edenhofer, O. (2009). Mitigation costs in a globalized world: climate policy analysis with REMIND-R. *Environ. Model. Assess.* 15, 155–173.

Leuthold, F. U., Weigt, H., and von Hirschhausen, C. (2012). A large-scale spatial optimization model of the European electricity market. *Network Spatial Econ.* 12, 75–107. doi:10.1007/s11067-010-9148-1

Levitt, S., and List, J. (2007). What do laboratory experiments measuring social preferences reveal about the real world? *J. Econ. Perspect.* 21, 153–174. doi:10.1257/jep.21.2.153

Lindbeck, A., Nyberg, S., and Weibull, J. W. (1999). Social norms and economic incentives in the welfare state. *Q. J. Econ.* 114, 1–35. doi:10.1162/003355399555936

Makri, M., and Lane, P. J. (2007). A search theoretic model of productivity, science and innovation. *RD Manag.* 37, 303–317. doi:10.1111/j.1467-9310.2007.00477.x

NEP. (2014). Available at: http://www.netzentwicklungsplan.de/

Nijkamp, P., Wang, S., and Kremers, H. (2005). Modeling the impacts of international climate change policies in a CGE context: the use of the GTAP-E model. *Econ. Model.* 22, 955–974. doi:10.1016/j.econmod.2005.06.001

Nordhaus, W. (2012). Available at: http://www.econ.yale.edu/~nordhaus/homepage/RICEmodels.htm

Nyborg, K., Howarth, R., and Brekke, K. (2006). Green consumers and public policy: on socially contingent moral motivation. *Resource Energy Econ.* 28, 351–366. doi:10.1016/j.reseneeco.2006.03.001

Oikonomou, V., Becchis, F., Stegc, L., and Russolillo, D. (2009). Energy saving and energy efficiency concepts for policy making. *Energy Policy* 37, 4787–4796. doi:10.1016/j.enpol.2009.06.035

Paltsev, S., Reilly, J. M., Jacoby, H. D., Eckaus, R. S., McFarland, J. R., Sarofim, M. C., et al. (2005). *The MIT Emissions Prediction and Policy Analysis (EPPA) Model: Version 4*. Cambridge: MIT Joint Program on the Science and Policy of Global Change.

Paul, A., Burtraw, D., and Palmer, K. (2009). Haiku documentation: RFF's electricity market model. *Resources Future* 52, 222.

Richter, J. (2011). *Dimension-A Dispatch and Investment Model for European Electricity Markets* (No. 11/03). Cologne: EWI Working Paper.

Schäefer, A., and Jacoby, H. D. (2006). Vehicle technology under CO 2 constraint: a general equilibrium analysis. *Energy Policy* 34, 975–985. doi:10.1016/j.enpol.2004.08.051

SFOE. (2012). *Consultation on the Energy Strategy 2050*. Berne: Swiss Federal Office of Energy.

SFOE. (2013). *Strategie Stromnetze; Detailkonzept im Rahmen der Energiestrategie 2050*. Berne: Swiss Federal Office of Energy.

Shang, J., and Croson, R. (2009). A field experiment in charitable contribution: the impact of social information on the voluntary provision of public goods. *Econ. J.* 119, 1422–1439. doi:10.1111/j.1468-0297.2009.02267.x

Traxler, C. (2010). Social norms and conditional cooperative taxpayers. *Eur. J. Polit. Econ.* 26, 89–103. doi:10.1016/j.ejpoleco.2009.11.001

Vöhringer, F. (2012). Linking the Swiss emissions trading system with the EU ETS: economic effects of regulatory design alternatives. *Swiss J. Econ. Stat.* 148, 167–196.

Weidlich, A., and Veit, D. (2008). A critical survey of agent-based wholesale electricity market models. *Energy Econ.* 30, 1728–1759. doi:10.1016/j.eneco.2008.01.003

Yang, H. (2010). Carbon-reducing taxes and income inequality: general equilibrium evaluation of alternative energy taxation in Taiwan. *Appl. Econ.* 32, 1213–1221. doi:10.1080/000368400404353

Conflict of Interest Statement: The authors declare that the research was conducted in the absence of any commercial or financial relationships that could be construed as a potential conflict of interest.

2

Development of Lithium-Stuffed Garnet-Type Oxide Solid Electrolytes with High Ionic Conductivity for Application to All-Solid-State Batteries

Ryoji Inada*, Satoshi Yasuda, Masaru Tojo, Keiji Tsuritani, Tomohiro Tojo and Yoji Sakurai

Department of Electrical and Electronic Engineering, Toyohashi University of Technology, Toyohashi, Japan

Edited by:
Jeff Sakamoto,
University of Michigan, USA

Reviewed by:
Hui Xia,
Nanjing University of Science and Technology, China
Candace K. Chan,
Arizona State University, USA

*Correspondence:
Ryoji Inada
inada@ee.tut.ac.jp

Specialty section:
This article was submitted to Energy Storage, a section of the journal Frontiers in Energy Research

Citation:
Inada R, Yasuda S, Tojo M, Tsuritani K, Tojo T and Sakurai Y (2016) Development of Lithium-Stuffed Garnet-Type Oxide Solid Electrolytes with High Ionic Conductivity for Application to All-Solid-State Batteries. Front. Energy Res. 4:28.

All-solid-state lithium-ion batteries are expected to be one of the next generations of energy storage devices because of their high energy density, high safety, and excellent cycle stability. Although oxide-based solid electrolyte (SE) materials have rather lower conductivity and poor deformability than sulfide-based ones, they have other advantages, such as their chemical stability and ease of handling. Among the various oxide-based SEs, lithium-stuffed garnet-type oxide, with the formula of $Li_7La_3Zr_2O_{12}$ (LLZ), has been widely studied because of its high conductivity above 10^{-4} S cm^{-1} at room temperature, excellent thermal performance, and stability against Li metal anode. Here, we present our recent progress for the development of garnet-type SEs with high conductivity by simultaneous substitution of Ta^{5+} into the Zr^{4+} site and Ba^{2+} into the La^{3+} site in LLZ. Li^{+} concentration was fixed to 6.5 per chemical formulae, so that the formula of our Li garnet-type oxide is expressed as $Li_{6.5}La_{3-x}Ba_xZr_{1.5-x}Ta_{0.5+x}O_{12}$ (LLBZT) and Ba contents x are changed from 0 to 0.3. As a result, all LLBZT samples have a cubic garnet structure without containing any secondary phases. The lattice parameters of LLBZT decrease with increasing Ba^{2+} contents $x \leq 0.10$ while increase with x from 0.10 to 0.30, possibly due to the simultaneous change of Ba^{2+} and Ta^{5+} substitution levels. The relative densities of LLBZT are in a range between 89 and 93% and are not influenced in any significant way by the compositions. From the AC impedance spectroscopy measurements, the total (bulk + grain) conductivity at 27°C of LLBZT shows its maximum value of 8.34×10^{-4} S cm^{-1} at $x = 0.10$, which is slightly higher than the conductivity ($= 7.94 \times 10^{-4}$ S cm^{-1}) of LLZT without substituting Ba ($x = 0$). The activation energy of the conductivity tends to become lower by Ba substation, while excess Ba substitution degrades the conductivity in LLBZT. LLBZT has a wide electrochemical potential window of 0–6 V vs. Li^{+}/Li, and Li^{+} insertion and extraction reactions of $TiNb_2O_7$ film electrode formed on LLBZT by aerosol deposition are demonstrated at 60°C. The results indicate that LLBZT can potentially be used as a SE in all-solid-state batteries.

Keywords: garnet-type oxide, solid electrolyte, ionic conductivity, all-solid-state battery, aerosol deposition

INTRODUCTION

Lithium-ion batteries (LIBs) consist of a graphite negative electrode, organic liquid electrolyte, and lithium transition-metal oxide. They were first commercialized in 1991, and since then such batteries have been widely distributed globally as a power source for mobile electronic devices, such as cell phones and laptop computers. Nowadays, large-scale LIBs have been developed for application to automotive propulsion and stationary load-leveling for intermittent power generation from solar or wind energy (Tarascon and Armand, 2001; Armand and Tarascon, 2008; Scrosati and Garche, 2010; Goodenough and Kim, 2011). However, increasing battery size creates more serious safety issues for LIBs; one reason being the increased amount of flammable organic liquid electrolytes.

All-solid-state LIBs are expected to be one of the next generations of energy storage devices because of their high energy density, high safety, and excellent cycle stability (Fergus, 2010; Takada, 2013; Tatsumisago et al., 2013). The materials used for solid electrolyte (SE) must have not only high lithium-ion conductivity above 10^{-3} S cm^{-1} at room temperature but also possess chemical stability against electrode materials, air, and moisture. Although oxide-based SE materials have rather lower conductivity and poor deformability compared to sulfide-based ones, they have other advantages such as their chemical stability and ease of handling (Knauth, 2009; Ren et al., 2015a). Furthermore, the formation of solid–solid interface with low resistance between SE and electrode is another challenging issue in order to achieve better electrochemical performance in solid-state batteries with an oxide-based SE.

Among the various oxide-based SE materials, lithium-stuffed garnet-type oxide with the formula of Li$_7$La$_3$Zr$_2$O$_{12}$ (LLZ) has been widely studied because of its high conductivity above 10^{-4} S cm^{-1} at room temperature, excellent thermal performance, and stability against Li metal anode (Murugan et al., 2007a). LLZ has two different crystal phases, one is the cubic phase (Awaka et al., 2011) and the other is tetragonal (Awaka et al., 2009a; Geiger et al., 2011), but high conductivity above 10^{-4} S cm^{-1} at room temperature is mostly confirmed in cubic phase sintered at high temperature (1100–1200°C) in Al$_2$O$_3$ crucible or with Al$_2$O$_3$ substitution (Murugan et al., 2007a; Jin and McGinn, 2011; Kotobuki et al., 2011; Kumazaki et al., 2011; Li et al., 2012a; Rangasamy et al., 2012). During high temperature sintering, Al^{3+} enters from the crucible and/or substituted Al$_2$O$_3$ into the LLZ pellet and works as sintering aid. In addition, it has been pointed out that some amount of Al^{3+} enters into the LLZ lattice, occupies the part of Li$^+$ site and modifies the Li$^+$ vacancy concentration for charge compensation, and stabilizes the cubic phase. Ga^{3+} has a similar effect as Al^{3+} for cubic phase stabilization and enhancement of ionic conducting properties (Wolfenstine et al., 2012; Shinawi and Janek, 2013; Bernuy-Lopez et al., 2014; Jalem et al., 2015).

Partial substitution of the Zr^{4+} site in LLZ by other higher valence cations, such as Nb^{5+} (Ohta et al., 2011; Kihira et al., 2013), Ta^{5+} (Buschmann et al., 2012; Li et al., 2012b; Logéat et al., 2012; Wang and Wei, 2012; Inada et al., 2014a; Thompson et al., 2014, 2015; Ren et al., 2015a,b), W^{6+} (Dhivya et al., 2013; Li et al., 2015), and Mo^{6+} (Bottke et al., 2015) is also reported to be effective in stabilizing the cubic garnet phase, and their conductivity at room temperature is greatly enhanced up to ~1 × 10^{-3} S cm^{-1} by controlling the contents of dopants and optimizing Li$^+$ concentration in the garnet framework. Although a demonstration of a solid-state battery with Nb-doped LLZ as SE has been already reported (Ohta et al., 2012, 2013, 2014), it has also been reported that the chemical stability against a Li metal electrode of Ta-doped LLZ is much superior to a Nb-doped one (Nemori et al., 2015). Partially substituted W^{6+} in LLZ could potentially become a redox center as well as Nb^{5+} at relatively high potential against Li$^+$/Li (Xie et al., 2012).

In this paper, we synthesized garnet-type SEs with simultaneous substitution of Ta^{5+} into the Zr^{4+} site and Ba^{2+} into the La^{3+} site in LLZ, by a conventional solid state reaction method. Li$^+$ concentration was fixed to 6.5 per chemical formulae to stabilize the cubic garnet phase, so that the chemical formula of our samples is expressed as Li$_{6.5}$La$_{3-x}$Ba$_x$Zr$_{1.5-x}$Ta$_{0.5+x}$O$_{12}$ (LLBZT) and Ba contents x are changed from 0 to 0.30. Many researchers have pointed out that the Li$^+$ conducting property in cubic garnet-type oxide is influenced not only by the Li$^+$ concentration but also by the lattice constant (Thangadurai and Weppner, 2005; Murugan et al., 2007b; Awaka et al., 2009b; Kihira et al., 2013; Thangadurai et al., 2014, 2015). Since the ionic radii of Ba^{2+} (142 pm) is greater than La^{3+} (116 pm), the lattice sizes of LLBZT with fixed Li$^+$ concentration per chemical formulae will be changed by the Ba substitution level, which will modify the ionic conducting property. Although it has been already confirmed that LLZ with substitution of both Ta and Ba can be fabricated (Tong et al., 2015), the influence of Ba substitution levels on the properties have not been fully discussed. In our study, the influence of Ba substitution level on the crystal phase, microstructure, and ionic conductivity for LLBZT was investigated. Furthermore, in order to investigate the feasibility for all-solid-state battery application, a TiNb$_2$O$_7$ film electrode was formed on garnet-type LLBZT without any thermal treatment by an aerosol deposition (AD) method (Akedo, 2006) and its electrochemical properties were evaluated.

MATERIALS AND METHODS

Synthesis and Characterization of LLBZT

LLBZT oxides with fixed Li content of 6.5 but different Ba content $x = 0$, 0.05, 0.1, 0.20, and 0.30 were prepared by a conventional solid state reaction method. It should be noted that, in this work, we did not use an Al$_2$O$_3$ crucible for sample synthesis because Al^{3+} contamination from the crucible into the LLBZT lattice during high temperature sintering may have an influence on Li$^+$ concentration and that this contamination level is very difficult to control. In addition, we had already succeeded in obtaining cubic garnet-type Ta-doped LLZ without containing Al by using a Pt–Au alloy crucible and sintering at 1100°C for 15 h. The highest room temperature conductivity attains to 6.1 × 10^{-4} S cm^{-1} at Li$^+$ concentration of 6.5 (Inada et al., 2014a), so we believe that Al^{3+} is not always required to stabilize the cubic phase and achieve high ionic conductivity.

Stoichiometric amounts of LiOH·H$_2$O (Kojundo chemical laboratory, 99%, 10% excess was added to account for the evaporation of lithium at high temperatures), La(OH)$_3$ (Kojundo chemical laboratory, 99.99%), BaCO$_3$ (Kojundo chemical laboratory, 99.9%), ZrO$_2$ (Kojundo chemical laboratory, 98%), and Ta$_2$O$_5$ (Kojundo chemical laboratory, 99.9%) were ground and mixed by planetary ball milling (Nagao System, Planet M2-3F) with zirconia balls and ethanol for 3 h and then calcined at 900°C for 6 h in a Pt-Au 5% alloy crucible. The calcined powders were ground again by planetary ball milling for 1 h, and then pressed into pellets at 300 MPa by cold isostatic pressing (CIP). Finally, all LLBZT pellets were sintered at 1150°C for 15 h in air using a Pt–Au 5% alloy crucible. During the sintering stage, the pellet was covered with the same mother powder to suppress excess Li loss and the formation of secondary phase, such as La$_2$Zr$_2$O$_7$.

As the characterization of obtained LLBZT samples, the crystal structure of the samples was evaluated by X-ray diffraction (XRD, Rigaku Multiflex) using CuKα radiation ($\lambda = 0.15418$ nm), with a measurement range of 2$\theta = 5$–90° and a step interval of 0.004°. Using the XRD data for all LLBZT samples, the lattice constants were calculated by Rigaku PDXL XRD analysis software. Scanning electron microscope (SEM) observation and energy dispersive X-ray (EDX) analysis were performed using a field-emission SEM (SU8000 Type II, Hitachi) to investigate the fractured surface microstructure of sintered LLBZT and the distribution of La, Ba, Zr, Ta, and O elements.

The conductivity of each LLBZT sample was evaluated with AC impedance measurements using a chemical impedance meter (3532-80, Hioki) and an LCR tester (3532-50, Hioki) at a temperature from 27 to 100°C, a frequency from 5 Hz to 5 MHz, and an applied voltage amplitude of 0.1 V. The former was used for measurement from 5 Hz to 1 MHz while the latter was used for measurement above 1 MHz, respectively. Both parallel surfaces of the pellet were sputtered with Li$^+$ blocking Au electrodes for the conductivity measurement. The electrochemical windows of the samples were evaluated by cyclic voltammetry (CV) using a potentio-galvanostat (VersaSTAT 3, Princeton Applied Research) at a scanning rate of 0.2 mV s^{-1} between −0.4 and 6 V vs. Li$^+$/Li. An Au electrode and lithium metal were attached to both faces of the pellet as working and counter electrodes, respectively.

TNO Film Electrode Fabrication on LLBZT and Its Characterization

In order to study the feasibility of LLBZT for application to solid state batteries, TiNb$_2$O$_7$ (TNO) film electrode was formed on LLBZT pellet by AD method (Akedo, 2006). AD has several advantages compared to other conventional thin film deposition methods, including the deposition of a crystallized film without any heat treatments and a fast deposition rate. A film is deposited through impact and adhesion of fine particles on substrate at room temperature. This phenomenon is known as "room temperature impact consolidation (RTIC)." Consequently, as-deposited film has similar properties to the raw powder material, such as its crystal structure, composition, and physical property. By addressing these features, several papers have been reported for the feasibility of the AD method as a battery electrode or SE

fabrication process (Popovici et al., 2011; Kim et al., 2013; Inada et al., 2014b, 2015; Iwasaki et al., 2014; Ahn et al., 2015; Kato et al., 2016).

TNO is known to be an insertion-type electrode material operating at a potential around 1.6–1.7 V vs. Li$^+$/Li and shows reversible charge and discharge capacities of around 250 mAh g^{-1} and reasonably good cycle stability (Han et al., 2011; Guo et al., 2014; Ashisha et al., 2015), so that we considered TNO as suitable for the feasibility study for an all-solid-state battery application. TNO powders used in this work were prepared by a conventional solid-state reaction method. Stoichiometric amount of anatase TiO$_2$ (Kojundo Chemical Laboratory Co., Ltd., 99%) and Nb$_2$O$_5$ (Kojundo Chemical Laboratory Co., Ltd., 99.9%) were ground and mixed in ethanol by planetary ball milling (Nagao System, Planet M2-3F) with zirconia balls for 1 h. The mixture was annealed at 1100°C for 24 h in air. From the XRD measurements, we confirmed that TNO was successfully obtained without forming any impurity phases.

It has been reported that both the size and morphology of raw powder are important factors for the structure and property of the film formed by AD (Akedo, 2006; Popovici et al., 2011; Inada et al., 2014b, 2015). In order to prepare TNO powder with suitable particle size for dense film fabrication by AD, TNO powder was pulverized using planetary ball milling (Nagao System, Planet M2-3F) with ethanol and zirconia balls. The rotation speed of planetary ball milling was fixed to 300 rpm but milling time was changed in order to control particle size.

The AD apparatus consists of a carrier gas supply system, an aerosol chamber, a deposition chamber equipped with a motored X–Y–Z stage, and a nozzle with a thin rectangular shaped-orifice with a size of 10 mm × 0.5 mm (Inada et al., 2014b, 2015). Deposition starts with evacuating the deposition chamber. A pressure difference between the carrier gas system and the deposition chamber is generated as a power source for film deposition. A carrier N$_2$ gas flows out from the gas supply system to the aerosol chamber. In the aerosol chamber, powder is dispersed into carrier gas and then the aerosol flows into the deposition chamber through a tube and is sprayed onto LLBZT as substrate. The deposition chamber was evacuated to a low vacuum state around 30 Pa, and deposition was carried out for several periods of 10 min. During the deposition process, the stage was moved uni-axially with a back-and-forth motion length of 50 mm. Distance between the tip of the nozzle and the LLBZT substrate was set to be 10 mm, and mass flow of N$_2$ carrier gas was fixed at 20 L min^{-1}. Before depositing the film electrode, an end surface of the LLBZT pellet was polished using sandpaper in order to form a smooth interface with the electrolyte.

Scanning Electron Microscope observation was performed using a field-emission SEM (SU8000 Type II, Hitachi) to reveal the morphology of the TNO powder and both the surface and cross-sectional microstructure of the TNO film electrode formed on LLBZT. EDX analysis was also performed during SEM observation to investigate the distribution of Ti and Nb elements contained in the film electrode.

An all-solid-state cell was fabricated by attaching a Li metal foil on the opposite end surface of a LLBZT pellet covered by a TNO film electrode in an Ar-filled grove box. TNO film is used as the working electrode and Li metal foil as the counter one. CV measurements (three cycles) of TNO film/LLBZT/Li all-solid-state cell were carried out using a potentio-galvanostat (VersaSTAT 3, Princeton Applied Research) at a temperature of 60°C and a scanning rate of 0.1 mV s^{-1} between 1.0 and 2.5 V vs. Li$^+$/Li. After the CV measurement, the solid-state cell was charged and discharged over a cell voltage range between 1.0 and 2.5 V at a constant current density of 2 µA cm^{-2} (corresponding to 5 mA g^{-1} per TNO film) and 60°C, using a battery test system (TOSCAT-3100, Toyo System).

RESULTS AND DISCUSSION

Characterization of LLBZT with Different Ba Contents

Figure 1 shows the XRD patterns of sintered LLBZT with different Ba content $x = 0$–0.30. Calculated diffraction peak patterns for cubic LLZ are also plotted as the reference (Awaka et al., 2011). All of the peaks for each LLBZT sample are well indexed as cubic garnet-type structures with a space group $Ia\bar{3}d$, and no other secondary phases were observed. Enlarged XRD patterns at $2\theta = 50.0$–54.5° are shown in **Figure 2**. As can be seen, the peaks shifted toward a higher angle as x increases from 0 to 0.10, while at $x \geq 0.10$, the peaks shifted toward a lower angle with increasing x. It is worth noting that this trend is confirmed in other diffraction peaks, indicating that the lattice constants of LLBZT are not changed monotonically with x.

Using measured XRD data, we calculated the lattice constant for each LLBZT, and the results are summarized in **Table 1**. Among all the samples, LLBZT with $x = 0.10$ has the smallest lattice constant (=12.944 Å). As confirmed in the chemical formulae of LLBZT prepared in this work, Ta contents $y = 0.5 + x$ increase with Ba contents x and the ionic radii of Ta^{5+} (64 pm) is smaller than Zr^{4+} (78 pm). Therefore, the complex change of the lattice constants in LLBZT, depending on the composition, could be caused by the simultaneous change of Ba and Ta substitution

levels. At $x \leq 0.10$, increase of Ta^{5+} substitution contents influences more on the lattice size of LLBZT, but at $x > 0.10$, the influence of Ba^{2+} substitution contents on the lattice size become dominant. Consequently, the lattice constants of LLBZT are decreased with $x \leq 0.10$, while increased with x from 0.10 to 0.30.

The calculated lattice constants for all LLBZT are also plotted as a function of Ba contents x (lower horizontal axis) and Ta contents $y = 0.5 + x$ (upper horizontal axis) in **Figure 3**. For comparison, the calculated lattice constants for Ba-free Li$_{7-y}$La$_3$Zr$_{2-y}$Ta$_y$O$_{12}$ (LLZT) with $y = 0.50, 0.75$, and 1.00 prepared in our previous work (Inada et al., 2014a) are also plotted against Ta contents y, which is almost linearly decreased with y. With increasing x, it is expected that the volume expansion of dodecahedral (La/Ba)O$_8$ and the volume contraction of octahedral (Ta/Zr)O$_6$ occur in a garnet framework of LLBZT, resulting in complex change of lattice constants depending on x. At Ba contents $x \leq 0.10$ and Ta contents $y \leq 0.60$, the lattice constants for LLBZT nearly fall on the y dependence of lattice constants in LLZT, so that, for LLBZT with $x \leq 0.10$ and $y \leq 0.60$, the contribution of a smaller Ta^{5+} substitution to the Zr^{4+} site is dominant. On the other hand, the lattice constants for LLBZT with $x = 0.20$ and 0.30 and $y = 1.70$ and 1.80 become larger than those for LLZT with the same y levels and the deviation becomes more remarkable with increasing x, suggesting that the contribution of a larger Ba^{2+} substitution to the La^{3+} site becomes dominant.

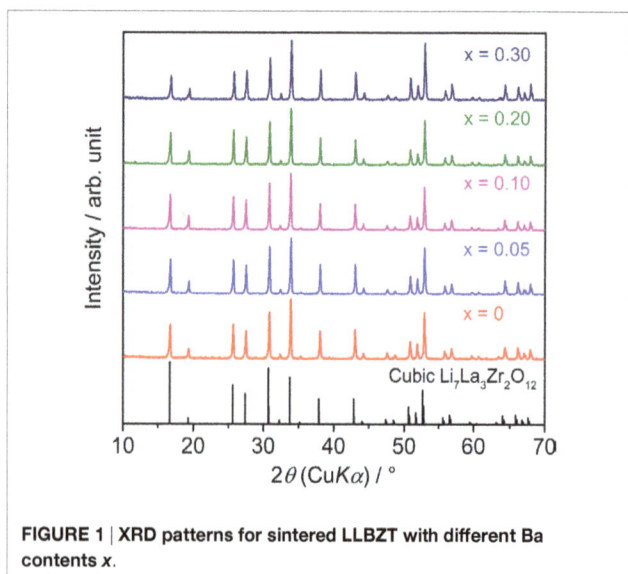

FIGURE 2 | Enlarged XRD patterns for sintered LLBZT with different Ba contents x.

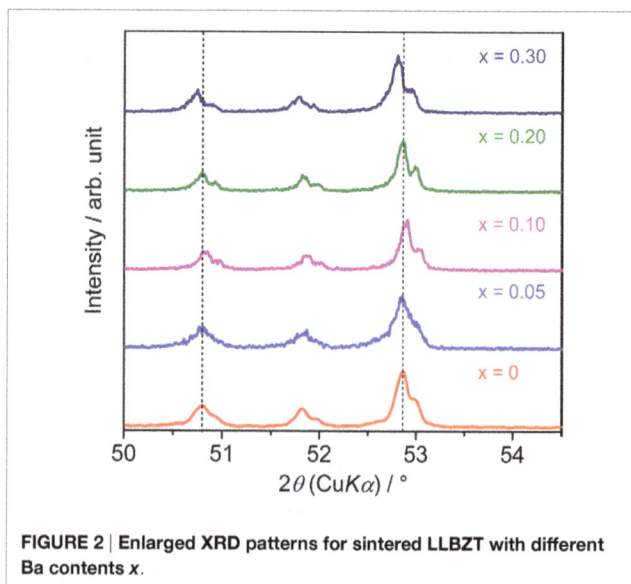

FIGURE 1 | XRD patterns for sintered LLBZT with different Ba contents x.

TABLE 1 | Expected compositions, lattice constants, relative densities of sintered LLZT and LLBZT with different Ba contents x.

Ba content x	Expected composition	Lattice constant/Å	Relative density %
0	Li$_{6.5}$La$_3$Zr$_{1.5}$Ta$_{0.5}$O$_{12}$	12.954	92.8
0.05	Li$_{6.5}$La$_{2.95}$Ba$_{0.05}$Zr$_{1.45}$Ta$_{0.55}$O$_{12}$	12.949	92.6
0.10	Li$_{6.5}$La$_{2.9}$Ba$_{0.1}$Zr$_{1.4}$Ta$_{0.6}$O$_{12}$	12.944	90.4
0.20	Li$_{6.5}$La$_{2.8}$Ba$_{0.2}$Zr$_{1.3}$Ta$_{0.7}$O$_{12}$	12.952	88.8
0.30	Li$_{6.5}$La$_{2.7}$Ba$_{0.3}$Zr$_{1.2}$Ta$_{0.8}$O$_{12}$	12.964	92.0

Figure 4 shows the comparison of SEM images for fractured cross sections of all LLBZT. Average grain size in the sintered LLBZT samples was confirmed to be around 5 μm, and all grains are in good contact with each other. The density of each sintered pellet was determined from their weight and physical dimensions. The relative density (measured density normalized by the theoretical one) of LLBZT with different compositions are summarized in **Table 1**. Here, the theoretical density for each sample was calculated from the lattice constant and expected compositions also shown in **Table 1**. The relative density of each sample was estimated to be in a range between 89 and 93%, indicating

that the difference in relative density among the sintered LLBZT with different composition is not so large. Enlarged SEM images and corresponding elementary mapping for La, Ba, Zr, and Ta and O in LLBZT with different Ba contents $x = 0.05$–0.30 are also shown in **Figure 5**. It can be seen that all elements show similar distribution in the sample, indicating that La^{3+} and Zr^{4+} are successfully substituted by Ba^{2+} and Ta^{5+}, respectively.

Ionic Conductivity and Electrochemical Stability of LLBZT

The conductivity of LLBZT with different compositions was examined by AC impedance spectroscopy using a Li-ion-blocking Au electrode. **Figure 6** shows Nyquist plots of AC impedance measured at 27 and 50°C for all samples. For direct comparison among the samples with different sizes, real and imaginary parts of impedance Z and Z'' multiplied by a factor of AL^{-1} are plotted, where A and L are surface area and thickness of each pellet. For all samples, a part of the semicircle and linear portion data were obtained in high and low frequency regions, indicating that the conducting nature is primarily ionic. Only a part of one semicircle was confirmed and the intercept point of the linear tail in the low frequency range with a real axis corresponding to the sum of the bulk- and grain-boundary resistance. Total conductivity σ for each sample can be calculated by the inverse of the total (bulk and grain boundary) resistance.

The lattice constant and conductivity σ at 27°C of LLBZT are plotted against Ba contents x in **Figures 7A,B**. The sample without Ba substitution shows $\sigma = 7.94 \times 10^{-4}$ S cm^{-1}, which is higher than LLZT with the same composition reported in our previous work (Inada et al., 2014a). This is mainly due to the higher sintering temperature ($=1150°C$) used in this work than that used in previous work ($=1100°C$). At $x \leq 0.1$, σ of LLBZT slightly increases, while the lattice constants slightly decrease with

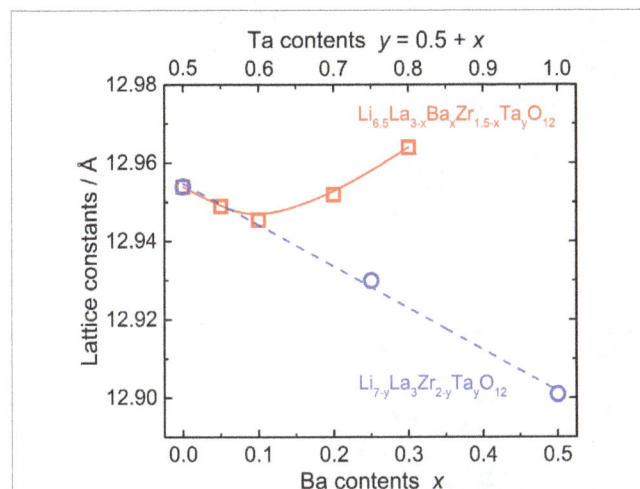

FIGURE 3 | Lattice constants of cubic garnet-type LLBZT plotted against Ba contents x and Ta contents y = 0.5 + x. Lattice constants of cubic garnet-type $Li_{7-y}La_3Zr_{2-y}Ta_yO_{12}$ without Ba substitution (Inada et al., 2014a,b) are plotted as a function of Ta contents y for comparison.

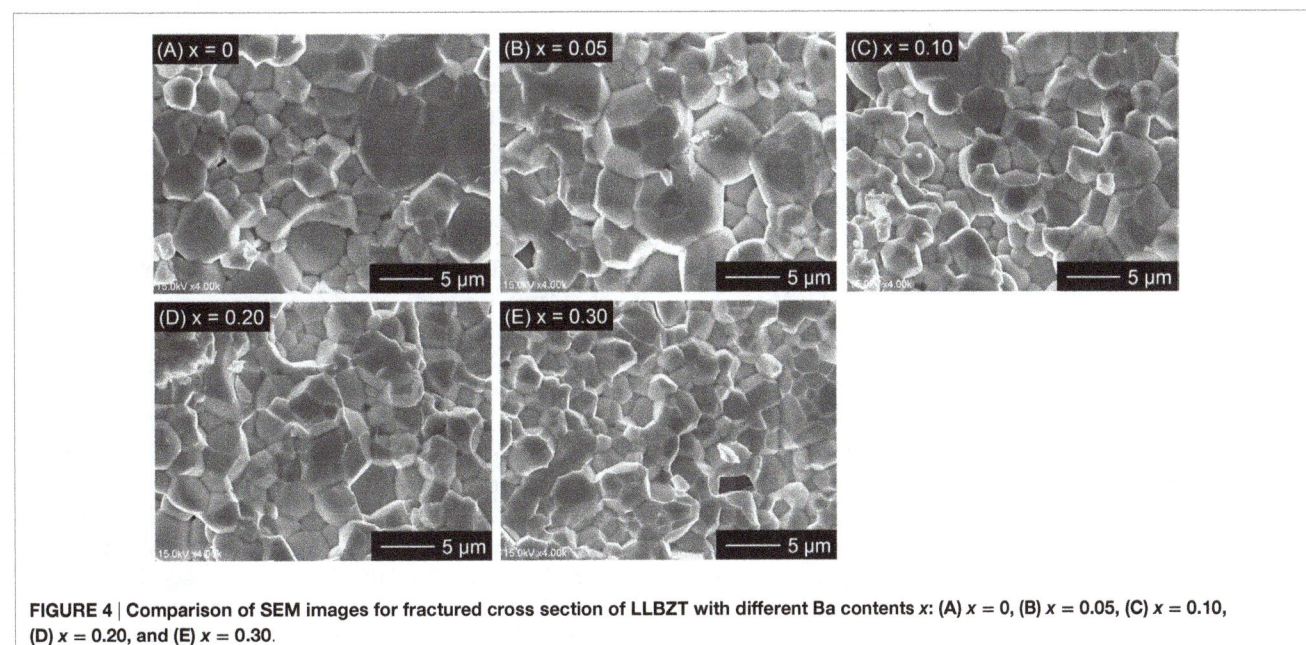

FIGURE 4 | Comparison of SEM images for fractured cross section of LLBZT with different Ba contents x: (A) x = 0, (B) x = 0.05, (C) x = 0.10, (D) x = 0.20, and (E) x = 0.30.

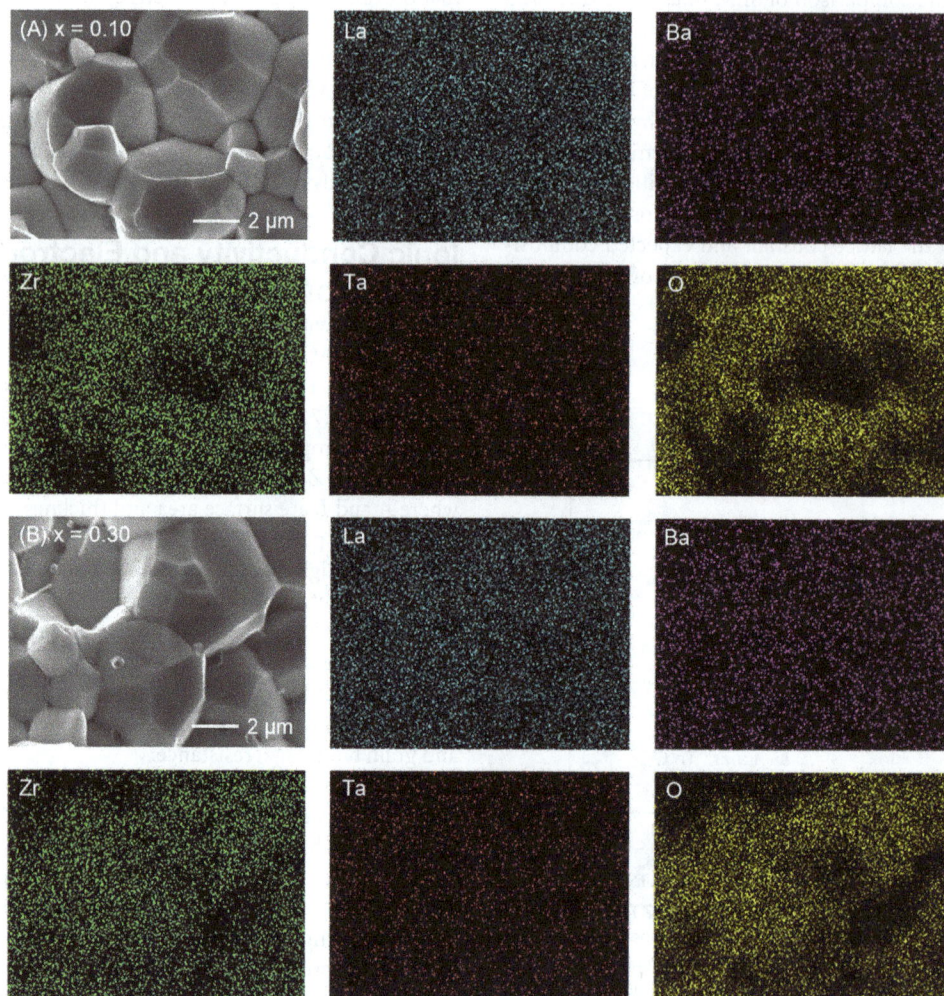

FIGURE 5 | SEM images and corresponding elementary mapping for La, Ba, Zr, Ta, and O for LLBZT with (A) x = 0.10 and (B) x = 0.30.

increasing x. The highest $\sigma = 8.34 \times 10^{-4}$ S cm^{-1} was obtained in LLBZT with $x = 0.1$ and the smallest lattice constant (=12.944 Å) among all samples. The relative density of LLBZT with $x = 0.1$ is slightly smaller than that for the sample with $x = 0$ and 0.05. Unfortunately, we could not obtain the bulk ionic conductivity of LLBZT, quantitatively, because of the limitation of frequency ranges in our experimental set-up, but it is expected that the bulk conductivity of LLBZT shows its maximum at $x = 0.1$. At $x > 0.1$, σ monotonically decreases while the lattice constants increase with x. The lowest $\sigma = 4.92 \times 10^{-4}$ S cm^{-1} was confirmed in LLBZT with $x = 0.3$ and the largest lattice constant among all samples.

The conductivity of LLBZT with fixed Li$^+$ concentration (=6.5) shows its maximum at a lattice constant of around 12.94–12.95 Å. Kihira et al. (2013) have reported their systematical investigation for the properties of cubic garnet-type LLZ with substituting Nb^{5+} to Zr^{4+} site and alkali earth metal cations, such as Sr^{2+} and Ca^{2+}, to La^{3+} site simultaneously, to modify the lattice constant. They also tried to substitute Mg^{2+} and Ba^{2+} to La^{3+} site in Nb-doped LLZ but some impurity phases, such as MgO and BaZrO$_3$, were formed.

Interestingly, regardless of the substitution element, the highest conductivity in their samples is observed at nearly the same lattice constant of 12.94–12.96 Å, which is close to our results for LLBZT with $x = 0$–0.10. However, LLBZT with $x = 0.20$ has lower room temperature conductivity than LLBZT with $x = 0.05$, although both samples have nearly the same lattice size.

Since the difference in the relative densities among all samples is not so large, the contribution of grain-boundary resistivity to the total ionic conductivity does not depend on the composition of LLBZT. We believe that the composition dependence of σ is mainly caused by the change of Li$^+$ conduction property in LLBZT bulk. The bulk Li$^+$ conductivity σ_{Li} is described as $\sigma_{Li} = ne\mu$, where n is the number of carriers, e is the charge and μ is the carrier mobility. Since the charge value will be constant, σ_{Li} will be influenced by the number of carriers and the mobility. As mentioned above, it is expected that in LLBZT garnet, the volume of the dodecahedral (La/Ba)O$_8$ is expanded while the volume of octahedral (Ta/Zr)O$_6$ is contracted with increasing Ba contents x. These polyhedrons are adjacent to the tetrahedral LiO$_4$ and the

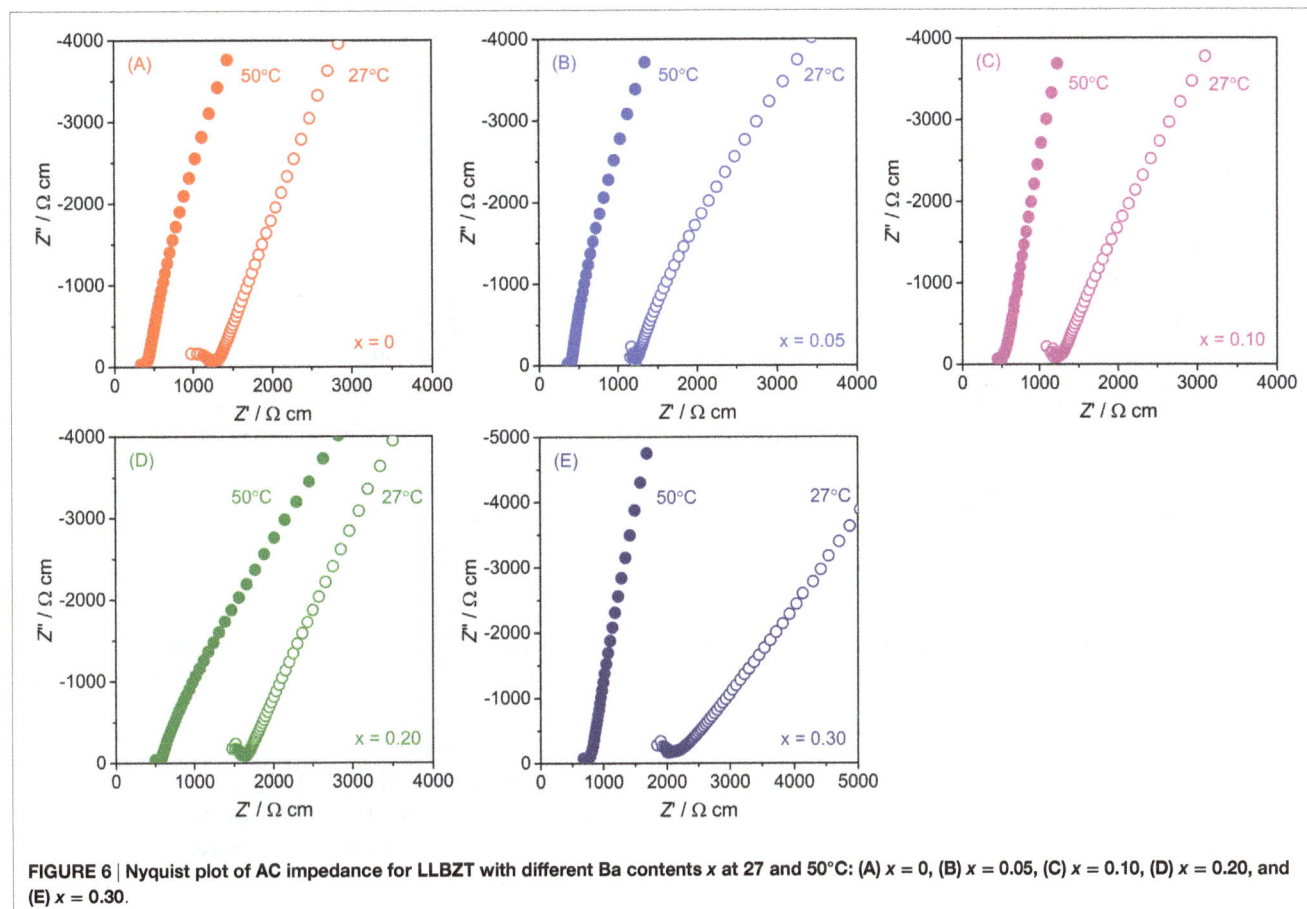

FIGURE 6 | Nyquist plot of AC impedance for LLBZT with different Ba contents x at 27 and 50°C: (A) $x = 0$, (B) $x = 0.05$, (C) $x = 0.10$, (D) $x = 0.20$, and (E) $x = 0.30$.

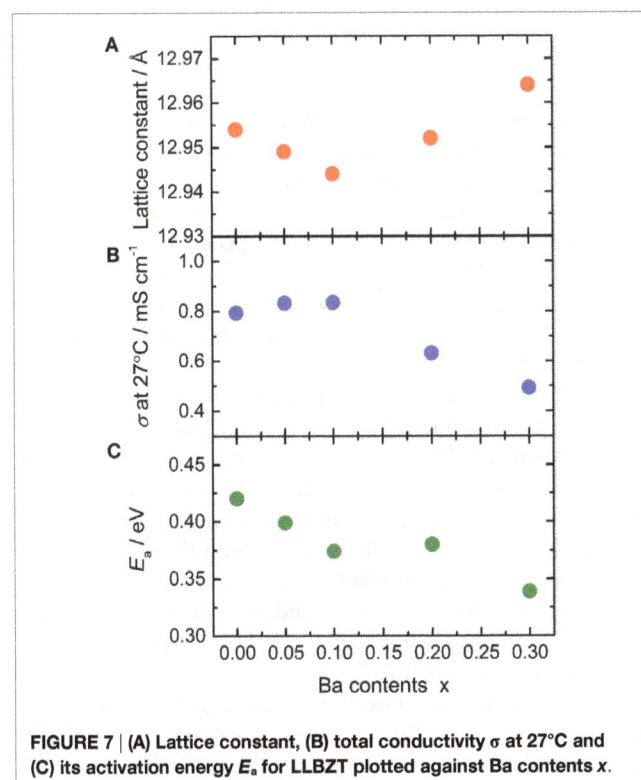

FIGURE 7 | (A) Lattice constant, (B) total conductivity σ at 27°C and (C) its activation energy E_a for LLBZT plotted against Ba contents x.

octahedral LiO_6 in a garnet framework, so that the volumes of the LiO_4 and LiO_6 polyhedrons also show complex change depending on x. It has been pointed out that the Li^+ conductivity of garnet-type oxides is strongly influenced by the site occupancy in both LiO_4 ($24d$ site) and LiO_6 ($96h$ site) polyhedrons. The $24d$ and $96h$ site occupancies effect on the number of carriers, and higher σ_{Li} is achieved by well-balanced site occupancy (Li et al., 2012b; Kihira et al., 2013; Thangadurai et al., 2014, 2015). We believe that the $24d$ and $96h$ site occupancies are influenced not only by the Li^+ contents in a garnet framework but also by the sizes of both LiO_4 and LiO_6 polyhedrons. The highest conductivity at room temperature for LLBZT with x from 0.05 to 0.10 would be attributed to the well-balanced $24d$ and $96h$ site occupancies, while the deterioration of the conductivity with increasing $x > 0.20$ would be caused by the change in the site occupancies depending on the sizes of both LiO_4 and LiO_6 polyhedrons.

Temperature dependence of the conductivity for all LLBZT was also evaluated in the temperature range from 27 to 100°C. **Figure 8** shows the variation of σ for all sintered samples as a function of an inverse of temperature $1000T^{-1}$. The temperature dependence of σ is expressed by the Arrhenius equation as $\sigma = \sigma_0 \exp[-E_a/(k_B T)]$, where σ_0 is constant, E_a is activation energy of conductivity, and k_B is Boltzmann constant ($= 1.381 \times 10^{-23}$ J/K). The E_a of each sample are estimated from the slope of σT data plotted in **Figure 8** and summarized in **Table 2**, together with σ at 27°C. The sample without Ba

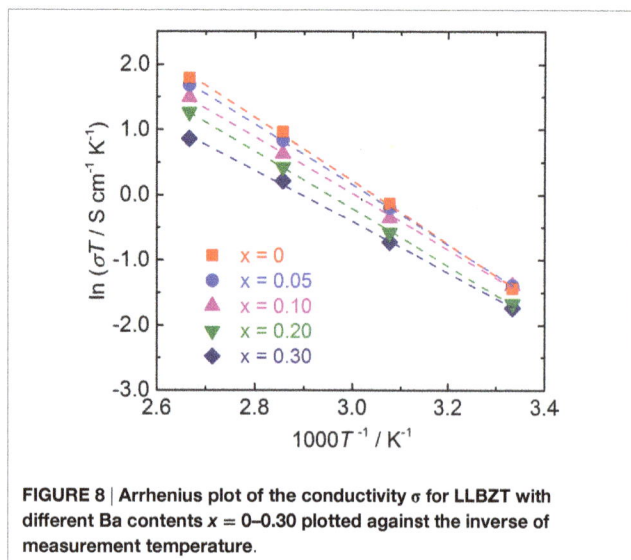

FIGURE 8 | Arrhenius plot of the conductivity σ for LLBZT with different Ba contents x = 0–0.30 plotted against the inverse of measurement temperature.

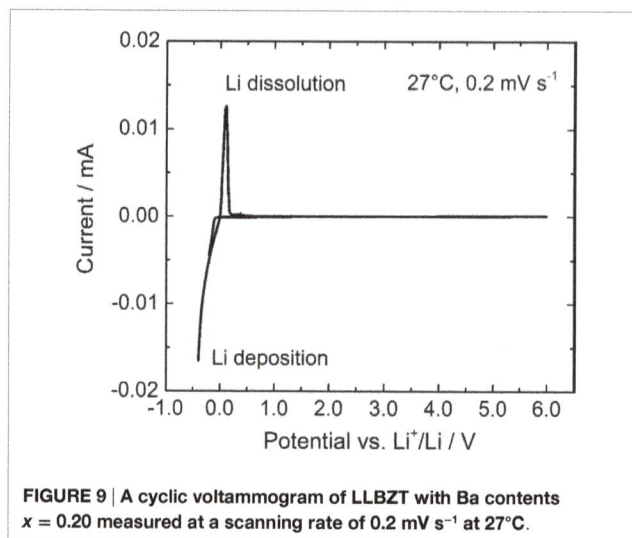

FIGURE 9 | A cyclic voltammogram of LLBZT with Ba contents x = 0.20 measured at a scanning rate of 0.2 mV s^{-1} at 27°C.

TABLE 2 | Total conductivity σ at different measurement temperature and activation energy E_a for LLBZT with different Ba contents x = 0–0.30.

Ba content x	Total conductivity σ/S cm^{-1}				E_a/eV
	27°C	50°C	75°C	100°C	
0	7.94×10^{-4}	2.70×10^{-3}	7.5×10^{-3}	16.0×10^{-3}	0.420
0.05	8.32×10^{-4}	2.50×10^{-3}	6.6×10^{-3}	14.4×10^{-3}	0.399
0.10	8.34×10^{-4}	2.20×10^{-3}	5.40×10^{-3}	11.9×10^{-3}	0.374
0.20	6.31×10^{-4}	1.80×10^{-3}	4.50×10^{-3}	10.6×10^{-3}	0.380
0.30	4.92×10^{-4}	1.30×10^{-3}	3.10×10^{-3}	5.40×10^{-3}	0.339

substitution has the highest $E_a = 0.42$ eV, which is higher than an Al-doped LLZT sample with a similar composition (Li et al., 2012b) but nearly the same as the Al-free sample prepared by hot pressing (Thompson et al., 2015). The E_a of LLBZT is in the range of 0.34–0.40 eV and tends to decrease with increasing Ba substitution levels as shown in **Figure 7C**.

Ba substitution levels influence on both the ionic conductivity and its activation energy. At $x \leq 0.10$, the lattice size of LLBZT is reduced with increasing x, while σ at room temperature is increased slightly, suggesting that the number of carriers for Li$^+$ conduction in LLBZT are increased with x from 0 to 0.10. Therefore, the reduction of E_a by Ba substitution at $x \leq 0.10$ would be mainly attributed to the increase in the number of carriers by the slight modification of the 24d and 96h site occupancies. On the other hand, the lattice size of LLBZT is increased with x, while σ at room temperature is lowered with increasing x from 0.10 to 0.30, suggesting that the number of carriers for Li$^+$ conduction is decreased with x while the bottleneck size for Li$^+$ conduction expands. It is expected that the expansion of bottleneck size result in enhancing the mobility, so the reduction of E_a at $x > 0.10$ would be mainly caused by the enhancement of the mobility.

In order to evaluate the electrochemical stability of LLBZT, a cyclic voltammogram of LLBZT is shown in **Figure 9**. Li deposition and dissolution peaks near 0 V vs. Li$^+$/Li are clearly observed, but no other reactions are confirmed up to 6 V vs. Li$^+$/Li. Therefore, LLBZT has a wide electrochemical potential window and various electrode materials can potentially be used to construct all-solid-state batteries with LLBZT as the SE.

Characterization of TNO Film Electrode Formed on LLBZT

For the feasibility study of all-solid-state battery application, TNO film electrode was formed on the surface of sintered LLBZT pellet by the AD method, and its electrochemical property was evaluated. We used pulverized TNO powder by planetary ball milling as raw material for fabrication of the film electrode. As can be seen in **Figure 10A**, raw TNO powder dominantly includes particles with a size of 0.5–0.8 μm. SEM images of the surface and cross section of TNO film electrode are shown in **Figures 10B,C**, together with a schematic illustration of the film formation by AD in **Figure 10D**. It is confirmed that strongly deformed and/or fractured TNO particles form a dense film electrode on LLBZT without any thermal treatment. The thickness of the TNO film electrode is confirmed to be ~1.5 μm. The interface morphology between the TNO film and the LLBZT pellet seems to be very smooth. **Figure 11** shows elementary mapping results of Ti and Nb on the surface and the cross section of the TNO film electrode formed on LLBZT. On the film surface, both Ti and Nb are detected in the whole observed area, while in the cross section, Ti and Nb are dominantly detected in the film electrode area.

A cyclic voltammogram of a TNO film/LLBZT/Li solid-state cell measured at 60°C and a scan rate of 0.1 mV s^{-1} is shown in **Figure 12**. After the cell construction, the open-circuit voltage of the cell was confirmed to be around 1.7 V. Except for the first scanning toward anodic direction, both the oxidation and reduction reaction of the TNO film electrode on LLBZT seems to be reversible. Broad oxidation and reduction reaction peaks are confirmed at the cell voltages (nearly corresponding to the potential vs. Li$^+$/Li) of 1.8 V and 1.6 V, which corresponds to the Li$^+$ extraction and insertion reaction potential for TNO (Ashisha et al., 2015). **Figure 13** shows the charge and discharge properties of a TNO film/LLBZT/Li solid-state cell measured at

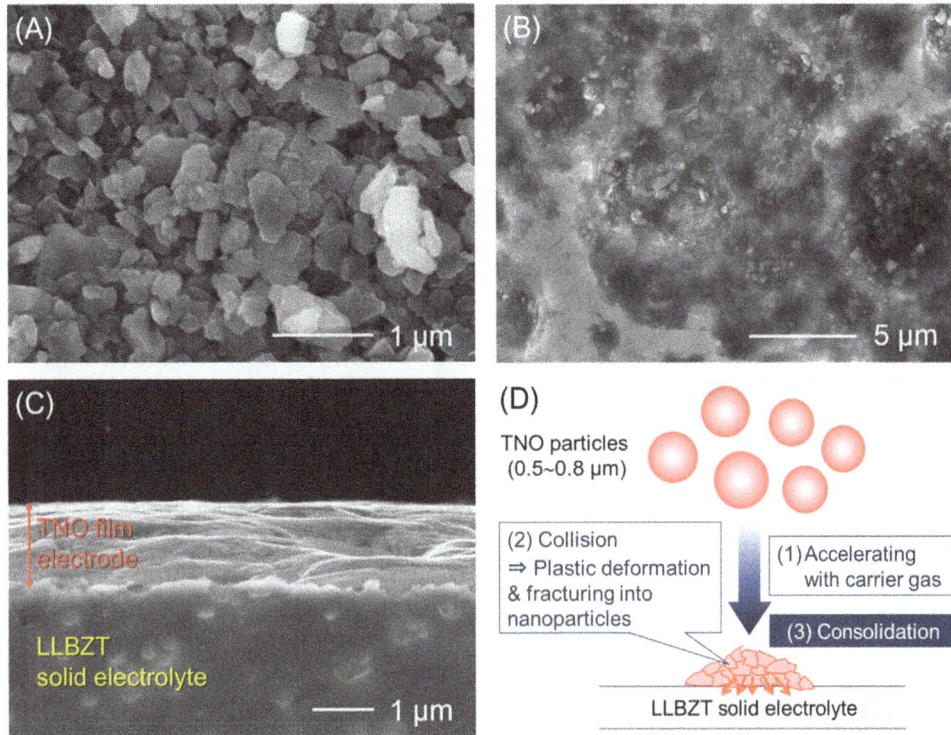

FIGURE 10 | SEM images for (A) TNO powder used as raw materials for film electrode fabrication by AD, (B) the surface of TNO film electrode, and (C) polished cross section of TNO film electrode. Plastically deformed and/or fractured TNO particles form the dense film on LLBZT *via* impact consolidation as shown in **(D)**.

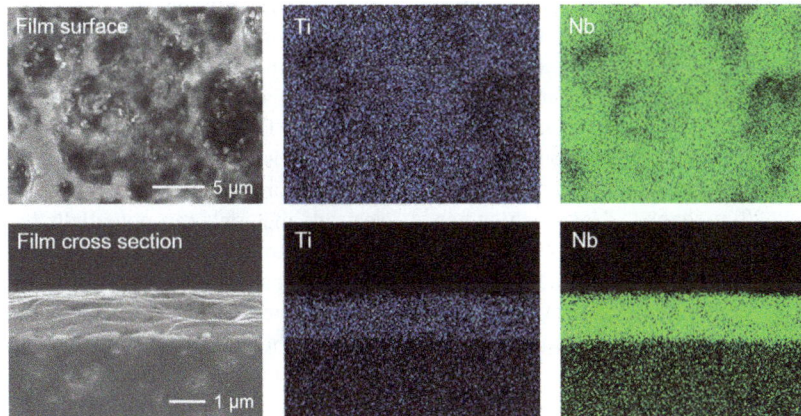

FIGURE 11 | SEM images and corresponding elementary mapping for Ti and Nb on the surface and polished cross section of TNO film electrode formed on LLBZT.

60°C and a current density of 2 μA cm⁻². As can be seen, TNO film electrode formed on LLBZT by AD showed capacities of 220 mAh g⁻¹ for Li⁺ insertion reaction and 170 mAh g⁻¹ for Li⁺ extraction reaction. Electrode reactions occur at an averaged cell voltage of 1.6 V, which is nearly the same as the TNO composite electrode with conducting additive and binder in an organic liquid electrolyte (Han et al., 2011; Guo et al., 2014; Ashisha et al., 2015). Although the coulombic efficiency of TNO film electrode is far below 100% at present, the results shown in **Figures 12** and **13** indicate that LLBZT works as the Li⁺ conduction path between the TNO film as the working electrode and the Li metal foil as the counter one.

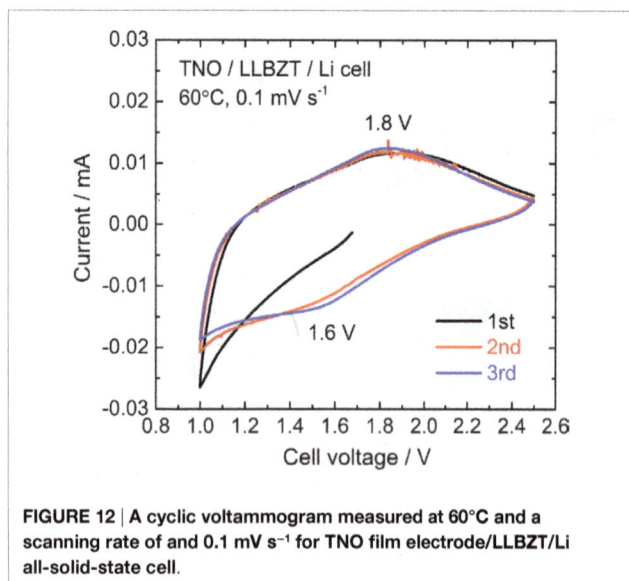

FIGURE 12 | A cyclic voltammogram measured at 60°C and a scanning rate of and 0.1 mV s⁻¹ for TNO film electrode/LLBZT/Li all-solid-state cell.

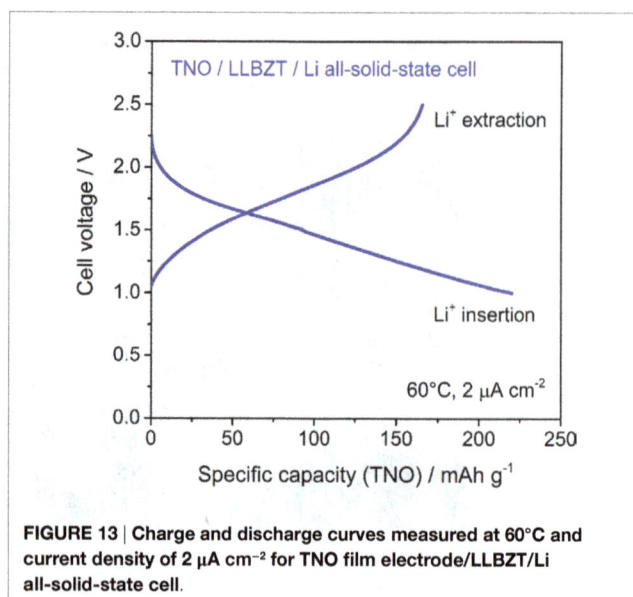

FIGURE 13 | Charge and discharge curves measured at 60°C and current density of 2 μA cm⁻² for TNO film electrode/LLBZT/Li all-solid-state cell.

As a future prospect for solid state battery development with garnet-type LLBZT SE, both increasing the thickness of the electrode and reducing the thickness of the SE layer should be indispensable for enhancing the volumetric energy density of a solid-state battery. In a thicker electrode layer with a thickness of several 10 μm, a composite structure with electrode active material and SE must be needed to increase the solid–solid interface among them for the high utilization of active material in a composite electrode. We are considering that AD is one of the potential processing methods to form composite electrode on LLBZT sheet because composite powders with active material and Li⁺ conducting SE can be directly used (Iwasaki et al., 2014; Kato et al., 2016) as raw material to form the composite electrode on SE sheet. Since thermal treatment is not always needed to form the electrode on SE sheet by AD, undesired reactions between the electrode and SE can be greatly suppressed. In addition, AD is also potentially applicable to the fabrication of film-shaped SE layer, which has been already demonstrated in NASICON-type Li⁺ conducting oxide-based SE (Popovici et al., 2011; Inada et al., 2015). We are now trying to fabricate composite thick film electrode on LLBZT by AD, and progress on this will be reported in a forthcoming paper.

CONCLUSION

In this paper, we investigate the properties of garnet-type Li⁺ conducting oxide SEs with simultaneous substitution of Ta⁵⁺ into the Zr⁴⁺ site and Ba²⁺ into the La³⁺ site in LLZ. Li⁺ concentration was fixed to 6.5 per chemical formulae, so that the composition of our samples is expressed as $Li_{6.5}La_{3-x}Ba_xZr_{1.5-x}Ta_{0.5+x}O_{12}$ (LLBZT), and Ba contents x are changed from 0 to 0.30. As a result, all LLBZT samples have a cubic garnet-type structure without any secondary phases. The lattice constants of LLBZT decrease with increasing Ba²⁺ contents $x \leq 0.10$ while increase with x from 0.10 to 0.30, possibly caused by the simultaneous change of smaller

Ta⁵⁺ and larger Ba²⁺ substitution levels. On the other hand, the relative densities of LLBZT are not influenced greatly by x. The total conductivity at 27°C of LLBZT has its maximum value of 8.34×10^{-4} S cm⁻¹ at $x = 0.10$, which is slightly higher than the conductivity of LLZT without substituting Ba (= 7.94×10^{-4} S cm⁻¹). This improvement would be attributed to the modification of the number of carriers by tuning the $24d$ and $96h$ site occupancies. A larger lattice size created by excess Ba substitution is not suitable to achieve high ionic conductivity in LLBZT. The activation energy of the conductivity tends to become lower than the sample without Ba substitution. However, the influence of Ba contents on the Li⁺ occupation site and carrier concentration in garnet-type LLBZT cannot be examined at present so that a detailed structural analysis would be necessary to clarify the ionic conducting property in LLBZT.

LLBZT is confirmed to be electrochemically stable at the potential range of 0–6 V vs. Li⁺/Li so that various kinds of electrode materials can potentially be used for constructing an all-solid-state battery. In order to investigate the feasibility for solid-state battery application, TNO film electrode was formed on LLBZT without any thermal treatment by an AD method, and its electrochemical properties were evaluated. The Li⁺ insertion and extraction reaction of the film electrode formed on LLBZT could be confirmed, indicating that LLBZT can be used as a SE in an all-solid-state battery.

AUTHOR CONTRIBUTIONS

RI was in charge of planning and performing all the experiment (containing both material preparation and characterization) in this work, together with summarizing the paper preparation. SY and KT were in charge of the preparation and characterization of oxide solid electrolyte samples. MT was in charge of the fabrication of the film-type electrode on solid electrolyte material and its characterization. TT and YS were in charge of the discussion for all experimental results.

ACKNOWLEDGMENTS

RI would like to thank Prof. Dr. Venkataraman Thangadurai of University of Calgary, Prof. Dr. Jeff Sakamoto of University of Michigan, Prof. Dr. Palani Balaya of National University of Singapore, and Prof. Dr. Ashutosh Tiwari of The University of Utah for their kind and technical advices in our sample preparation and characterization during his visiting in their laboratory on March 2015. This work was partly supported by JSPS KAKENHI [Challenging Exploratory Research and Scientific Research (C)] grant numbers 26630111 and 16K06218 from the Japan Society for the Promotion of Science (JSPS), Research Foundation for the Electrotechnology of Chubu (R-25209) and TOYOAKI Scholarship Foundation.

REFERENCES

Ahn, C.-W., Choi, J.-J., Ryu, J., Hahn, B.-D., Kim, J.-W., Yoon, W.-Ha, et al. (2015). Microstructure and electrochemical properties of iron oxide film fabricated by aerosol deposition method for lithium ion battery. *J. Power Sources* 275, 336–340. doi:10.1016/j.jpowsour.2014.11.033

Akedo, J. (2006). Aerosol deposition of ceramic thick films at room temperature: densification mechanism of ceramic layers. *J. Am. Ceramic Soc.* 89, 1834–1839. doi:10.1111/j.1551-2916.2006.01030.x

Armand, M., and Tarascon, J.-M. (2008). Building better batteries. *Nature* 451, 652–657. doi:10.1038/451652a

Ashisha, A. G., Arunkumara, P., Babua, B., Manikandana, P., Saranga, S., and Shaijumon, M. M. (2015). TiNb$_2$O$_7$/graphene hybrid material as high performance anode for lithium-ion batteries. *Electrochim. Acta* 176, 285–292. doi:10.1016/j.electacta.2015.06.122

Awaka, J., Kijima, N., Hayakawa, H., and Akimoto, J. (2009a). Synthesis and structure analysis of tetragonal Li$_7$La$_3$Zr$_2$O$_{12}$ with the garnet-related type structure. *J. Solid State Chem.* 182, 2046–2052. doi:10.1016/j.jssc.2009.05.020

Awaka, J., Kijima, N., Takahashi, Y., Hayakawa, H., and Akimoto, J. (2009b). Synthesis and crystallographic studies of garnet-related lithium-ion conductors Li$_6$CaLa$_2$Ta$_2$O$_{12}$ and Li$_6$BaLa$_2$Ta$_2$O$_{12}$. *Solid State Ionics.* 180, 602–606. doi:10.1016/j.ssi.2008.10.022

Awaka, J., Takashima, A., Kataoka, K., Kijima, N., Idemoto, Y., and Akimoto, J. (2011). Crystal structure of fast lithium-ion-conducting cubic Li$_7$La$_3$Zr$_2$O$_{12}$. *Chem. Lett.* 40, 60–62. doi:10.1246/cl.2011.60

Bernuy-Lopez, C., Manalastas, W. Jr., Lopez, J. M., Aguadero, A., Aguesse, F., and Kilner, J. A. (2014). Atmosphere controlled processing of Ga-substituted garnets for high Li-ion conductivity ceramics. *Chem. Mater.* 26, 3610–3617. doi:10.1021/cm5008069

Bottke, P., Rettenwander, D., Schmidt, W., Amthauer, G., and Wilkening, M. (2015). Ion dynamics in solid electrolytes: NMR reveals the elementary steps of Li$^+$ hopping in the garnet Li$_{6.5}$La$_3$Zr$_{1.75}$Mo$_{0.25}$O$_{12}$. *Chem. Mater.* 27, 6571–6582. doi:10.1021/acs.chemmater.5b02231

Buschmann, H., Berendts, S., Mogwitz, B., and Janek, J. (2012). Lithium metal electrode kinetics and ionic conductivity of the solid lithium ion conductors "Li$_7$La$_3$Zr$_2$O$_{12}$" and Li$_{7-x}$La$_3$Zr$_{2-x}$Ta$_x$O$_{12}$ with garnet-type structure. *J. Power Sources* 206, 236–244. doi:10.1016/j.jpowsour.2012.01.094

Dhivya, L., Janani, N., Palanivel, B., and Murugan, R. (2013). Li$^+$ transport properties of W substituted Li$_7$La$_3$Zr$_2$O$_{12}$ cubic lithium garnets. *AIP Adv.* 3, 082115. doi:10.1063/1.4818971

Fergus, J. W. (2010). Ceramic and polymeric solid electrolytes for lithium-ion batteries. *J. Power Sources* 195, 4554–4569. doi:10.1016/j.jpowsour.2010.01.076

Geiger, C. A., Alekseev, E., Lazic, B., Fisch, M., Armbruster, T., Langner, R., et al. (2011). Crystal chemistry and stability of "Li$_7$La$_3$Zr$_2$O$_{12}$" garnet: a fast lithium-ion conductor. *Inorg. Chem.* 50, 1089–1097. doi:10.1021/ic101914e

Goodenough, J. B., and Kim, Y. (2011). Challenges for rechargeable batteries. *J. Power Sources* 196, 6688–6694. doi:10.1016/j.jpowsour.2010.11.074

Guo, B. K., Yu, X. Q., Sun, X. G., Chi, M. F., Qiao, Z. A., Liu, J., et al. (2014). A long-life lithium-ion battery with a highly porous TiNb$_2$O$_7$ anode for large-scale electrical energy storage. *Energy Environ. Sci.* 7, 2220–2226. doi:10.1039/C4EE00508B

Han, J.-T., Huan, Y.-H., and Goodenough, J. B. (2011). New anode framework for rechargeable lithium batteries. *Chem. Mater.* 23, 2027–2029. doi:10.1021/cm200441h

Inada, R., Ishida, K., Tojo, M., Okada, T., Tojo, T., and Sakurai, Y. (2015). Properties of aerosol deposited NASICON-type Li$_{1.5}$Al$_{0.5}$Ge$_{1.5}$(PO$_4$)$_3$ solid electrolyte thin films. *Ceramics Int.* 41, 11136–11142. doi:10.1016/j.ceramint.2015.05.062

Inada, R., Kusakabe, K., Tanaka, T., Kudo, S., and Sakurai, Y. (2014a). Synthesis and properties of Al-free Li$_{7-x}$La$_3$Zr$_{2-x}$Ta$_x$O$_{12}$ garnet related oxides. *Solid State Ionics.* 262, 568–572. doi:10.1016/j.ssi.2013.09.008

Inada, R., Shibukawa, K., Masada, C., Nakanishi, Y., and Sakurai, Y. (2014b). Characterization of as-deposited Li$_4$Ti$_5$O$_{12}$ thin film electrode prepared by aerosol deposition method. *J. Power Sources* 244, 646–651. doi:10.1016/j.jpowsour.2013.12.084

Iwasaki, S., Hamanaka, T., Yamakawa, T., West, W. C., Yamamoto, K., Motoyama, M., et al. (2014). Preparation of thick-film LiNi$_{1/3}$Co$_{1/3}$Mn$_{1/3}$O$_2$ electrodes by aerosol deposition and its application to all-solid-state batteries. *J. Power Sources* 272, 1086–1090. doi:10.1016/j.jpowsour.2014.09.038

Jalem, R., Rushton, M. J. D., Manalastas, W. Jr., Nakayama, M., Kasuga, T., Kilner, J. A., et al. (2015). Effects of Gallium doping in garnet-type Li$_7$La$_3$Zr$_2$O$_{12}$ solid electrolytes. *Chem. Mater.* 27, 2821–2831. doi:10.1021/cm5045122

Jin, Y., and McGinn, P. J. (2011). Al-doped Li$_7$La$_3$Zr$_2$O$_{12}$ synthesized by a polymerized complex method. *J. Power Sources* 196, 8683–8687. doi:10.1016/j.jpowsour.2011.05.065

Kato, T., Iwasaki, S., Ishii, Y., Motoyama, N., West, W. C., Yamamoto, Y., et al. (2016). Preparation of thick-film electrode-solid electrolyte composites on Li$_7$La$_3$Zr$_2$O$_{12}$ and their electrochemical properties. *J. Power Sources* 303, 65–72. doi:10.1016/j.jpowsour.2015.10.101

Kihira, Y., Ohta, S., Imagawa, H., and Asaoka, T. (2013). Effect of simultaneous substitution of alkali earth metals and Nb in Li$_7$La$_3$Zr$_2$O$_{12}$ on lithium-ion conductivity. *ECS Electrochem. Lett.* 2, A56–A59. doi:10.1149/2.001307eel

Kim, I., Park, J., Nam, T.-H., Kim, K.-W., Ahn, J.-H., Park, D.-S., et al. (2013). Electrochemical properties of an as-deposited LiFePO$_4$ thin film electrode prepared by aerosol deposition. *J. Power Sources* 243, 181–186. doi:10.1016/j.jpowsour.2012.12.108

Knauth, P. (2009). Inorganic solid Li ion conductors: an overview. *Solid State Ionics.* 180, 911–916. doi:10.1016/j.ssi.2009.03.022

Kotobuki, M., Kanamura, K., Sato, Y., and Yoshida, T. (2011). Fabrication of all-solid-state lithium battery with lithium metal anode using Al$_2$O$_3$-added Li$_7$La$_3$Zr$_2$O$_{12}$ solid electrolyte. *J. Power Sources* 196, 7750–7754. doi:10.1016/j.jpowsour.2011.04.047

Kumazaki, S., Iriyama, Y., Kim, K.-H., Murugan, R., Tanabe, K., Yamamoto, K., et al. (2011). High lithium ion conductive Li$_7$La$_3$Zr$_2$O$_{12}$ by inclusion of both Al and Si. *Electrochem. commun.* 13, 509–512. doi:10.1016/j.elecom.2011.02.035

Li, Y., Han, J.-T., Wang, C.-A., Vogel, S. C., Xie, H., Xu, M., et al. (2012a). Ionic distribution and conductivity in lithium garnet Li$_7$La$_3$Zr$_2$O$_{12}$. *J. Power Sources* 209, 278–281. doi:10.1016/j.jpowsour.2012.02.100

Li, Y., Han, J.-T., Wang, C.-A., Xie, H., and Goodenough, J. B. (2012b). Optimizing Li$^+$ conductivity in a garnet framework. *J. Mater. Chem.* 22, 15357–15361. doi:10.1039/C2JM31413D

Li, Y., Wang, Z., Cao, Y., Du, F., Chen, C., Cui, Z., et al. (2015). W-doped Li$_7$La$_3$Zr$_2$O$_{12}$ ceramic electrolytes for solid state Li-ion batteries. *Electrochim. Acta* 180, 37–42. doi:10.1016/j.electacta.2015.08.046

Logéat, A., Köhler, T., Eisele, U., Stiaszny, B., Harzer, A., Tovar, M., et al. (2012). From order to disorder: the structure of lithium-conducting garnets Li$_{7-x}$La$_3$Ta$_x$Zr$_{2-x}$O$_{12}$ (x = 0–2). *Solid State Ionics.* 206, 33–38. doi:10.1016/j.ssi.2011.10.023

Murugan, R., Thangadurai, V., and Weppner, W. (2007a). Fast lithium ion conduction in garnet-type Li$_7$La$_3$Zr$_2$O$_{12}$. *Angew. Chem. Int. Ed.* 46, 7778–7781. doi:10.1002/anie.200701144

Murugan, R., Weppner, W., Schmid-Beurmann, P., and Thangadurai, V. (2007b). Structure and lithium ion conductivity of bismuth containing lithium garnets Li$_5$La$_3$Bi$_2$O$_{12}$ and Li$_6$SrLa$_2$Bi$_2$O$_{12}$. *Mater. Sci. Eng. B* 143, 14–20. doi:10.1016/j.mseb.2007.07.009

Nemori, H., Matsuda, Y., Mitsuoka, S., Matsui, M., Yamamoto, O., Takeda, Y., et al. (2015). Stability of garnet-type solid electrolyte Li$_x$La$_3$A$_{2-y}$B$_y$O$_{12}$ (A=Nb or Ta, B = Sc or Zr). *Solid State Ionics.* 282, 7–12. doi:10.1016/j.ssi.2015.09.015

Ohta, S., Kobayashi, T., and Asaoka, T. (2011). High lithium ionic conductivity in the garnet-type oxide $Li_{7-x}La_3(Zr_{2-x}, Nb_x)O_{12}$ (X = 0–2). *J. Power Sources* 196, 3342–3345. doi:10.1016/j.jpowsour.2010.11.089

Ohta, S., Komagata, S., Seki, J., and Asaoka, T. (2012). Electrochemical performance of an all-solid-state lithium ion battery with garnet-type oxide electrolyte. *J. Power Sources* 202, 332–335. doi:10.1016/j.jpowsour.2011.10.064

Ohta, S., Komagata, S., Seki, J., Saeki, T., Morishita, S., and Asaoka, T. (2013). All-solid-state lithium ion battery using garnet-type oxide and Li_3BO_3 solid electrolytes fabricated by screen-printing. *J. Power Sources* 238, 53–56. doi:10.1016/j.jpowsour.2013.02.073

Ohta, S., Seki, J., Yagi, Y., Kihira, Y., Tani, T., and Asaoka, T. (2014). Co-sinterable lithium garnet-type oxide electrolyte with cathode for all-solid-state lithium ion battery. *J. Power Sources* 265, 40–44. doi:10.1016/j.jpowsour.2014.04.065

Popovici, D., Nagai, H., Fujisima, S., and Akedo, J. (2011). Preparation of lithium aluminum titanium phosphate electrolytes thick films by aerosol deposition method. *J. Am. Ceramic Soc.* 94, 3847–3850. doi:10.1111/j.1551-2916.2011.04551.x

Rangasamy, E., Wolfenstine, J., and Sakamoto, J. (2012). The role of Al and Li concentration on the formation of cubic garnet solid electrolyte of nominal composition $Li_7La_3Zr_2O_{12}$. *Solid State Ionics* 206, 28–32. doi:10.1016/j.ssi.2011.10.022

Ren, Y., Chen, K., Chen, R., Liu, T., Zhang, Y., and Nan, C.-W. (2015a). Oxide electrolytes for lithium batteries. *J. Am. Ceramic Soc.* 98, 3603–3623. doi:10.1111/jace.13844

Ren, Y., Deng, H., Chen, R., Shen, Y., Lin, Y., and Nan, C.-W. (2015b). Effects of Li source on microstructure and ionic conductivity of Al-contained $Li_{6.75}La_3Zr_{1.75}Ta_{0.25}O_{12}$ ceramics. *J. Eur. Ceramic Soc.* 35, 561–572. doi:10.1016/j.jeurceramsoc.2014.09.007

Scrosati, B., and Garche, J. (2010). Lithium batteries: status, prospects and future. *J. Power Sources* 195, 2419–2430. doi:10.1016/j.jpowsour.2009.11.048

Shinawi, H. E., and Janek, J. (2013). Stabilization of cubic lithium-stuffed garnets of the type "$Li_7La_3Zr_2O_{12}$" by addition of gallium. *J. Power Sources* 225, 13–19. doi:10.1016/j.jpowsour.2012.09.111

Takada, K. (2013). Progress and prospective of solid-state lithium batteries. *Acta Mater.* 61, 759–770. doi:10.1016/j.actamat.2012.10.034

Tarascon, J.-M., and Armand, M. (2001). Issues and challenges facing rechargeable lithium batteries. *Nature* 414, 359–367. doi:10.1038/35104644

Tatsumisago, M., Nagao, M., and Hayashi, A. (2013). Recent development of sulfide solid electrolytes and interfacial modification for all-solid-state rechargeable lithium batteries. *J. Asian Ceramic Soc.* 1, 117–125. doi:10.1016/j.jascer.2013.03.005

Thangadurai, V., Narayanan, S., and Pinzaru, D. (2014). Garnet-type solid-state fast Li ion conductors for Li batteries: critical review. *Chem. Soc. Rev.* 43, 4714–4727. doi:10.1039/c4cs00020j

Thangadurai, V., Pinzaru, D., Narayanan, S., and Baral, A. K. (2015). Fast solid-state Li ion conducting garnet-type structure metal oxides for energy storage. *J. Phys. Chem. Lett.* 6, 292–299. doi:10.1021/jz501828v

Thangadurai, V., and Weppner, W. (2005). $Li_6ALa_2Ta_2O_{12}$ (A = Sr, Ba): novel garnet-like oxides for fast lithium ion conduction. *Adv. Funct. Mater.* 15, 107–112. doi:10.1002/adfm.200400044

Thompson, T., Sharafi, A., Johannes, M. D., Huq, A., Allen, J. L., Wolfenstine, J., et al. (2015). Tale of two sites: on defining the carrier concentration in garnet-based ionic conductors for advanced Li batteries. *Adv. Energy Mater.* 5: 1500096. doi:10.1002/aenm.201500096

Thompson, T., Wolfenstine, J., Allen, J. L., Johannes, M., Huq, A., Davida, I. N., et al. (2014). Tetragonal vs. cubic phase stability in Al-free Ta doped $Li_7La_3Zr_2O_{12}$ (LLZO). *J. Mater. Chem. A* 2, 13431–13436. doi:10.1039/c4ta02099e

Tong, X., Thangadurai, V., and Wachslnan, E. D. (2015). Highly conductive Li garnets by a multielement doping strategy. *Inorg. Chem.* 54, 3600–3607. doi:10.1021/acs.inorgchem.5b00184

Wang, Y., and Wei, L. (2012). High ionic conductivity lithium garnet oxides of $Li_{7-x}La_3Zr_{2-x}Ta_xO_{12}$ compositions. *Electrochem. Solid State Lett.* 15, A68–A71. doi:10.1149/2.024205esl

Wolfenstine, J., Ratchford, J., Rangasamy, E., Sakamoto, J., and Allen, J. L. (2012). Synthesis and high Li-ion conductivity of Ga-stabilized cubic $Li_7La_3Zr_2O_{12}$. *Mater. Chem. Phys.* 134, 571–575. doi:10.1016/j.matchemphys.2012.03.054

Xie, H., Park, K.-S., Song, J., and Goodenough, J. B. (2012). Reversible lithium insertion in the garnet framework of $Li_3Nd_3W_2O_{12}$. *Electrochem. commun.* 19, 135–137. doi:10.1016/j.elecom.2012.03.014

Conflict of Interest Statement: The authors declare that the research was conducted in the absence of any commercial or financial relationships that could be construed as a potential conflict of interest.

3

Improving NASICON Sinterability through Crystallization under High-Frequency Electrical Fields

Ilya Lisenker and Conrad R. Stoldt*

Department of Mechanical Engineering, University of Colorado Boulder, Boulder, CO, USA

The effect of high-frequency (HF) electric fields on the crystallization and sintering rates of a lithium aluminum germanium phosphate (LAGP) ion conducting ceramic was investigated. LAGP with the nominal composition $Li_{1.5}Al_{0.5}Ge_{1.5}(PO_4)_3$ was crystallized and sintered, both conventionally and under effect of electrical field. Electrical field application, of 300 V/cm at 1 MHz, produced up to a 40% improvement in sintering rate of LAGP that was crystallized and sintered under the HF field. Heat sink effect of the electrodes appears to arrest thermal runaway and subsequent flash behavior. Sintered pellets were characterized using X-ray diffraction, scanning electron microscope, TEM, and electrochemical impedance spectroscopy to compare conventionally and field-sintered processes. The as-sintered structure appears largely unaffected by the field as the sintering curves tend to converge beyond initial stages of sintering. Differences in densities and microstructure after 1 h of sintering were minor with measured sintering strains of 31 vs. 26% with and without field, respectively. Ionic conductivity of the sintered pellets was evaluated, and no deterioration due to the use of HF field was noted, though capacitance of grain boundaries due to secondary phases was significantly increased.

Keywords: NASICON, field-assisted sintering, high frequency, sintering, grain boundary capacitance

Edited by:
Jeff Sakamoto,
University of Michigan, USA

Reviewed by:
Fatih A. Cetinel,
BASF SE, Germany
Juchen Guo,
University of California Riverside, USA

*Correspondence:
Conrad R. Stoldt
stoldt@colorado.edu

Specialty section:
This article was submitted to Energy Storage, a section of the journal Frontiers in Energy Research

Citation:
Lisenker I and Stoldt CR (2016) Improving NASICON Sinterability through Crystallization under High-Frequency Electrical Fields. Front. Energy Res. 4:13.

INTRODUCTION

Sodium superionic conducting (NASICON) ceramics are a family of sodium- or lithium-based solids that have been investigated as potential solid-state electrolytes since the 1970s. They offer relatively high room temperature conductivity, improved safety due to their intrinsic stability, and the possibility of assembling mechanically robust monolithic battery cells (Birke et al., 1997; Nagata and Nanno, 2007). Furthermore, as a high modulus and hardness ceramic solid electrolyte, these materials are proposed to inhibit dendrite formation during cycling, thus preventing the possibility of a short circuit, thermal runaway, and ultimately battery cell failure. The leading candidates among this family of ceramics are aluminum substituted lithium titanium and lithium germanium phosphates. Trivalent ion substitution within the NASICON structure requires insertion of additional lithium to maintain charge neutrality. The additional lithium ions, which force a rearrangement of the lithium sublattice, have the effect of dramatically increasing lithium mobility and therefore conductivity. In particular, the $Li_{1+x}Al_xGe_{2-x}(PO_4)_3$ [lithium aluminum germanium phosphate (LAGP)], $x = 0.45$ appears to provide the highest conductivity for Li^+ ions (Li et al., 1988; Francisco et al., 2015), with higher x values resulting in secondary phase formation (Cretin and Fabry, 1999). At a minimum, a dense structure with no interconnected porosity is required to inhibit dendrite propagation in lithium metal battery cells; however, the sintering of LAGP to a consistently high density remains a challenge with densities of 70% to just over 90% of theoretically reported despite sometimes lengthy sintering cycles (Cretin and Fabry, 1999; Yang et al., 2015).

A number of sintering approaches can be utilized to increase densification rates. Raising sintering temperatures is the simplest approach, but for LAGP above 800–825°C, phase separation occurs with formation of insulating phases (Thockchom and Kumar, 2010; Mariappan et al., 2011). The use of high heating rates, for example, by spark plasma sintering (SPS), to increase sintering rates has been well described. In addition to the applied mechanical pressure, the material is in intimate contact with a resistively heated graphite die resulting in heating rates in the hundreds of degree Celsius per minute. In materials for which the bulk diffusion activation energy, responsible for densification, is higher than the activation energy for surface diffusion, responsible for grain coarsening, rapid densification occurs (Stanciu et al., 2001). Resistively heating the sample itself, a process limited to electronically conductive materials is even more effective than traditional SPS but requires an SPS furnace with additional complexity of the die structure (Zapata-Solvas et al., 2015). SPS sintering of LAGP has been reported, but the achieved density was only 87% of theoretical, and the 20°C conductivity was measured to be 3.3×10^{-5} S cm^{-1} (Kubanska et al., 2014), a much lower value than what is commonly reported for such a density (Mariappan et al., 2011). It is notable that an additional, though low intensity, peak appears in the X-ray diffraction (XRD) pattern of the sample after SPS, corresponding to the most intense peak of $AlPO_4$, an insulating secondary phase, though a causal connection was not asserted.

Microwave sintering is another effective approach for rapid sintering of ceramic materials. The alternating electrical field component in microwave processing is used to polarize the ceramic, an inherently lossy process, thereby transferring energy to the crystal lattice with every reversal while rapidly heating the material (Sudiana et al., 2013). Notably, the higher frequencies above about 28 GHz that are needed to prevent formation of standing waves in the chamber and resultant hot spots in the material (Sudiana et al., 2013) may prevent its use in a co-sintering application of a battery device, since the bulk of the device would be effectively shielded by any electronically conductive materials (Bokhan et al., 1995).

Flash sintering methods use DC or lower frequency (up to 1000 Hz) AC electrical current passing directly through the sample to achieve remarkably rapid densification (Cologna et al., 2010). However, the charge carrier drift speed at the currents involved would inevitably cause NASICON phase breakdown due to local changes in stoichiometry resulting from lithium extraction near the anode and lithium accumulation near the cathode, similar phenomena to what has been shown in work with zirconia (Downs, 2013) and thoria (Goldwater, 1961). In addition, flash sintering, with an ohmic conducting path from the electrodes to the component and a dielectric breakdown-like thermal initiation mechanism (Todd et al., 2015; Zhang et al., 2015), presents a formidable engineering challenge in assuring an even current distribution across essentially 2D objects, such as battery components.

Finally, we note that a number of investigations have shown that the manner of material synthesis, and especially crystallization parameters, can substantially influence the conductivity of the final product (Thockchom and Kumar, 2010; Yang et al., 2015). Subjecting amorphous materials to alternating electrical fields, microwave irradiation in particular, has been used to promote uniform crystallization (Mahmoud, 2007).

To the authors' knowledge, frequencies between 1 kHz and 2.4 GHz represent an unexplored frequency range for field-assisted sintering. In this work, we evaluate the use of high-frequency (HF) electrical fields during synthesis, crystallization, and sintering as a means of enhancing sintering rate beyond that found in conventional thermal sintering processes for NASICON type ceramics. The choice of frequencies in the megahertz range represents a compromise between the shallow penetration of microwaves and electrolysis-inducing direct current. We further demonstrate the ability of a HF field to capacitively couple to the sample, eliminating both hot spots, and the need for platinum paints used in flash sintering to achieve a reliable electrical contact to the ceramic specimen.

EXPERIMENTAL PROCEDURE

Experimental Apparatus

A custom built set-up, as shown in **Figure 1**, was designed and fabricated to allow for the application of a variable frequency and variable voltage AC electrical field across the pellet during both crystallization and sintering. The set-up was contained inside a vertical tube furnace (Mellen Co., Concord, NH, USA). The furnace temperature was controlled using a k-type thermocouple embedded within one of the electrodes, approximately 1 mm below the sample, in an attempt to mitigate temperature gradients and sample self-heating. Any sample self-heating was thereby

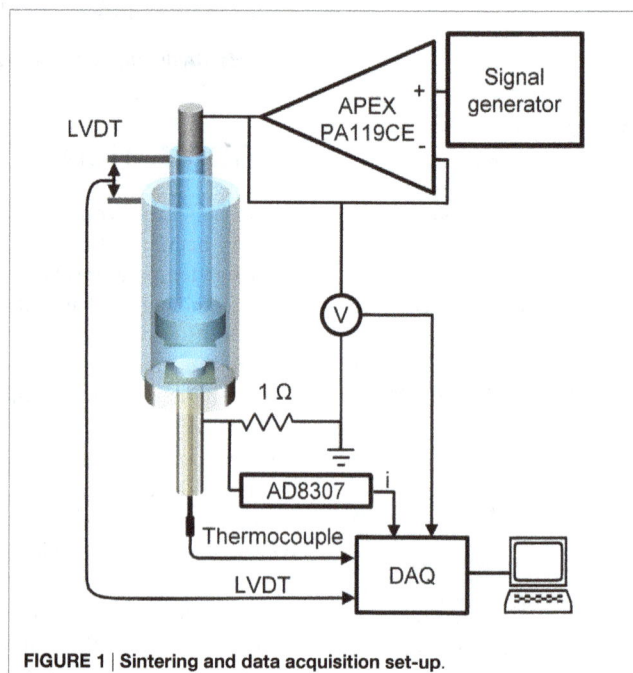

FIGURE 1 | Sintering and data acquisition set-up.

compensated by the furnace controller ensuring a stable sample temperature to within 2°C of the target.

Inconel 625 electrodes were separated from the sample pellet by 0.2-mm thick wafers made of Hexoloy SA silicon carbide (Saint-Gobain Ceramics, France) in order to prevent chromium oxide, deriving from Inconel surface oxidation, from contaminating the pellets. A custom built, all-quartz rod-in-tube dilatometer, based on a GTX-1000 LVDT and S7AC transducer (RDP Group, England), was used to monitor shrinkage in real time. The weight of the dilatometer components exerted an effective pressure of 0.2 MPa on the pellet during crystallization and sintering, thus ensuring intimate contact between electrodes and the pellet. The pressure exerted by the dilatometer was sufficiently low to prevent any sinter-forging of the sample, but was sufficient to prevent radial shrinkage of the pellets. An up to 30-V peak-to-peak, 1-MHz AC field can be applied across the pellet using a custom built amplifier based on APEX PA119CE (Apex Microtechnology, Tucson, AZ, USA) op-amp driven by Agilent 33220A signal generator (Keysight Technologies, Santa Rosa, CA, USA), resulting in a field approximately 300 V/cm for a typical 1-mm thick pellet. Note that the capacitor formed by the electrode and the pellet faces acts as a short circuit, inducing a current flow in the sample without any conductive coatings (e.g., platinum) on the pellet faces, as is required for DC and low frequency AC flash sintering experiments. Voltage and current monitoring were provided by a Schottky diode peak detector circuit and AD8307A (Analog Devices, Norwood, MA, USA) logarithmic amplifier to convert high-frequency signal to 0–10 V analog signal. Data acquisition was performed using U3-HV DAQ (Labjack Corp. Lakewood, CO, USA) at 1-s intervals.

Material Synthesis and Crystallization

The entire sample preparation process can be followed by referencing the flowchart in **Figure 2**. Single phase $Li_{1.5}Al_{0.5}Ge_{1.5}(PO_4)_3$

was prepared by coprecipitation from an aqueous mixture of $LiNO_3$ (Alfa Aesar, 99%), $AlNO_3 \cdot 9H_2O$ (Alfa Aesar, 98–102%), $(NH_4)H_2PO_4$ (Sigma-Aldrich, 99.99%), and an ethanol solution of $Ge(EtOH)_4$ (Gelest, 99.99%). The resulting precipitate was thoroughly dried by heating with stirring for 8 h at 80°C followed by an additional 8 h at 120°C. The precipitate was then reground in a mortar and heat treated in an alumina crucible in air for 8 h at 450°C to drive out most of the volatiles, including water and ammonia. The powder was then uniaxially pressed at a pressure of 220 MPa into 70-mg pellets with approximate dimensions of 6.35 mm diameter by 1.3 mm thickness. No binder was used for the pressing.

Two subsequent material crystallization routes were pursued. In both routes, the pellets were heat treated in the experimental furnace according to the schedule shown in **Figure 3**. Half of the material was heat treated under an applied high-frequency (HF) electrical field of 30 V peak-to-peak amplitude and 1.0 MHz frequency. Control samples were heat treated under identical conditions without HF field. The evolution of pellet thickness during the heat treatment is also illustrated in **Figure 3** (note that here positive displacement corresponds to reduction in thickness). The measured volume change coincides with LAGP crystallization peak reported previously by differential scanning calorimetry (Kubanska et al., 2014; Yang et al., 2015). The crystallization process resulted in highly porous bodies of agglomerated crystalline powder that were subsequently reground by hand in a mortar.

Sample Sintering

After crystallization, the respective LAGP powders were again uniaxially pressed (without binder) at a pressure of 220 MPa into 70-mg pellets with approximate dimensions of 6.35 mm diameter by 1.0 mm thickness. All pellets were sintered in a vertical tube furnace in air at 800°C for 1 h with a 10°C/min heating ramp under a 300 V/cm, 1-MHz HF field. The field was applied at the beginning of the heating ramp and was turned off after 4 h, once the sample had cooled below approximately 100°C. A total of five

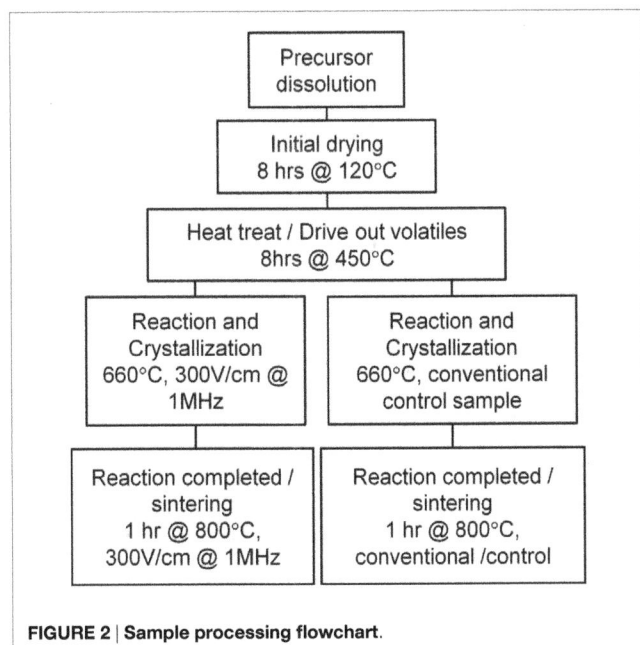

FIGURE 2 | Sample processing flowchart.

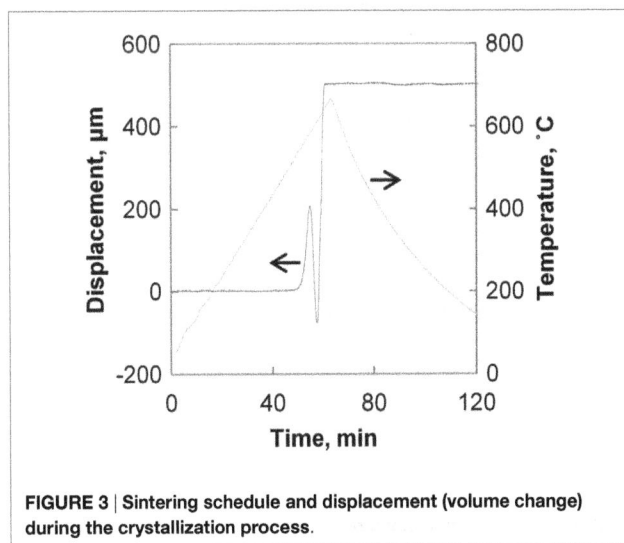

FIGURE 3 | Sintering schedule and displacement (volume change) during the crystallization process.

replicates from the two different synthesis runs were made. An additional five replicates using the same batch of material were used as controls with no field applied. Post-sintering sample shrinkage was confirmed using digital calipers.

The shorter time interval, the schedule of which is shown in **Figure 4**, while not allowing for full sintering, serves to accentuate the differences in initial densification rates for the methods used.

Characterization

Powder at every step of preparation and the sintered samples was characterized by XRD using a Bruker D2 Phaser (Bruker GmbH, Germany) using Cu-K_α radiation. Raman spectra, not reported here, on pellet faces was acquired using a Jasco NRS-3100 system (Jasco Corp., Tokyo, Japan) at a 532-nm excitation wavelength with a laser power level of approximately 65 mW and 50× effective magnification to non-destructively confirm material structure. Sample porosity was determined based on geometric measurement of pellet volume using digital calipers and the LAGP theoretical density as calculated from unit cell parameters.

Sample faces were subsequently sputter coated with approximately 500 nm of pure gold and electrochemical impedance spectroscopy (EIS) measurements were conducted using a Solartron 1250B potentiostat with a Solartron 1287 electrochemical interface (Ametek, Inc., Farnborough, UK) with a 100-mV excitation amplitude over a frequency range from 0.1 Hz to 1 MHz. EIS measurements were performed in an airtight enclosure under flowing dry nitrogen in a range of temperatures from room temperature down to −70°C. Fracture surfaces of two of the pellets were sputter coated with carbon and imaged using JEOL JSM-7401F field emission scanning electron microscope (SEM).

Additionally, two samples were prepared using FIB and imaged using a Talos F200S high-resolution transmission electron microscope (HRTEM) (FEI, Hillsboro, OR, USA) with energy dispersive X-ray spectroscopy (EDS) capability for elemental mapping.

RESULTS AND DISCUSSION

A comparison of XRD patterns from as-crystallized powders is shown in **Figure 5**. The powder crystallized under electrical field shows a slightly higher degree of crystallinity with approximately 10% larger grain size as determined by the Scherrer formula, specifically 69 and 63 nm, respectively. With no apparent differences in the timing and temperature of the onset of the crystallization,

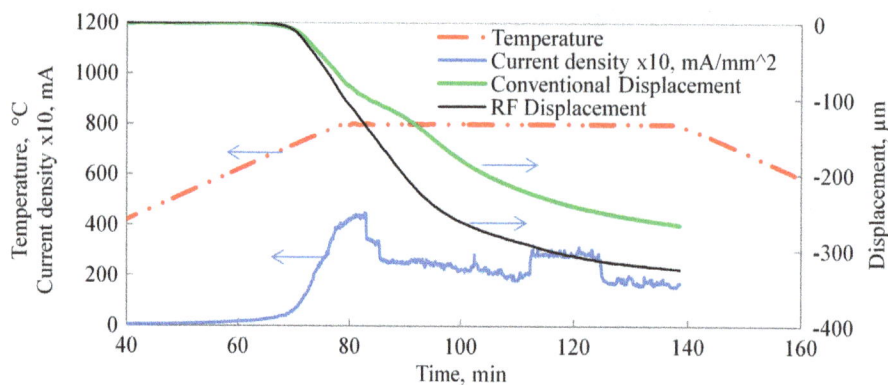

FIGURE 4 | **Conventional vs. HF sintering displacement curves**. Note that 100-μm displacement corresponds to 10% strain.

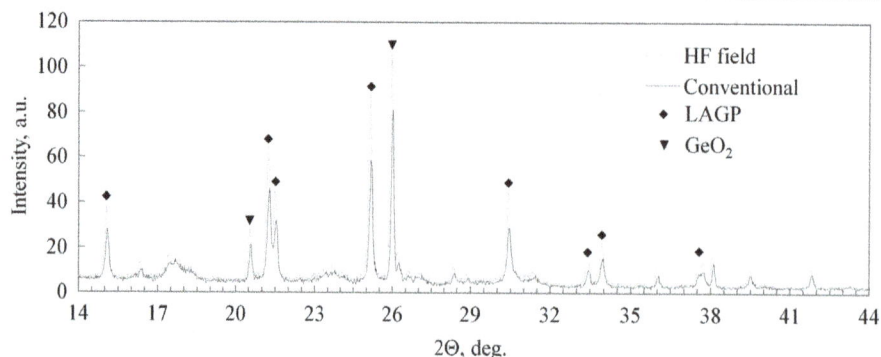

FIGURE 5 | **XRD of as-crystallized powders**.

it is reasonable to conclude that the field promotes crystallization through a mild non-thermal effect.

A comparison of the conventional and field-assisted final sintering curves is shown in **Figure 4**. Both curves are averages of five samples each. Sintering commences at approximately 630°C with the sintering rate accelerating and rapidly slowing down as some grain growth occurs, thus reducing sintering pressure. Residual thermal expansion of the dilatometer is subtracted from the data shown. The typical current profile is overlaid on the chart in **Figure 4** and illustrates the expected rise in electrical conductivity with temperature. The significant rise in measured current illustrates the effectiveness of using HF fields to achieve capacitive coupling with the material and obviates the need for any added conductive coating. The dramatic fluctuations in current are assignable to secondary phases melting and providing intermittent highly conductive paths until they react with the matrix forming the primary phase. This process can repeat itself many times during HF sintering.

It is notable that any effect on sintering rate is minor. Despite the slightly larger starting grain size, the sintering rate under HF electrical field is only increased by approximately 30% before and 40% after reaching steady state temperature. This brings into question the significance of non-thermally activated matter diffusion mechanisms in the LAGP system. Such effects are widely claimed for a number of electrically assisted sintering processes in other material systems. However, most of the theoretical treatment and experimental work is done on systems where charge carriers simultaneously belong to the crystal lattice. This is not the case for LAGP, and it is possible that polarization of the lattice is a requirement for distinct field effects. The final density of the samples was approximately 80.3 vs. 77.9%, and sintering strain of 30.5 vs. 26.2% for field processed vs. conventionally sintered samples, respectively. With the applied field being in the hundreds of volt per centimeter, a question of why the sample does not flash sinter has to be addressed. Few experimental geometries allow for direct control of the sample temperature by virtue of aspect ratio and electrical connections utilized. By comparison, our experiments embed a shielded thermocouple within the electrode, in intimate contact with the sample itself. This configuration allows for accurate temperature measurement that is not compromised by the applied field. Another benefit is our ability to control the furnace from the same sensor, in effect using the sample temperature for control. Thus, the sample is actively prevented from proceeding into the thermal runaway condition. This is further aided by the very low aspect ratio of the sample, effectively reducing internal thermal gradients and maximizing the intimate thermal contact with the electrodes, thus allowing the electrodes to serve as effective heat sinks. A similar effect has been reported for the flash sintering of zirconia, where the onset of flash was significantly delayed in low aspect ratio samples (Bichaud et al., 2015). To verify that the absence of flash sintering is a result of the thermal sink effect of the electrode configuration in our apparatus, a single sample was pressed from the same powder in the shape of a dog bone using method taught by Francis et al. (2012) and an attempt was made to sinter the sample using the flash sintering technique. The sample flashed immediately upon application of 300 V/cm DC

field at a comparatively low temperature of 400°C proving that the material is indeed amenable to flash if allowed to experience thermal runaway. The sample did not densify measurably; however, no further investigation was undertaken since this did not further the current study. Thus, thermal runaway in flash sintering appears to behave similarly to thermal dielectric breakdown (O'Dwyer, 1964).

Typical SEM photographs of fracture surfaces, as shown in **Figure 6**, demonstrate very similar microstructure, consistent with our density measurements. Grain size can be visually estimated to average at 0.3 μm in samples sintered with and without field. It appears that any difference in material that results from crystallization under field is effectively diminished after 1 h at 800°C. No abnormal grain growth was observed.

The XRD spectra, as shown in **Figure 7**, for the two processing methods appear nearly identical with slightly better definition of the weaker peaks, such as [006] and [202] in the HF field sintered material. Besides the primary LAGP phase peaks, only minor features are detected at 26.3° and 20.6°, identified as [101] and [100] peaks of the GeO_2 secondary phase. There is a near consensus in the LAGP community that $AlPO_4$ and some version of lithium phosphate are almost invariably present in amorphous form, and therefore do not produce diffraction peaks in XRD spectra. In order to identify the presence of additional amorphous impurities, HRTEM EDS images of HF field sintered material are shown in **Figure 8**, and contrasted with HRTEM EDS images of conventionally sintered material shown in **Figure 9**. These highlight misalignment of nearly perfectly formed tetragonal crystals

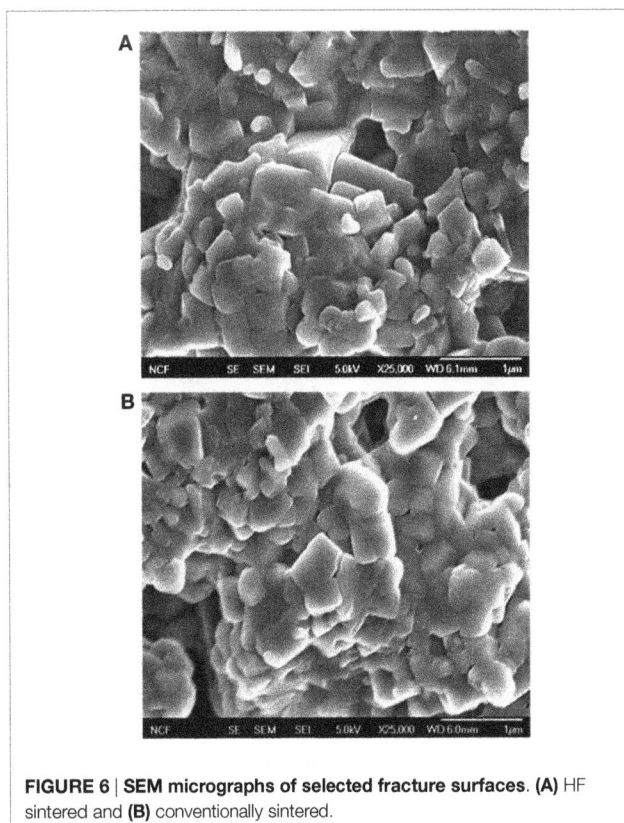

FIGURE 6 | **SEM micrographs of selected fracture surfaces. (A)** HF sintered and **(B)** conventionally sintered.

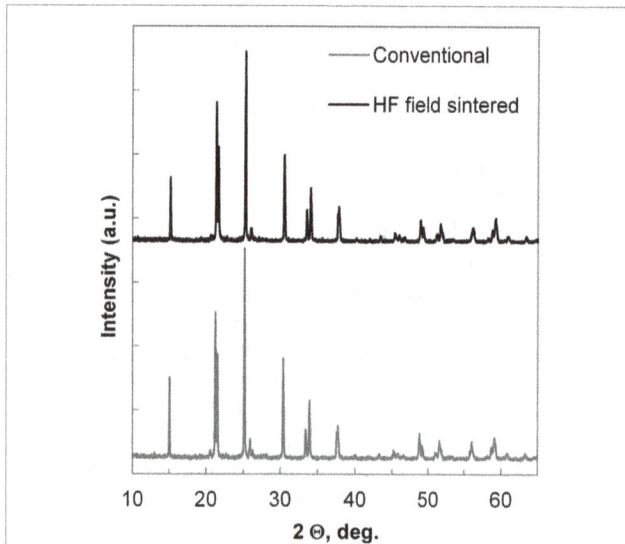

FIGURE 7 | X-ray diffraction patterns for HF (top) and conventionally (bottom) sintered LAGP.

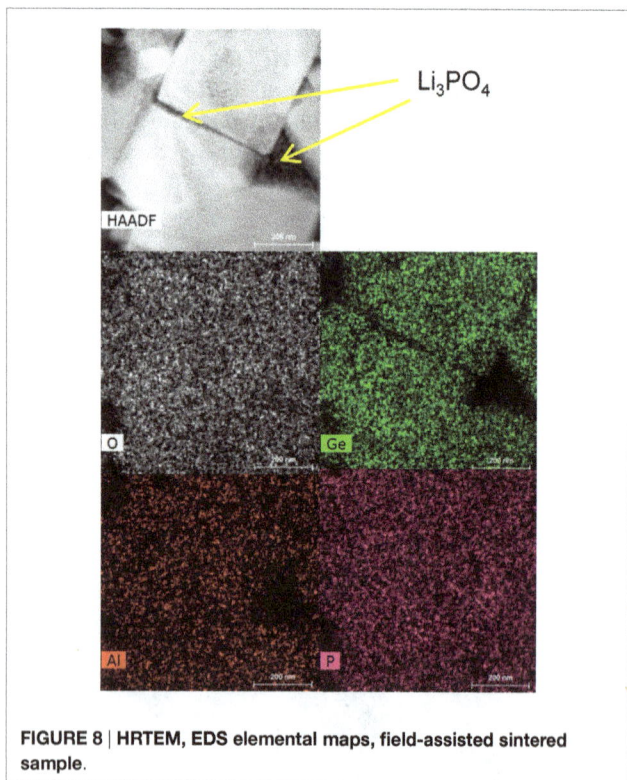

FIGURE 9 | HRTEM, EDS elemental maps, conventionally sintered sample.

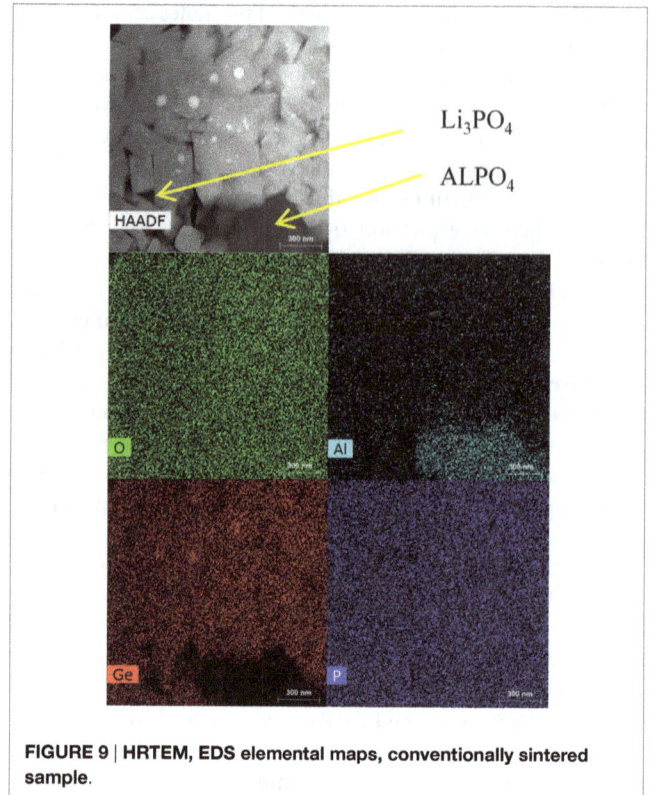

FIGURE 8 | HRTEM, EDS elemental maps, field-assisted sintered sample.

of LAGP with relatively little direct grain-to-grain contact. The amorphous phases between grains are identified by elemental mapping as either lithium phosphate, by absence of Al or Ge, or as AlPO₄. Precise phase identification is challenging as these phases appear extremely vulnerable to electron beam damage, consistent with previous reports (Mariappan et al., 2011). In

general, AlPO₄ seems to be more likely to exist as large inclusions or in inter-grain regions with large grain misalignment, whereas lithium phosphate appears to be present in thin layers between mostly aligned grain facets. Due to the much higher 1800°C melting point, AlPO₄ has a tendency to segregate and form larger inclusions as shown in the image of conventionally sintered LAGP, **Figure 9**.

Impedance spectroscopy results show essentially identical behavior for samples derived from the two processing methods. A typical Nyquist plot is presented in **Figure 10**. The expected contributions from bulk and grain boundary conductivity are clearly visible. An RC network model, as shown in the inset in **Figure 10**, was fit to the data using Zview software (Scribner Associates). Conductivity of the conventional and HF processed material was calculated according to

$$\sigma = \frac{1}{R} \times \frac{t}{A}$$

Although the data exhibit some variation among the five samples of each processing type, the distributions overlap between conventionally and field-assisted sintered samples. Typical results for the bulk and grain boundary conductivity are presented in **Figure 11** in the form of an Arrhenius plot. The calculated activation energies for the grain and grain boundary conductivity are 0.38 and 0.44 eV, respectively, and overall conductivities are on the order of 1.5×10^{-4} S/cm, consistent with values reported in the literature (Kotobuki et al., 2010).

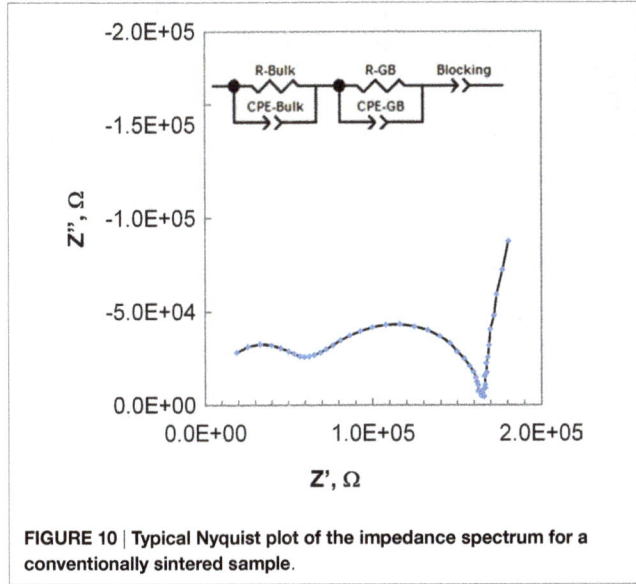

FIGURE 10 | Typical Nyquist plot of the impedance spectrum for a conventionally sintered sample.

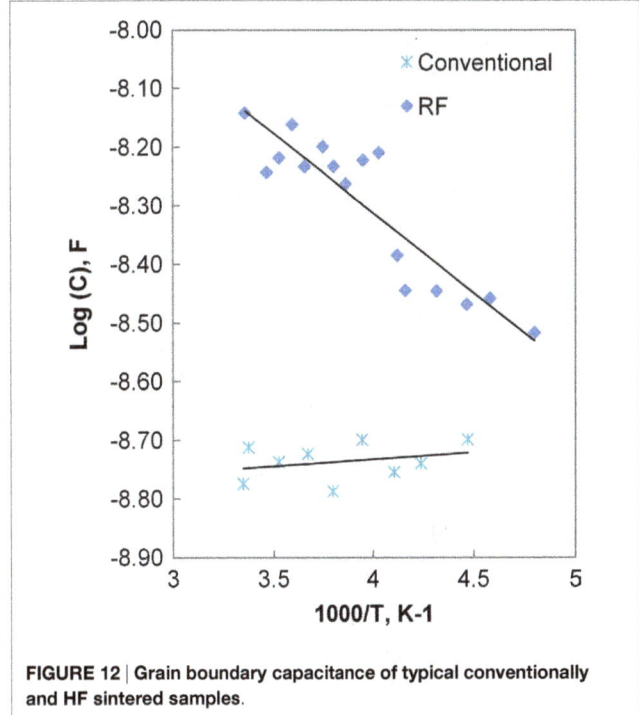

FIGURE 11 | Arrhenius plot of bulk and grain boundary conductivity for conventional and HF sintered samples.

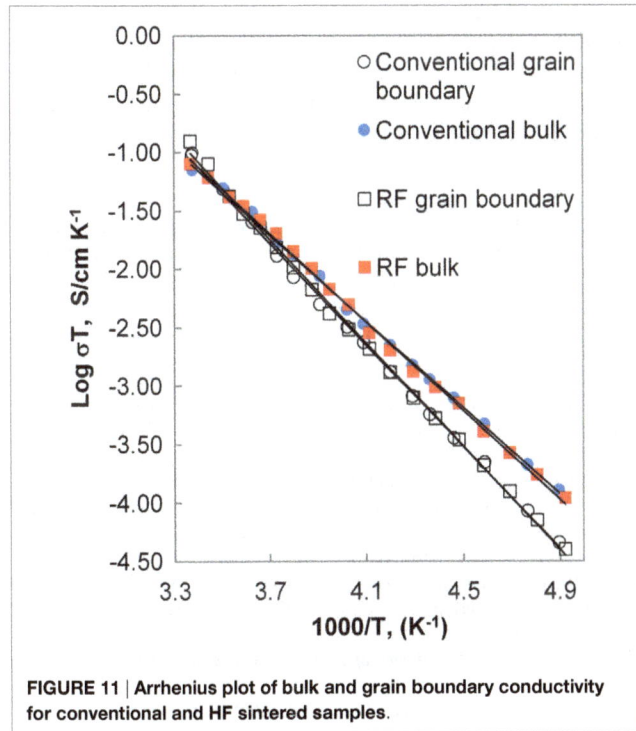

FIGURE 12 | Grain boundary capacitance of typical conventionally and HF sintered samples.

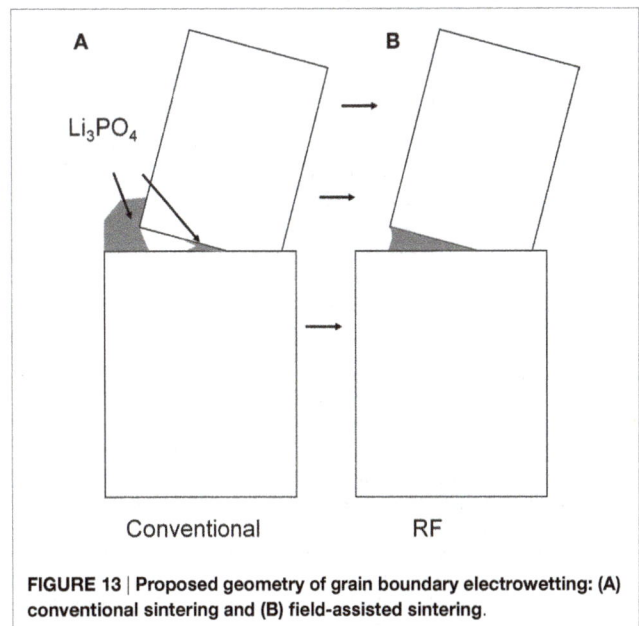

FIGURE 13 | Proposed geometry of grain boundary electrowetting: (A) conventional sintering and (B) field-assisted sintering.

A wholly unexpected effect is the significantly increased grain boundary capacitance of the samples processed under electrical field, as illustrated in **Figure 12**. It may be tempting to ascribe this effect to a thinner grain boundary, as thinner grain boundaries were convincingly demonstrated in flash sintered zirconia (M'Peko et al., 2013). However, the nearly identical grain boundary conductivity values between field-assisted and conventional sintering would suggest otherwise. We propose that the cause of the increased grain boundary capacitance is an increased contact area of an insulating intergranular phase adjacent to the actual conducting grain boundary. At least one of the secondary phases, e.g., lithium phosphate, melts at 847°C and might form a lower melting eutectic. If melting of these phases occurs at grain boundaries, application of electrical field would lead to electrowetting, effectively pulling these liquid phases into regions of highest field intensity where they subsequently act as a capacitive element in parallel with the grain boundary resistance, as illustrated in **Figure 13**.

The HRTEM images from our study and other researchers suggest that LAGP itself is fairly resistant to densification with actual grain contact area being substantially lower than the grain facet area (Mariappan et al., 2011). Thus, the classic brick layer model may be inadequate to explain grain boundary behavior in these materials with abundant secondary phases. In addition, since NASICON crystal conductivity is anisotropic, the widely reported grain conductivity is, at best, the average of a randomly oriented ensemble. The presence of intergranular spaces with a random distribution of insulating, e.g., Ge_2O and $AlPO_4$, and ionically conductive phases in addition to substantial porosity complicates the ability to rigorously characterize conduction mechanisms. Our future work will focus on separating the contribution of each phase to sinterability and conductivity with an eventual goal of engineering higher performing materials through intelligent grain boundary control.

CONCLUSION

High-frequency electrical fields by themselves do not appear to have a pronounced effect on sintering rate of LAGP ceramic. However, by allowing the starting material to crystallize under the same field, LAGP does show a moderately increased (+40%) sintering rate when sintered under field, as compared to powders that were conventionally calcined. Providing efficient heat sinks for Joule heating appears to arrest any tendency toward thermal

runaway and flash sintering. Microstructure of the ceramic and its ionic conduction properties also show no negative effects from application of HF electric fields. Finally, we demonstrate a HF field's ability to efficiently couple capacitively into ionically conductive ceramics, preserving stoichiometry and foregoing the need for electrically conductive coatings prior to field-assisted sintering.

AUTHOR CONTRIBUTIONS

CS provided guidance in experiment design and sample characterization, contributed to and edited the manuscript, and provided final approval for submission. IL designed and built the experimental set-up, manufactured samples for characterization, performed data analysis, assembled the manuscript, and provided final approval for submission.

ACKNOWLEDGMENTS

Support for instrumentation development was provided under ARPA-E 0869-1584 through a subcontract from Solid Power, Inc. (Louisville, CO, USA). Support for materials characterization was provided under the NSF Sustainable Energy Pathways Program (Project No. DMR-1231048). Additionally, the authors acknowledge the assistance of Dr. David Diercks for the HRTEM analyses and Dr. Brian Francisco for numerous discussions.

REFERENCES

Bichaud, E., Chaix, J. M., Carry, C., Kleitz, M., and Steil, M. C. (2015). Flash sintering incubation in Al_2O_3/TZP composites. *J. Eur. Ceram. Soc.* 35, 2587–2592. doi:10.1016/j.jeurceramsoc.2015.02.033

Birke, P., Salam, F., Döring, S., and Weppner, W. (1997). A first approach to a monolithic all solid state inorganic lithium battery. *Solid State Ionics* 118, 149–157. doi:10.1016/S0167-2738(98)00462-7

Bokhan, Yu. I., Komar, V. G., Misyuvyanets, V. Z., Mikhnevich, V. V., and Saraseko, M. N. (1995). High frequency sintering of multilayer ceramic capacitors. *J. Eng. Phys. Thermophys.* 68, 135–137. doi:10.1007/BF00854378

Cologna, M., Rashkova, B., and Raj, R. (2010). Flash sintering of nanograin zirconia in <5sec at 850 degrees C. *J. Am. Ceram. Soc.* 93, 3556–3559. doi:10.1111/j.1551-2916.2010.04089.x

Cretin, M., and Fabry, P. (1999). Comparative study of lithium ion conductors in the system $LiAlA(PO_4)_3$ with A=Ti or Ge and $0≤x≤0.7$ for use as Li+ sensitive membranes. *J. Eur. Ceram. Soc.* 19, 2931–2940. doi:10.1016/S0955-2219(99)00055-2

Downs, J. A. (2013). *Mechanisms of Flash Sintering in Cubic Zirconia*. Ph.D. dissertation, University of Trento, Trento.

Francis, J. S. C., Cologna, M., and Raj, R. (2012). Particle size effects in flash sintering. *J. Eur. Ceram. Soc.* 32, 3129–3136. doi:10.1016/j.jeurceramsoc.2012.04.028

Francisco, B. E., Stoldt, C. R., and M'Peko, J.-C. (2015). Energetics of ion transport in NASICON-type electrolytes. *J. Phys. Chem. C* 119, 16432–16442. doi:10.1021/acs.jpcc.5b03286

Goldwater, D. L. (1961). The electrolysis of thorium oxide crystals. *J. Phys. Chem. Solids* 18, 259–260. doi:10.1016/0022-3697(61)90172-X

Kotobuki, M., Hoshina, K., Isshiki, Y., and Kanamura, K. (2011). Preparation of Li1.5Al0.5Ge1.5(PO4)3 solid electrolyte by sol-gel method. *Phosphorus Res. Bull.* 25, 061–063.

Kubanska, A., Castro, L., Tortet, L., Schäf, O., Dollé, M., and Bouchet, R. (2014). Elaboration of controlled size $Li_{1.5}Al_{1.5}(PO_4)_3$ crystallites from glass-ceramics. *Solid State Ionics* 266, 44–50. doi:10.1016/j.ssi.2014.07.013

Li, S. C., Cai, J. Y., and Lin, Z. X. (1988). Phase relationships and electrical conductivity of LAGP and LACP systems. *Solid State Ionics* 2, 1265–1270. doi:10.1016/0167-2738(88)90368-2

Mariappan, C. R., Yada, C., Rosciano, F., and Roling, B. (2011). Correlation between micro-structural properties and ionic conductivity of $Li_{1.5}Al_{0.5}Ge_{1.5}(PO_4)_3$ ceramics. *J. Power Sources* 196, 6456–6464. doi:10.1016/j.jpowsour.2011.03.065

Mahmoud, M. M. (2007). *Crystallization of Lithium Disilicate Glass Using Variable Frequency Microwave Processing*. Master's thesis, Virginia Polytechnic Institute and State University, Blacksburg, VA.

M'Peko, J.-C., Francis, J. S. C., and Raj, R. (2013). Impedance spectroscopy and dielectric properties of flash versus conventionally sintered yttria-doped zirconia electroceramics viewed at the microstructural level. *J. Am. Ceram. Soc.* 96, 3760–3767. doi:10.1111/jace.12567

Nagata, N., and Nanno, T. (2007). All solid battery with phosphate compounds made through sintering process. *J. Power Sources* 174, 832–837. doi:10.1016/j.jpowsour.2007.06.227

O'Dwyer, J. J. (1964). *The Theory of Dielectric Breakdown*. Oxford: Clarendon Press.

Stanciu, L. A., Kodash, V. Y., and Groza, J. R. (2001). Effects of heating rate on densification and grain growth during field-assisted sintering of Al_2O_3 and $MoSi_2$. *Metall. Mater. Trans. A* 32, 2633–2638. doi:10.1007/s11661-001-0053-6

Sudiana, I. N., Ito, R., Inagaki, S., Kuwayama, K., Sako, K., and Mitsudo, S. (2013). Densification of alumina ceramics sintered by using submillimeter wave gyrotron. *J. Infrared, Millim. Terahertz Waves* 34, 627–638. doi:10.1007/s10762-013-0011-6

Thockchom, J. S., and Kumar, B. (2010). The effect of crystallization parameters of the ionic conductivity of a lithium aluminum germanium phosphate glass-ceramic. *J. Power Sources* 195, 2870–2876. doi:10.1016/j.jpowsour.2009.11.037

Todd, R. I., Zapata-Solvas, E., Bonilla, R. S., Sneddon, T., and Wilshaw, P. R. (2015). Electrical characteristics of flash sintering: thermal. *J. Eur. Ceram. Soc.* 35, 1865–1877. doi:10.1016/j.jeurceramsoc.2014.12.022

Yang, J., Huang, Z., Huang, B., Zhou, J., and Xu, X. (2015). Influence of phosphorus sources on lithium ion conducting performance in the system of Li_2O-Al_2O_3-GeO_2-P_2O_5 glass-ceramics. *Solid State Ionics* 270, 61–65. doi:10.1016/j.ssi.2014.12.013

Zapata-Solvas, E., Gomez-Garcia, D., Dominguez-Rodriguez, A., and Todd, R. I. (2015). Ultra-fast and energy-efficient sintering of ceramics by electric current concentration. *Sci. Rep.* 5, 1–7. doi:10.1038/srep08513

Zhang, Y., Jung, J., and Luo, J. (2015). Thermal runaway, flash sintering and asymmetrical microstructural development of ZnO and ZnO–Bi_2O_3 under direct currents. *Acta. Mater.* 94, 87–100. doi:10.1016/j.actamat.2015.04.018

Conflict of Interest Statement: The corresponding author CS is a founder and member of the Board of Solid Power, Inc., one of the two sponsors of this research work. IL has no conflict of interest to declare.

4

The Electrochemical Characteristics and Applicability of an Amorphous Sulfide-Based Solid Ion Conductor for the Next-Generation Solid-State Lithium Secondary Batteries

Yuichi Aihara[1]*, Seitaro Ito[1], Ryo Omoda[1], Takanobu Yamada[1], Satoshi Fujiki[1], Taku Watanabe[1], Youngsin Park[2] and Seokgwang Doo[2]

[1] Samsung R&D Institute Japan, Minoo-shi, Japan, [2] Samsung Advanced Institute of Technology, Samsung Electronics Co., Ltd, Suwon-si, South Korea

Edited by:
Fuminori Mizuno,
Toyota Research Institute of North
America, USA

Reviewed by:
Yongguang Zhang,
Hebei University of Technology, China
Xiao-Liang Wang,
Seeo Inc., USA

*Correspondence:
Yuichi Aihara
yuichi.aihara@samsung.com

Specialty section:
This article was submitted to
Energy Storage,
a section of the journal
Frontiers in Energy Research

Citation:
Aihara Y, Ito S, Omoda R, Yamada T,
Fujiki S, Watanabe T, Park Y and
Doo S (2016) The Electrochemical
Characteristics and Applicability of an
Amorphous Sulfide-Based Solid Ion
Conductor for the Next-Generation
Solid-State Lithium
Secondary Batteries.
Front. Energy Res. 4:18.

Sulfide-based solid electrolytes (SEs) are of considerable practical interest for all solid-state batteries due to their high ionic conductivity and pliability at room temperature. In particular, iodine containing lithium thiophosphate is known to exhibit high ionic conductivity, but its applicability in solid-state battery remains to be examined. To demonstrate the possibility of the iodine-doped SE, $LiI–Li_3PS_4$, was used to construct two different types of test cells: $Li/SE/S$ and $Li/SE/LiNi_{0.80}Co_{0.15}Al_{0.05}$ cells. The SE, $LiI–Li_3PS_4$, showed a high ionic conductivity approximately 1.2 mS cm^{-1} at 25°C. Within 100 cycles, the capacity retention was better in the $Li/SE/S$ cell, and no redox shuttle was observed due to physical blockage of SE layer. The capacity fade after 100 cycles in $Li/SE/S$ cell was approximately 4% from the maximum capacity observed at 10th cycle. In contrast, the capacity fade was much larger in $Li/SE/LiNi_{0.80}Co_{0.15}Al_{0.05}$ cell, probably due to the decomposition of the electrolyte at the operating potential range. Nevertheless, both the $Li/SE/LiNi_{0.80}Co_{0.15}Al_{0.05}$ and $Li/SE/S$ cells exhibited high coulombic efficiencies above 99.6 and 99.9% during charge–discharge cycle test, respectively. This indicates that a high energy density can be achieved without an excess lithium metal anode. In addition, it was particularly interesting that the SE showed a reversible capacity about 260 mAh g$^{-1}_{SE}$ (the value calculated using the net solid electrolyte weight). This electrolyte may behave not only as an ionic conductor but also as a catholyte.

Keywords: solid-state battery, solid electrolyte, lithium–sulfur batteries, lithium secondary batteries, sulfide electrolyte, catholyte

INTRODUCTION

Nowadays, requirements for batteries have been increasingly stringent in terms of energy density and safety because of the diversity of functions of electronic devices and their usage as a power source especially in transportation. Lithium ion batteries (LIBs) have been in the market over two decades because of their high energy density, but even better batteries have been sought to meet ever stricter requirements (Scrosati and Garche, 2010). Solid-state lithium batteries have received considerable attention since the inception of primary and secondary lithium batteries (Weppner, 1981). From viewpoints of (1) theoretical capability and (2) physical properties, solid-state lithium batteries have been believed to be next-generation power sources. Unfortunately, the commercial applications of all solid-state batteries are limited to a few cases, such as the Li–I primary battery, and thin film type solid-state batteries because of serious technical challenges (Bates et al., 1993).

Recently, fast lithium ion conductors have been found in sulfide compounds (Kamaya et al., 2011). A large difference between typical oxide- and sulfide-based solid electrolytes (SEs) is in their physicomechanical properties, especially on their deformability. In particular, a large grain boundary resistance, observed in oxide SEs, does not unambiguously exist in many sulfide SEs because of the pliable nature of the sulfide compounds; seamless reaction interfaces can be easily formed by a cold press without high temperature sintering (Tatsumisago et al., 2002). Recently, excellent cell characteristics have been reported using a small pellet cell (Ogawa et al., 2012). Unfortunately, most of these studies adopted a Li–In alloy, because of its stable cycle due to the solid solution, unlike a dissolution–deposition reaction. As well as a conventional LIB, a rocking chair type all-solid-state battery (Ogawa et al., 2012) and Li–S battery (Nagata and Chikusa, 2014) have also been proposed. We have shown the possibility of a practical 2-Ah class all-solid LIB using a sulfide electrolyte (Ito et al., 2014). We also have investigated the anode kinetics using Li_3PS_4, i.e., $Li_2S:P_2S_5 = 75:25$ (Yamada et al., 2015). Use of the solid-state electrolyte greatly facilitates the design of a highly save battery unlike the ones using flammable organic liquid solvents.

Among many sulfide compounds, lithium thiophosphates containing halogens exhibit relatively high ionic conductivities. Mercier et al. (1981) reported a structure and property relation of $Li_2S–P_2S_5–LiI$ systems. They reported that the addition of LiI to $Li_2S–P_2S_5$ lowered the glass transition temperature, and the material had a high ionic conductivity of approximately 10^{-3} S cm^{-1} at 25°C when the molar ratio of $LiI:0.66Li_2S–0.33P_2S_5$ is 0.45. Ohtomo et al. (2013) investigated the electrochemical stability and ionic conductivity of $LiI–Li_2O–Li_2S–P_2S_5$ glass. They found that $30LiI–70 (0.07Li_2O–0.68Li_2S–0.25P_2S_5)$ is stable over 0–10 V and exhibited a high ionic conductivity of 1.3×10^{-3} S cm^{-1}. The stability and ionic conductivity of $Li_7P_2S_8I$ was reported by Rangasamy et al. (2015). They observed that the addition of iodine to the β-Li_3PS_4-based structure resulted in good electrochemical stability against an Li metal anode while maintaining a conductivity of 6.3×10^{-4} S cm^{-1} at room temperature. These literatures indicate that iodine-containing lithium thiophosphates have great potential and that there is considerable practical interest in investigating the applicability of this class of materials as SE in all solid-state batteries.

In this paper, we report the cell cycle characteristics longer than 100 cycles with the theoretical sulfur redox capacity plus the electrolyte redox capacity, using an amorphous $LiI–Li_3PS_4$ electrolyte, referring with $LiNi_{0.80}Co_{0.15}Al_{0.05}$ (NCA) cathode. To clarify the origin of the excess cell capacity from a net sulfur specific capacity, the electrolyte was also tested as a cathode. It was found that amorphous $LiI–Li_3PS_4$ had a reversible capacity of about 260 mAh g^{-1} within the potential range of 1.3–2.7 V versus Li/Li$^+$. The possibility of using a sulfide-based solid electrolyte in solid-state lithium secondary batteries using was also discussed.

EXPERIMENTAL

Materials

The solid-state electrolyte (SE), $LiI–Li_3PS_4$, was prepared by the mechanical milling method in an Ar-gas-filled vessel was used for above preparation to prevent degradation of the sample. Specifically, 35 mol% of LiI (Aldrich, 99.999%) and 65mol% of $0.75Li_2S$ (Alfa, 99.9%)–$0.25P_2S_5$ (Aldrich, 99%) were mixed at 380 rpm for 35 h (70 cycles of 20 min of operation and 10 min of interval) by using Fritsch (Idar-Oberstein) P-5 grinding bowl fasteners. The synthesized electrolyte was characterized using Raman spectroscopy (JASCO, NRS-3100, Tokyo) and X-ray diffraction (XRD, CuKα, 45 kV, 40 mA, Panalytical, Empyrean XRD, Almelo).

$LiNi_{0.80}Co_{0.15}Al_{0.05}$ (NCA) was coated with $Li_2O–ZrO_2$ (LZO) and was used for the reference cathode material. The preparation of LZO–NCA was given in our previous report (Ito et al., 2014). An NCA cathode composite was prepared by mixing the LZO–NCA, $LiI–Li_3PS_4$ and a vapor grown carbon fiber (VF, the fiber diameter: approximately 200–500 nm, the length: approximately 5–10 μm) in the weight ratio of 0.60:0.35:0.05 using a mortar (hand mixing) in an Argon box (MBraun, LABmaster dp, $H_2O < 0.1$ ppm, $O_2 < 0.1$ ppm).

For preparing the lithium-sulfur (Li–S) cell, the carbon-sulfur (C–S) composite was first prepared using the ball milling method. Dried sulfur powder (Alfa Aesar) was mixed with an activated carbon (MAXSORB® MSC-30, Kansai Coke and Chemicals Co., Ltd.) in the weight ratio of 26:74 wt% in an argon-filled 45-mL zirconia vessel by using high-energy ball milling at 370 rpm for 10 h (20 cycles of 20 min operation and 10 min interval). Then, the C–S composite was further mixed with $LiI–Li_3PS_4$ (SE) in the weight ratio 0.5:0.5 to make a carbon-sulfur-solid electrolyte (C–S–SE) composite, using the same high energy ball milling method. The mixing procedure is the same as for the C–S preparation. The weight fraction of the prepared sulfur cathode composite was 0.37:0.13:0.50 for sulfur:carbon:LiI–Li_3PS_4. The C–SE composite (0.21:0.79) was similarly prepared using the ball milling method. Lithium foil ($t = 30$ μm, Honjo Metal, Osaka) was used for the anode.

Differential scanning calorimetry (DSC) was performed on a SII X-DSC7000 calorimeter (Seiko SII, Tokyo) using an Ar-filled closed pan for the prepared LiI, Li_3PS_4, $LiI–Li_3PS_4$, C–S–SE, and C–SE composites. The measurements were performed between room temperature and 400°C with scanning rate of 2°min^{-1}.

For the evaluation of the applicability of cathode materials, the in-house electrochemical cell (two electrodes cell, φ 13 mm) was adopted. The cell preparation method and the schematic structure are also described in our previous paper (Ito et al., 2014). The details of the prepared cells are listed in **Table 1**.

Electrochemical Measurements

The ionic conductivity of the SE was determined from the cell constant and bulk resistance observed in electrochemical

TABLE 1 | Specification of the test cells.

	S-cell	NCA-cell
Cathode active material weight fraction (wt%)	37	60
Cathode composite loading (mg · cm⁻²)	2.7	11.3
Active material loading (mg · cm⁻²)	1.0	6.8
Active material/carbon ratio (w/w)	2.9	(12.0)
Cell capacity (mAh)	2.1	1.2
Cathode area specific capacity (mAh · cm⁻²)	1.6	0.9
Anode area specific capacity (mAh · cm⁻²)	23.2	23.2
Anode/cathode capacity ratio	14.5	25.8
Utilization of anode at DOD = 100% (%)	6.9	4.0

DOD, depth of discharge.

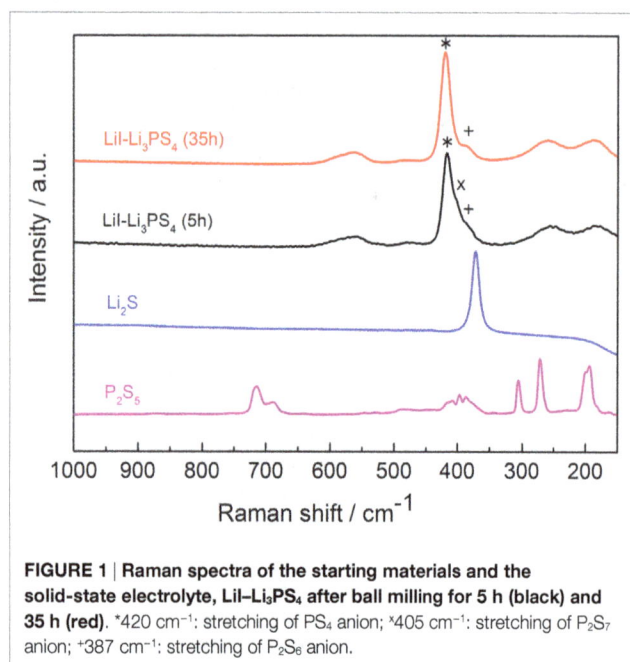

FIGURE 1 | Raman spectra of the starting materials and the solid-state electrolyte, LiI–Li₃PS₄ after ball milling for 5 h (black) and 35 h (red). *420 cm⁻¹: stretching of PS₄ anion; ˣ405 cm⁻¹: stretching of P₂S₇ anion; ⁺387 cm⁻¹: stretching of P₂S₆ anion.

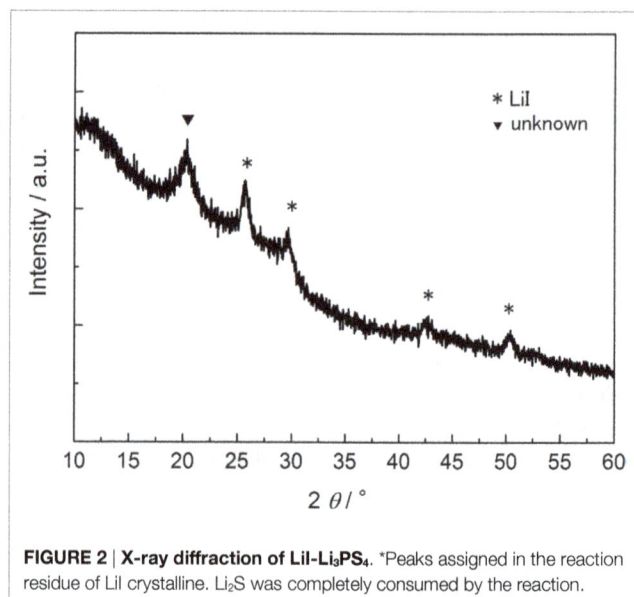

FIGURE 2 | X-ray diffraction of LiI–Li₃PS₄. *Peaks assigned in the reaction residue of LiI crystalline. Li₂S was completely consumed by the reaction.

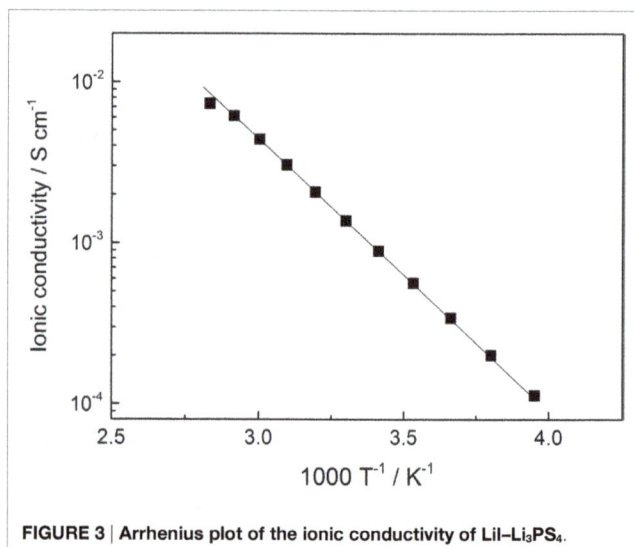

FIGURE 3 | Arrhenius plot of the ionic conductivity of LiI–Li₃PS₄.

impedance spectroscopy (EIS). The EIS was performed using an AUTOLAB PGSTAT30 with an frequency response analysis FRA module (Metrohm Autolab, Utrecht) controlled by a personal computer. Linear sweep voltammetry (LSV) was performed using the same equipment. An asymmetric two electrodes cell with a Pt working electrode and a lithium counter (reference) electrode was prepared for LSV. The galvanostatic charge/discharge profiles were obtained using a battery test station, TOSCAT3100 (Toyo system, Iwaki) at 25°C position in a temperature chamber.

RESULTS

Characterization of LiI–Li₃PS₄

The crystallinity and anion structure of the SE, LiI–Li₃PS₄, was characterized by Raman spectroscopy (**Figure 1**) and XRD (**Figure 2**). It is clear from the Raman spectra that the reaction proceeded because of the high-energy ball milling within the first 5 h. After 35 h of ball milling, the intensities of the P₂S₇ dimer diminished in the spectrum. However, even after 35 h, the P₂S₆ peak was not completely suppressed. Furthermore, some reaction

residue of LiI was found in the XRD pattern. The XRD pattern and Raman spectrum did not show any obvious change after 35 h, and so synthesis was discontinued at this point. From the above results, the synthesized electrolyte was mostly amorphous containing a small fraction of P₂S₆ dimer and crystalline LiI. The main anion frame structure was ortho - PS_4^- (and I⁻).

The temperature dependence of the ionic conductivity for LiI–Li₃PS₄ is plotted in **Figure 3**. The temperature dependence was a typical Arrhenius type, and the activation energy of the ionic conductivity was 29.1 ± 1.1 kJ mol⁻¹. The ionic conductivity was 1.2 mS cm⁻¹ at 25°C, and it is approximately 10 times than that of amorphous Li₃PS₄ SE (Yamada et al., 2015).

The electrochemical stability of the electrolyte was determined using LSV (**Figure 4**). The potential was scanned from the open circuit potential to 5.0 V versus co-counter/reference electrodes of Li/Li⁺. At 2.8 V, a small spike peak was observed for LiI–Li₃PS₄.

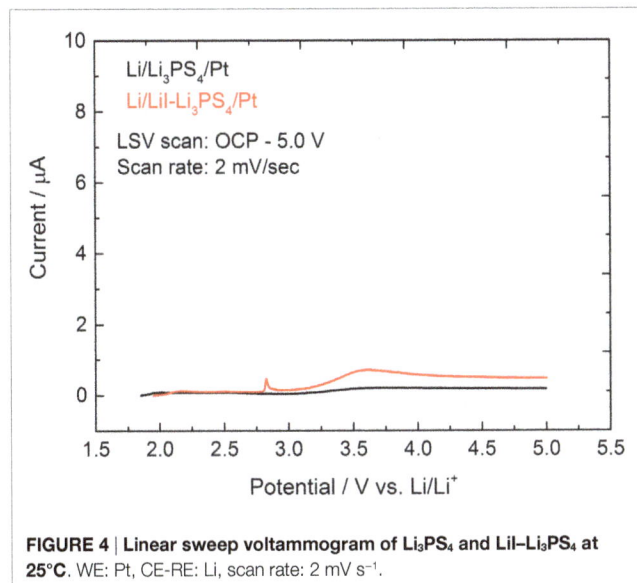

FIGURE 4 | Linear sweep voltammogram of Li₃PS₄ and LiI–Li₃PS₄ at 25°C. WE: Pt, CE-RE: Li, scan rate: 2 mV s⁻¹.

This peak might originate from isolated lithium iodide and its anodic oxidation. Although the anodic decomposition current was not significant, the current profile indicated a small anodic oxidation above 3.1 V for LiI–Li₃PS₄. On the contrary, no anodic current was observed in Li₃PS₄ cell.

Li/LiI–Li₃PS₄/LZO–NCA Cell (NCA Cell)

The discharge profiles and capacity attenuation during 100 cycles for Li/LiI–Li₃PS₄/LZO–NCA cell (NCA cell) is shown in **Figures 5A,B**. The cell showed initial discharge capacities of approximately 134 and 94 mAh g⁻¹ at 0.05 and 0.5°C, respectively. These cathode specific capacities were larger than in our previous study (Ito et al., 2014). However, the specific capacity decreased from 95 to 70 mAh g⁻¹ after 100 cycles with the charge/discharge condition of 0.1 C CC–CV/0.5 C CC. The capacity fade was much faster than that in Graphite/Li₂S–P₂S₅ (80:20 mol%)/LZO–NCA (Ito et al., 2014). The initial coulombic efficiency was 69.4% and gradually increased to 99.0% by the sixth cycle. The average coulombic efficiency during the last 50 cycles was 99.6%. After 100 cycles, the average cell closed potential decreased due to the increase of iR drop up in the first 2 s right after the galvanostatic charge/discharge started. These performance characteristics indicate that the cell-specific capacity gradually decreased due to the increase of cell resistance with the same cutoff potential.

Li/LiI–Li₃PS₄/C–S Cell (S Cell)

The discharge profiles and capacity attenuation during 100 cycles for the Li/LiI–Li₃PS₄/S (S cell) is shown in **Figures 6A,B**. On account of the expected poor rate capability of the S cell, the discharge rate was fixed at 0.15 C (0.25 mA cm⁻²) with the charge/discharge cut off potentials between 3.1 and 1.3 V (constant current charge/discharge). In **Figure 6A**, the discharge profiles including 1st, 10th, and 100th are shown at the same discharge rate. The initial cell capacity was only about 1300 mAh g⁻¹₋ₛ (the value calculated using the net sulfur weight), lower than what is

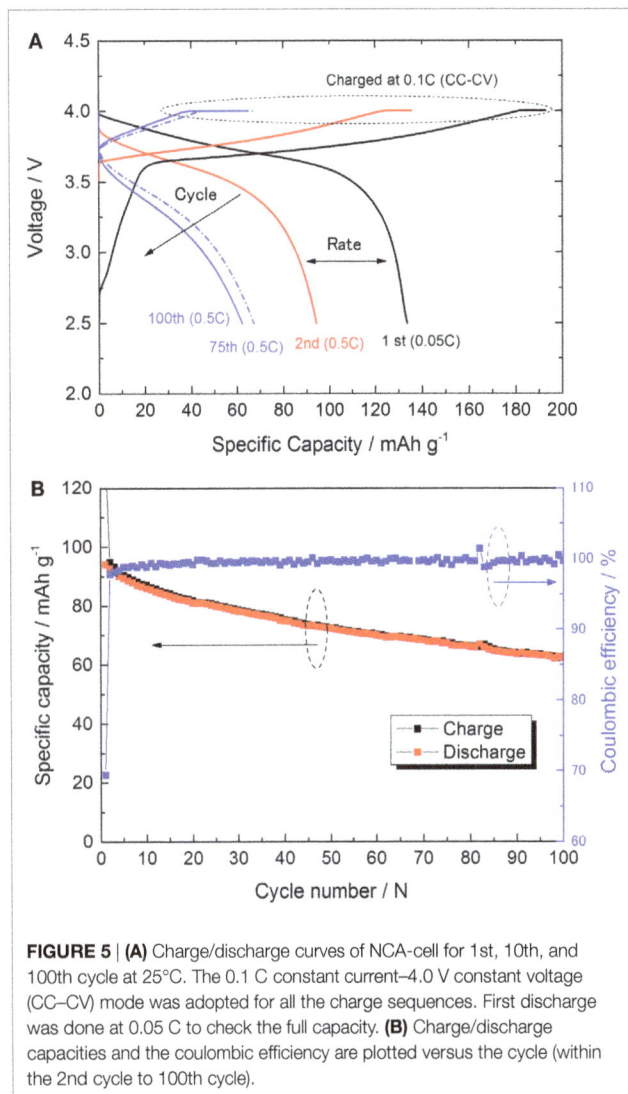

FIGURE 5 | (A) Charge/discharge curves of NCA-cell for 1st, 10th, and 100th cycle at 25°C. The 0.1 C constant current–4.0 V constant voltage (CC–CV) mode was adopted for all the charge sequences. First discharge was done at 0.05 C to check the full capacity. **(B)** Charge/discharge capacities and the coulombic efficiency are plotted versus the cycle (within the 2nd cycle to 100th cycle).

expected for a Li–S cell of this type. The specific capacity increased with the cycle and achieved a maximum value of 1688 mAh g⁻¹₋ₛ (calculated using the net sulfur weight) at the 10th cycle.

After the 10th cycle, the specific capacity gradually decreased to 1622 mAh g⁻¹₋ₛ by 100th cycle. The coulombic efficiency reached beyond 99.9% after the 30th cycle. From the maximum capacity obtained at the 10th cycle, the specific capacity decreased only 4% in the last 90 cycles. Also, it is clear that no redox shuttle was observed. This result is consistent with our previous report on the all-solid-state Li–S cell (Yamada et al., 2015). Unfortunately, the cycle performance was lower than the expectation. Nevertheless, in this paper, we changed the SE and the carbon support used in the cathode. Both the material change positively influenced the cell performance. The results are summarized and compared with the former NCA cell in **Table 2**.

Li/LiI–Li₃PS₄/C–LiI–Li₃PS₄ (E Cell)

Because of the strange charge plateau above 2.5 V, the possibility of a reversible redox reaction of the electrolyte was investigated.

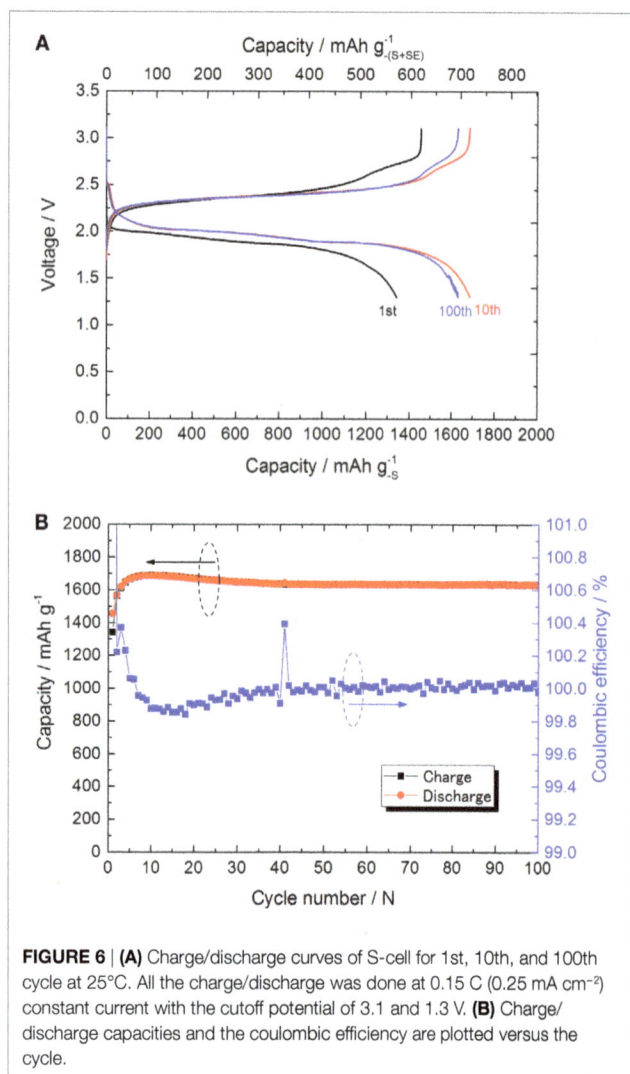

FIGURE 6 | (A) Charge/discharge curves of S-cell for 1st, 10th, and 100th cycle at 25°C. All the charge/discharge was done at 0.15 C (0.25 mA cm⁻²) constant current with the cutoff potential of 3.1 and 1.3 V. (B) Charge/discharge capacities and the coulombic efficiency are plotted versus the cycle.

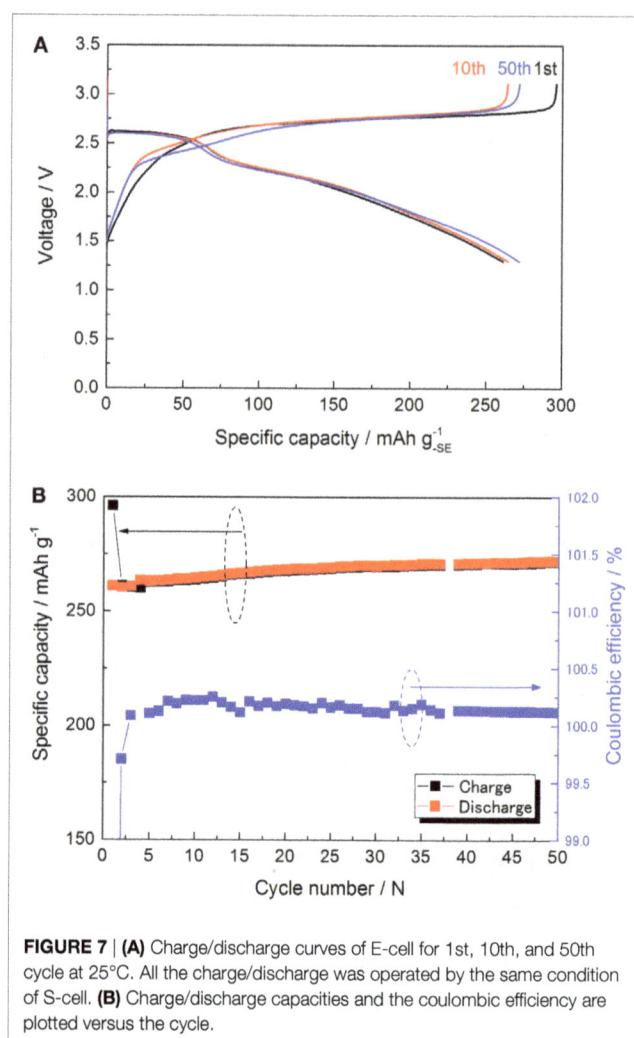

FIGURE 7 | (A) Charge/discharge curves of E-cell for 1st, 10th, and 50th cycle at 25°C. All the charge/discharge was operated by the same condition of S-cell. (B) Charge/discharge capacities and the coulombic efficiency are plotted versus the cycle.

TABLE 2 | Summary of the cell characteristics.

	S-cell	NCA-cell
Cell capacity (mAh)[a]	2.1	1.2
$V_{1/2Q}$ at 1st cycle (V)[b]	1.9	3.7
Cathode active material energy density (Wh/kg)	3000	493
Cathode composite energy density (Wh/kg)	1111	259
Average coulombic efficiency within last 50 cycles (%)	100.0	99.6
Capacity retention at 100th cycle (%)[c]	96	74

[a]Maximum capacities are adopted (S-cell: the discharge capacity at 10th cycle and NCA-cell: the discharge capacity at 1st cycle).
[b]The average discharge voltage was determined at the half capacity.
[c]The capacity retention was determined from 1st/100th capacity (S-cell) and 2nd/100th capacity (NCA-cell) to be fixed as the discharge rate.

Recently, the reversible reaction of Li_3PS_4 and "sulfur-rich lithium polysulfidophosphates" has been suggested (Lin et al., 2013; Hakari et al., 2015). Reactivity of the carbon–electrolyte (C–SE) composite was investigated using a similar cell configuration to the S-cell, but it did not contain sulfur in the cathode composite. The galvanostatic charge/discharge profile and the capacity

attenuation are plotted in **Figures 7A,B**. The redox capacity calculated using the weight of the electrolyte in the cathode composite was 261 mAh g_{-SE}^{-1} at 10th cycle. The charge/discharge capacities were quite stable during the cycle (the coulombic efficiency was always nearly equal to 100%). Although the actual reaction scheme is unclear, two potential plateaus were clearly observed in the discharge profile.

To examine the condition of the electrolyte in the C–SE composite, we measured Raman spectroscopy of our sample. However, after the ball milling, the Raman shift related to the electrolyte (420 cm⁻¹, PS_4^{3-} anion structure) was not observable. Other observed peaks were assigned to D/G bands of graphite (1570–1360 cm⁻¹). DSC was performed (**Figure 8**) and the results are summarized in **Table 3**. There was a significant difference between the profiles of Li_3PS_4 and $LiI–Li_3PS_4$. $LiI–Li_3PS_4$ exhibited a large peak at 160.9°C. This exothermic peak might be related to the reorientation and crystallization of LiI. In the scan, the glass transition appeared just before the crystallization of LiI. The heat flow decreases slightly in the SE and C–SE samples. Furthermore, the other peaks related to Li_3PS_4 (crystallizing and transformation to the thio-LISICON phase) showed high temperature shifts.

The peak pattern of the C–SE composite was unchanged after ball milling with activated carbon. However, the sharp peak shifted to high temperature in comparison with LiI–Li$_3$PS$_4$. The reason for this high temperature shift is obscure. In the C–S–SE sample, the peak pattern was completely changed by the ball milling. The sulfur (S$_8$) generally does not show any exothermic peak, and only endothermic melting peaks were observed before 300°C. However, the C–S–SE composite showed only a small endothermic peak around 107°C and two exothermic peaks. The presence of excess sulfur certainly changes the chemical environment and must be promoting alternate chemical reactions. Unfortunately, it was still not clear whether the materials were changed by the heating during the DSC or by the mechanical milling.

DISCUSSION

Characteristic of LiI–Li$_3$PS$_4$

In this paper, we have demonstrated two types of lithium metal secondary cells using a solid-state electrolyte based on an amorphous LiI–Li$_3$PS$_4$. This electrolyte consists of 48 mol%

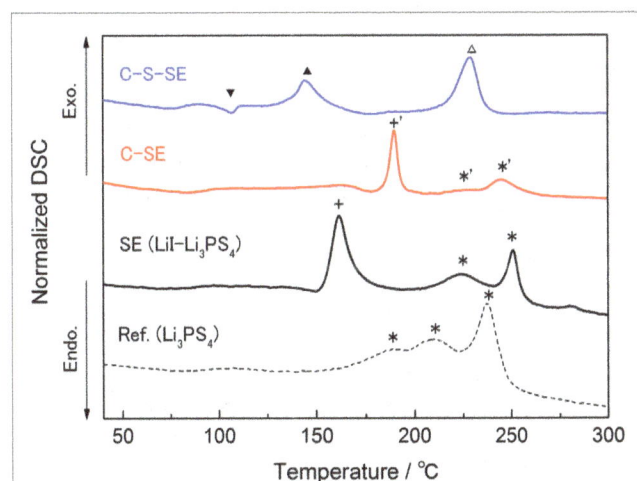

FIGURE 8 | DSC chart for the electrolytes and composites at the first heating. The heat flow was normalized by the sample weight. DSC scan rate was 2°C min^{-1}.

ortho-Li$_3$PS$_4$, and the 52 mol% of LiI. Although LiI is an ionic conductive material (3.2 µS cm^{-1}) (Rao et al., 1978), it is impossible to explain the improvement of ionic conductivity by the addition of LiI to Li$_3$PS$_4$, because of the poorer conductivity of LiI than that of Li$_3$PS$_4$. We assumed that the carrier number did not significantly change upon 52 mol% substitution by LiI compared to amorphous Li$_3$PS$_4$. The PS$_4$ anion actually has a symmetrical tetrahedron structure, and substitution with LiI replaces this highly symmetric spherical anion in the framework. This substitution influences the probability of the ion exchange between anions and Li$^+$ in the mixed electrolytes and increases the total ionic conductivity. However, the details of the phenomenon are still unclear, requiring further investigation. The LiI–Li$_3$PS$_4$ showed a high ionic conductivity, and this improved the cell performance, especially for the S-cell in comparison with our previous study (Ito et al., 2014). Unfortunately, LiI–Li$_3$PS$_4$ is unstable at high potentials (>2.7 V) probably due to oxidation of I$^-$. In the NCA cell, the capacity fade in the cycle test is much larger compared to previous results although the anode is different. On the other hand, the S cell showed very stable cycles (even though the utilization of the anode is deeper in the S cell); the electrolyte is at least as stable against the Li/Li$^+$ redox reaction. Although LiI–Li$_3$PS$_4$ may not be applicable to a 4 V system, it could be used for a metal lithium secondary battery with a high coulombic efficiency.

At this point, one may wonder whether the electrolyte material in the separate layer plays any role on the charge–discharge capacities. Hakari et al. (2015) have already demonstrated that the cathode composite without sufficient electronic conductivity does not allow any observable redox reactions. In our system, the electrical conductivity is given to the cathode composite by ball milling the SE with carbon to form complexes. Therefore, the SE in the separator region has no contribution in the capacity of the battery.

Exact Reactions in S Cell

The charge/discharge curves nearly emulated the theoretical redox capacity of sulfur. However, as described above, the charge/discharge capacities must be based on the summation of redox capacities of sulfur and SE, because the electrolyte shows reversible redox capacity in the same potential range 3.1–1.3 V versus Li/Li$^+$. The cathode composite consists of 50 wt% (SE) and 37 wt% (sulfur). Considering the reversible capacity of the electrolyte,

TABLE 3 | Summary of the DSC analysis of the samples.

Sample	Peak top/°C [heat capacity (J/g)]			
	Sulfur related	LiI related	Li$_3$PS$_4$ related	Unknown
Ref. (Li$_3$PS$_4$)			*188.8 (−17.1)^x *208.3 (−17.8)^x *237.3 (−28.6)^x	
SE (Li$_3$PS$_4$–LiI)		+160.9 (−18.5)	*224.2 (−9.1)^x *250.6 (−9.1)^x	
C–SE (SE = 80%)		+190.0 (−7.2)	*226.8 (−1.2)^x *244.5 (−4.8)^x	
C–S–SE (S = 37%, SE = 50%)	▼107 (1.0)			▲145 (−7.1) Δ 229 (−15.7)

^{x}Vertical partitioning.
The symbols before the values correspond to the symbols in **Figure 8**.

FIGURE 9 | dQ/dV plot of the charge process for S-cell (bottom) and E-cell (upper). Gaussian peak fitting was performed onto the dQ/dV data with $R^2 > 0.998$. Each peak consists of I: main red-ox reaction of sulfur, II: red-ox capacity with poly-condensation of Li3PS4, and III: red-ox capacity of LiI.

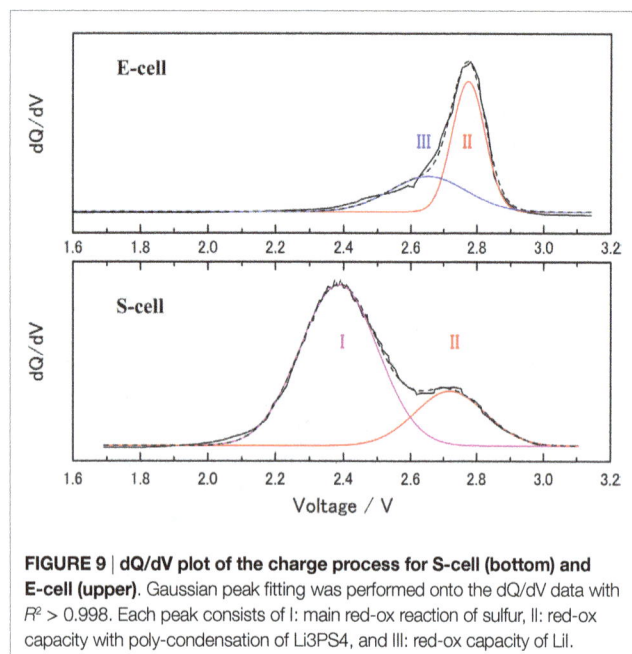

the net capacity of sulfur was estimated. For estimating the each value, dQ/dV profiles of S- and E-cell were examined and are presented in **Figure 9**. The main peak in region I (around 2.37 V) including a shoulder peak around 2.19 V represents the majority of the reaction in S-cell. Region II (around 2.75 V) is not generally observed in a typical Li–S cell. In the dQ/dV of the S-cell, the areas assigned in regions I and II were separated by Gaussian peak fitting ($R^2 > 0.998$). The peak area of region II in the S-cell was about 25% of the total integrated area between 1.8 and 3.0 V. If it is assumed that the extra capacity originated from the reaction of the electrolyte at 2.75 V, the net sulfur redox capacity was about 75% of the apparent sulfur specific capacity (1600 mAh g_{-S}^{-1}) and was equal to 1200 mAh g_{-S}^{-1}. For the E-cell, the peak could be separated in two peaks. One of them gives the same potential with region II. The second peak (region III) appeared at 2.65 V. The ratio of the peak areas for region II:III is 61:39. We have calculated and estimated the reaction based on the electrolyte composition. There was only one possibility that the summation of the redox capacity of LiI and Li3PS4 (higher than two electrons reaction) can provide the capacity of 250 mA g^{-1} for the SE.

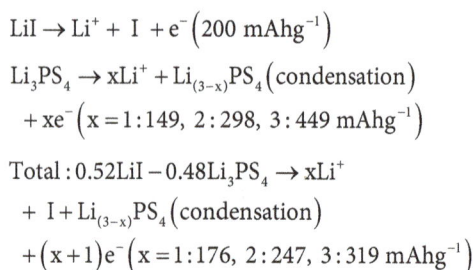

$$LiI \rightarrow Li^+ + I + e^- \left(200 \text{ mAhg}^{-1}\right)$$

$$Li_3PS_4 \rightarrow xLi^+ + Li_{(3-x)}PS_4 \left(\text{condensation}\right)$$
$$+ xe^- \left(x = 1:149, \ 2:298, \ 3:449 \text{ mAhg}^{-1}\right)$$

$$\text{Total}: 0.52LiI - 0.48Li_3PS_4 \rightarrow xLi^+$$
$$+ I + Li_{(3-x)}PS_4 \left(\text{condensation}\right)$$
$$+ \left(x+1\right)e^- \left(x = 1:176, \ 2:247, \ 3:319 \text{ mAhg}^{-1}\right)$$

Although there are some unclear points (e.g., the exact formation of the anions in the cathode composite), LiI and two electron redox of Li_3PS_4 ($3 \geq x \geq 2$) were suitable for considering the reversible capacity of the LiI–Li_3PS_4, which was used in this study. Because of the unusual DSC peaks observed in the C–S–SE composite, carbon and sulfur might also contribute to the highly complicated active material. Although further investigation is required to clarify this redox system, it is clear that the solid-state electrolyte, LiI–Li_3PS_4, has scope for use in real cell reactions.

CONCLUSION

A sulfide-based SE, amorphous LiI–Li_3PS_4, was prepared by mechanical milling. The ionic conductivity was determined to be 1.2 mS cm^{-1} at 25°C with an activation energy of 29.1 kJ mol^{-1}. We have demonstrated the successful fabrication of two types of lithium secondary cells using LiI–Li_3PS_4. In the NCA-cell, the cell performance improved in comparison to the Li_3PS_4 adopted cell in our previous study, except for the cycle ability. Due to its poor electrochemical stability of LiI–Li_3PS_4, the cycle performance was unsatisfactory in the 4-V system. On the contrary, the cycle performance was significantly improved in the S-cell using the same electrolyte compared to our previous study. The high coulombic efficiency above 99.9% during charge–discharge cycle test was verified in the S-cell. This indicates that the cell can basically operate 1000 cycles without reserve lithium capacity in terms of the figure of merit.

Interestingly, the C–SE composite electrochemically reacted and clearly showed a reversible capacity in the potential range of 3.1–1.3 V versus Li/Li$^+$. Although the exact reaction needs to be clarified, the stable capacity was about 260 mAh g^{-1}. Also, a cooperative reaction between sulfur, LiI, and Li_3PS_4 was observed in the C–S–SE composite. The coulombic efficiency was extremely high, even though some complex reaction was expected. Also, no evidence for a polysulfide redox shuttle was observed within 100 cycles. These facts indicate that the solid-state electrolyte and its adopted system is one of the most promising approaches for producing high-capacity lithium secondary batteries.

AUTHOR CONTRIBUTIONS

YA and SI proposed the idea and strategy for the experimental work. SI, SF, RO, and TY designed the experiments and performed the synthesis and characterization of the samples in addition to the electrochemical measurements. The experimental data were analyzed and the manuscript was written by YA and TW. The entire project was coordinated by YP and SD.

ACKNOWLEDGMENTS

This work is financially supported by Samsung Electronics. The authors acknowledge Prof. Machida (Konan University) for valuable discussion on the solid electrolyte and also express their sincere gratitude to Prof. Price (Western Sydney University) for improving the manuscript.

REFERENCES

Bates, J. B., Dudney, N. J., Gruzalski, G. R., Zuhr, R. A., Choudhury, A., Luck, C. F., et al. (1993). This volume contains the Proceedings of the 6th International Meeting on Lithium Batteries Fabrication and characterization of amorphous lithium electrolyte thin films and rechargeable thin-film batteries. *J. Power Sources* 43, 103–110. doi:10.1016/0378-7753(93)80106-Y

Hakari, T., Nagao, M., Hayashi, A., and Tatsumisago, M. (2015). All-solid-state lithium batteries with Li_3PS_4 glass as active material. *J. Power Sources* 293, 721–725. doi:10.1016/j.jpowsour.2015.05.073

Ito, S., Fujiki, S., Yamada, T., Aihara, Y., Park, Y., Kim, T. Y., et al. (2014). A rocking chair type all-solid-state lithium ion battery adopting Li_2O-ZrO_2 coated $LiNi_{0.8}Co_{0.15}Al_{0.05}O_2$ and a sulfide based electrolyte. *J. Power Sources* 248, 943–950. doi:10.1016/j.jpowsour.2013.10.005

Kamaya, N., Homma, K., Yamakawa, Y., Hirayama, M., Kanno, R., Yonemura, M., et al. (2011). A lithium superionic conductor. *Nat. Mater.* 10, 682–686. doi:10.1038/nmat3066

Lin, Z., Liu, Z., Fu, W., Dudney, N. J., and Liang, C. (2013). Lithium polysulfidophosphates: a family of lithium-conducting sulfur-rich compounds for lithium-sulfur batteries. *Angew. Chem. Int. Ed.* 52, 7460–7463. doi:10.1002/anie.201300680

Mercier, R., Malugani, J.-P., Fahys, B., and Robert, G. (1981). Proceedings of the International Conference on Fast Ionic Transport in Solids Superionic conduction in Li_2S-P_2S_5-LiI glasses. *Solid State Ionics* 5, 663–666. doi:10.1016/0167-2738(81)90341-6

Nagata, H., and Chikusa, Y. (2014). A lithium sulfur battery with high power density. *J. Power Sources* 264, 206–210. doi:10.1016/j.jpowsour.2014.04.106

Ogawa, M., Kanda, R., Yoshida, K., Uemura, T., and Harada, K. (2012). High-capacity thin film lithium batteries with sulfide solid electrolytes. *J. Power Sources* 205, 487–490. doi:10.1016/j.jpowsour.2012.01.086

Ohtomo, T., Hayashi, A., Tatsumisago, M., and Kawamoto, K. (2013). All-solid-state batteries with Li_2O-Li_2S-P_2S_5 glass electrolytes synthesized by two-step mechanical milling. *J. Sol. St. Electrochem.* 17, 2551–2557. doi:10.1007/s10008-013-2149-5

Rangasamy, E., Liu, Z., Gobet, M., Pilar, K., Sahu, G., Zhou, W., et al. (2015). An iodide-based $Li_7P_2S_8I$ superionic conductor. *J. Am. Chem. Soc.* 137, 1384–1387. doi:10.1021/ja508723m

Rao, B. M. L., Silbernagel, B. G., and Jacobson, A. J. (1978). Evaluation of solid electrolytes for high temperature lithium batteries: a preliminary study. *J. Power Sources* 3, 59–66. doi:10.1016/0378-7753(78)80005-6

Scrosati, B., and Garche, J. (2010). Lithium batteries: status, prospects and future. *J. Power Sources* 195, 2419–2430. doi:10.1016/j.jpowsour.2009.11.048

Tatsumisago, M., Hama, S., Hayashi, A., Morimoto, H., and Minami, T. (2002). New lithium ion conducting glass-ceramics prepared from mechanochemical Li_2S-P_2S_5 glasses. *Solid State Ionics* 15, 635–640. doi:10.1016/S0167-2738(02)00509-X

Weppner, W. (1981). Proceedings of the International Conference on Fast Ionic Transport in Solids Trends in new materials for solid electrolytes and electrodes. *Solid State Ionics* 5, 3–8. doi:10.1016/0167-2738(81)90186-7

Yamada, T., Ito, S., Omoda, R., Watanabe, T., Aihara, Y., Agostini, M., et al. (2015). All solid-state lithium-sulfur battery using a glass-type P_2S_5-Li_2S electrolyte: benefits on anode kinetics. *J. Electrochem. Soc.* 162, A646–A651. doi:10.1149/2.0441504jes

Conflict of Interest Statement: The authors declare that the research was conducted in the absence of any commercial or financial relationships that could be construed as a potential conflict of interest.

5

Novel Solid Electrolytes for Li-Ion Batteries: A Perspective from Electron Microscopy Studies

Cheng Ma and Miaofang Chi*

Oak Ridge National Laboratory, Center for Nanophase Materials Sciences, Oak Ridge, TN, USA

Edited by:
Shyue Ping Ong,
University of California San Diego,
USA

Reviewed by:
Paulina Półrolniczak,
Institute of Non-Ferrous Metals
Division in Poznań Central Laboratory
of Batteries and Cells, Poland
Yan E. Wang,
MIT, USA

*Correspondence:
Miaofang Chi
chim@ornl.gov

Specialty section:
This article was submitted
to Energy Storage,
a section of the journal
Frontiers in Energy Research

Citation:
Ma C and Chi M (2016) Novel Solid
Electrolytes for Li-Ion Batteries:
A Perspective from Electron
Microscopy Studies.
Front. Energy Res. 4:23.

Solid electrolytes can simultaneously overcome two of the most formidable challenges of Li-ion batteries: the severe safety issues and insufficient energy densities. However, before they can be implemented in actual batteries, the ionic conductivity needs to be improved and the interface with electrodes must be optimized. The prerequisite for addressing these issues is a thorough understanding of the material's behavior at the microscopic and/or the atomic level. (Scanning) transmission electron microscopy is a powerful tool for this purpose, as it can reach an ultrahigh spatial resolution. Here, we review recent electron microscopy investigations on the ion transport behavior in solid electrolytes and their interfaces. Specifically, three aspects will be highlighted: the influence of grain interior atomic configuration on ionic conductivity, the contribution of grain boundaries, and the behavior of solid electrolyte/electrode interfaces. Based on this, the perspectives for future research will be discussed.

Keywords: lithium battery, solid electrolyte, electron microscopy, atomic resolution analysis, interface

INTRODUCTION

With the exhaustion of fossil fuels, high-performance energy storage devices have received significant attention in recent years (Quartarone and Mustarelli, 2011; Bruce et al., 2012). While the Li-ion battery (LIB) is a very promising alternative power source, the safety concerns and insufficient energy density have hindered its implementation in heavy-duty applications, e.g., electric vehicles and grid energy storage (Quartarone and Mustarelli, 2011; Bruce et al., 2012). Fortunately, these issues can be addressed by integrating novel solid electrolytes (Quartarone and Mustarelli, 2011; Takada, 2013; Wang et al., 2015). On the one hand, these solid materials are typically non-flammable and impossible to leak, which circumvents the safety issues associated with conventional organic liquid electrolytes. This is a prerequisite for large-scale application. On the other hand, the energy density can also be effectively improved. The much larger electrochemical window allows for the use of advanced electrode materials that are incompatible with conventional liquid electrolytes. In addition, by eliminating the need for bulky safety mechanisms, the battery size can be greatly reduced. Because of these advantages, solid electrolytes have received tremendous interest in recent years.

However, two grand challenges must be overcome before solid electrolytes can be used in commercial batteries. First, their ionic conductivity is typically low, preventing a fast charge and discharge (Takada, 2013; Wang et al., 2015). Second, forming a stable conductive interface between the solid electrolyte and electrode is difficult (Zhu et al., 2015 & 2016; Richards et al., 2016). Overcoming the first challenge requires a mechanistic understanding of the interplay between Li migration and the atomic framework of the material. For the second challenge, the correlation between interface structure/chemistry and ionic transport must first be systematically established. Clearly, both tasks demand structural and chemical analysis with an ultrahigh spatial resolution.

Transmission electron microscopy (TEM), most notably aberration-corrected scanning transmission electron microscopy (STEM), is an ideal tool for gaining critical atomic level insight. It is not only capable of directly visualizing the atomic configurations but can also elucidate chemical information at a sub-angstrom spatial resolution using electron energy loss spectroscopy (EELS) and energy-dispersive X-ray spectroscopy (EDS) (Pennycook, 1992; Muller et al., 2008; Chi et al., 2011; Yabuuchi et al., 2011; Wu et al., 2015). However, STEM investigations of solid electrolytes pose numerous challenges as the high Li mobility and poor electronic conductivity make these materials highly vulnerable to electron irradiation damage (Egerton et al., 2004). Fortunately, with the significantly improved capabilities for imaging and TEM specimen preparation, this issue has been greatly alleviated in recent years. Several beam-sensitive materials that could not be studied previously can now be analyzed at the atomic scale (Ma et al., 2015), and many of these studies have made significant contributions toward the research on solid electrolytes.

The present mini-review will highlight electron microscopy investigations for three important factors governing the behavior of solid electrolytes: (1) influence of grain interior atomic configuration on ionic conductivity, (2) impacts of grain boundaries, and (3) behavior of solid electrolyte–electrode interfaces. Based on this, opportunities, challenges, and perspectives for future research will be discussed.

INFLUENCE OF GRAIN INTERIOR ATOMIC CONFIGURATION ON IONIC CONDUCTIVITY

Li migration within the crystalline lattice is dictated by the atomic framework, which forms channels for Li transport. A precise understanding of the atomic structure is required to explain ionic transport within the lattice. With its ultrahigh spatial resolution and sensitivity to subtle differences in diffraction, (S)TEM not only complements X-ray and neutron scattering studies but also provides unique insights at the atomic level. Recent microscopy studies mainly focused on two systems: $Li_7La_3Zr_2O_{12}$ (LLZO) and $Li_{3x}La_{2/3-x}TiO_3$ (LLTO).

$Li_7La_3Zr_2O_{12}$ is currently the most promising oxide solid electrolyte due to the coexistence of an excellent stability against Li metal and a relatively high conductivity (Murugan et al., 2007; Cussen, 2010). It crystallizes in the garnet structure with two polymorphs (Cussen, 2010): a cubic phase with a relatively high conductivity (c-LLZO) and a less conductive tetragonal phase (t-LLZO). Distinguishing these two phases is critical to properly interpret the ionic transport behavior. The precession electron diffraction (PED) study by Buschmann et al. (2011) successfully differentiated these two phases by circumventing the influence from double diffraction. This result further confirmed that Al-doping is critical for stabilizing the cubic phase. When combined with neutron diffraction, it was found that the Li sites in c-LLZO, unlike those in t-LLZO, are partially filled. The high

concentration of vacancies in c-LLZO gives rise to a higher Li mobility and superior conductivity. Besides the PED study, Buschmann et al. also tried to perform high-resolution TEM (HRTEM), but detailed analysis was prevented by the electron beam-irradiation damage. Recently, this issue was successfully alleviated by Ma et al. (2015). The careful selection of imaging and specimen preparation conditions enabled a high-quality, atomic resolution (S)TEM/EELS analysis (**Figures 1A,B**). Ma's research demonstrated that c-LLZO maintains its cubic crystal structure even in an aqueous environment with a pH >7. Such a high structural stability indicates that c-LLZO offers a robust atomic framework for Li transport. Given the high ionic conductivity, Li compatibility, and desirable structural stability against aqueous solutions with a broad range of pH values, LLZO is a promising candidate for the separator in novel aqueous Li batteries.

Another system that has been extensively interrogated by electron microscopy is LLTO, which has a perovskite-type structure (Stramare et al., 2003). By varying the composition and/or processing conditions, multiple polymorphs with different ionic conductivity may be obtained. Regardless, most of them exhibit an alternate stacking between La-rich and La-poor A-site layers, and Li migration is favored by the La-poor layers. The highest bulk conductivity is 10^{-3} S cm^{-1}, approaching that of conventional liquid electrolytes (10^{-2} S cm^{-1}) (Takada, 2013). Therefore, an in-depth understanding of the origin for such exceptional performances is crucial for designing highly conductive solid electrolytes. The (S)TEM studies made important contributions toward this cause. Taking advantage of the sensitivity of annular-bright-field (ABF) STEM imaging to light elements such as Li, Gao et al. (2013) directly visualized the variation of Li positions in different LLTO polymorphs. Li was observed to reside at the O4 window for the Li-poor composition $La_{0.62}Li_{0.16}TiO_3$, but near the A-site position for the Li-rich composition $La_{0.56}Li_{0.33}TiO_3$. The Li content, valence state of cations, and geometry of the oxygen octahedra in the La-rich and La-poor layers were also revealed by EELS. Furthermore, domain structures associated with the ordering between La-rich and La-poor layers were examined (Gao et al., 2014). With La blocking the Li pathways, the domain boundaries were found to impede ionic transport. Beyond this, structural features that cannot be easily detected by diffraction methods may also be visualized. As mentioned above, the Li transport within LLTO relies on the La-poor layers. However, none of the previous diffraction studies detected such crucial features in the most conductive polymorph, the 1350°C-quenched $La_{0.56}Li_{0.33}TiO_3$ (Stramare et al., 2003). As a result, its ionic transport mechanism remained unclear for years. Recently, an atomic resolution STEM study directly visualized the previously overlooked short-range-ordered Li pathways in this material (Ma et al., 2016). The coherence length of the ordering was found to be at the mesoscopic scale (below 10 nm), which prevented it from being detected by most diffraction methods. In combination with molecular dynamics (MD) simulations, this observation indicated that such an elusive mesoscopic framework can most effectively maximize the number of Li transport pathways, leading to a high conductivity. The discovery not only reconciled the long-existing structure–property inconsistency but also pointed out a new angle on improving ionic conductivity.

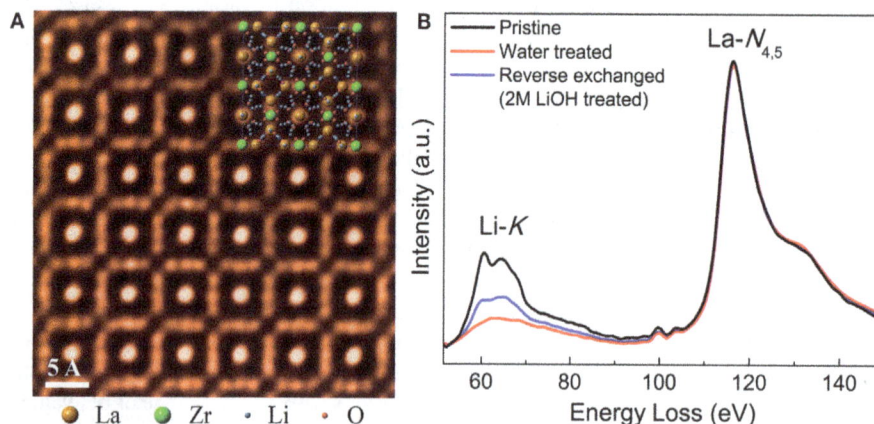

FIGURE 1 | (A) The atomic structure of the electron beam-sensitive solid electrolyte LLZO successfully visualized by high-angle annular dark-field (HAADF) STEM imaging. **(B)** EELS data of LLZO after the Li+/H+ exchange with different aqueous solutions. The Li content can be precisely monitored. Reproduced with permission from (Ma et al., 2015).

Although atomic resolution (S)TEM greatly benefited the fundamental understanding of ionic transport, current studies are limited to oxides. In comparison, sulfide solid electrolytes, regardless of their higher conductivity (Takada, 2013), are rarely investigated. Microscopy studies on these materials are extremely challenging because of (1) the vulnerability of weak Li–S bonds to electrons and (2) their sensitivity to ambient atmosphere. If these issues can be mitigated, (S)TEM will play an even more critical role in the research of solid electrolytes.

IMPACTS OF GRAIN BOUNDARIES

Although the research on solid electrolytes primarily focuses on the grain interior, grain boundaries are frequently the actual bottleneck. While the bulk conductivity of many solid electrolytes is already comparable to those of conventional liquid electrolytes, their large grain boundary resistance typically lowers the total conductivity by orders of magnitude (Takada, 2013). Due to the absence of a proper understanding on the grain boundary Li conduction mechanism, a targeted optimization is not yet possible.

Grain boundaries in solids are often confined to a very small length scale with widths of only a few unit cells. Therefore, STEM, with its sub-angstrom resolution, appears to be an ideal tool to study them. Ma et al. (2014) successfully utilized atomic resolution STEM/EELS to unravel the atomic-scale origin of the large grain boundary resistance in LLTO. Most grain boundaries were observed to show darker Z-contrast than the adjacent grains, suggesting that the average atomic number at the grain boundary is lower. Further atomic-scale analysis showed that the grain boundary atomic configuration significantly deviated from that of the grain interior (**Figures 2A,B**). Instead of the ABO$_3$ perovskite structure, such reconstructed grain boundaries are essentially a binary Ti–O layer, forbidding the abundance of the charge carrier Li+. Therefore, they act as internal barriers for Li transport. This topic has also been investigated by HRTEM and EDS. In addition, Gellert et al. (2012) studied the grain boundaries in lithium aluminum titanium phosphate (LATP).

Depending on the relative orientation between the neighboring grains, two types of grain boundaries were observed. If the orientations are similar, a thick crystalline grain boundary will be present. Its high degree of crystallinity was believed to allow for a relatively facile ionic transport. If the orientations differ greatly, a thinner but amorphous layer forms, which was believed to be highly resistive.

Unlike the two materials discussed above, LLZO exhibits a grain boundary resistance comparable to that of the grain interior (Murugan et al., 2007). However, the origin of this benign behavior remains unknown. Several research groups have attempted to study LLZO grain boundaries using electron microscopy, but the results are inconsistent. Kumazaki et al. (2011) observed amorphous Li–Al–Si–O and nanocrystalline LiAlSiO$_4$ at the LLZO grain boundaries. In contrast, clean grain boundaries, which are free of any second phase or compositional variations, were reported by Wolfenstine et al. (2012). Systematic investigations with higher spatial resolution are necessary to reach a conclusive explanation.

These studies demonstrate that the grain boundaries, despite their highly localized nature, can be effectively investigated using (S)TEM combined with local analytical methods, such as EELS and EDS. However, current efforts in this area are very limited. Before a systematic understanding and a rational optimization of Li transport at grain boundaries can be realized, further in-depth investigations are needed.

BEHAVIOR OF ELECTROLYTE-ELECTRODE INTERFACES

The stable and conductive electrode/electrolyte interface is a prerequisite for the long-term operation of solid electrolyte-based batteries (Zhu et al., 2015 & 2016; Richards et al., 2016). Nevertheless, due to the absence of a mechanistic understanding to guide the rational improvement, it is still very difficult to form such interfaces. As the first step toward this goal, a direct experimental observation of the interfaces is essential.

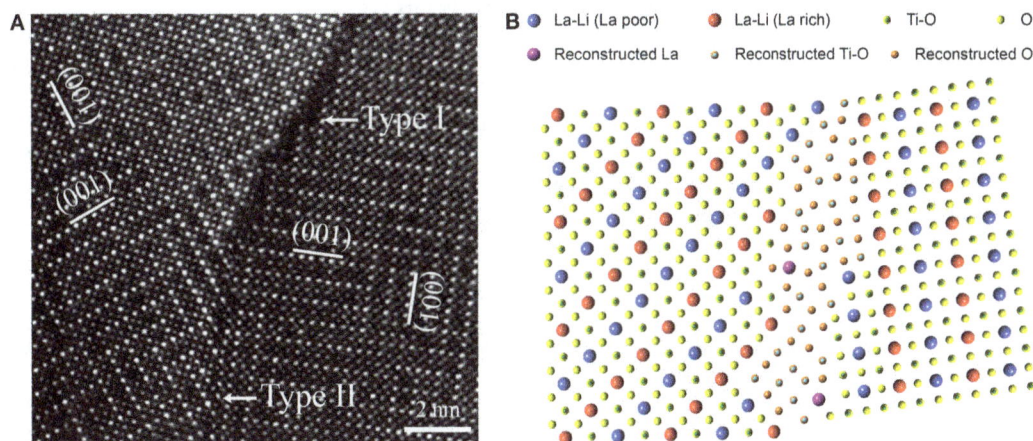

FIGURE 2 | (A) Atomic resolution HAADF-STEM image of a grain boundary in LLTO. **(B)** Atomic model of the Li-deficient LLTO grain boundary based on a comprehensive STEM/EELS study. Reproduced with permission from (Ma et al., 2014).

Although no atomic resolution electron microscopy studies have been reported to date, interfaces between cathode materials and several solid electrolytes have been examined *via* nano-electron diffraction (NED), STEM, and EDS. Kim et al. (2011) investigated the interfacial stability between LLZO and $LiCoO_2$ (LCO). An LCO thin film was grown on the polished surface of the LLZO ceramic through pulsed laser deposition at 937 K. TEM observations revealed the existence of an interface reaction layer of ~50 nm thick. EDS line profile measurements and NED acquired in the vicinity of the interface suggested that this reaction layer consisted of La_2CoO_4, which is believed to hinder Li diffusion. In addition, the interface between LCO and $Li_2S–P_2S_5$, a prototypical sulfide electrolyte, was studied by Sakuda et al. (2009). The interface was simply formed by mechanical grinding. After charging, an interfacial layer associated with the mutual diffusion of Co, P, and S emerged, and this layer gave rise to a large resistance. A similar behavior was observed between $LiMn_2O_4$ and $Li_2S–P_2S_5$ (Kitaura et al., 2010). An interfacial layer resulting from the diffusion of Mn into the solid electrolyte was observed and believed to yield large resistance. These electron microscopy studies suggest that a reaction layer may frequently form between the solid electrolyte and cathode due to inter-diffusion. Unlike the solid electrolyte interface (SEI) in conventional LIBs, reaction layers at solid electrolyte/electrode interfaces are usually detrimental rather than beneficial, as they typically impede the ionic transport (Qian et al., 2015).

Beyond these experimentally observed reaction layers, highly localized interfacial decomposition at the solid electrolyte and electrode interfaces was frequently speculated, although they show a certain degree of stability in electrochemical measurements (Zhu et al., 2015 & 2016; Richards et al., 2016). However, most of such speculations are proposed based on theoretical calculations. The experimental verification is quite challenging due to the extremely small length scale of the speculated thickness and the high volatility/instability of Li metal (Wenzel et al., 2015, 2016). (S)TEM, which can probe local features at an extremely high spatial resolution down to sub-angstrom level, presents excellent opportunities to interrogate these intriguing interfacial behaviors.

SUMMARY AND PROSPECTS

In this mini-review, we discussed recent progress of (S)TEM studies on solid electrolytes for Li batteries. With the success in alleviating the challenges caused by electron beam irradiation damage, more and more investigations that could not be performed previously are reported. These studies clarified several long-standing confusions regarding the structure–property relationship, provided first experimental insights into the large grain boundary resistance, and contributed insights regarding the reaction layer at cathode/SEIs.

Regardless, further challenges remain. For the ionic transport within the grain, sulfide electrolytes, which often exhibit a higher conductivity than oxides, demand a thorough study at atomic scale. Their vulnerary to electron beam, due to the weak bonds of Li with S in the structure and limited electron conductivity, significantly limit reliable measurements of their atomic and electronic structures in TEMs. In order to understand the role of grain boundaries in solid electrolytes, a wide range of materials need to be investigated in order to establish the systematic understanding. In particular, the materials with benign grain boundaries deserve special attention, as they may inspire the design of materials with conductive grain boundaries. For the solid electrolyte/electrode interface, one of the most pressing tasks is to verify the highly localized interfacial reaction layers that are proposed recently by theoretical works. Furthermore, the variation of these interfaces with composition, processing conditions, and cycling also remained to be investigated. It must be emphasized that recently developed *in situ* TEM techniques, such as *in situ* heating, and *in situ* electrochemical cycling with a desirable spatial resolution, will greatly facilitate these studies (Gu et al., 2013; Chi et al., 2015; Zeng et al., 2015). Their capability of real-time, high-resolution structural/chemical analysis will provide unique insights that cannot be acquired otherwise. With the recent remarkable developments in microscopy instrumentations, such as fast cameras and detectors, low voltage TEMs, and multi-functional specimen stages, these challenges should be overcome in the near future, and electron microscopy is expected to play an increasingly significant role in the research of Li-ion-conducting solid electrolytes.

AUTHOR CONTRIBUTIONS

All authors listed have made substantial, direct, and intellectual contribution to the work and approved it for publication.

REFERENCES

Bruce, P. G., Freunberger, S. A., Hardwick, L. J., and Tarascon, J.-M. (2012). Li-O$_2$ and Li-S batteries with high energy storage. *Nat. Mater.* 11, 19–29. doi:10.1038/nmat3191

Buschmann, H., Dolle, J., Berendts, S., Kuhn, A., Bottke, P., Wilkening, M., et al. (2011). Structure and dynamics of the fast lithium ion conductor "Li$_7$La$_3$Zr$_2$O$_{12}$". *Phys. Chem. Chem. Phys.* 13, 19378–19392. doi:10.1039/c1cp22108f

Chi, M. F., Mizoguchi, T., Martin, L. W., Bradley, J. P., Ikeno, H., Ramesh, R., et al. (2011). Atomic and electronic structures of the SrVO$_3$-LaAlO$_3$ interface. *J. Appl. Phys.* 110, 046104. doi:10.1063/1.3601870

Chi, M. F., Wang, C., Lei, Y. K., Wang, G. F., Li, D. G., More, K. L., et al. (2015). Surface faceting and elemental diffusion behaviour at atomic scale for alloy nanoparticles during *in situ* annealing. *Nat. Commun.* 6, 8925. doi:10.1038/ncomms9925

Cussen, E. J. (2010). Structure and ionic conductivity in lithium garnets. *J. Mater. Chem.* 20, 5167–5173. doi:10.1039/b925553b

Egerton, R. F., Li, P., and Malac, M. (2004). Radiation damage in the TEM and SEM. *Micron* 35, 399–409. doi:10.1016/j.micron.2004.02.003

Gao, X., Fisher, C. A. J., Kimura, T., Ikuhara, Y. H., Kuwabara, A., Moriwake, H., et al. (2014). Domain boundary structures in lanthanum lithium titanates. *J. Mater. Chem. A* 2, 843–852. doi:10.1039/C3TA13726K

Gao, X., Fisher, C. A. J., Kimura, T., Ikuhara, Y. H., Moriwake, H., Kuwabara, A., et al. (2013). Lithium atom and A-site vacancy distributions in lanthanum lithium titanate. *Chem. Mater.* 25, 1607–1614. doi:10.1021/cm3041357

Gellert, M., Gries, K. I., Yada, C., Rosciano, F., Volz, K., and Roling, B. (2012). Grain boundaries in a lithium aluminum titanium phosphate-type fast lithium ion conducting glass ceramic: microstructure and nonlinear ion transport properties. *J. Phys. Chem. C* 116, 22675–22678. doi:10.1021/jp305309r

Gu, M., Parent, L. R., Mehdi, B. L., Unocic, R. R., McDowell, M. T., Sacci, R. L., et al. (2013). Demonstration of an electrochemical liquid cell for operando transmission electron microscopy observation of the lithiation/delithiation behavior of Si nanowire battery anodes. *Nano Lett.* 13, 6106–6112. doi:10.1021/nl403402q

Kim, K. H., Iriyama, Y., Yamamoto, K., Kumazaki, S., Asaka, T., Tanabe, K., et al. (2011). Characterization of the interface between LiCoO$_2$ and Li$_7$La$_3$Zr$_2$O$_{12}$ in an all-solid-state rechargeable lithium battery. *J. Power Sources* 196, 764–767. doi:10.1016/j.jpowsour.2010.07.073

Kitaura, H., Hayashi, A., Tadanaga, K., and Tatsumisago, M. (2010). All-solid-state lithium secondary batteries using LiMn$_2$O$_4$ electrode and Li$_2$S-P$_2$S$_5$ solid electrolyte. *J. Electrochem. Soc.* 157, A407–A411. doi:10.1149/1.3298441

Kumazaki, S., Iriyama, Y., Kim, K.-H., Murugan, R., Tanabe, K., Yamamoto, K., et al. (2011). High lithium ion conductive Li$_7$La$_3$Zr$_2$O$_{12}$ by inclusion of both Al and Si. *Electrochem. Commun.* 13, 509–512. doi:10.1016/j.elecom.2011.02.035

Ma, C., Chen, K., Liang, C. D., Nan, C. W., Ishikawa, R., More, K., et al. (2014). Atomic-scale origin of the large grain-boundary resistance in perovskite Li-ion-conducting solid electrolytes. *Energy Environ. Sci.* 7, 1638–1642. doi:10.1039/c4ee00382a

Ma, C., Cheng, Y., Chen, K., Li, J., Sumpter, B., Nan, C.-W., et al. (2016). Mesoscopic framework enables facile ionic transport in solid electrolytes for Li batteries. *Adv. Energy Mater.* doi:10.1002/aenm.201600053

Ma, C., Rangasamy, E., Liang, C. D., Sakamoto, J., More, K. L., and Chi, M. F. (2015). Excellent stability of a lithium-ion-conducting solid electrolyte upon reversible Li$^+$/H$^+$ exchange in aqueous solutions. *Angew. Chem. Int. Ed.* 54, 129–133. doi:10.1002/anie.201410930

Muller, D. A., Kourkoutis, L. F., Murfitt, M., Song, J. H., Hwang, H. Y., Silcox, J., et al. (2008). Atomic-scale chemical imaging of composition and bonding by aberration-corrected microscopy. *Science* 319, 1073–1076. doi:10.1126/science.1148820

Murugan, R., Thangadurai, V., and Weppner, W. (2007). Fast lithium ion conduction in garnet-type Li$_7$La$_3$Zr$_2$O$_{12}$. *Angew. Chem. Int. Ed.* 46, 7778–7781. doi:10.1002/anie.200701144

Pennycook, S. J. (1992). Z-contrast transmission electron microscopy – direct atomic imaging of materials. *Ann. Rev. Mater. Sci.* 22, 171–195. doi:10.1146/annurev.ms.22.080192.001131

Qian, D., Ma, C., More, K. L., Meng, Y. S., and Chi, M. (2015). Advanced analytical electron microscopy for lithium-ion batteries. *NPG Asia Mater.* 7, e193. doi:10.1038/am.2015.50

Quartarone, E., and Mustarelli, P. (2011). Electrolytes for solid-state lithium rechargeable batteries: recent advances and perspectives. *Chem. Soc. Rev.* 40, 2525–2540. doi:10.1039/c0cs00081g

Richards, W. D., Miara, L. J., Wang, Y., Kim, J. C., and Ceder, G. (2016). Interface stability in solid-state batteries. *Chem. Mater.* 28, 266–273. doi:10.1021/acs.chemmater.5b04082

Sakuda, A., Hayashi, A., and Tatsumisago, M. (2009). Interfacial observation between LiCoO$_2$ electrode and Li$_2$S-P$_2$S$_5$ solid electrolytes of all-solid-state lithium secondary batteries using transmission electron microscopy. *Chem. Mater.* 22, 949–956. doi:10.1021/cm901819c

Stramare, S., Thangadurai, V., and Weppner, W. (2003). Lithium lanthanum titanates: a review. *Chem. Mater.* 15, 3974–3990. doi:10.1021/cm0300516

Takada, K. (2013). Progress and prospective of solid-state lithium batteries. *Acta Mater.* 61, 759–770. doi:10.1016/j.actamat.2012.10.034

Wang, Y., Richards, W. D., Ong, S. P., Miara, L. J., Kim, J. C., Mo, Y., et al. (2015). Design principles for solid-state lithium superionic conductors. *Nat. Mater.* 14, 1026–1031. doi:10.1038/nmat4369

Wenzel, S., Leichtweiss, T., Krüger, D., Sann, J., and Janek, J. (2015). Interphase formation on lithium solid electrolytes – an *in situ* approach to study interfacial reactions by photoelectron spectroscopy. *Solid State Ionics* 278, 98–105. doi:10.1016/j.ssi.2015.06.001

Wenzel, S., Weber, D. A., Leichtweiss, T., Busche, M. R., Sann, J., and Janek, J. (2016). Interphase formation and degradation of charge transfer kinetics between a lithium metal anode and highly crystalline Li$_7$P$_3$S$_{11}$ solid electrolyte. *Solid State Ionics* 286, 24–33. doi:10.1016/j.ssi.2015.11.034

Wolfenstine, J., Sakamoto, J., and Allen, J. L. (2012). Electron microscopy characterization of hot-pressed Al substituted Li$_7$La$_3$Zr$_2$O$_{12}$. *J. Sci. Mater.* 47, 4428–4431. doi:10.1007/s10853-012-6300-y

Wu, Y., Ma, C., Yang, J. H., Li, Z. C., Allard, L. F., Liang, C. D., et al. (2015). Probing the initiation of voltage decay in Li-rich layered cathode materials at the atomic scale. *J. Mater. Chem. A* 3, 5385–5391. doi:10.1039/C4TA06856D

Yabuuchi, N., Yoshii, K., Myung, S. T., Nakai, I., and Komaba, S. (2011). Detailed studies of a high-capacity electrode material for rechargeable batteries, Li$_2$MnO$_3$-LiCo$_{1/3}$Ni$_{1/3}$Mn$_{1/3}$O$_2$. *J. Am. Chem. Soc.* 133, 4404–4419. doi:10.1021/ja108588y

Zeng, Z. Y., Zhang, X. W., Bustillo, K., Niu, K. Y., Gammer, C., Xu, J., et al. (2015). *In situ* study of lithiation and delithiation of MoS$_2$ nanosheets using electrochemical liquid cell transmission electron microscopy. *Nano Lett.* 15, 5214–5220. doi:10.1021/acs.nanolett.5b02483

Zhu, Y., He, X., and Mo, Y. (2015). Origin of outstanding stability in the lithium solid electrolyte materials: insights from thermodynamic analyses based on first-principles calculations. *ACS Appl. Mater. Interfaces* 7, 23685–23693. doi:10.1021/acsami.5b07517

FUNDING

This work was sponsored by the U.S. Department of Energy (DOE), Office of Science, Office of Basic Energy Sciences, Materials Sciences and Engineering Division. Materials characterization was performed as part of a user proposal at the Center for Nanophase Materials Sciences, which is a U.S. DOE Office of Science User Facility.

Zhu, Y., He, X., and Mo, Y. (2016). First principles study on electrochemical and
chemical stability of the solid electrolyte-electrode interfaces in all-solid-state
Li-ion batteries. *J. Mater. Chem. A.* 4, 3253–3266. doi:10.1039/C5TA08574H

Conflict of Interest Statement: The authors declare that the research was conducted in the absence of any commercial or financial relationships that could be construed as a potential conflict of interest.

6

Exploring societal preferences for energy sufficiency measures in Switzerland

Corinne Moser[1,2]*, Andreas Rösch[2] and Michael Stauffacher[2,3]

[1] Institute of Sustainable Development, School of Engineering, Zurich University of Applied Sciences, Winterthur, Switzerland, [2] Natural and Social Science Interface, Institute for Environmental Decisions, Department of Environmental Systems Science, ETH Zürich, Zürich, Switzerland, [3] Transdisciplinarity Laboratory, Department of Environmental Systems Science, ETH Zürich, Zürich, Switzerland

Edited by:
Tobias Brosch,
University of Geneva, Switzerland

Reviewed by:
Anya Skatova,
University of Nottingham, UK
Andrea Tabi,
University of St. Gallen, Switzerland

*Correspondence:
Corinne Moser,
Institute of Sustainable Development,
School of Engineering, Zurich
University of Applied Sciences,
Technoparkstrasse 2, Winterthur
8401, Switzerland
corinne.moser@zhaw.ch

Specialty section:
This article was submitted to Energy
Systems and Policy, a section of the
journal Frontiers in Energy Research

Citation:
Moser C, Rösch A and Stauffacher M
(2015) Exploring societal preferences
for energy sufficiency measures in
Switzerland.
Front. Energy Res. 3:40.

Many countries are facing a challenging transition toward more sustainable energy systems, which produce more renewables and consume less energy. The latter goal can only be achieved through a combination of efficiency measures and changes in people's lifestyles and routine behaviors (i.e., sufficiency). While research has shown that acceptance of technical efficiency is relatively high, there is a lack of research on societal preferences for sufficiency measures. However, this is an important prerequisite for designing successful interventions to change behavior. This paper analyses societal preferences for different energy-related behaviors in Switzerland. We use an online choice-based conjoint analysis ($N = 150$) to examine preferences for behaviors with high technical potentials for energy demand reduction in the following domains: mobility, heating, and food. Each domain comprises different attributes across three levels of sufficiency. Respondents were confronted with trade-off situations evoked through different fictional lifestyles that comprised different combinations of attribute levels. Through a series of trade-off decisions, participants were asked to choose their preferred lifestyle. The results revealed that a vegetarian diet was considered the most critical issue that respondents were unwilling to trade off, followed by distance to workplace and means of transportation. The highest willingness to trade off was found for adjustments in room temperature, holiday travel behaviors, and living space. Participants' preferences for the most energy-sufficient lifestyles were rather low. However, the study showed that there were lifestyles with substantive energy-saving potentials that were well accepted among respondents. Our study results suggest that the success of energy-sufficiency interventions might depend strongly on the targeted behavior. We speculate that they may face strong resistance (e.g., vegetarian diet). Thus, it seems promising to promote well-balanced lifestyles, rather than extremely energy-sufficient lifestyles, as potential role models for sufficiency.

Keywords: energy, sufficiency, societal preferences, routine behavior, lifestyles, conjoint analysis

Introduction

The Importance of Energy Sufficiency for Switzerland's Energy Transition

Countries worldwide are facing challenging transitions of their energy systems with regard to fighting climate change and declining availability of fossil fuels. Switzerland has adopted a new energy strategy (Energy Strategy 2050) that promotes the implementation of new renewables, the stepwise phase-out of nuclear power, and sets ambitious reduction targets for per capita energy consumption (Swiss Federal Council, 2013). This goal shall be achieved primarily through increased energy efficiency, i.e., through the implementation of technologies that require less energy to maintain current levels of services. Examples of such energy-efficient technologies include cars that use less fuel per kilometer and well-insulated buildings that require less heat.

However, there are technological and economical limitations to energy efficiency. Furthermore, increased energy efficiency often causes rebound effects (Herring, 2006; Darby, 2007), which at least partly offset the saved resources (e.g., energy, time, money). For example, although many appliances, such as fridges or TVs, are more energy efficient than ever before, these appliances have also increased in size and/or in number over time. Along these lines, researchers have also found that people rely on symbols of energy efficiency, which may lead to paradoxical effects (Sütterlin and Siegrist, 2014). For example, in an experiment, participants judged a person driving an energy-efficient car (i.e., a Prius) over longer distances to be more energy conscious than an SUV driver who covered shorter distances – and so, in total, consumed less energy than the Prius driver (Sütterlin and Siegrist, 2014).

Thus, in order to guarantee an absolute reduction in energy consumption, efficiency needs to be complemented by more sustainable consumption patterns. This requires behavioral changes on the part of energy consumers. This perspective was confirmed by a recent study in which Notter et al. (2013) estimated Switzerland's potential to become a 2000-Watt/1 ton CO_2 society[1]. The authors conclude that this goal is only realistic "when assuming a pronounced technological increase in efficiency combined with a smart sufficiency strategy" (Notter et al., 2013, p. 4019).

Sufficiency can be understood as a process of changing existing consumption patterns for more sustainable ones. The literature distinguishes two different approaches to sufficiency (for an overview, see Jenny, 2014). First, in a narrow sense, sufficiency can be understood as a necessary complement to energy efficiency and renewable energy sources in order to reach political goals regarding climate targets, resource use, or per capita energy consumption. Second, it can also be understood as a critique of our consumer society and our growth-based economic system, as well as of respective attempts to change these systems (Linz et al., 2002; Linz, 2012). In this study, we focus on the former, narrower understanding of sufficiency: that is, while energy *efficiency* refers to technological means to minimize resource input, energy *sufficiency* refers to changes in individual behaviors that lead to lower demand for energy services. In accordance with Breukers et al. (2013), we understand energy-sufficient behavior to involve changes in routine behaviors and lifestyles that lead to lower energy consumption. Examples of energy-sufficient behaviors include line-drying laundry instead of using a tumble dryer, eating vegetarian food instead of meat, commuting by bike instead of by private car, and so on.

With respect to private energy consumption in Switzerland, over the entire lifecycle of products and services, the domains of mobility, heating, and food are the most energy-demanding (Notter et al., 2013). A study by Jungbluth and Itten (2012) indicated substantial potential for energy-sufficient behaviors in these domains. In their study, the following reduction potentials in primary energy consumption were found: nutrition at around 8% (i.e., eating vegetarian and seasonal food); mobility at around 17% (traveling by bike/walking) or around 11% (traveling by public transport); and living at around 12% (i.e., lowering the room temperature, reducing the living space per person).

To unlock these sufficiency potentials, private consumers are key agents of change. However, private energy consumption patterns are strongly shaped by habits, norms, and cultural, social, and technological contexts and are, therefore, difficult to change (e.g., Owens and Driffill, 2008). Under certain conditions, interventions have the potential to induce such changes (Thøgersen, 2005).

Background: Private Energy-Saving Behaviors and Interventions to Change Behavior

There exists a broad range of research that empirically tests or reviews interventions to change energy-relevant behaviors in different national contexts (Abrahamse et al., 2005; Steg and Vlek, 2009; Mourik and Rotmann, 2013). Steg and Vlek (2009) suggested a general framework for encouraging pro-environmental behavior, which is also relevant in the context of promoting energy-sufficient behavior. To design successful interventions, it is important to identify the relevant behaviors to be changed and to understand how they are influenced. This means that (i) those behaviors that actually have an impact on energy consumption should be identified (Gardner and Stern, 2008; Huddart Kennedy et al., 2015) and (ii) there is a need to better understand "the feasibility of various behavior changes and the acceptability of its consequences" (Steg and Vlek, 2009, p. 310). While (i) can be assessed from a technical perspective, (ii) requires a thorough understanding of people's current energy-saving behavior and their preferences regarding behavior change. In other words, a purely technical approach is not enough to design a successful intervention; it needs to be combined with social-scientific knowledge on behavior and behavior change.

Many studies in Switzerland and internationally that take a social-scientific perspective on energy saving differentiate between curtailment behaviors and efficiency decisions (Gardner and Stern, 2008; Karlin et al., 2012). As the topic of this study is energy sufficiency, we focus on the former aspect. A representative survey in Switzerland analyzed Swiss people's current energy-saving behaviors in the domains of housing, food, and mobility (Sütterlin et al., 2011). People on average perform energy-saving behaviors in the domain of housing very often (e.g., turning the

---

[1] The 2000-Watt/1 ton CO_2 society is a Swiss energy vision that envisages a more equal distribution of global energy consumption by setting per capita consumption goals, it is very popular among Swiss authorities and academics (http://www.2000watt.ch).

TV off when not watching it, filling the washing machine to its capacity, ventilating briefly but intensely during winter). By contrast, energy-saving behaviors in the domain of food (e.g., avoiding buying foods that are flown in, buying seasonal fruits and vegetables) and, in particular, energy-saving behaviors in the domain of mobility (e.g., going on holidays by train, covering short distances by foot or bicycle) are performed less frequently on average (see **Table 1** in Sütterlin et al., 2011). This study did not cover meat consumption, although this is a crucial factor with respect to energy consumption in the domain of food (Dutilh and Kramer, 2000). The results of this Swiss study are in line with those of a Dutch study (Poortinga et al., 2003) that analyzed the acceptability of different energy-saving measures and found that such behaviors as switching off the lights or appliances are well accepted, while such behaviors as going on holidays by train or altering food patterns were somewhat contested. A study in nine OECD countries focusing on housing identified that turning off lights in unused rooms and fully loading washing machines and dishwashers were the most commonly performed energy-saving behaviors, while switching off stand-by modes in appliances seemed less popular (Urban and Ščasný, 2012). The reported results[2] indicate that private energy-saving behavior is domain dependent and, similarly, that people's preferences for energy-saving measures differ for different domains.

What is more, there is a need to better understand contextual influences on energy consumption to design successful interventions (Steg and Vlek, 2009). Barr et al. (2011) noted with some criticism that studying and promoting sustainable consumption is often focused on isolated behaviors in the everyday "home" context where many pro-environmental behaviors (e.g., turning off lights) are socially desired and require no or only small adjustments in lifestyle. With changing contexts, for example when traveling to holidays, pro-environmental behaviors (e.g., not flying to a distant country) often ask for substantial adjustments in lifestyle that leads to conflicts and trade-offs. From focus group discussions, the authors conclude that, "in short, holidays were 'off limits' to sustainability" (Barr et al., 2011, p. 717). This means that designing successful interventions that actually have an impact on energy consumption requires a comprehensive approach that takes into account different contexts where energy is consumed, such as the home, in transit (e.g., from home to work), and while traveling (e.g., on vacation). At the same time, behavior in everyday situations (e.g., commuting) as well as in extraordinary situations (e.g., traveling to holidays) should be considered.

Goal of this Study

The literature review has revealed that (i) for realizing the energy transition in Switzerland, private consumers are key agents of change, (ii) there exist considerable energy-saving potentials through more energy-sufficient behavior, (iii) appropriate interventions may help unlock these potentials, (iv) an important prerequisite for designing successful interventions is knowing

what behaviors have the most impact on energy consumption, how they are influenced by context and what people's preferences are regarding behavior change. While many studies exist that analyze current energy-saving behaviors in Switzerland and in other countries, our study focuses on people's preferences for behaviors that differ in energy-sufficiency. Our approach considers behaviors that have a considerable impact on private energy consumption. Furthermore, behaviors are evaluated together as lifestyles, which are characterized by certain behavior patterns in different domains (everyday mobility, holiday travel, housing, food consumption). Thereby, different contexts for energy-sufficient behavior are considered. In other words, we aim to analyze what people think about energy sufficiency in different domains of life and to which energy-sufficient behaviors they can relate.

The goal of this paper is to identify societal preferences in Switzerland concerning different energy-related behaviors in order to reveal barriers and opportunities related to the promotion of energy sufficiency. Such knowledge provides an important basis for designing successful energy-sufficiency interventions by identifying potential levers and "no-go" areas for such interventions. More concretely, we investigate the following research questions:

- Which energy-sufficiency-related domains and behaviors do people prefer when evaluating different lifestyles?
- Which energy-sufficient lifestyles are perceived to be attractive by the public?

Context Information about Switzerland

As our analysis is focused on Switzerland, we briefly provide some key figures on private energy consumption as well as some context information about the domains we are looking at, that is, commuting, holiday travel, housing, and meat consumption. Swiss households demand 29% of final energy – mostly for heating and hot water – and mobility/transport demands 35% (Swiss Federal Office of Energy, 2014). In the domain of mobility and transport, 74% of final energy is demanded for transporting people on the road, that is, mostly for private mobility (Swiss Federal Office of Energy, 2013). Although Switzerland has an excellent public transport system, approximately half of the inhabitants own a car (536 cars per 1000 inhabitants in 2014; Swiss Federal Statistical Office, 2015d). However, the level of motorization is usually lower in bigger cities compared to rural areas. On average, Swiss commuters commute 14.3 km from home to the workplace (one way). Of these commuters, 53% commute by car, 30% use public transport (train, tram, and bus), 9% bike, and 6% walk (Swiss Federal Statistical Office, 2014). In 2012, Swiss people (older than 6 years) completed a total of 20,300,000 trips with at least one overnight stay, which is roughly three trips per person. The purpose of 65% of these trips was holidays, which people spend abroad (2/3 of cases, 1/3 in Switzerland). Around 50% of trips were made by car and 27% by plane (Swiss Federal Statistical Office, 2013b). The average living space per capita was 45 m^2 in 2013. Around 60% of Swiss people rent their home, while around 40% are homeowners (Swiss Federal Statistical Office, 2015a). In 2012, around 25% of Swiss people ate meat almost every day (6–7 days per week); around 50% ate meat 3–5 days per week, around 20% ate meat 1–2 times a week, and 3% never ate meat (Swiss Federal Office of Public Health, 2014).

---

[2]While the cited papers all use a quantitative approach, there are various studies on the issue of energy consumption that use more qualitative approaches, such as focus groups (e.g., Barr et al., 2011) or ethnographic research (e.g., Higginson et al., 2014).

Materials and Methods

Conjoint Analysis

We apply a conjoint analysis to determine societal preferences for different fictional lifestyles that are characterized by different levels of energy-sufficient behaviors in relevant domains. Conjoint analysis is a method for studying complex decisions that are characterized by trade-offs among different attributes. This method has classically been used in consumer and marketing research (Green and Srinivasan, 1978) and has recently been applied to energy and infrastructure-related decisions (e.g., Dohle et al., 2010; Krütli et al., 2012; Rudolf et al., 2014). Participants are confronted with decision situations composed of sets of attributes. For example, decisions regarding future energy systems may be characterized by different prices and production technologies. Each attribute is associated with different levels (e.g., levels for price: different prices per kilowatt hour of electricity; levels of energy production technologies: solar, nuclear, wind, and hydropower). Participants are then asked to evaluate the decision situation by providing rankings, which requires them to consider combinations of different attribute levels jointly to make a decision. Next, the relative importance values of the different attributes and the part-worth utilities of all the levels can be assessed.

An advantage of conjoint analysis is that it reflects real-world decisions, which are usually characterized by combinations of criteria. Furthermore, it measures preferences indirectly, thus minimizing the potential for respondents to give socially desired responses (Sattler and Hensel-Börner, 2001). From a methodological perspective, another advantage is that not all combinations of levels need to be evaluated empirically; instead, utilities of all combinations can be estimated based on a limited set of choices.

In this study, a choice-based conjoint (CBC) was applied. The main difference between a CBC and other conjoint procedures is that, in a CBC, rather than ranking or rating different options, participants choose their preferred option (Sawtooth Software, 2008). For assessing choices at an individual level, a hierarchical Bayesian estimation was applied.

Attributes and Levels

For the study at hand, only domains with high-energy-saving potentials were chosen (based on Jungbluth and Itten, 2012; Notter et al., 2013). Specifically, we selected the domains of mobility, heating, and food. For each of these domains, a set of attributes was selected:

- Mobility: distance to workplace, means of transport when commuting, holiday travel behavior.
- Heating: amount of heated living space per person, room temperature.
- Food: weekly meat consumption.

The basis for selecting these attributes was a study by Notter et al. (2013), who analyzed and quantified private behaviors based on their cumulative energy demand (CED), global warming potential (GWP) and environmental impact (EI99) by considering the entire lifecycles surrounding these behaviors. Private car use was the most important influencing factor for all three indicators (38% of CED, 31% of GWP, and 29% of EI99). Additional important influencing factors were heating (26% of CED, 25% of GWP, and 18% of EI99) and food (6% of CED, 15% of GWP, and 20% of EI99). Private aviation accounted for 7% of CED, 5% of GWP, and 6% of EI99 (data based on Notter et al., 2013). Also Jungbluth and Itten (2012) identified substantial reduction potentials for primary energy consumption in the domains of mobility, living, and nutrition.

For each attribute, three different levels were defined based on the literature or on thorough discussions among the authors of this paper, such that Level 1 is set as the least energy-sufficient level and Level 3 is the most energy-sufficient level. In contrast to most existing research, this research makes no explicit reference to energy consumption in its descriptions of lifestyles. Rather, the focus is on concrete social practices, which seems to be a more appropriate measurement for the embedded character of energy consumption. Furthermore, this approach also serves the purpose of describing lifestyles realistically, without the use of extensive technical jargon. **Table 1** provides an overview of the selected domains, attributes, and levels.

Design of Conjoint Analysis and Procedure

The defined attributes and levels served as a basis to describe fictional characters and their lifestyles. These lifestyles were composed randomly by combining different levels (one per attribute, full-profile CBC). In each decision situation, participants were presented with three different lifestyles and then asked to choose their preferred lifestyle (see **Table 2**). The study used a full-profile design, meaning that all attributes (with different levels each time) were represented in every option. For each option, the sequence of attributes was kept constant in order to maintain consistency and to better enable comparisons of levels across options. The study was a forced-choice situation; that is, there was no possibility to not choose an option.

Each participant made 10 choices in total (i.e., 10 decision situations): eight randomized tasks and two fixed holdout tasks (all participants evaluated the same two holdout tasks). The holdout tasks were used to validate the conjoint model (Orme et al., 1997; see Chapter Model Fit). Sawtooth Software was used to conduct the experiments and analyze the results (Sawtooth Software, 2008). Three sample lifestyles are presented in **Table 2**.

The data were collected as part of the second author's master's thesis (Rösch, 2013) in autumn of 2013 in the German-speaking part of Switzerland. Participants were recruited from an online panel and received a small incentive for participation. Potential participants were invited to the study by e-mail. The participants first responded to the 10 CBC tasks described above. Afterwards, they answered questions on their personal energy-related behaviors, as well as socio-demographic questions. On average, participants required 12.5 min to complete the survey. All participants who completed the survey were included in the statistical analyses.

Sample

In total, $N = 150$ participants took part in the study. On average, the participants were 47.7 years old ($SD = 12.67$ years), with youngest participant being 18 and the oldest being 66 years old. 52% ($n = 78$) of respondents were female. A total of 50%

TABLE 1 | Domains, attributes, and levels for the conjoint analysis.

Domain	Attribute	Level	Level description
Mobility	Distance to workplace (Swiss Federal Statistical Office, 2012)	1	100 km from home to workplace (100 km)
		2	10 km from home to workplace (10 km)
		3	2 km from home to workplace (2 km)
	Means of transport	1	Car or motorcycle (car)
		2	Public transport or park and rail (public transport)
		3	Public transport or bike (public transport/bike)
	Holiday travel behavior	1	Short trips in Europe, vacations on another continent, solely air travel (World)
		2	Short trips to cities in adjacent countries, vacations within Europe, air travel for vacations, trains for short trips (Europe)
		3	Short trips and vacations in Switzerland or adjacent countries, train whenever possible or car otherwise (Switzerland)
Heating	Living space (Swiss Federal Statistical Office, 2015a)	1	60 m² per person (60 m²)
		2	50 m² per person (50 m²)
		3	40 m² per person (40 m²)
	Room temperature (Stadt Zürich, 2006)	1	T-shirt can be worn even if cold outside (high)
		2	Thin pullover and trousers are worn if cold outside (medium)
		3	Thick pullover and warm socks are worn if cold outside (low)
Food	Weekly meat consumption (Notter et al., 2013)	1	Meat at least once a day (daily)
		2	Meat 3–4 times a week (3–4 times)
		3	Vegetarian or vegan diet (never)

Terms in brackets indicate the short labels for the levels.

TABLE 2 | Exemplary description of three lifestyles, as presented in the study.

The lifestyles of different people are presented below. Please read through them carefully and choose the lifestyle that appeals to you the most. Click the respective button at the end. Even if it is difficult for you to choose, please select one. (1 of 10 decisions)

Work	The person lives in Zürich and works in Bern. He or she commutes daily (**100 km** one way). He or she would consider moving if a new job involved a commute of more than 130 km each way (e.g., Zürich–Fribourg) He or she commutes to work by **public transport or park and rail**	The person lives and works in Zürich. His or her place of work is **2 km** away from home. He or she would consider moving if a new job involved a commute of more than 10 km each way He or she commutes to work by **car**	The person lives in Zürich and works in Thalwil. He or she commutes daily (**10 km** one way). He or she would consider moving if a new job involved a commute of more than 50 km each way (e.g., Zürich–Olten) He or she commutes to work by **public transport or park and rail**
Travel	The person regularly goes on short trips **within Europe** (e.g., a weekend trip to Rome, London, or Barcelona). At least once a year, he or she travels to **another continent** for vacation (e.g., Maldives, the USA, or Brazil). For short trips and longer vacations, he or she usually takes the **plane**	The person regularly goes on short trips **within Switzerland** (e.g., a weekend trip to Ticino or to the Alps). He or she spends vacations in **Switzerland** or in adjacent countries (e.g., France or Germany). For short trips and longer vacations, he or she usually **uses public transport whenever possible or car otherwise**	The person regularly goes on short trips in **adjacent countries** (e.g., a weekend trip to Paris or Berlin). At least once a year, he or she travels to a more **distant country in Europe** (e.g., Spain or Norway) or to a close country on another continent (e.g., Egypt). For vacation, he or she takes a **plane**, and for short trips, he or she uses **public transport**
Housing	The person's flat offers **50 m²** per person The colder it gets, the more clothes he or she wears at home to keep warm. On days that are particularly cold, he or she wears **thick clothes and warm socks**	The person's flat offers **40 m²** per person Even if it is less than 0°C outside, he or she only wears **thin clothes** at home because the rooms are comfortably warm	The person's flat offers **60 m²** per person Even if it is less than 0°C outside, he or she only wears **thin clothes** at home because the rooms are comfortably warm
Food	The person **consumes meat three to four times per week**	The person **consumes meat daily**	The person does not eat meat; he or she is a **vegetarian or vegan**
	○	○	○

Bolded phrases were depicted in red, and the original descriptions were in German.

of participants had concluded vocational training, 20% had completed higher education (e.g., university, PhD), 16% had completed senior high school, 5% had completed higher vocational training, 4% had completed compulsory school, and the rest did not specify their education level.

Regarding political attitudes, most participants positioned themselves in the center of a left wing-right wing scale [from 1 (left) to 7 (right); $M = 3.98$, $SD = 1.13$]. Most participants (43%) lived in two-people households, 24% lived in single-person households, 16% lived in three-people households, and the remainder lived in households larger than three people.

Table 3 summarizes key characteristics of our sample and compares them to Swiss average data (where comparable Swiss data are available; Swiss Federal Statistical Office, 2013a, 2015b,c).

TABLE 3 | Key characteristics of our sample in comparison with Swiss population statistics.

Key characteristics	Study sample (N = 150)	Swiss population
Gender	52% females, 48% males	51% females, 49% males (Swiss Federal Statistical Office, 2015c)
Age (mean)	47.7 years	41.8 years (Swiss Federal Statistical Office, 2015c)
Education	50% vocational training 20% higher education (e.g., university, PhD) 16% senior high school 5% higher vocational training 4% compulsory school Rest: other	44% vocational training 18% higher education (e.g., university, PhD) 9% senior high school 13% higher vocational training 15% compulsory school (Swiss Federal Statistical Office, 2013a)
Household size	24% one person 43%: two people 16%: three people Rest: larger households	35% one person 33%: two people 13%: three people Rest: larger households (Swiss Federal Statistical Office, 2015b)

The sample is approximately representative of Switzerland's population with regard to gender. With respect to vocational and university education, the sample is roughly comparable to the Swiss population, though, in our sample, fewer people had completed only compulsory school and fewer people had completed higher vocational training than the Swiss average. Regarding age, our sample is slightly older than the Swiss average; however, this could be due to the fact that only participants 18 or older were invited to participate in the survey. Regarding household size, more people in our sample lived in two-people households, and fewer people lived in single households than in the Swiss population.

Finally, participants were asked about their personal behaviors in relation to their energy consumption in the domains of mobility, housing, and food. Around 65% of participants lived 10 km away from their workplace or closer, around 30% lived between 10 and 50 km from their workplace, and around 5% lived more than 50 km from their workplace. Most participants (41%) used their car to commute to work, 35% used public transport, 9% used bikes, and 15% walked to their workplace. Over the last 5 years, participants had flown, on average, around nine times ($M = 9.33$, $SD = 10.84$). In terms of living space, 34% of participants used 40 m² per person or less, 25% of participants used around 50 m² per person, 19% used around 60 m² per person, and the remainder used more than 60 m² per person. Participants also indicated the general room temperature they used during the heating season in their apartments: around 33% of participants wore only light clothing during the winter time, 37% wore thin pullovers and trousers, and 30% wore thick clothes and warm socks. With regard to food consumption patterns, 21% of participants ate meat at least once a day, 63% of participants ate meat three to four times a week, 12% ate meat one to four times a month, and 4% were vegetarians or vegans.

Results

Model Fit

The validity of the conjoint model is assessed by observing how well part-worth utilities of the levels can predict the evaluations of the two fixed holdout tasks (Orme et al., 1997). We ran a simulation to estimate participants' choices regarding both holdout

TABLE 4 | Importance of attributes ordered by importance (relative importance values sum to 100%).

Attribute	Relative attribute importance (rounded) (%)
Weekly meat consumption	32
Distance to workplace	22
Means of transport	13
Room temperature	12
Holiday travel behavior	11
Living space	10

tasks based on the individual part-worth utilities derived from the eight random tasks. The simulated results for both holdout tasks were then compared with the actually observed choices regarding the two holdout tasks. A mean absolute error (MAE) test was used to calculate the fitness of the model. This means that, for every holdout task, the difference between the predicted and the observed choices was calculated, with a smaller MAE indicating a better model fit. The MAE for both holdout tasks was 4.24, which is a good result for holdout tasks with three options, according to Orme, President of Sawtooth Software (personal communication). The holdout tasks were only used for this analysis; all further analyses include only the eight randomized tasks.

Importance of Attributes

All participants who finished the survey were included in the subsequent statistical analysis ($N = 150$). As a first step, the attribute importance values were calculated using Sawtooth Software. Attribute importance is a relative measure that allows a relative comparison among the different attributes used in a study. Relative attribute importance is calculated by dividing the range of part-worth utilities for each attribute by the total utility range for all attributes and multiplying the result by 100%. Therefore, when interpreting importance values, it is important to note that these values are relative to the other attributes in the study and that they depend upon the chosen attribute levels (Orme, 2010). Analyses reveal that meat consumption is the most important attribute for participants, followed by distance to workplace (see **Table 4**). Means of transport, room temperature, holiday travel behaviors, and living space can be considered less important attributes, such that meat consumption is about three times as important as those attributes.

Average part-worth utilities of the levels were calculated and displayed in **Figure 1**. The part-worth utilities of each attribute sum to 0. The negative part-worth utilities were not necessarily disliked by participants; however, compared to the other levels, they were preferred less (all else being equal; Orme, 2010). **Figure 1** again indicates the greater importance of the two attributes – meat consumption and distance to workplace – but also shows the part-worth utilities represented by the different attribute levels. This allows the preferences for different attribute levels to be identified. Regarding distance to workplace, larger commuting distances (100 km) were clearly less preferred over shorter commuting distances (2 km or 10 km). For commuting, bike and public transport were the most preferred means of transportation, while, for travel behavior, no clear preference pattern emerged. Regarding living space, larger apartments were preferred to smaller ones, and lower room temperatures were preferred to higher temperatures. In terms of food, eating

meat daily or several times a week was clearly preferred to a vegetarian diet.

Lifestyles that are Sufficiency-Oriented and Perceived as Being Attractive

Based on the aggregated part-worth utilities of all the levels, the overall utilities for all composed lifestyles can be calculated (adding up all part-worth utilities for different combinations of attribute levels; in total, $3^6 = 729$ lifestyle combinations are possible). As can be expected based on the results presented in **Figure 1**, the lifestyle with the highest utility is that of a person who lives around 10 km from work, commutes to work by bike or public transport, takes holiday trips to Europe, lives in quite a large apartment that is not extensively heated during the winter, and eats meat three to four times a week (first rank, overall utility = 123.49). By contrast, the lowest utility is provided by the following lifestyle: a person who lives around 100 km from work, commutes to work by car, makes holiday trips to Switzerland, lives in quite a small apartment that is medium heated during the

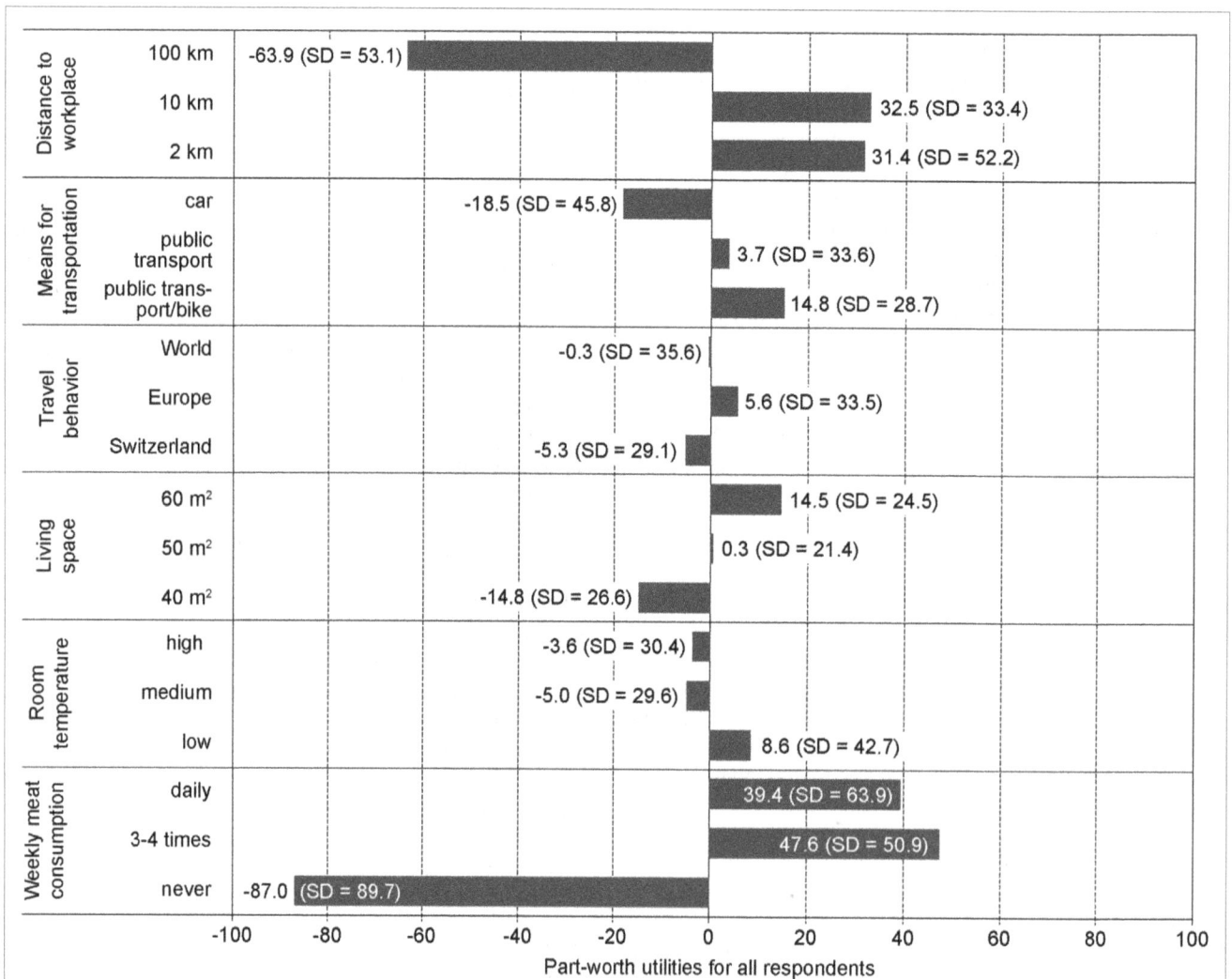

FIGURE 1 | Aggregated average part-worth utilities of all attribute levels. The part-worth attributes for each attribute sum to 0 ($N = 150$).

winter and never eats meat (last rank, overall utility = −194.50; see **Table 5**).

A visual inspection of **Table 5** indicates several important findings: the 10 most popular lifestyles do not include a strictly vegetarian diet, while the 10 least preferred lifestyles all do. Similarly, the 10 most popular lifestyles are characterized by short commuting distances that are traveled by bike or by car, while the 10 least popular lifestyles are characterized by large commuting distances that are traveled by car. Regarding travel, Switzerland does not seem to be a popular destination, and small apartments are not found among any of the top 10 lifestyles. Regarding room temperature, there is a preference for low temperatures.

In order to make judgments about the sufficiency levels of the various lifestyle concepts, we calculated an additive sufficiency index (S-index) based on the attribute levels (i.e., for the most sufficient level, three points were calculated; for the mid-sufficient level, two points; and for the least sufficient level, one point). We are aware that this is an estimate that does not account for differences in energy-saving potentials among attributes, since all attributes are weighted equally. However, calculating the exact amount of energy per level is difficult, since some of the levels are described quite vaguely (e.g., in the case of the level "Europe" in the domain of travel behavior, the exact number of trips taken is not specified; moreover, the exact destination location within Europe is not specified). Thus, our S-index is a very rough estimate of potential savings. It does not reflect differences in energy-saving potentials between domains (e.g., commuting

2 km to work is weighted the same as never eating meat), and it gives different weights to levels within domains. The S-index takes a value between $S_{min} = 6$ and $S_{max} = 18$, with a higher index representing a more sufficient lifestyle. In **Table 5**, the S-index is displayed for the 10 most preferred and the 10 least preferred lifestyles. On average, the S-index for the 10 most preferred lifestyles is slightly higher than the S-index for the 10 least preferred lifestyles, indicating that the most preferred lifestyles are not less sufficient than the least preferred ones; rather, the contrary seems to be the case.

The lifestyle composed of the most sufficient attribute levels is represented by a person who lives around 2 km from work, commutes to work by bike or public transport, makes holiday trips to Switzerland, lives in quite a small apartment that is not heated a lot during winter, and never eats meat. This most sufficient lifestyle is not popular at all – mainly because of its strictly vegetarian diet. It has an overall utility of −52.35 (rank 549 out of 729; see **Table 6**). Thus, the question of interest concerns which sufficiency-oriented lifestyles – specifically, ones that may be less extreme – trigger broad social support and, thus, have the potential to be promising models for energy-sufficient lifestyles for individuals.

Table 6 indicates two key things: first, reasonably energy-sufficient lifestyles can be found among the 50 most preferred lifestyles (11 of the top 50-ranked lifestyles have S-indexes of 14 or 15; $S_{max} = 18$). However, the most energy-sufficient lifestyles do not seem to be very popular. Second, our results indicate the biggest potentials for support for sufficiency in the domains

TABLE 5 | Ranking of lifestyles (10 top-ranked and lowest-ranked lifestyles), ordered by overall utility.

	Overall utility	Rank	Distance to workplace (km)	Means of transport	Travel behavior	Living space (m²)	Room temperature	Weekly meat consumption	S-index
10 most preferred lifestyles	123.49	1	10	Bike/PT	Europe	60	Low	3–4 times	13
	122.47	2	2	Bike/PT	Europe	60	Low	3–4 times	14
	117.62	3	10	Bike/PT	World	60	Low	3–4 times	12
	116.60	4	2	Bike/PT	World	60	Low	3–4 times	13
	115.30	5	10	Bike/PT	Europe	60	Low	Daily	12
	114.28	6	2	Bike/PT	Europe	60	Low	Daily	13
	112.55	7	10	Bike/PT	CH	60	Low	3–4 times	14
	112.45	8	10	PT	Europe	60	Low	3–4 times	12
	111.53	9	2	Bike/PT	CH	60	Low	3–4 times	15
	111.43	10	2	PT	Europe	60	Low	3–4 times	13
10 least preferred lifestyles	−175.87	720	100	Car	World	40	Low	Never	12
	−177.99	721	100	Car	CH	50	High	Never	11
	−179.40	722	100	Car	CH	50	Medium	Never	12
	−180.94	723	100	Car	CH	40	Low	Never	14
	−182.15	724	100	Car	Europe	40	High	Never	11
	−183.56	725	100	Car	Europe	40	Medium	Never	12
	−188.02	726	100	Car	World	40	High	Never	10
	−189.43	727	100	Car	World	40	Medium	Never	11
	−193.09	728	100	Car	CH	40	High	Never	12
	−194.50	729	100	Car	CH	40	Medium	Never	13

Levels are colored according to sufficiency: white = most sufficient levels; light gray = mid-sufficient levels; dark gray = least sufficient levels. S-Index, sufficiency index, which is calculated by adding points according to level: most sufficient level, 3; mid-sufficient level, 2; least sufficient level, 1. N = 150.

TABLE 6 | Display of those lifestyles within the 50 most preferred lifestyles with the highest energy sufficiency, ranked by overall utility.

Overall utility	Utility ranking	Distance to workplace (km)	Means of transport	Travel behavior	Living space (m²)	Room temperature	Weekly meat consumption	S-index
122.47	2	2	Bike/PT	Europe	60	Low	3–4 times	14
112.55	7	10	Bike/PT	CH	60	Low	3–4 times	14
111.53	9	2	Bike/PT	CH	60	Low	3–4 times	15
109.26	15	10	Bike/PT	Europe	50	Low	3–4 times	14
108.24	18	2	Bike/PT	Europe	50	Low	3–4 times	15
103.34	27	2	Bike/PT	CH	60	Low	Daily	14
102.37	31	2	Bike/PT	World	50	Low	3–4 times	14
100.49	37	2	PT	CH	60	Low	3–4 times	14
100.05	40	2	Bike/PT	Europe	50	Low	Daily	14
98.32	46	10	Bike/PT	CH	50	Low	3–4 times	15
97.97	48	2	Bike/PT	CH	60	Medium	3–4 times	14
97.30	51	2	Bike/PT	CH	50	Low	3–4 times	16
82.20	124	2	Bike/PT	CH	40	Low	3–4 times	17
−52.35	549	2	Bike/PT	CH	40	Low	Never	18

Below the straight lines, we show the first occurrences of lifestyles with S-Indices of 16, 17, and 18. The last row displays the most sufficient lifestyle (S-index = 18). Levels are colored according to sufficiency: white = most sufficient levels; light gray = mid-sufficient levels; and dark gray = least sufficient levels. N = 150.

of distance to workplace, means of transport, and room temperature. By contrast, there did not seem to be any support for lifestyles promoting a strictly vegetarian diet and reduced living space per person.

Discussion

Key Findings

This study investigates the societal potentials for sufficiency interventions by investigating people's preferences for different lifestyles using a conjoint analysis. Our first research question is: *Which energy-sufficiency-related domains and behaviors do people prefer when evaluating different lifestyles?* Based on the data from the conjoint analysis, the following patterns are suggested: distance to workplace and meat consumption are considered to be the most important factors when participants make choices regarding their preferred lifestyle. More specifically, participants strongly preferred shorter commuting distances over longer distances and eating meat several times a week over a vegetarian diet.

Our second research question is: *Which energy-sufficient lifestyles are perceived to be attractive by the public?* Our data suggest that there are lifestyles that are reasonably energy-sufficient and, at the same time, able to attract broad public support. These lifestyles are characterized by short commuting distances, using bikes and public transport for commuting and lowered room temperatures during the heating season. Lifestyles characterized by a strictly vegetarian diet and reduced living spaces per person were the least preferred ones.

Discussion of Key Findings and Potential Implications

As demonstrated in our conjoint analysis, there is a disparity between the most energy-sufficient and the most preferred lifestyles. However, this does not mean that energy sufficiency and

popular lifestyles must necessarily conflict. There are lifestyles that are both widely preferred and relatively energy-sufficient. Our research has shown that people weigh such domains as mobility, heating, and food differently when making choices about their preferred lifestyles, indicating that they make different trade-offs between these domains. In the following, we separately discuss all of the domains, their potentials for energy sufficiency and possible implications for practice (e.g., for interventions). However, it must be kept in mind that these results were established in an integrated and, thus, indirect way. Also, results are strongly influenced by the attributes and levels chosen and are situated in the context of the German-speaking part of Switzerland.

Shorter Commuting Distances and Mode of Transport

Our results suggest a preference for shorter commuting distances. This is likely because shorter commuting distances provide significant benefits to individuals in the sense that shorter commutes give employees more leisure time. Similarly, our results suggest a preference for biking to work. This preference may also relate to individual benefits, since "active commuting" (i.e., biking to work) is positively related to physical well-being (Humphreys et al., 2013). Our results reflect a trend in Swiss cities (e.g., Zürich) in which an increasing number of households refrain from having a car, instead opting to use bikes, public transport, or car sharing. In Zürich, 48% of households do not own a car, and the rate of motorization has declined since the 1990s to around 350 cars per 1000 inhabitants (Stadt Zürich, 2012). This is almost certainly due to the city's excellent public transport system, which offers regular, punctual, and modern means of transport. Our results indicate a high social acceptance for commuting by bike or public transport – and, as such, suggest the potential for interventions to reduce energy consumption through commuting. For example, campaigns that promote biking to work (Bike to Work, 2013), sharing offers (e.g., car sharing, bike sharing), or even car-free lifestyles (e.g., car-free residential areas) may be effective. As the level of motorization in the rural areas of Switzerland is still high compared to that in urban areas (Swiss Federal Statistical Office, 2015d), it might be particularly promising to develop respective interventions for rural areas. One interesting example is the bike-4car campaign, which encourages car owners to give up driving and try out e-biking for free for a period of 2 weeks[3]. It is important to note that the success of interventions to change commuting behavior depends strongly on the available infrastructure, such as quality of public transport and spatial separation of activities (e.g., shopping, sports, daycare; Thøgersen, 2005). This implies that our results are very context-sensitive and might look very different in another country.

Travel Behavior

Our study indicates that, relative to the other attributes, holiday travel behavior was not a very important attribute. Our study does not indicate large social support for more local travel behavior; however, this option was perceived as only slightly less attractive than vacations and short trips in Europe or on other continents. We can only speculate about why this might be the case. One reason could be that, although locations varied, all levels seemed to include many short trips and holiday opportunities over the year. Therefore, the different levels might have been perceived as equally attractive, leading only to a small spread across the levels.

Living Space and Room Temperature

Our study indicates a preference for large living spaces, since 60 m^2 per person is clearly preferred to 40 m^2 per person. Moreover, our results follow a clear trend in Switzerland toward larger living spaces (Swiss Federal Statistical Office, 2015a). One crucial question is how to address the need for more personal space within cities, where such space is particularly scarce. One option would be to complement a limited amount of personal space with shared spaces (e.g., shared guest rooms, workshops, office spaces, and common rooms). Such shared rooms offer benefits on several different levels: (i) in total, less space needs to be heated, lighted, etc., thus reducing energy consumption; (ii) people can profit from shared infrastructure; and (iii) shared rooms address people's need to connect with other people, in that they offer opportunities to meet, exchange, and learn from each other. Since this approach (i.e., increasing the usage of shared spaces while limiting personal space) addresses different needs and does not only focus on reduced energy consumption, it is particularly promising in terms of attracting broad social support (International Energy Agency, 2014; Moser et al., 2014). Regarding room temperature, the study indicates a certain potential for interventions to reduce room temperature, since lower room temperatures were preferred to higher temperatures. However, it should be noted here that the study was conducted in autumn, before the heating season; thus, the results may be framed by the time of year. Interestingly, the results of our study indicate paradoxical effects regarding living space and room temperature that are similar to those determined by Sütterlin and

Siegrist (2014). They found that participants regard a person who has more living space but lower room temperatures to be more environmentally conscious than a person with less living space but higher room temperatures, although the latter actually uses less energy for heating.

Meat Consumption

Our results clearly show that a vegetarian or vegan diet is not a viable option for the vast majority of participants. For most participants, it was important to eat meat several times per week or even daily. However, the results also showed that daily meat consumption is not the most preferred option, indicating that many people are ready to refrain from meat consumption from time to time. A link to personal health may play an important role here. Based on our results, we may speculate that restrictions on eating meat might trigger strong reactions and protests. Instead, campaigns promoting vegetarian dishes from time to time may be more effective. Furthermore, people might be nudged into less energy-intense dietary habits through attractive alternatives in canteens and restaurants (Bucher et al., 2011).

Critical Reflections, Limitations and Further Research

In our study, we see three particular limitations related to: the choice of levels, context influences, and the construction of the S-index.

In conjoint analyses, the importance values of attributes are vastly influenced by the levels and the spread associated with the attributes. Although Orme (2002) suggests that levels should spread across the full range of possibilities, this might have been too extreme for the meat consumption attribute. As shown in **Table 4** and **Figure 1**, the importance of the meat consumption attribute was significantly influenced by the negative part-worth utility of the level of vegetarian diet (never). In comparison with the other levels (i.e., daily meat consumption and three to four times meat consumption per week), this option is more extreme because it suggests a strictly vegetarian or vegan diet. It is possible that a less extreme level (e.g., meat consumption once a month) would not have triggered such extreme reactions by participants. On the other hand, our responses are in line with the intense public response that followed the proposal by some Swiss canteens to launch "vegetarian days." This announcement resulted in large protests in social media and through online comments to media articles, indicating that meat consumption may be non-negotiable for many Swiss people. Although daily consumption is not necessarily desired, many people in Switzerland are not ready to completely give up meat consumption. In 2012, only around 3% of Switzerland's population never ate meat or sausage products (Swiss Federal Office of Public Health, 2014). For a future study, it would be interesting to add another level between never and three to four times a week, such as once a month or on special occasions. A similar logic regarding the spread of levels can be found in the attribute commuting distance, since the commuting distances of 2 and 10 km are quite close together, while 100 km is more extreme. It would be interesting to include a less extreme option in a future study, thereby allowing the identification of tipping points in preferences with respect to commuting distances.

---

[3] www.bike4car.ch

It is likely that our study results are affected by contextual influences. These influences may have manifested, in particular, in participant responses to the attribute room temperature. Here, many participants preferred lifestyles in which the protagonists wore warm clothes during winter to keep warm in their apartments, instead of turning up the thermostat. This result may have been influenced by contextual effects, since the data were collected in September and not during the heating season. Furthermore, the amount of clothing worn serves only as a proxy for drawing conclusions about the actual room temperature (in terms of absolute values). It could be that people prefer warm clothing, but simultaneously heat their living spaces to higher temperatures. However, we would argue that it is easier for participants to imagine the types of clothes implied by room temperature than to imagine a particular room temperature in degree celsius. Similarly, our results regarding travel behavior could be influenced by the season in which the study was conducted. If the study had been conducted before summer, when many people in Switzerland usually plan longer vacations, preferences for flying to distant countries might have been more distinct. In general, actual behaviors cannot be inferred directly from the revealed preferences, as there might be additional constraints in people's lives. For example, although a participant might prefer a commuting distance of 2 km, he or she may not be able to move closer to the workplace or change jobs due to his or her family situation.

Although our S-index provides certain indications regarding the energy-saving potentials of the presented lifestyles, these indications offer only a very rough estimate, which does not account for differences in the potential savings of the attributes. For future research, it would be interesting to describe the levels more precisely. More finely grained data would facilitate the calculation of the actual energy-saving potentials of the presented lifestyles (e.g., based Life Cycle Assessment databases), thus allowing us to draw precise conclusions about the energy-saving potentials of the most preferred lifestyles.

Conclusion

Our study results suggest that the success of energy-sufficiency interventions might depend strongly on the targeted behavior.

References

Interventions to change certain behaviors (e.g., meat consumption) seem likely to face strong public resistance. As such, our results have implications for the promotion of energy-sufficient lifestyles through, for example, energy-saving campaigns. Specifically, our results show that extremely energy-sufficient lifestyles are not perceived as attractive – or, more technically speaking, they are characterized by a negative overall utility. We thus speculate that the promotion of such extremely energy-sufficient lifestyles might backfire, potentially evoking resistance or resignation.

Our study results could be interpreted to suggest that well-balanced lifestyles with substantive (but not extreme) energy-saving potentials might better serve as social models for energy sufficiency (compared to extremely sufficient lifestyles). As role models, such well-balanced lifestyles may motivate people to change their routine behaviors and lifestyles in order to consume less energy. A similar effect was found in a study on scenarios for urban development, in which the most sustainable scenarios were unable to trigger consensus among different stakeholders (e.g., investors, urban planners, housing target groups), whereas more balanced scenarios were able to gain broader support (Bügl et al., 2012). However, the study at hand is exploratory; thus, questions concerning exactly how the public reacts when confronted with extreme energy-sufficient lifestyles or respective interventions, what types of emotions these lifestyles trigger and how well different groups identify with them remain unanswered. Field experiments could be a promising approach to investigate these questions.

Acknowledgments

This paper is based on the MSc thesis of the second author, supervised by the first and third author. The work was supported by the National Research Programme "Managing Energy Consumption" (NRP 71) of the Swiss National Science Foundation (SNSF). Further information on the National Research Programme can be found at www.nrp71.ch. According to ETH ethical guidelines, no ethical approval was required for this study. We would like to thank the participants for taking part in the study. Furthermore, we would like to thank Yann Blumer, Merla Kubli, the two reviewers Anya Skatova and Andrea Tabi for providing feedback on earlier versions of this paper, as well as Sandro Bösch for designing **Figure 1**.

Abrahamse, W., Steg, L., Vlek, C., and Rothengatter, T. (2005). A review of intervention studies aimed at household energy conservation. *J. Environ. Psychol.* 25, 273–291. doi:10.1016/j.jenvp.2005.08.002

Barr, S., Gilg, A., and Shaw, G. (2011). "Helping people make better choices": exploring the behaviour change agenda for environmental sustainability. *Appl. Geogr.* 31, 712–720. doi:10.1016/j.apgeog.2010.12.003

Bike to Work. (2013). *Ergebnisse Befragung TeilnehmerInnen, TeamchefInnen [Survey Results of Teamleaders and Participants]*. Bern: Bike to Work.

Breukers, S., Mourik, R., and Heiskanen, E. (2013). "Changing energy demand behavior: potential of demand-side-management," in *Handbook of Sustainable Engineering*, eds Kauffmann J. and Lee K.-M. (Dordrecht: Springer), 773–792.

Bucher, T., van der Horst, K., and Siegrist, M. (2011). Improvement of meal composition by vegetable variety. *Public Health Nutr.* 14, 1357–1363. doi:10.1017/S136898001100067X

Bügl, R., Stauffacher, M., Kriese, U., Pollheimer, D. L., and Scholz, R. W. (2012). Identifying stakeholders' views on sustainable urban transition: desirability,

utility and probability assessments of scenarios. *Eur. Plann. Stud.* 20, 1667–1687. doi:10.1080/09654313.2012.713332

Darby, S. (2007). "Enough is as good as a feast: sufficiency as policy," in *ECEEE 2007 Summer Study*, 111–119.

Dohle, S., Keller, C., and Siegrist, M. (2010). Conjoint measurement of base station siting preferences. *Hum. Ecol. Risk Assess.* 16, 825–836. doi:10.1080/10807039.2010.501250

Dutilh, C. E., and Kramer, K. J. (2000). Energy consumption in the food chain. *Ambio* 29, 98–101. doi:10.1579/0044-7447-29.2.98

Gardner, G. T., and Stern, P. C. (2008). The short list: the most effective actions US households can take to curb climate change. *Environment* 50, 12–24. doi:10.3200/envt.50.5.12-25

Green, P. E., and Srinivasan, V. (1978). Conjoint analysis in consumer research: issues and outlook. *J. Consum. Res.* 5, 103–123. doi:10.1086/208721

Herring, H. (2006). Energy efficiency: a critical view. *Energy* 31, 10–20. doi:10.1016/j.energy.2004.04.055

Higginson, S., Thomson, M., and Bhamra, T. (2014). "For the times they are a-changin": the impact of shifting energy-use practices in time and space. *Local Environ.* 19, 520–538. doi:10.1080/13549839.2013.802459

Huddart Kennedy, E., Krahn, H., and Krogman, N. T. (2015). Are we counting what counts? A closer look at environmental concern, pro-environmental behaviour, and carbon footprint. *Local Environ.* 20, 220–236. doi:10.1080/13 549839.2013.837039

Humphreys, D. K., Goodman, A., and Ogilvie, D. (2013). Associations between active commuting and physical and mental wellbeing. *Prev. Med.* 57, 135–139. doi:10.1016/j.ypmed.2013.04.008

International Energy Agency. (2014). *Capturing the Multiple Benefits of Energy Efficiency.* Paris: OECD/IEA.

Jenny, A. (2014). *Suffizienz auf individueller Ebene – Literaturanalyse zu psychologischen Grundlagen der Suffizienz. Zwischenbericht Nr. 18, Forschungsprojekt FP-1.7.* Zürich: Energieforschung Stadt Zürich.

Jungbluth, N., and Itten, R. (2012). *Umweltbelastungen des Konsums in der Schweiz und in der Stadt Zürich: Grundlagendaten und Reduktionspotenziale.* Zürich: Energieforschung Stadt Zürich.

Karlin, B., Davis, N., Sanguinetti, A., Gamble, K., Kirkby, D., and Stokols, D. (2012). Dimensions of conservation: exploring differences among energy behaviors. *Environ. Behav.* 46, 423–452. doi:10.1177/0013916512467532

Krütli, P., Stauffacher, M., Pedolin, D., Moser, C., and Scholz, R. W. (2012). The process matters: fairness in repository siting for nuclear waste. *Soc. Justice Res.* 25, 79–101. doi:10.1007/s11211-012-0147-x

Linz, M. (2012). *Weder Mangel noch Übermass: Warum Suffizienz unentbehrlich ist.* München: Oekom Verlag.

Linz, M., Bartelmus, P., Hennicke, P., Jungkeit, R., Sachs, W., Scherhorn, G., et al. (2002). "Von nichts zu viel: Suffizienz gehört zur Zukunftsfähigkeit," in *Wuppertal Working Papers*, Wuppertal: Wuppertal Institut, 125.

Moser, C., von Wirth, T., Adler, C., Fonseca, J., Schlüter, A., and Stauffacher, M. (2014). *Nachhaltige Arealentwicklung: Der Fall Siemensareal in Zug [Sustainable Urban Development: The Case of the Siemens Site in Zug].* Zürich: ETH-UNS TdLab.

Mourik, R., and Rotmann, S. (2013). *Most of the Time What We Do is What We Do Most of the Time. And Sometimes We Do Something New: Analysis of Case Studies IEA DSM Task 24 Closing the Loop – Behaviour Change in DSM: From Theory to Practice.* Paris: International Energy Agency.

Notter, D. A., Meyer, R., and Althaus, H.-J. (2013). The western lifestyle and its long way to sustainability. *Environ. Sci. Technol.* 47, 4014–4021. doi:10.1021/ es3037548

Orme, B. K. (2002). *Formulating Attributes and Levels in Conjoint Analysis.* Sequim, WA: Sawtooth Software.

Orme, B. K. (2010). *Getting Started with Conjoint Analysis: Strategies for Product Design and Pricing Research,* 2nd Edn. Madison, WI: Research Publishers LLC.

Orme, B. K., Alpert, M. I., and Christensen, E. (1997). *Assessing the Validity of Conjoint Analysis - Continued,* Vol. 98382. Sequim, WA: Sawtooth Software.

Owens, S., and Driffill, L. (2008). How to change attitudes and behaviours in the context of energy. *Energy Policy* 36, 4412–4418. doi:10.1016/j.enpol.2008.09.031

Poortinga, W., Steg, L., Vlek, C., and Wiersma, G. (2003). Household preferences for energy-saving measures: a conjoint analysis. *J. Econ. Psychol.* 24, 49–64. doi:10.1016/s0167-4870(02)00154-x

Rösch, A. (2013). *Analysis of Energy Sufficiency: Barriers and Opportunities on the Path Towards an Energy Sufficient Consumer Society.* Zürich: ETH Zürich.

Rudolf, M., Seidl, R., Moser, C., Krütli, P., and Stauffacher, M. (2014). Public preferences of electricity options before and after Fukushima. *J. Integr. Environ. Sci.* 11, 1–15. doi:10.1080/1943815X.2014.881887

Sattler, H., and Hensel-Börner, S. (2001). "A comparison of conjoint measurement with self-explicated approaches," in *Conjoint Measurement,* eds Gustafsson A., Herrmann A., and Huber F. (Berlin: Springer), 121–133.

Sawtooth Software. (2008). *CBCv6.0 Technical Paper.* Orem: Sawtooth Software, Inc.

Stadt Zürich. (2006). *Raumtemperatur-Richtlinie.* Zürich: Stadt Zürich.

Stadt Zürich. (2012). *Mobilität in Zahlen [Mobility Figures].* Zürich: Stadt Zürich.

Steg, L., and Vlek, C. (2009). Encouraging pro-environmental behaviour: an integrative review and research agenda. *J. Environ. Psychol.* 29, 309–317. doi:10.1016/j.jenvp.2008.10.004

Sütterlin, B., Brunner, T. A., and Siegrist, M. (2011). Who puts the most energy into energy conservation? A segmentation of energy consumers based on energy-related behavioral characteristics. *Energy Policy* 39, 8137–8152. doi:10.1016/j. enpol.2011.10.008

Sütterlin, B., and Siegrist, M. (2014). The reliance on symbolically significant behavioral attributes when judging energy consumption behaviors. *J. Environ. Psychol.* 40, 259–272. doi:10.1016/j.jenvp.2014.07.005

Swiss Federal Council. (2013). *Botschaft zum ersten Massnahmenpaket der Energiestrategie 2050 und zur Volksinitiative «Für den geordneten Ausstieg aus der Atomenergie (Atomausstiegsinitiative)».* Bern: Swiss Federal Council.

Swiss Federal Office of Energy. (2013). *Analyse des schweizerischen Energieverbrauchs 2000-2012 nach Verwendungszwecken [Energy Consumption in Switzerland According to Purpose 2000-2012].* Bern: Federal Department of the Environment, Transport, Energy and Communications.

Swiss Federal Office of Energy. (2014). *Schweizerische Gesamtenergiestatistik 2013 [Swiss Energy Statistics].* Ittigen: Swiss Federal Office of Energy.

Swiss Federal Office of Public Health. (2014). *Schweizerische Gesundheitsbefragung 2012: Standardtabellen [Swiss Health Survey 2012: Tables].* Available at: http:// www.portal-stat.admin.ch/sgb2012/

Swiss Federal Statistical Office. (2012). *Taschenstatistik der Schweiz 2012 ["Pocket Statistics" Switzerland 2012].* Neuchâtel: Swiss Federal Statistical Office.

Swiss Federal Statistical Office. (2013a). *Lebenslanges Lernen in der Schweiz: Ergebnisse des Mikrozensus Aus- und Weiterbildung 2011 [Lifelong Learning in Switzerland: Results from the Education Census 2011].* Neuchâtel: Swiss Federal Statistical Office.

Swiss Federal Statistical Office. (2013b). *Reisen der Schweizer Wohnbevölkerung 2012 [Trips of Swiss People 2012].* Neuchâtel: Swiss Federal Statistical Office.

Swiss Federal Statistical Office. (2014). *Pendlermobilität in der Schweiz: 2012 [Commuting in Switzerland: 2012].* Neuchâtel: Swiss Federal Statistical Office.

Swiss Federal Statistical Office. (2015a). *Bau- und Wohnungswesen: Die wichtigsten Zahlen [Construction and Housing: Key Figures].* Available at: http://www.bfs. admin.ch/bfs/portal/de/index/themen/09/01/key.html

Swiss Federal Statistical Office. (2015b). *Families, Households – Data, Indicators: Households by Household Size 2013.* Available at: http://www.bfs.admin.ch/bfs/ portal/en/index/themen/01/04/blank/key/01/05.html

Swiss Federal Statistical Office. (2015c). *Population: Key Figures 2013.* Available at: http://www.bfs.admin.ch/bfs/portal/en/index/themen/01/01/key.html

Swiss Federal Statistical Office. (2015d). *Transport Infrastructure and Vehicles - Vehicles Stock, Level of Motorisation.* Available at: http://www.bfs. admin.ch/bfs/portal/en/index/themen/11/03/blank/02/01/01.html

Thøgersen, J. (2005). How may consumer policy empower consumers for sustainable lifestyles? *J. Consum. Policy* 28, 143–177. doi:10.1007/s10603-005-2982-8

Urban, J., and Ščasný, M. (2012). Exploring domestic energy-saving: the role of environmental concern and background variables. *Energy Policy* 47, 69–80. doi:10.1016/j.enpol.2012.04.018

Conflict of Interest Statement: The authors declare that the research was conducted in the absence of any commercial or financial relationships that could be construed as a potential conflict of interest.

7

Comparison of Microalgae Cultivation in Photobioreactor, Open Raceway Pond, and a Two-Stage Hybrid System

Rakesh R. Narala[†], Sourabh Garg[†], Kalpesh K. Sharma, Skye R. Thomas-Hall, Miklos Deme, Yan Li and Peer M. Schenk*

Algae Biotechnology Laboratory, School of Agriculture and Food Sciences, The University of Queensland, Brisbane, QLD, Australia

Edited by:
*Arumugam Muthu,
Council of Scientific and Industrial
Research, India*

Reviewed by:
*Maria Gonzalez Alriols,
University of the Basque
Country, Spain
Jennifer Stewart,
University of Delaware, USA*

*Correspondence:
*Peer M. Schenk
p.schenk@uq.edu.au*

*[†]Rakesh R. Narala and
Sourabh Garg contributed equally.*

Specialty section:
*This article was submitted to
Bioenergy and Biofuels,
a section of the journal
Frontiers in Energy Research*

Citation:
*Narala RR, Garg S, Sharma KK,
Thomas-Hall SR, Deme M, Li Y and
Schenk PM (2016) Comparison of
Microalgae Cultivation in
Photobioreactor, Open Raceway
Pond, and a Two-Stage Hybrid
System.
Front. Energy Res. 4:29.*

In the wake of intensive fossil fuel usage and CO_2 accumulation in the environment, research is targeted toward sustainable alternate bioenergy that can suffice the growing need for fuel and also that leaves a minimal carbon footprint. Oil production from microalgae can potentially be carried out more efficiently, leaving a smaller footprint and without competing for arable land or biodiverse landscapes. However, current algae cultivation systems and lipid induction processes must be significantly improved and are threatened by contamination with other algae or algal grazers. To address this issue, we have developed an efficient two-stage cultivation system using the marine microalga *Tetraselmis* sp. M8. This hybrid system combines exponential biomass production in positive pressure air lift-driven bioreactors with a separate synchronized high-lipid induction phase in nutrient deplete open raceway ponds. A comparison to either bioreactor or open raceway pond cultivation system suggests that this process potentially leads to significantly higher productivity of algal lipids. Nutrients are only added to the closed bioreactors, while open raceway ponds have turnovers of only a few days, thus reducing the issue of microalgal grazers.

Keywords: biodiesel, lipid induction, microalgae cultivation, microalgal oil, open pond, photobioreactor

INTRODUCTION

Microalgae are considered a promising feedstock for next-generation biofuel production because they are potentially 10–20 times more productive than any other biofuel crop and their large-scale cultivation does not need to compete for arable land or precious biodiverse landscapes (Hannon et al., 2010; Mata et al., 2010; Ahmad et al., 2011; Ndimba et al., 2013). Importantly, they are also able to grow in saline and even wastewater (Brennan and Owende, 2010; Christenson and Sims, 2011; Abou-Shanab et al., 2013). Microalgal oil is a valuable commodity, for example, when used as a substitute for omega-3-rich fish oil. However, commercial cultivation of microalgae for biodiesel, a relatively low-value product, requires further improvements to reach economical feasibility.

Microalgae accumulate large amounts of lipid bodies containing triacylglycerides under adverse conditions, such as during nutrient deprivation (Hu et al., 2008; Sharma et al., 2012). Under these circumstances, microalgae stop dividing but are still able to perform photosynthesis and the accumulation of triacylglycerides is considered a survival strategy to endure adverse conditions (Schenk et al., 2008; Breuer et al., 2012; Liu and Benning, 2013). Cultivation of microalgae for biodiesel production, however, aims at maximizing lipid productivity (or lipid yields) which takes both, growth rates and lipid contents, into consideration. In batch cultivation systems, microalgae are first grown exponentially to increase their biomass that is then followed by a lipid induction process, usually by omitting nutrient supply toward the end of the growth phase. Other lipid induction techniques are also available and their combination may lead to improved lipid contents (Sharma et al., 2012).

The two most common methods of microalgae cultivation are open cultivation systems, such as open ponds, tanks, and raceway ponds, and controlled closed cultivation systems using different types of bioreactors. One of the first attempts to scale up and cultivate microalgae was achieved using open raceway ponds (Johnson et al., 1988). Since then, extensive research has been carried out to cultivate microalgae in open cultivation systems. Some of the major advantages of an open cultivation system are minimal capital and operating costs, and a lower energy requirement for culture mixing. On the downside, open systems require large areas to scale up and are susceptible to contamination (e.g., introduced by birds) and adverse weather conditions. Although rarely reported, we know from our own experience that contamination with other microalgae, high bacterial loads, and grazers is common in large-scale open pond cultivation systems. In particular, we have experienced the damaging effects of the occurrence of rotifers to cultures of *Tetraselmis*, *Chlorella*, *Nannochloropsis*, and *Scenedesmus*, and the damage of amoeba to diatoms, which can be consumed in as little as 2 days after visual detection of contamination. Also in open pond systems, it is difficult to have control over growth parameters, such as evaporation, culture temperature, etc. (Oyler, 2009; Rupprecht, 2009; Mata et al., 2010).

Closed cultivation systems, here referred to as closed photobioreactors (PBRs), are more efficient in terms of quality as they can be operated at highly controlled conditions and, therefore, can overcome the disadvantages of an open cultivation system. PBRs can be designed and optimized in accordance with the strain of choice. This closed system utilizes relatively little space, while increasing the light availability and greatly decreasing the contamination issues. However, PBRs also have some disadvantages, such as bio-fouling, overheating, benthic algae growth, cleaning issues and high build-up of dissolved oxygen resulting in growth limitation, and, more importantly, very high capital costs for designing and operating (Molina Grima et al., 1999; Chisti, 2008).

The design and principle of cultivation systems change based on the specific needs (Schenk et al., 2008). Open ponds built in a wastewater treatment plant can be circular in shape or driven by gravity flow. Similarly, the basic tubular design of PBRs has been improved over the past decade to facilitate better light availability and culture mixing to produce a range of pharmaceutical products to high value nutritional products. PBRs were found to be more efficient when operated with continuous cultures (Otero and Fábregas, 1997; Mata et al., 2010). Continuous cultures in closed systems can be used for higher biomass productivity, but

cannot be used for lipid induction by stress mechanisms (nutrient starvation) to produce biodiesel.

Although extensive studies have been carried out on open and closed cultivation methods, limited work has been carried out on two-stage hybrid cultivation systems. Two-phase hybrid cultivation systems have been proposed as an advantageous microalgae cultivation system, as they are able to essentially separate biomass growth from the lipid accumulation phase (Olaizola, 2000; Schenk et al., 2008; Su et al., 2011). Recently, a life cycle analysis demonstrated a considerably reduced environmental impact when comparing various open and closed cultivation systems with hybrid cultivation (Adesanya et al., 2014). A recent study on large-scale hybrid cultivation concluded that PBRs were economical to provide a continuous and consistent inoculum for short-period batch open pond cultures that prevented biological system crashes compared to longer term open pond cultures (Huntley et al., 2015).

To test whether productivity may also vary for these different systems, a side-by-side comparison was carried out in the present study, using the same algal culture. A new airlift-driven low-cost tubular PBR was designed and used for continuous growth phase of the microalgal culture while an open raceway pond was used for stress induction and synchronized lipid accumulation. A pilot-scale hybrid cultivation system has been constructed where microalgal growth for lipid production was compared to cultivation in either closed airlift-driven tubular PBR or open raceway ponds. The proposed hybrid cultivation system allows for a separate lipid accumulation phase where one or more efficient stress induction techniques can be carried out, while effectively avoiding the issue of contamination. Key parameters, such as sunlight, nutrients, CO_2 and water, which affect outdoor cultivation of microalgae, were examined and monitored over time to understand the importance of these various factors for growth and lipid productivity.

MATERIALS AND METHODS

Microalgae Cultivation and Analyses

Strain *Tetraselmis* sp. M8 had previously been shown to accumulate significant amounts of lipids after nutrient deprivation (Lim et al., 2012). It was collected in an intertidal rock pool at Maroochydore, Australia (26°39′39″S 153°6′18″E). Pure *Tetraselmis* sp. M8 cultures were grown in f/2 medium (AlgaBoost™) with the omission of silica (f/2 (−Si); Guillard, 1975) in autoclaved seawater (collected at Cleveland, Brisbane, QLD, Australia) using laboratory culturing conditions described previously (Lim et al., 2012). A hemocytometer (Bright Line, Sigma-Aldrich, St. Louis, MO, USA) was used to count microalgal cells manually. A total of 100 μL of culture was transferred into an Eppendorf tube and 0.1 μL of acetic acid was added to ensure that the algae lose their motility. Growth rate and doubling time were calculated using the formulae mentioned below (Lim et al., 2012)

$$\text{Growth rate } (K') = Ln\left(\frac{N2/N1}{t2 - t1}\right)$$

$$\text{Doubling time} = \frac{1}{k' / Ln(2)}$$

where, K' = growth rate,

$N1$ and $N2$ = biomass at time1 ($t1$) and time2 ($t2$), respectively.

Mean growth rates for the entire length of each growing cycle were determined based on cell concentrations at start and finish of each cycle. Nitrate and phosphate concentrations were measured in seawater using API Nutrient testing kits (API Fishcare, UK) according to the manufacturer's instructions.

Culture Scale-Up and Monitoring

A 50 mL preculture of pure *Tetraselmis* sp. M8 was used to inoculate a 2-L glass bottle and culture was made up to 1 L with fresh autoclaved seawater containing f/2 (−Si) medium. Filtered air was supplied through a Millipore syringe filter for uniform mixing of the culture and to prevent stagnation at the bottom of the bottle. The set-up was undisturbed for 4 days. At the end of the fourth day, 250 mL of the culture was transferred into a clear hanging polyethylene bag (80 cm × 50 cm) which contained 4.75 L of fresh f/2 (−Si) medium in autoclaved seawater. Filtered air was supplied through a Millipore filter for mixing. The set-up was undisturbed for 4 days.

Four polyethylene bags were set up under outdoor conditions (roof structure of a three-storey building; Goddard building 8, University of Queensland St Lucia campus, Brisbane, QLD, Australia) and were each filled with 19 L of fresh f/2 (−Si) medium in seawater. The 5 L culture grown in the laboratory was mixed uniformly and 1 L was added to each 19 L bag. These outdoor cultures were left under direct sunlight for 4 days to achieve maximum cell density. The algal cultures of the four bags were then used for inoculation of a closed PBR that contained 1,200 L fresh f/2 (−Si) medium in seawater. CO_2 (100%) was supplied to the PBR to control the pH at around 8.4 using a Weipro pH 2010 controller. This culture served as the starting culture for all subsequent experiments described in the section "Results".

Cell densities and nutrients were monitored on a daily basis using 5 mL samples of culture from the closed PBR or raceway ponds. Daily sampling was carried out at 04:00 p.m. (AEST). f/2 (−Si) nutrients were added when required (nutrients depleted to less than 10%). In the PBR and raceway pond, starvation phase was started once the cells attained a density of 1.5×10^6 cells/mL. Half the volume was harvested after the nutrients were completely used and the cell density exceeding 2.0×10^6 cells/mL. In hybrid system cultivation, when the desired PBR cell density of over 2×10^6 cells/mL was reached, typically half the volume of the PBR (600 L) was transferred into the open raceway pond for lipid induction for 3–4 days and the removed volume of the PBR was replaced with fresh f/2 (−Si) medium in seawater. Apart from this sampling, climate data were obtained from Australia's Bureau of Meteorology, including temperature and solar exposure. PBR and raceway pond cultivation systems were tested simultaneously, but separately from hybrid cultivation. Both cultivation systems were cleaned and sanitized with bleach before cultivation. Cell count, temperature, and solar exposure values were plotted on graphs for the open raceway pond, the closed PBR and the hybrid cultivation system. To avoid biofilm formation in the PBR that

could cause light limitation, the polyethylene tubes were occasionally (once a week) slapped to loosen any benthic cells. Nile red staining was performed as described previously (Lim et al., 2012). Data were analyzed statistically using one-way ANOVA followed by Tukey's test.

RESULTS

Design and Construction of Pilot-Scale Photobioreactor and Raceway Ponds

A closed tubular PBR and an open raceway pond were designed for side-by-side pilot-scale outdoor algae cultivation using an airlift mechanism for mixing. The volume of the PBR was 1,200 L and the raceway pond held 1,000 L, as shown in **Figures 1** and **2**. The surface area of the PBR and each raceway pond was 4.6 and 6 m^2, respectively. For a direct comparison of both cultivation systems, PBR and open raceway pond were used for simultaneous side-by-side algae cultivation. This included cycles of growth and lipid induction before harvesting of biomass. Synchronized lipid induction was verified by Nile red staining before harvesting (**Figure 3**) and only when lipid-rich biomass was produced it was harvested. The following paragraphs describe the cultivation cycles applied and monitored for each cultivation system.

Outdoor Photobioreactor Cultivation

Cell counts (cells/milliliter) and all other parameters of PBR-grown *Tetraselmis* sp. M8 culture were recorded for 33 days,

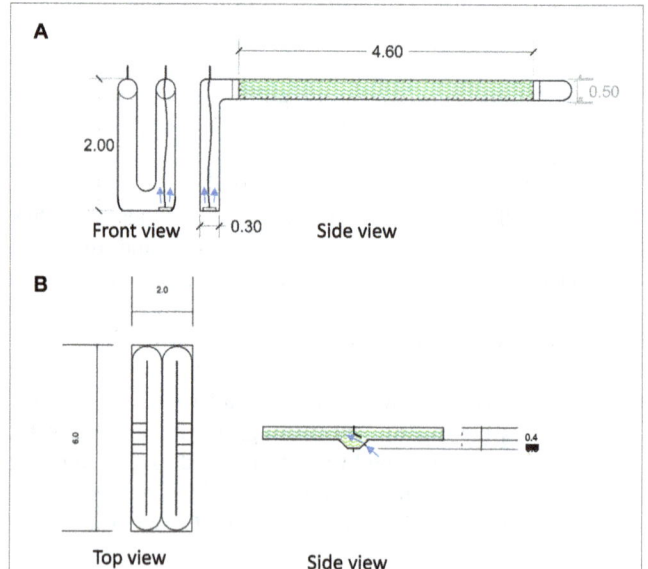

FIGURE 1 | Design and specifications of pilot-scale (A) tubular PBR (1:35 scale) and (B) two open raceway ponds (1:75 scale). A 2-m-long U-shaped PVC pipe with 0.3 m diameter was used as the airlift for the PBR that was connected to clear 4.6 m polyethylene flexible tubes that were also connected to a PVC bend at the end. Two 6-m-long open raceway ponds were constructed using fiberglass-coated plywood. Mixing was achieved by aeration with pressurized air using airlifts (arrows). This included an aeration disk with 6000 exit holes for the PBR and a single exit point for pressured air at a lowered section of the raceway ponds. Specifications are shown in meters.

FIGURE 2 | Photograph of pilot-scale two-stage microalgae cultivation system. Individual modules were used for single testing of photobioreactor or open ponds.

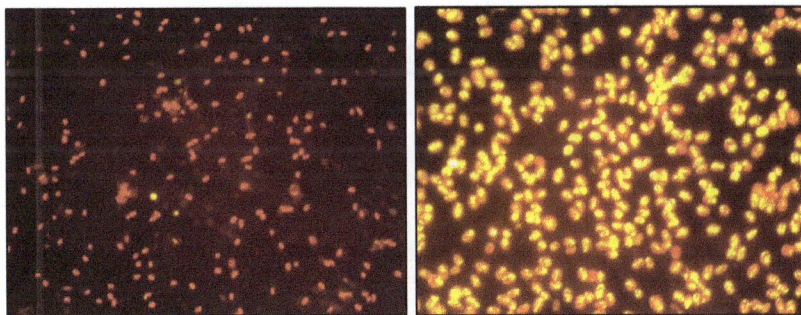

FIGURE 3 | Two Nile red-stained samples of *Tetraselmis* sp. M8 culture (40× magnification) before (left) and after nutrient deprivation stress at the time of harvesting (right).

which underwent several growth and harvesting cycles (**Figure 4**). Only biomass verified to be rich in lipids was harvested. The first cycle started with exponential growth after day 6 and nutrient stress set in on day 8. Nile red staining was performed on the following days to monitor the lipid induction. Low sunlight was recorded on day 10 followed by a decrease in lipids and cell numbers on day 11. Fresh f/2 (−Si) medium was added on day 12 and half of the culture volume (600 L) was harvested on day 13. Similarly, nutrient stress set in on day 20 during the second growth cycle and half of the culture was harvested again on day 23. The third growth cycle achieved maximum cell density on day 32 with accumulated lipid content, but declined in cell numbers on day 33. The entire culture was harvested.

Open Raceway Pond Cultivation

Cultivation in the raceway pond led to three harvesting events on days 13, 23, and 33 over the same time period (**Figure 4**). Similar to the first growth cycle in the PBR, nutrient stress was measured on day 8 and the culture reached a density of 2×10^6 cells/mL followed by a decline in density on day 12. f/2 (−Si) nutrients were added and half the volume (500 L) was harvested. On day 23, the cell density reached 2.3×10^6 cells/mL after nutrient stress from days 20 to 22. Maximum cell densities were observed on days 27 and 28. The culture was nutrient-stressed during the following 2 days and the cell density was 2.8×10^6 cells/mL on day 32 with substantial lipid accumulation. Hence, all the culture was harvested.

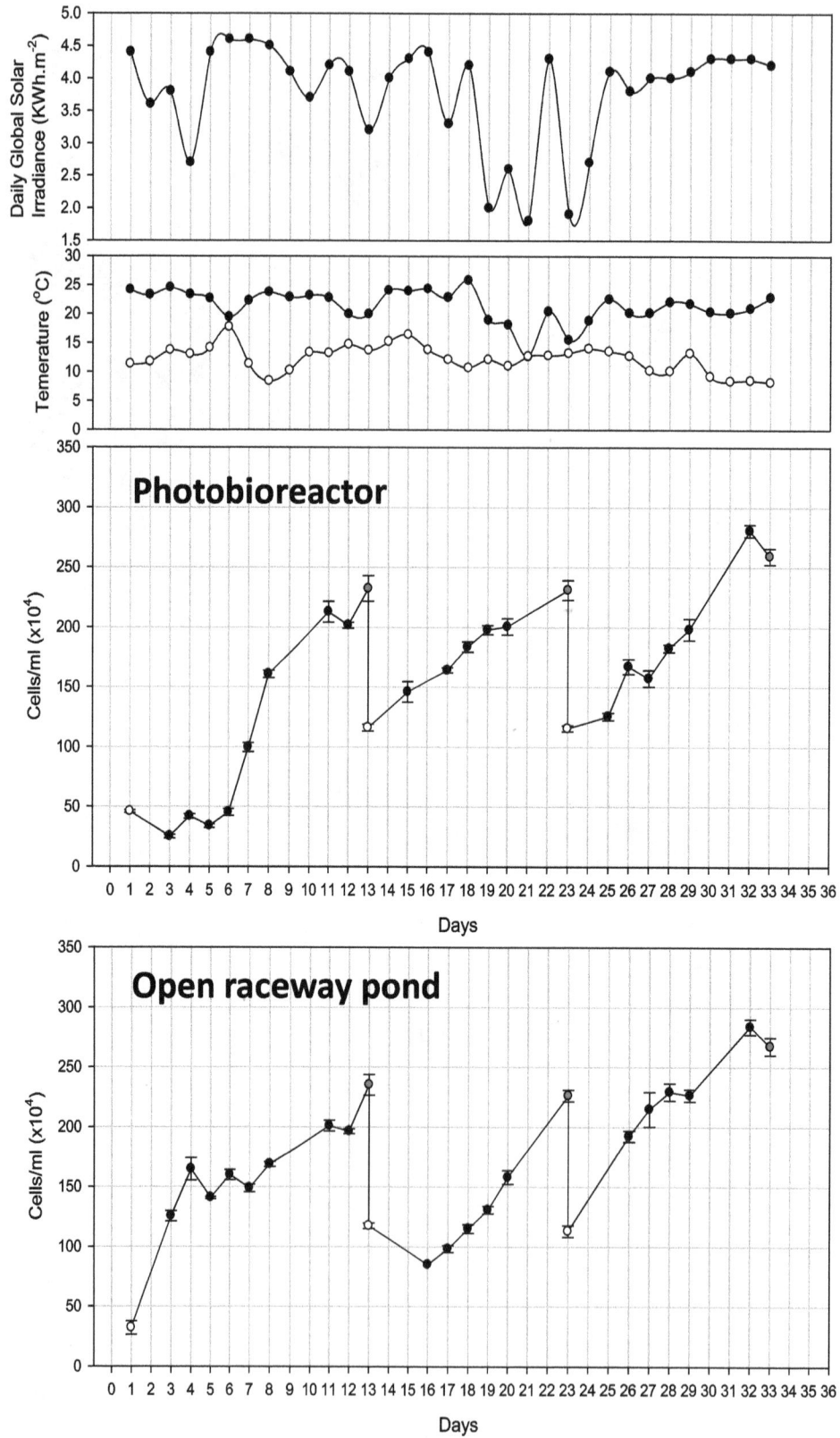

FIGURE 4 | Observations of photobioreactor and open raceway pond cultivation for 33 days. Also shown is daily global solar irradiance, as well as minimum and maximum air temperature recorded for the respective day. White and red dots represent the start and finish of each cycle.

Two-Stage Hybrid Cultivation System

A two-stage hybrid cultivation system was applied by transferring a portion of rapidly growing cells from the PBR to open raceway ponds where then nutrients diminished and algae were harvested upon lipid accumulation. **Figure 5** shows the cell density and various harvesting points in the PBR as part of the two-stage hybrid cultivation approach. During these harvesting events, at least half of the PBR culture volume was transferred to one of the raceway ponds, where lipid biosynthesis and accumulation was stimulated by nutrient depletion. The initial cell concentration of the PBR was 1.2×10^6 cells/mL. Initially it took 8 days for the medium to get exhausted and, during this cycle, the highest cell density was monitored (up to 3×10^6 cells/mL). On day 8, based on the high cell density, 900 L of the culture was transferred into the raceway pond for lipid induction. The PBR was refilled with medium for the second cycle. The duration of the second cycle was 7 days in the PBR followed by raceway pond cultivation for lipid induction. For the third cycle, as the cell density was 1.9×10^6 cells/mL and the nutrient concentration was below detection limit, half (600 L) of the culture was transferred into the raceway pond only after

4 days. During the last three cycles, microalgae were cultured in the closed PBR for 5, 6, and 6 days, respectively, followed by 3–4 days each of starvation in the raceway pond.

Comparison of Individual Open Pond or Closed PBR Cultivation with Two-Stage Hybrid Cultivation System

Both, closed PBR and open raceway pond cultivation resulted in three main growth cycles and harvesting events (**Figure 4**). Only biomass verified to be rich in lipids was harvested. Considering the growth curves for closed PBR and open raceway pond (**Figure 4**), a possible forth harvest appeared possible when cell concentrations reached 2×10^6/mL at days 28 and 29, respectively, but the biomass during this high irradiation phase was not rich in lipids and had, therefore, be harvested a few days later. During the same amount of time, the hybrid cultivation system led to six main growth cycles and harvesting events (**Figure 5**). Accordingly, the average growth rate of the hybrid system was significantly higher than that of both single

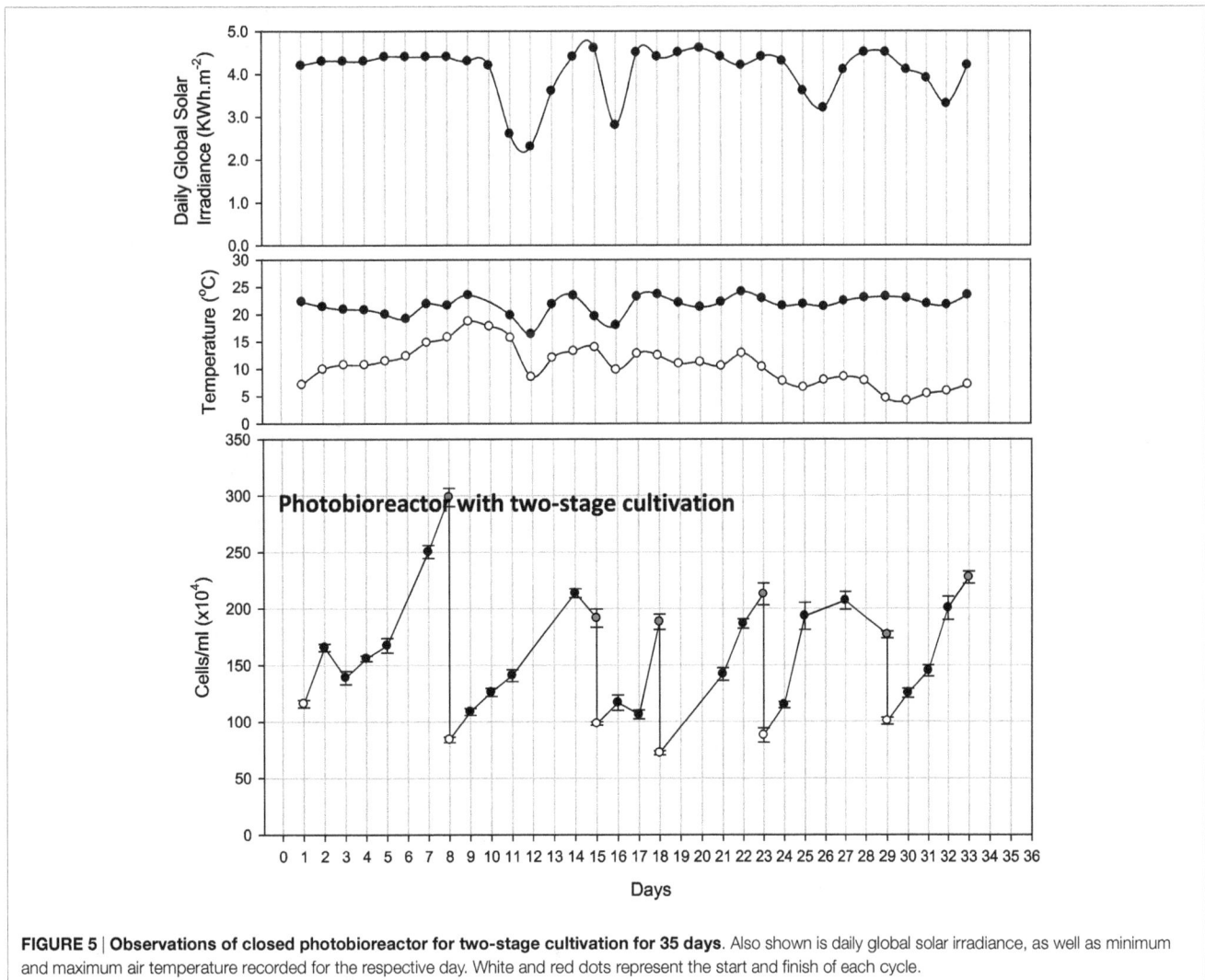

FIGURE 5 | Observations of closed photobioreactor for two-stage cultivation for 35 days. Also shown is daily global solar irradiance, as well as minimum and maximum air temperature recorded for the respective day. White and red dots represent the start and finish of each cycle.

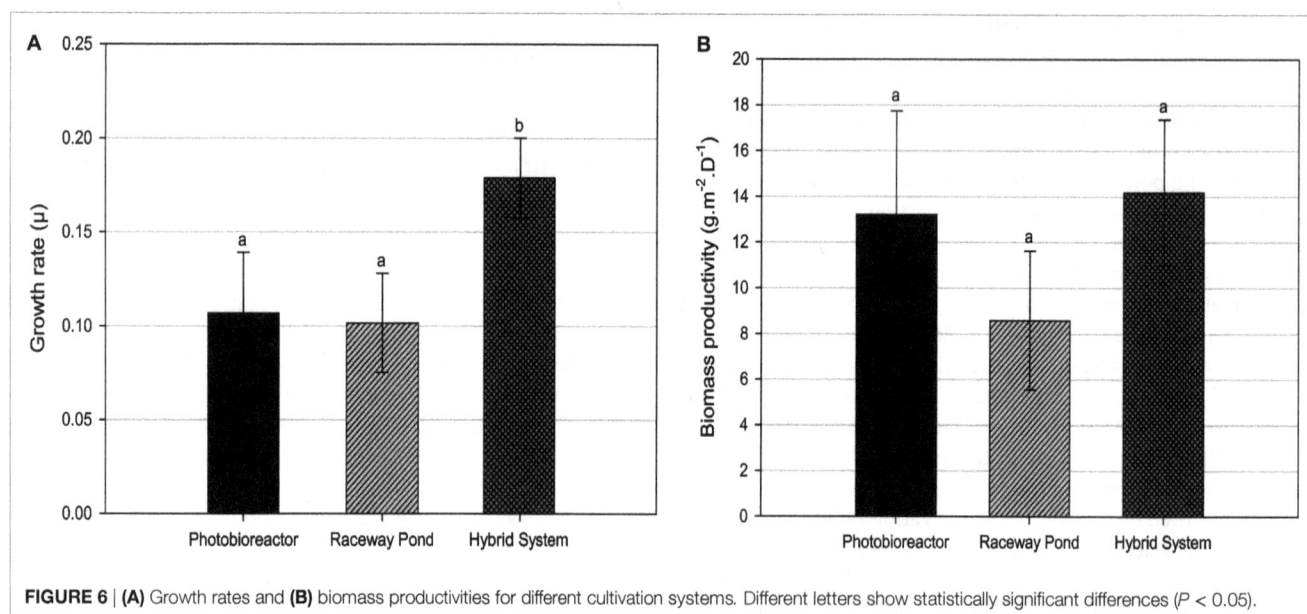

FIGURE 6 | (A) Growth rates and (B) biomass productivities for different cultivation systems. Different letters show statistically significant differences ($P < 0.05$).

systems (**Figure 6A**). The main reason for this appears to be that biomass growth and lipid induction phases are essentially independent from each other when using the hybrid cultivation system. This enables to keep the culture in rapid growth at a very high cell density. On the other hand, microalgal cultures in individual cultivation systems (PBR or open raceway pond) go through phases of exponential growth and nutrient starvation to enable lipid accumulation, followed by a brief lag phase before the next growth phase, leading to reduced overall growth rates (**Figure 4**). To determine productivity of all systems, biomass harvests (presented as gram/square meter/day) over the duration of the experiments are shown in **Figure 6B**.

Since the average growth rate of hybrid cultivation cannot be directly compared to that of either closed PBR or raceway pond due to different weather conditions, especially solar irradiance, further analysis was carried out to normalize the areal productivity of lipid-rich biomass to the total solar irradiance that occurred during the testing periods (**Figure 7**). The biomass productivity relative to solar irradiance was significantly higher for the hybrid cultivation system compared to open ponds.

DISCUSSION

In the present study, three growth and harvesting cycles were carried out for a closed PBR and an open raceway pond, while for a similar duration, six cultivation cycles were achieved using a two-stage hybrid cultivation system. The higher number of growth and harvesting cycles in the hybrid system was based on the use of separate cultivation systems for biomass growth and lipid induction phases. As the culture in growth phase never ran critically low in nutrients, two separate systems helped maintain higher growth rates and lowered the chance of culture dormancy or contamination. It was hypothesized that a separation of growth phase and lipid induction would be advantageous because microalgae typically either divide to increase the cell numbers (usually

during ideal nutrient replete conditions) or lipid biosynthesis will be initiated as a means to increase survival capability during adverse conditions, such as nutrient deprivation. The lipid accumulation capacity of *Tetraselmis* sp. M8 under these conditions has been previously described (Lim et al., 2012; Li et al., 2014; Sharma et al., 2014). Although lipids were not quantified in the present study, only lipid-rich biomass (checked by Nile red staining) after nutrient depletion phases was harvested to allow comparisons between cultivation systems. We found that nutrient deplete cultures during the lipid accumulation phase still underwent cell divisions, although at a lower rate.

Tetraselmis sp. M8 grown in the hybrid system had the highest average growth rate and, thus, the lowest doubling time (**Figure 6A**), while the increased biomass harvests compared to the open raceway pond was not significant (**Figure 6B**). To enable a direct comparison that takes into account the different weather conditions, the areal productivity of lipid-rich biomass was normalized to the total solar irradiance that occurred during the testing periods (**Figure 7**). This showed that the cultures in the hybrid cultivation produced significantly ($P < 0.05$) more lipid-rich biomass than the raceway cultivation system when normalized to solar exposure, while there was no significant difference in areal productivity between hybrid system and single-stage PBR. If assuming that the amount and quality of lipids did not differ between the cultivation systems and that all systems are at an equal level of the technological learning curve, this suggests, in principle, that hybrid cultivation systems should be preferred for the production of lipid-rich biomass. This should be weighed up against the increased capital expenditure and labor costs compared to simpler open pond systems and other possible advantages and disadvantages contrasted in **Table 1**.

A major advantage for hybrid systems can be expected for longer term cultivation where contamination by other algae or predators becomes a major concern. These problems occur

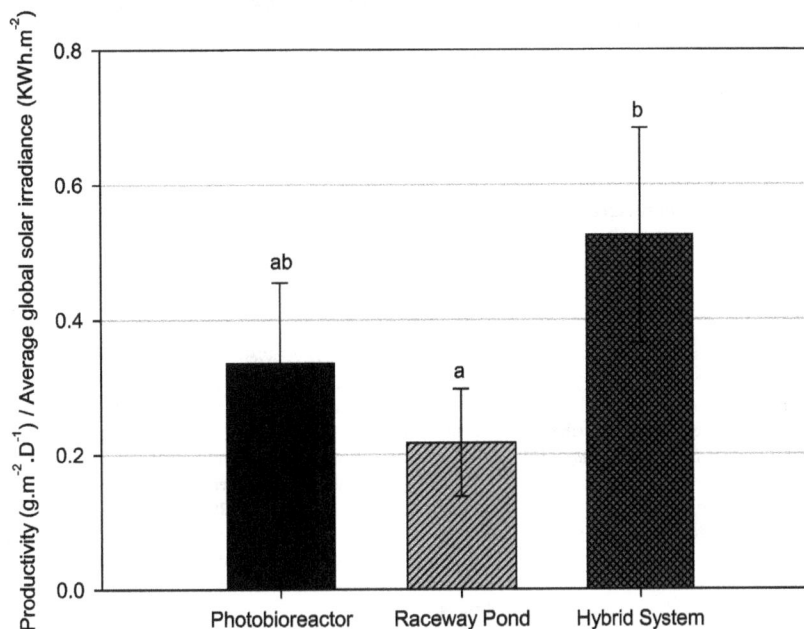

FIGURE 7 | Biomass productivity normalized to daily global solar exposure for different cultivation systems. Different letters show statistically significant differences ($P < 0.05$).

TABLE 1 | Comparison of various microalgae cultivating methods considering findings from the present and previous studies.

Factor	Photobioreactor	Raceway pond	Hybrid system
Space required	Moderate	High	High
Evaporation loss	Low	High	Moderate
CO_2 sparging efficiency	High	Low	Moderate
Maintenance	Difficult	Easy	Moderate
Contamination risk	Low	High	Low
Biomass quality	Reproducible	Variable	Reproducible
Energy input for mixing	High	Low	Moderate
Operation type	Batch	Batch	Continuous
Setup cost	High	Low	Moderate
Maintaining continuous exponential phase	Difficult	Difficult	Easy

(Borowitzka, 1999; Barbosa et al., 2003; Moheimani and Borowitzka, 2006; Chisti, 2007; Huntley and Redalje, 2007; Eriksen, 2008; Ugwu et al., 2008; Brennan and Owende, 2010; Harun et al., 2010; Mata et al., 2010).

especially in open pond systems or during times of reduced algae growth (Moheimani and Borowitzka, 2006; Wang et al., 2013). During hybrid cultivation in the present study, the continuously grown culture was contained in a closed PBR that rarely experiences phases of reduced growth or stagnation, while contamination-prone open ponds only ever held cultures for a few days before being cleaned, therefore avoiding this critical issue.

The present study was a pilot study aimed at identifying the most suitable system for algae cultivation for lipid production. Productivities were relatively low under these unoptimized conditions. Higher yields can be expected if cultivation conditions are further improved, for example, by automated nutrient supply and automated harvesting based on cell density. Carefully

dosed removal of cells during harvests is required to stay within the optimal window of exponential growth. Similarly, it may be advisable to harvest high cell density cultures when irradiance is expected to be low for the following days based on weather forecasts. This will avoid light limitations that may lead to reduced or stagnant growth or even partial culture death. Future studies should focus on long-term monitoring of these systems and use carefully optimized larger-scale demonstration facilities to enable reduction of operating costs, energy input, and environmental impact.

CONCLUSION

The comparison of microalgae cultivation systems, including open pond, closed PBR, and hybrid cultivation, suggests that the hybrid system is superior for the production of lipid-rich microalgae. The evaluated hybrid cultivation enables a separation of biomass growth and lipid induction phases, so that exponential biomass production and one or more efficient stress induction techniques can be carried out simultaneously, while effectively avoiding the issue of contamination. Techno-economic analyses of large-scale production in hybrid cultivation mode will reveal whether this system is also economically more viable.

AUTHOR CONTRIBUTIONS

RN performed bioreactor and open pond experiments, collected and interpreted data, and contributed to the writing of this manuscript. SG performed bioreactor and open pond experiments, collected and interpreted data, and contributed to the writing of this manuscript. KS performed *Tetraselmis* laboratory experiments, collected and interpreted data, and contributed to the writing of this manuscript. ST-H designed experiments, interpreted data, and contributed of the writing of this manuscript. MD designed and constructed photobioreactor and open ponds, performed experiments, interpreted data, and contributed to the writing of this manuscript. YL designed experiments, interpreted data, and contributed to the writing of this manuscript. PS designed experiments, interpreted data, and contributed to the writing of this manuscript.

REFERENCES

Abou-Shanab, R. A. I., Ji, M.-K., Kim, H.-C., Paeng, K.-J., and Jeon, B.-H. (2013). Microalgal species growing on piggery wastewater as a valuable candidate for nutrient removal and biodiesel production. *J. Environ. Manage.* 115, 257–264. doi:10.1016/j.jenvman.2012.11.022

Adesanya, V. O., Cadena, E., Scott, S. A., and Smith, A. G. (2014). Life cycle assessment on microalgal biodiesel production using a hybrid cultivation system. *Bioresour. Technol.* 163, 343–355. doi:10.1016/j.biortech.2014.04.051

Ahmad, A. L., Yasin, N. H. M., Derek, C. J. C., and Lim, J. K. (2011). Microalgae as a sustainable energy source for biodiesel production: a review. *Renew. Sustain. Energy Rev.* 15, 584–593. doi:10.1016/j.rser.2010.09.018

Barbosa, M. J., Albrecht, M., and Wijffels, R. H. (2003). Hydrodynamic stress and lethal events in sparged microalgae cultures. *Biotechnol. Bioeng.* 83, 112–120. doi:10.1002/bit.10657

Borowitzka, M. B. (1999). Commercial production of microalgae: ponds, tanks, tubes and fermenters. *J. Biotechnol.* 70, 313–321. doi:10.1016/S0168-1656(99)00083-8

Brennan, L., and Owende, P. (2010). Biofuels from microalgae – a review of technologies for production, processing, and extractions of biofuels and co-products. *Renew. Sustain. Energy Rev.* 14, 557–577. doi:10.1016/j.rser.2009.10.009

Breuer, G., Lamers, P. P., Martens, D. E., Draaisma, R. B., and Wijffels, R. H. (2012). The impact of nitrogen starvation on the dynamics of triacylglycerol accumulation in nine microalgae strains. *Bioresour. Technol.* 124, 217–226. doi:10.1016/j.biortech.2012.08.003

Chisti, Y. (2007). Biodiesel from microalgae. *Biotechnol. Adv.* 25, 294–306. doi:10.1016/j.biotechadv.2007.02.001

Chisti, Y. (2008). Biodiesel from microalgae beats bioethanol. *Trends Biotechnol.* 26, 126–131. doi:10.1016/j.tibtech.2007.12.002

Christenson, L., and Sims, R. (2011). Production and harvesting of microalgae for wastewater treatment, biofuels, and bioproducts. *Biotechnol. Adv.* 29, 686–702. doi:10.1016/j.biotechadv.2011.05.015

Eriksen, N. (2008). The technology of microalgal culturing. *Biotechnol. Lett.* 30, 1525–1536. doi:10.1007/s10529-008-9740-3

Guillard, R. R. L. (1975). "Culture of phytoplankton for feeding marine invertebrates," in *Culture of Marine Invertebrate Animals*, eds W. L. Smith and M. H. Chanley (New York: Plenum Press), 29–60.

Hannon, M., Gimpel, J., Tran, M., Rasala, B., and Mayfield, S. (2010). Biofuels from algae: challenges and potential. *Biofuels* 1, 763–784. doi:10.4155/bfs.10.44

Harun, R., Singh, M., Forde, G. M., and Danquah, M. K. (2010). Bioprocess engineering of microalgae to produce a variety of consumer products. *Renew. Sustain. Energy Rev.* 14, 1037–1047. doi:10.1016/j.rser.2009.11.004

Hu, Q., Sommerfeld, M., Jarvis, E., Ghirardi, M., Posewitz, M., Seibert, M., et al. (2008). Microalgal triacylglycerols as feedstocks for biofuel production: perspectives and advances. *Plant J.* 54, 621–639. doi:10.1111/j.1365-313X.2008.03492.x

Huntley, M., and Redalje, D. (2007). CO_2 mitigation and renewable oil from photosynthetic microbes: a new appraisal. *Mitigat. Adapt. Strat. Global Change* 12, 573–608. doi:10.1007/s11027-006-7304-1

Huntley, M. E., Johnson, Z. I., Brown, S. L., Sills, D. L., Gerber, L., Archibald, I., et al. (2015). Demonstrated large-scale production of marine microalgae for fuels and feed. *Algal Res.* 10, 249–265. doi:10.1016/j.algal.2015.04.016

ACKNOWLEDGMENTS

We wish to thank the Australian Research Council and North Queensland & Pacific Biodiesel for financial support.

FUNDING

This work was financially supported by a linkage grant of the Australian Research Council and North Queensland & Pacific Biodiesel (LP0990558).

Johnson, D. A., Weissman, J., and Goebel, R. (1988). *An Outdoor Test Facility for the Large-scale Production of Microalgae (No. SERI/TP-231-3325; CONF-880215-4).* Golden, CO, Fairfield, CA: Solar Energy Research Inst., Microbial Products, Inc.

Li, Y., Naghdi, F. G., Garg, S., Adarme-Vega, T. C., Thurecht, K. J., Ghafor, W. A., et al. (2014). A comparative study: the impact of different lipid extraction methods on current microalgal lipid research. *Microb. Cell Fact.* 13, 14. doi:10.1186/1475-2859-13-14

Lim, D. K., Garg, S., Timmins, M., Zhang, E. S., Thomas-Hall, S. R., Schuhmann, H., et al. (2012). Isolation and evaluation of oil-producing microalgae from subtropical coastal and brackish waters. *PLoS ONE* 7:e40751. doi:10.1371/journal.pone.0040751

Liu, B., and Benning, C. (2013). Lipid metabolism in microalgae distinguishes itself. *Curr. Opin. Biotechnol.* 24, 300–309. doi:10.1016/j.copbio.2012.08.008

Mata, T. M., Martins, A. A., and Caetano, N. S. (2010). Microalgae for biodiesel production and other applications: a review. *Renew. Sustain. Energy Rev.* 14, 217–232. doi:10.1016/j.rser.2009.07.020

Moheimani, N. R., and Borowitzka, M. A. (2006). The long-term culture of coccolithophore *Pleurochrysis carterae* (Haptophyta) in outdoor raceway ponds. *J. Appl. Phycol.* 18, 703–712. doi:10.1007/s10811-006-9075-1

Molina Grima, E., Fernández, F. G. A., García Camacho, F., and Chisti, Y. (1999). Photobioreactors: light regime, mass transfer, and scale-up. *J. Biotechnol.* 70, 231–247. doi:10.1016/S0168-1656(99)00078-4

Ndimba, B. K., Ndimba, R. J., Johnson, T. S., Waditee-Sirisattha, R., Baba, M., Sirisattha, S., et al. (2013). Biofuels as a sustainable energy source: an update of the applications of proteomics in bioenergy crops and algae. *J. Proteomics* 93, 234–244. doi:10.1016/j.jprot.2013.05.041

Olaizola, M. (2000). Commercial production of astaxanthin from *Haematococcus pluvialis* using 25,000-liter outdoor photobioreactors. *J. Appl. Phycol.* 12, 499–506. doi:10.1023/A:1008159127672

Otero, A., and Fábregas, J. (1997). Changes in the nutrient composition of *Tetraselmis suecica* cultured semicontinuously with different nutrient concentrations and renewal rates. *Aquaculture* 159, 111–123. doi:10.1016/S0044-8486(97)00214-7

Oyler, J. R. (2009). *Integrated Processes and Systems for Production of Biofuels Using Algae.* Patent No. US20110136217 A1.

Rupprecht, J. (2009). From systems biology to fuel: *Chlamydomonas reinhardtii* as a model for a systems biology approach to improve biohydrogen production. *J. Biotechnol.* 142, 10–20. doi:10.1016/j.jbiotec.2009.02.008

Schenk, P. M., Thomas-Hall, S., Stephens, E., Marx, U., Mussgnug, J., Posten, C., et al. (2008). Second generation biofuels: high-efficiency microalgae for biodiesel production. *BioEnergy Res.* 1, 20–43. doi:10.1007/s12155-008-9008-8

Sharma, K., Li, Y., and Schenk, P. M. (2014). UV-C mediated lipid induction and settling, a step change towards economical microalgae biodiesel production. *Green Chem.* 16, 3539–3548. doi:10.1039/C4GC00552J

Sharma, K. K., Schuhmann, H., and Schenk, P. M. (2012). High lipid induction in microalgae for biodiesel production. *Energies* 5, 1532–1553. doi:10.3390/en5051532

Su, C.-H., Chien, L.-J., Gomes, J., Lin, Y.-S., Yu, Y.-K., Liou, J.-S., et al. (2011). Factors affecting lipid accumulation by *Nannochloropsis oculata* in a two-stage cultivation process. *J. Appl. Phycol.* 23, 903–908. doi:10.1007/s10811-010-9609-4

Ugwu, C. U., Aoyagi, H., and Uchiyama, H. (2008). Photobioreactors for mass cultivation of algae. *Bioresour. Technol.* 99, 4021–4028. doi:10.1016/j.biortech.2007.01.046

Wang, H., Zhang, W., Chen, L., Wang, J., and Liu, T. (2013). The contamination and control of biological pollutants in mass cultivation of microalgae. *Bioresour. Technol.* 128, 745–750. doi:10.1016/j.biortech.2012.10.158

Conflict of Interest Statement: The authors declare that the research was conducted in the absence of any commercial or financial relationships that could be construed as a potential conflict of interest.

8

High Reversibility of "Soft" Electrode Materials in All-Solid-State Batteries

Atsushi Sakuda*, Tomonari Takeuchi*, Masahiro Shikano, Hikari Sakaebe and Hironori Kobayashi

Department of Energy and Environment, Research Institute for Electrochemical Energy, National Institute of Advanced Industrial Science and Technology (AIST), Ikeda, Japan

Edited by:
Jeff Sakamoto,
University of Michigan, USA

Reviewed by:
Francois Aguey-Zinsou,
The University of New South Wales,
Australia
Jeff B. Wolfenstine,
Army Research Laboratory, USA

*Correspondence:
Atsushi Sakuda
a.sakuda@aist.go.jp;
Tomonari Takeuchi
takeuchi.tomonari@aist.go.jp

Specialty section:
This article was submitted to
Energy Storage,
a section of the journal
Frontiers in Energy Research

Citation:
Sakuda A, Takeuchi T, Shikano M,
Sakaebe H and Kobayashi H (2016)
High Reversibility of "Soft" Electrode
Materials in All-Solid-State Batteries.
Front. Energy Res. 4:19.

All-solid-state batteries using inorganic solid electrolytes (SEs) are considered to be ideal batteries for electric vehicles and plug-in hybrid electric vehicles because they are potentially safer than conventional lithium-ion batteries (LIBs). In addition, all-solid-state batteries are expected to have long battery life owing to the inhibition of chemical side reactions because only lithium ions move through the typically used inorganic SEs. The development of high-energy density (more than 300 Wh kg^{-1}) secondary batteries has been eagerly anticipated for years. The application of high-capacity electrode active materials is essential for fabricating such batteries. Recently, we proposed metal polysulfides as new electrode materials. These materials show higher conductivity and density than sulfur, which is advantageous for fabricating batteries with relatively higher energy density. Lithium niobium sulfides, such as Li_3NbS_4, have relatively high density, conductivity, and rate capability among metal polysulfide materials, and batteries with these materials have capacities high enough to potentially exceed the gravimetric-energy density of conventional LIBs. Favorable solid–solid contact between the electrode and electrolyte particles is a key factor for fabricating high performance all-solid-state batteries. Conventional oxide-based positive electrode materials tend to give rise to cracks during fabrication and/or charge–discharge processes. Here, we report all-solid-state cells using lithium niobium sulfide as a positive electrode material, where favorable solid–solid contact was established by using lithium sulfide electrode materials because of their high processability. Cracks were barely observed in the electrode particles in the all-solid-state cells before or after charging and discharging with a high capacity of approximately 400 mAh g^{-1} suggesting that the lithium niobium sulfide electrode charged and discharged without experiencing substantial mechanical damage. As a result, the all-solid-state cells retained more than 90% of their initial capacity after 200 cycles of charging and discharging at 0.5 mA cm^{-2}.

Keywords: all-solid-state battery, lithium niobium sulfide, electrode morphology, sulfide solid electrolyte, long cycle life

INTRODUCTION

Secondary batteries are one of the most important components of electric vehicles (EVs) and plug-in hybrid electric vehicles (PHEVs). Among many types of secondary batteries, lithium-ion batteries (LIBs) are typically used in current EVs and PHEVs because of their high-energy density and long life (Armand and Tarascon, 2008; Dunn et al., 2011; Etacheri et al., 2011). To achieve a long driving range, the batteries must be densely packed in the battery pack because of the limited space for batteries in EVs and PHEVs. However, this dense packing usually increases the operating temperature of batteries. Thus, the safety and cyclability of these batteries at high temperature must be improved for use in EVs and PHEVs; otherwise, large-volume cooling units will be required.

Conventional LIBs employ liquid electrolytes with organic solvents (Armand and Tarascon, 2008; Dunn et al., 2011). These batteries are associated with some safety risks and exhibit accelerated degradation when they are operated at high temperatures. All-solid-state batteries that use inorganic solid electrolytes (SEs) are considered to be ideal batteries for EVs and PHEVs because they are potentially safer than conventional LIBs (Minami et al., 2006; Kamaya et al., 2011; Sakuda et al., 2013a). In addition, these batteries are expected to have long battery life because of the inhibition of chemical side reactions as only the lithium ions move in the typically used inorganic SEs.

The development of high-energy density (more than 300 Wh kg^{-1}) secondary batteries has been eagerly anticipated for years. The application of high-capacity electrode-active materials is essential for the fabrication of such batteries. Lithium/sulfur batteries have attracted attention because of their high theoretical energy densities based on the high capacities of sulfur in positive electrode materials (Yamin and Peled, 1983; Ji et al., 2009; Bruce et al., 2012; Schuster et al., 2012). However, the use of sulfur-carbon composites is required to activate insulating sulfur, and the composites require large amounts of carbon to achieve high performance, which decreases their energy density. Recently, we proposed metal polysulfides as new electrode materials (Hayashi et al., 2012a; Matsuyama et al., 2012; Sakuda et al., 2013b, 2014a,b,c,d). These materials show higher conductivity and density than sulfur. Thus, they are expected to exhibit improved volumetric-energy densities because of their relatively low conductive carbon ratios and relatively high material densities.

Understanding and exploiting the materials' mechanical properties are important in all-solid-state batteries (Hayashi et al., 2012b; Sakuda et al., 2013a). The all-solid-state batteries with sulfide SEs can be fabricated through high-pressure pressing at room temperature because of the sulfide SEs unique mechanical properties. The sulfide SE powder becomes highly densified as the grain boundaries decrease during pressing at room temperature. We term this densification "room-temperature pressure sintering" (Sakuda et al., 2013a). This phenomenon enables the creation of intimate contact between the electrode and SE materials, which is an essential requirement for all-solid-state operation, *via* room-temperature processing. The mechanical properties of electrode materials are also important. When brittle materials are used as the active material, the fragmentation of the electrode particles may occur during the high-pressure pressing used for cell construction. This fragmentation is especially serious at the interfaces between brittle materials as a result of stress concentration during high-pressure pressing (Sakuda et al., 2016). When electrode materials with some plastic character are used, the morphology of the electrode layer differs from that when brittle materials are used. For instance, the fragmentation during high-pressure pressing is expected to be suppressed. Furthermore, a dense electrode layer with a favorable electrode–electrolyte interface can likely be constructed. Electrode materials with large

capacities usually show large volume changes during charging and discharging. The maintenance of intimate solid–solid contact is an essential requirement for long cycle life. The plasticity of the electrode materials is believed to affect the maintenance of this contact. Lithium-containing metal sulfides with three-dimensional (3D) structures are attractive model materials with some degree of plastic character based on an intermediate bond character between ionic and covalent. Lithium niobium sulfides, such as rock-salt Li_3NbS_4, have relatively high density, conductivity, and rate capability among metal polysulfide materials (Sakuda et al., 2014c,d). Additionally, the batteries that use these materials have capacities that are high enough to potentially exceed the gravimetric-energy density of conventional LIBs. Furthermore, Li_3NbS_4 is considered to have relatively soft mechanical nature. Its application in all-solid-state batteries is expected to enhance these batteries' cyclability.

Here, we report high reversibility of soft electrode materials in all-solid-state batteries. The Li_3NbS_4 is used as a model material of soft electrode materials. This material shows unique mechanical properties that it can be densified by pressing at room temperature, and charged and discharged without the fragmentation. As a result, a dense electrode layer with a favorable conducting pathway and electrode–electrolyte interface is constructed. Furthermore, the dense electrode layer is maintained during charging and discharging, and the all-solid-state cell created here has a long cycle performance.

MATERIALS AND METHODS

Li_3NbS_4 was mechanochemically synthesized using a planetary ball mill apparatus (P-5, Fritsch GmbH). In an argon-filled glove box, a mixture of lithium sulfide (Li_2S, 99.9%, Mitsuwa Pure Chemicals), niobium disulfide (NbS_2, 99%, High Purity Chemicals), and sulfur (S_8, 99.9%, Wako Pure Chemical Industries) was placed into a zirconia pot (500 mL) along with zirconia balls (4 mm in diameter, 1,000 g). The total weight of the mixture of the starting materials was 10 g. The rotation speed and time of ball milling were fixed at 250 rpm and 120 h (60 min × 120 times), respectively. The $75Li_2S·25P_2S_5$ (mol%) glassy SE (Hayashi et al., 2001) was used, because it is the most typical sulfide-based SE. The SE was mechanochemically prepared from Li_2S and phosphorus pentasulfide (P_2S_5, 99%, Sigma-Aldrich) *via* planetary ball milling. Heptane was used as the ball-milling solvent. Zirconia pots (500 mL) and zirconia balls (4 mm in diameter, 1,000 g) were used. The starting materials were weighed in an argon-filled glove box and milled in air-sealing pots for 20 h. The lithium-ion conductivity of the powder-compressed pellet of the as-prepared SE and smaller-sized SE particles exhibited almost the same value (approximately 4×10^{-4} S cm^{-1}) at 25°C. All-solid-state cells were constructed as follows. The Li_3NbS_4 and the SE were mixed at a weight ratio of 75:25 for 3 min using a vortex mixer (IKA® Lab Dancer test tube shakers) *via* a dry process to prepare the positive composite electrode. The conductive additive was not included in this study because electronic conductivity of the Li_3NbS_4 is sufficiently high (>2 × 10^{-3} S cm^{-1}) relative to the lithium-ion conductivity in the composite electrode (Sakuda et al., 2014c,d). The resistance component attributable to electronic resistance in the composite electrode was hardly observed by AC impedance measurements. A lithium–indium alloy was used as the counter/reference electrode for the two-electrode cell. The lithium–indium alloy has been reported to show a voltage plateau at 0.62 V vs. Li$^+$/Li, over a wide range of compositions (Takada et al., 1996). Bilayer pellets (φ = 10 mm) consisting of the positive composite electrodes (10 mg) and the SE (80 mg) were obtained by pressing under 330 MPa at room temperature; indium foil (t = 0.3 mm, φ = 9 mm) and lithium foil (t = 0.2 mm, φ = 8 mm) were then attached to the bilayer pellets by pressing under 100 MPa. The pellets were pressed using two stainless steel rods, which were used as current collectors for both the positive and negative electrodes. All the processes involved in preparing the all-solid-state cells were performed in an argon-filled glove box [(H$_2$O) < 1 ppm].

Powder XRD measurements were performed at room temperature over the 2θ angle range 10° ≤ 2θ ≤ 80° with a step size of 0.1° using a D8 ADVANCE (Bruker AXS) diffractometer with CuKα radiation. The particle size distribution was measured using a particle size analyzer (SALD-7500nano, Shimadzu) with heptane/dibutyl ether as a solvent. Cross sections of the all-solid-state cells were prepared for scanning electron microscopy (SEM) observation using an argon-ion beam cross-section polisher (CP; IB-9020CP, JEOL) with an air-sealing holder. The samples were placed in the argon-filled glove box, transferred to the CP using the air-sealing holder, and milled for approximately 2 h. The acceleration voltage and argon gas pressure used were 6 kV and 3 × 10^{-3} Pa, respectively. During argon-ion milling, the stage was rocked ±30° to prevent beam striations and ensure uniform etching of the composite materials. Then, the obtained cross-section samples were transferred to the glove box and placed in the transfer vessel for SEM (JSM-6510A, JEOL) observation. The charge–discharge performances were evaluated at 50°C.

RESULTS

Figure 1 shows the XRD patterns of the Li$_3$NbS$_4$ prepared by mechanical milling (MM) with some reference materials. The peaks attributable to rock-salt type Li$_3$NbS$_4$ (Sakuda et al., 2014c,d) were confirmed after MM for 120 h, according to the preparation conditions used in this study. The average crystalline diameter of the Li$_3$NbS$_4$ estimated from the XRD peak widths was <10 nm, indicating that the sample obtained here included an amorphous phase.

Figure 2 shows the SEM images of the (**Figure 2A**) Li$_3$NbS$_4$ and (**Figure 2B**) 75Li$_2$S·25P$_2$S$_5$ glassy SE particles. The average particle sizes of the Li$_3$NbS$_4$ and 75Li$_2$S·25P$_2$S$_5$ glassy particles measured by the particle size analyzer were 5.7 and 15 μm, respectively. The Li$_3$NbS$_4$ particles are smaller than the SE particles in this study.

Figure 3 shows the cross-sectional SEM image of the positive electrode layer. The light- and dark-gray particles are Li$_3$NbS$_4$ and the SE, respectively. The dense pellet was obtained by pressing at room temperature. It should be noted that the grain boundaries are hardly visible at some interfaces between Li$_3$NbS$_4$ particles. No cracking or fragmentation can be seen in the Li$_3$NbS$_4$.

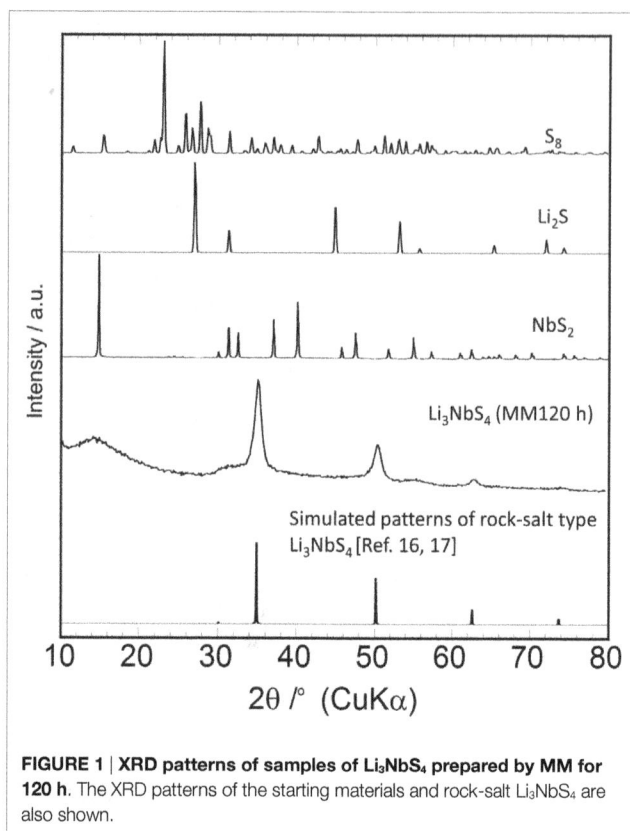

FIGURE 1 | XRD patterns of samples of Li$_3$NbS$_4$ prepared by MM for 120 h. The XRD patterns of the starting materials and rock-salt Li$_3$NbS$_4$ are also shown.

Figure 4 shows the dependence of the density of the Li$_3$NbS$_4$ pellet on the molding pressure. In this measurement, the Li$_3$NbS$_4$ was uniaxially compressed in a 10-mm-diameter mold. The relative densities were defined by the ratios between the pellets' densities and the powder true density, which was determined using a gas pycnometer (AccuPyc II 1340, Shimadzu), and the relative density was 2.92 g cm^{-3} for Li$_3$NbS$_4$. The relative density increases as the applied pressure increases and exceeds 90% when the Li$_3$NbS$_4$ powder is compressed by a pressure of over 500 MPa. Compressed Li$_3$NbS$_4$ with a remarkably high relative density can be obtained by pressing without heat treatment. Thus, Li$_3$NbS$_4$ is amenable to room-temperature pressure sintering.

Figure 5A shows the charge–discharge curves of the all-solid-state Li–In/Li$_3$NbS$_4$ cells. The current density used was 0.25 mA cm^{-2}. Cutoff voltages of 2.4 and 0.9 V (vs. Li–In) were used in this study and correspond to 3.0 and 1.5 V (vs. Li$^+$/Li). **Figure 5B** presents the charge–discharge curves, with the number of lithium atoms per formula unit shown on the x-axis. The initial charge and discharge capacities were 263 and 370 mAh g^{-1}, respectively. After the first cycle, a capacity of ca. 370 mAh g^{-1} was maintained for the charge and discharge process; however, a small increase in the capacity believed to be attributable to some sort of electrode activation was observed. This capacity corresponds to structures ranging from Li$_{0.6}$NbS$_4$ to Li$_{4.0}$NbS$_4$. The reversible capacity per unit area was 3.54 mAh cm^{-2}. Rock-salt Li$_3$NbS$_4$ has been reported to exhibit reversible charging and discharging for compositions ranging between Li$_{0.4}$NbS$_4$ and Li$_{3.9}$NbS$_4$ when charged and discharged with cutoff voltages of 3.0 and 1.5 V

FIGURE 2 | SEM images of (A) Li$_3$NbS$_4$ prepared by MM for 120 h and (B) 75Li$_2$S·25P$_2$S$_5$ glassy SE particles.

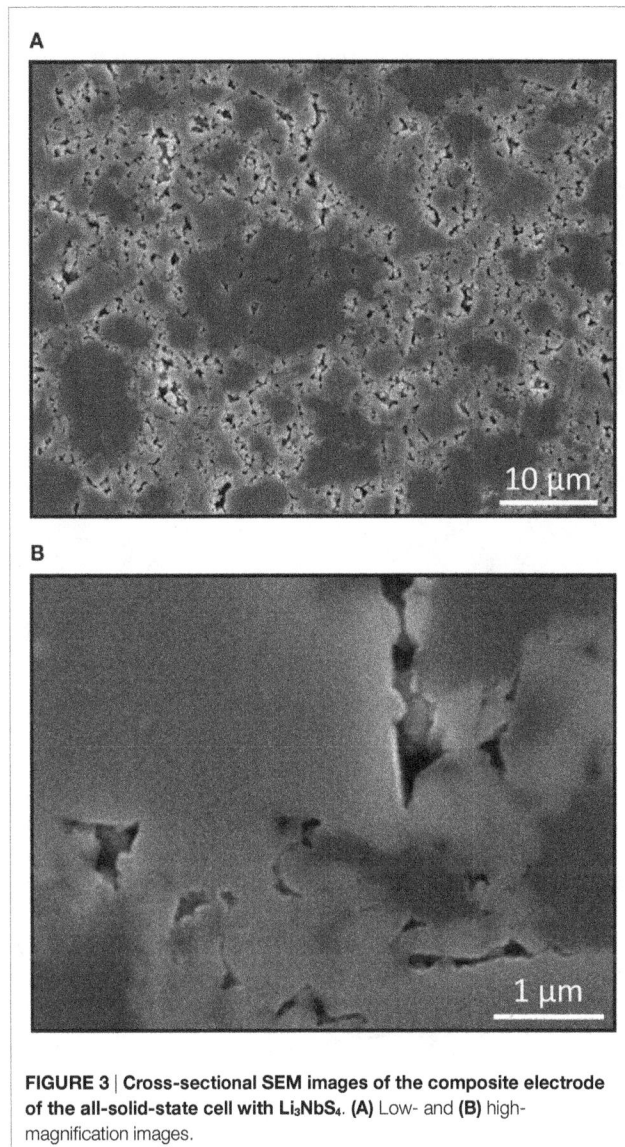

FIGURE 3 | Cross-sectional SEM images of the composite electrode of the all-solid-state cell with Li$_3$NbS$_4$. (A) Low- and (B) high-magnification images.

(vs. Li$^+$/Li), respectively (Sakuda et al., 2014c). Thus, Li$_3$NbS$_4$ can be used in all-solid-state cells as well as in the cells with carbonate-based liquid electrolytes. The potentials for charging and discharging were almost the same as those in cells with carbonate-based liquid electrolytes.

Figure 6 shows the cross-sectional SEM image of the positive composite electrode layer after the first charging (Figure 6A) and discharging (Figure 6B). The respective EDX mappings for Nb, P, and S are also shown in Figures 6C,D. The figures show that the electrode layers are free of cracks. The EDX mappings show that the light- and dark-gray particles are Li$_3$NbS$_4$ and the SE, respectively. The void volume increases after charging, and it decreases after discharging. It is noted that the Li$_3$NbS$_4$ particles are well connected in the composite electrode after discharging process. The fragmentation of Li$_3$NbS$_4$ is hardly observed.

Figure 7 shows the cycle performance of the all-solid-state cell. The current densities used for the cycle test were 0.25 mA cm^{-2} for the first through fifth cycles and 0.5 mA cm^{-2} after the sixth cycle. The capacity decreased only slightly when the current

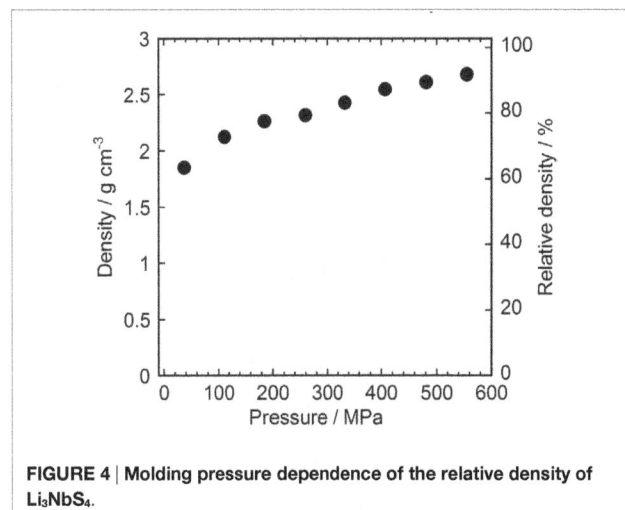

FIGURE 4 | Molding pressure dependence of the relative density of Li$_3$NbS$_4$.

density changed from 0.25 to 0.5 mA cm^{-2}. The all-solid-state cell exhibited high cyclability, retaining 92.3% of its capacity from the 7th to 200th cycle. The average cycle efficiency was 99.96% per cycle, and the average coulombic efficiency was 99.998%.

DISCUSSION

The experimental results show that dense electrode layer can be prepared solely by pressing the composite electrode with Li_3NbS_4 electrode and $75Li_2S \cdot 25P_2S_5$ SE particles. The grain boundaries

are hardly visible in the some interfaces between Li_3NbS_4 particles, indicating that this material sinters during pressing at room temperature. The Li_3NbS_4 particles are effectively deformed, make intimate contact, and create chemical bonds between the contacting particles. The cracking and fragmentation of electrode particles occur at the stress concentration point when typical lithium metal oxide electrodes, which are rather brittle materials, are used (Sakuda et al., 2016).

After charging and discharging, the electrode layer was still free of cracks (**Figures 6A,B**), despite the strong possibility of the electrode experiencing a large volume change during the charging and discharging cycle. The charge–discharge of Li_3NbS_4 involves large crystal structure changes, including amorphization. Therefore, it is difficult to determine the degree of volume change of the material by XRD measurements in charged and discharged electrodes. Thus, we estimated the volume change by measuring the molar volume of the mechanochemically prepared lithium niobium sulfides with the composition of Li_xNbS_4 ($x = 0, 1, 2,$ and 3) (Sakuda et al., 2014d) by measuring the powder true density using a gas pycnometer. **Table 1** summarizes the results of the powder true density measurements. These materials potentially include amorphous phase because of the preparation process. The powder true densities and molar volumes gradually changed as the lithium contents changed. The changes in the molar volume relative to Li_3NbS_4 range from 79% for $x = 0$ to 106% for $x = 4$. Thus, the molar volume of Li_3NbS_4 changes from ca. 82% ($x = 0.6$) to 106% ($x = 4$) during charging and discharging. In fact, the density of the electrode was largely changed during charging and discharging process (**Figures 6A,B**). Although an electrode that undergoes large volume changes usually exhibit fragmentation of the electrode particles, loss of interfacial contact between the electrode and electrolyte, and decreased packing density, these effects were not observed here. Instead, the electrode's density decreased by charging and increased by discharging relative to the as-prepared composite electrode layer (**Figure 3**). This result suggests that the room-temperature pressure sintering of the electrode particles occurred during discharging process, increasing the contact area of both electrode–electrolyte interface and electrode–electrode interface. The Nb-S bond in the Li_xNbS_4 structure has a relatively covalent bond character and is relatively strong. The Li–S bond in the Li_xNbS_4 structure has an intermediate bond character between ionic and covalent. The bond-dissociation energy of the Li–S bond in Li_3NbS_4 is relatively small compared to that of a typical ionic bond, such as Li–F, and a typical covalent bond, such as Si–O. This intermediate bond character and relatively small bond energy make these particles

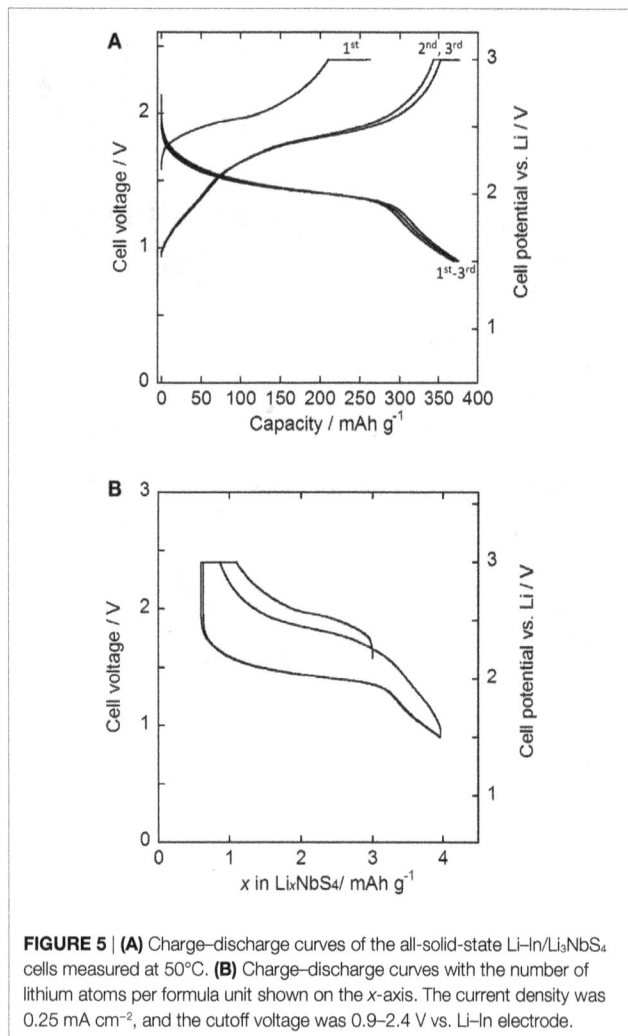

FIGURE 5 | **(A)** Charge–discharge curves of the all-solid-state Li–In/Li_3NbS_4 cells measured at 50°C. **(B)** Charge–discharge curves with the number of lithium atoms per formula unit shown on the x-axis. The current density was 0.25 mA cm^{-2}, and the cutoff voltage was 0.9–2.4 V vs. Li–In electrode.

TABLE 1 | Powder true density, molar mass, molar volume, mean atomic volume, and change in molar volume (vs. Li_3NbS_4) of lithium niobium sulfides (Li_xNbS_4) prepared by MM.

x in Li_xNbS_4	Composition	Observed crystalline phase	Powder true density (g cm^{-3})	Molar mass (g mol^{-1})	Molar volume (cm^3 mol^{-1})	Mean atomic volume (cm^3 mol^{-1})	Change in molar volume (%)
0	NbS_4	Amorphous	3.38	221.17	65.5	13.1	79.0
1	$LiNbS_4$	Amorphous	3.24	228.11	70.4	11.7	84.9
2	Li_2NbS_4	Rock salt	3.12	235.05	75.3	10.8	90.8
3	Li_3NbS_4	Rock salt	2.92	241.99	82.9	10.4	100.0
4	Li_4NbS_4	Antifluorite	2.82	248.93	88.1	9.8	106.3

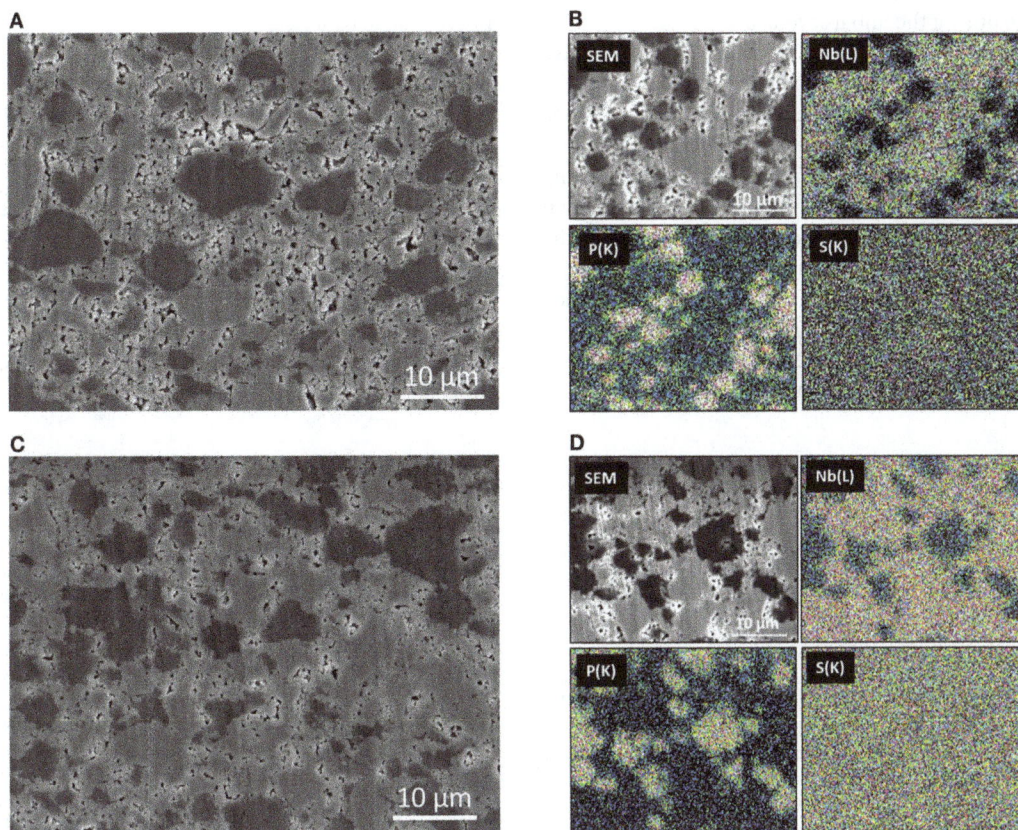

FIGURE 6 | Cross-sectional SEM images of the composite electrode of the all-solid-state cell after the first charging (A) and discharging (B). The EDX mappings for Nb–L, P–K, and S–K of the cross-sections of the composite electrode after first charging **(C)** and discharging **(D)** are also shown.

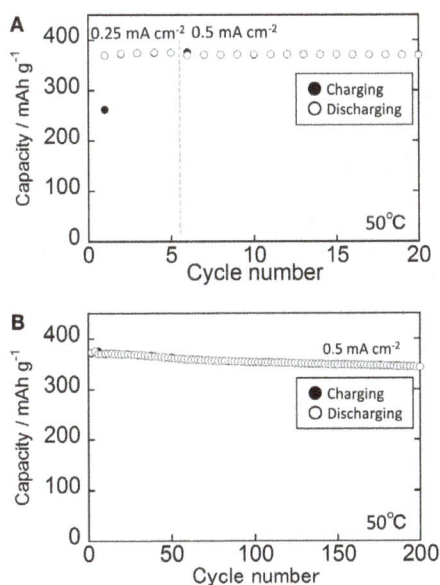

FIGURE 7 | Cycle performance of the all-solid-state Li–In/Li₃NbS₄ cell. **(A)** 1st to 20th cycle and **(B)** 1st to 200th cycles.

amenable to room-temperature pressure sintering by quasi-plastic deformation. Quasi-plastic deformation involves the breaking and reformation of chemical bonds, such as the Li–S bond. Furthermore, niobium's coordination structure is flexible in sulfide materials. Thus, the stress magnification factor of Li_xNbS_4 should be small, and the material should exhibit quasi-plastic deformation. The crystalline size of the Li_3NbS_4 in this study was very small; the crystalline diameter estimated based on the XRD pattern was <10 nm. The Li_3NbS_4 crystal domains are believed to be connected by an amorphous Li–Nb–S matrix. This structure is favorable for quasi-plastic deformation because amorphous phases usually have larger free volumes and smaller cohesive energies than crystal phases. Clearly, the application of some degree of pressure to the all-solid-state cells during operation is important for these cells' high performance. Under the applied pressure, the interfacial contact between the Li_3NbS_4 electrode and SE can be maintained despite relatively large volume changes.

The high cyclability of the all-solid-state batteries as shown in **Figure 7** is attributable to the unique mechanical properties of the Li_3NbS_4 particles, which are believed to deform effectively without fragmentation during charging and discharging. The composite electrode layer rather becomes dense during discharging process, although the electrode becomes less dense during charging. This densification is also likely related to the small capacity increase during the initial several cycles.

In this study, we presented a new, qualitative perspective concerning the usefulness of deformable electrode materials in all-solid-state batteries. The effective electrode–electrode intimate contacts can be maintained even if the electrode shows volume changes during charging and discharging by using the deformable soft electrode materials. The transition metal sulfides with sulfur-rich composition are model materials as the soft electrode materials with high capacity. In addition to the deformability of electrode materials, many other factors, such as elastic modulus and morphology of the electrode particles, and the pore volume of the composite electrode, also affect battery performance. Therefore, additional quantitative studies are required to better understand the requirements to achieve "ideal" all-solid-state batteries.

AUTHOR CONTRIBUTIONS

AS wrote the manuscript and TT, MS, HS, and HK revised the manuscript. AS, TT, and HK designed the work. AS conducted the experiments and AS, TT, MS, HS, and HK characterized the materials.

FUNDING

This work was partially supported by JSPS KAKENHI (Grant Number 15K17920).

REFERENCES

Armand, M., and Tarascon, J.-M. (2008). Building better batteries. *Nature* 451, 652–657. doi:10.1038/451652a

Bruce, P. G., Freunberger, S. A., Hardwick, L. J., and Tarascon, J.-M. (2012). Li-O$_2$ and Li-S batteries with high energy storage. *Nat. Mater.* 11, 19–29. doi:10.1038/nmat3191

Dunn, B., Kamath, H., and Tarascon, J.-M. (2011). Electrical energy storage for the grid: a battery of choices. *Science* 334, 928–935. doi:10.1126/science.1212741

Etacheri, V., Marom, R., Elazari, R., Salitra, G., and Aurbach, D. (2011). Challenges in the development of advanced Li-ion batteries: a review. *Energy Environ. Sci.* 4, 3243–3262. doi:10.1039/C1EE01598B

Hayashi, A., Hama, S., Morimoto, H., Tatsumisago, M., and Minami, T. (2001). Preparation of Li$_2$S-P$_2$S$_5$ amorphous solid electrolytes by mechanical milling. *J. Am. Ceram. Soc.* 84, 477–479. doi:10.1111/j.1151-2916.2001.tb00685.x

Hayashi, A., Matsuyama, T., Sakuda, A., and Tatsumisago, M. (2012a). Amorphous titanium sulfide electrode for all-solid-state rechargeable lithium batteries with high capacity. *Chem. Lett.* 41, 886–889. doi:10.1246/cl.2012.886

Hayashi, A., Noi, K., Sakuda, A., and Tatsumisago, M. (2012b). Superionic glass-ceramic electrolytes for room-temperature rechargeable sodium batteries. *Nat. Commun.* 3, 856–860. doi:10.1038/ncomms1843

Ji, X., Lee, K. T., and Nazar, L. F. (2009). A highly ordered nanostructured carbon-sulphur cathode for lithium-sulphur batteries. *Nat. Mater.* 8, 500–506. doi:10.1038/NMAT2460

Kamaya, N., Homma, K., Yamakawa, Y., Hirayama, M., Kanno, R. M., Yonemura, R., et al. (2011). A lithium superionic conductor. *Nat. Mater.* 10, 682–686. doi:10.1038/NMAT3066

Matsuyama, T., Sakuda, A., Hayashi, A., Togawa, Y., Mori, S., and Tatsumisago, M. (2012). Preparation of amorphous TiS$_x$ thin film electrodes by the PLD method and their application to all-solid-state lithium secondary batteries. *J. Mater. Sci.* 47, 6601–6606. doi:10.1007/s10853-012-6594-9

Minami, T., Hayashi, A., and Tatsumisago, M. (2006). Recent progress of glass and glass-ceramics as solid electrolytes for lithium secondary batteries. *Solid State Ionics.* 177, 2715–2720. doi:10.1016/j.ssi.2006.07.017

Sakuda, A., Hayashi, A., and Tatsumisago, M. (2013a). Sulfide solid electrolyte with favorable mechanical property for all-solid-state lithium battery. *Sci. Rep.* 3, 2261. doi:10.1038/srep02261

Sakuda, A., Taguchi, N., Takeuchi, T., Kobayashi, H., Sakaebe, H., Tatsumi, K., et al. (2013b). Amorphous TiS$_4$ positive electrode for lithium-sulfur secondary batteries. *Electrochem. commun.* 31, 71–75. doi:10.1016/j.elecom.2013.03.004

Sakuda, A., Taguchi, N., Takeuchi, T., Kobayashi, H., Sakaebe, H., Tatsumi, K., et al. (2014a). Composite positive electrode based on amorphous titanium polysulfide for application in all-solid-state lithium secondary batteries. *Solid State Ionics.* 31, 143–146. doi:10.1016/j.ssi.2013.09.044

Sakuda, A., Taguchi, N., Takeuchi, T., Kobayashi, H., Sakaebe, H., Tatsumi, K., et al. (2014b). Amorphous niobium sulfides as novel positive-electrode materials. *ECS Electrochem. Lett.* 3, A79–A81. doi:10.1149/2.0091407eel

Sakuda, A., Takeuchi, T., Okamura, K., Kobayashi, H., Sakaebe, H., Tatsumi, K., et al. (2014c). Rock-salt-type lithium metal sulphides as novel positive-electrode materials. *Sci. Rep.* 4, 4883. doi:10.1038/srep04883

Sakuda, A., Takeuchi, T., Kobayashi, H., Sakaebe, H., Tatsumi, K., and Ogumi, Z. (2014d). Preparation of novel electrode materials based on lithium niobium sulfides. *Electrochemistry* 82, 880–883. doi:10.5796/electrochemistry.82.880

Sakuda, A., Takeuchi, T., and Kobayashi, H. (2016). Electrode morphology in all-solid-state lithium secondary batteries consisting of LiNi$_{1/3}$Co$_{1/3}$Mn$_{1/3}$O$_2$ and Li$_2$S-P$_2$S$_5$ solid electrolytes. *Solid State Ionics.* 285, 112–117. doi:10.1016/j.ssi.2015.09.010

Schuster, J., He, G., Mandlmeier, B., Yim, T., Lee, K. T., Bein, T., et al. (2012). Spherical ordered mesoporous carbon nanoparticles with high porosity for lithium-sulfur batteries. *Angew. Chem. Int. Ed.* 51, 3591. doi:10.1002/anie.201107817

Takada, K., Aotani, N., Iwamoto, K., and Kondo, S. (1996). Solid state lithium battery with oxysulfide glass. *Solid State Ionics.* 86-88, 877–882. doi:10.1016/0167-2738(96)00199-3

Yamin, H., and Peled, E. (1983). Electrochemistry of a non-aqueous lithium sulfur cell. *J. Power Sources* 9, 281–287. doi:10.1016/0378-7753(83)87029-3

Conflict of Interest Statement: The authors declare that the research was conducted in the absence of any commercial or financial relationships that could be construed as a potential conflict of interest.

9

Grain Boundary Analysis of the Garnet-Like Oxides $Li_{7+X-Y}La_{3-X}A_XZr_{2-Y}Nb_YO_{12}$ (A = Sr or Ca)

Shingo Ohta*, Yuki Kihira and Takahiko Asaoka

Battery & Cell Division, Toyota Central R&D Labs. Inc., Nagakute, Japan

Garnet-like oxides having the formula $Li_{7+X-Y}La_{3-X}A_XZr_{2-Y}Nb_YO_{12}$ (A = Sr or Ca) were synthesized using a solid-state reaction, and their bulk and grain boundary resistivities were assessed by AC impedance measurements. A difference in grain boundary resistivity was identified between Sr and Ca materials, and so the grain boundaries were examined using electron probe microanalysis (EPMA). The difference in the grain boundary resistivities was attributed to the core–shell structure of the Sr-substituted samples. In contrast, the Ca-substituted materials exhibited accumulations of impurities at the grain boundaries.

Keywords: solid electrolytes, oxide, ion conductivity, grain boundary, all-solid-state, lithium ion battery

Edited by:
Shyue Ping Ong,
University of California,
San Diego, USA

Reviewed by:
Liqiang Mai,
Wuhan University of Technology,
China
Lincoln James Miara,
Samsung Electronics America,
Inc. (SEA)

*Correspondence:
Shingo Ohta
shingo_ohta@mail.toyota.co.jp,
sohta@mosk.tytlabs.co.jp

Specialty section:
This article was submitted
to Energy Storage,
a section of the journal
Frontiers in Energy Research

Citation:
Ohta S, Kihira Y and Asaoka T (2016)
Grain Boundary Analysis of the
Garnet-Like Oxides Li7+X-
YLa3-XAXZr2-YNbYO12 (A = Sr or Ca).
Front. Energy Res. 4:30.

INTRODUCTION

It is increasingly important to ensure the safety of lithium ion batteries as they are considered for use as power sources in electric vehicles or airplanes. These batteries typically use flammable organic liquids as the electrolyte, and so to improve the safety of lithium ion batteries, solid-state electrolytes have been developed to allow the fabrication of incombustible all-solid-state devices. Solid oxide electrolytes are believed to potentially have advantages over other inorganic electrolytes, such as sulfides, in terms of their chemical stability and lack of toxic degradation products (Ohtomo et al., 2013a,b). The garnet-like oxide $Li_7La_3Zr_2O_{12}$ (LLZ), which was reported by Prof. Murugan in 2007, is one of the most promising oxide electrolytes because of its excellent lithium ion conductivity ($\geq 10^{-4}$ Scm^{-1}) and wide potential window (Murugan et al., 2007). Following the discovery of LLZ, there were extensive efforts to improve its lithium ion conductivity, especially by elemental substitution leading to structural modification of the framework atoms (Rangasamy et al., 2013; Jalem et al., 2013; Thompson et al., 2014). This substitution modifies the lithium ion pathway by varying the bulk lithium ion conductivity, and many researchers have focused on improving the bulk lithium ion conductivity of garnet as a means of enhancing the conductivity of LLZ (Nyman et al., 2010; Ohta et al., 2011; Logeat et al., 2012; Allen et al., 2012). However, practical all-solid-state lithium ion batteries are likely to be fabricated by typical solid-state reactions, such that the primary contributor to the internal resistance of the solid oxide electrolyte will not be the bulk resistance, but rather the grain boundary resistivity (R_{gb}) of the polycrystalline electrolyte. As such, the total resistivity (R_{total}) of the electrolyte, including the R_{gb}, is important. Generally, the grain boundary resistance of a solid oxide electrolyte is high, meaning that the total ion conductivity has almost the same value as the grain boundary ion conductivity. Lithium lanthanum titanate, for example, is known to exhibit high ionic conductivity along with a high bulk conductivity of $\sigma_{bulk} = 1 \times 10^{-3}$ Scm^{-1}. However, the two-dimensional lithium ion paths in this material are interrupted by grain boundaries, resulting in a high R_{gb} with $\sigma_{grain\ boundary} \approx \sigma_{total} = 2 \times 10^{-5}$ Scm^{-1} (Inaguma et al., 1993; Stramare et al., 2003;

Yashima et al., 2005; Gao et al., 2014; Moriwake et al., 2015). In contrast, the grain boundary resistance of garnet-like oxides is relatively low compared to other ion-conducting oxides because of the good lithium ion connection between grains, forming three dimensional lithium ion paths. These materials thus show total lithium ion conductivity values as high as the bulk lithium ion conductivity (that is, $\sigma_{total} \approx \sigma_{bulk}$) (Ohta et al., 2011; Allen et al., 2012). However, the R_{gb} values of garnet-like oxides vary widely, from several percent to 40% of the R_{total}, depending on the composition (Li et al., 2012; Kihira et al., 2013). These differences in R_{gb} are believed to arise from the effects of substances at the grain boundaries. For this reason, the grain boundary resistivities induced by substitution with other elements should be assessed as a means of minimizing the R_{gb} and to determine the optimum substitution elements for garnet-like oxides.

In the present study, we simultaneously made substitutions at both the Zr and La sites to synthesize $Li_{7+X-Y}(La_{3-X}A_X)(Zr_{2-Y}Nb_Y)O_{12}$ (A-LLZNb; A = Ca, Sr). This was done to determine the optimum composition with regard to obtaining the highest lithium ion conductivity, based on differences in valency and ionic size: Nb^{5+} (0.64 Å), Zr^{4+} (0.72 Å), Ca^{2+} (1.34 Å), La^{3+} (1.36 Å), and Sr^{2+} (1.44 Å) (Shannon, 1976). Through these investigations, we found that the R_{gb} showed different characteristics between Ca and Sr substitutions. This study clarified that the ionic size and valency of the substitution element affect the grain and grain boundary morphology, and consequently the total lithium ion conductivity. This effect is discussed herein based on the grain boundaries of A-LLZNb incorporating Sr and Ca, as assessed *via* electron probe microanalysis (EPMA).

MATERIALS AND METHODS

$Li_{7+X-Y}(La_{3-X}A_X)(Zr_{2-Y}Nb_Y)O_{12}$ (A-LLZNb; A = Ca or Sr) bulk ceramic samples were synthesized by a conventional solid-state reaction in alumina crucibles. The starting materials were Li_2CO_3 (a 10% excess was added to allow for lithium losses by evaporation at high temperatures), $La(OH)_3$, ZrO_2, Nb_2O_5, and alkali earth metals in the form of carbonates.

The electrical conductivity of each sample was measured in air using a two-probe AC impedance method with an Agilent 4294A instrument over the frequency range of 110 MHz to 40 Hz at 25°C. Au electrodes were applied to both sides of the sample pellets using Au paste prior to these measurements, and the pellet and paste were subsequently annealed at 800°C for 0.5 h. The morphologies and the elemental distributions of the Ca or Sr-LLZNb bulk ceramics were measured by EPMA (JEOL Ltd. JXA-8500F). The crystal structures and impurity phases of the samples were determined by X-ray diffraction (XRD, Rigaku Co., Ltd., Ultima IV) using Cu Kα radiation.

RESULTS AND DISCUSSION

Figure 1 shows the AC impedance plots (Nyquist plots) of the synthesized A-LLZNb (A = Sr or Ca) bulk ceramic samples, where a = $Li_7La_2Sr_1Zr_1Nb_1O_{12}$ (1.0Sr-LLZNb), b = $Li_7La_{2.25}Ca_{0.75}Zr_{1.25}Nb_{0.75}O_{12}$ (0.75Ca-LLZNb), and c = $Li_7La_2Ca_1Zr_1Nb_1O_{12}$ (1.0Ca-LLZNb). When analyzing the impedance plot of a garnet

oxide, there will be two semi-circles indicating the bulk resistivity (R_{bulk}) at higher frequencies (each terminal frequencies are 0.3M ~ 0.1 MHz) and the R_{gb} at lower frequencies (each terminal frequencies are 5000 ~ 50 Hz), the sum of which results in the R_{total}. The R_{bulk} and R_{gb} of the A-LLZNb (A = Sr or Ca) specimens are listed in **Table 1**. The R_{bulk} values and terminal frequencies of all samples were almost similar (~10^3 Ωcm and ~0.1 MHz), indicating that the R_{bulk} was not significantly affected by the substitution element. In contrast, the R_{gb} values of samples a, b, and c demonstrate that the substitution element makes a large difference with regard to the R_{gb} range. Each terminal frequencies are different by two orders of magnitude (5000 ~ 50 Hz). As shown in **Figure 1D**, the R_{gb} to R_{total} ratio in the Sr-LLZNb series increased gradually upon elevating the substitution level. In contrast, the ratios of the Ca-LLZNb series remained low until the substitution level reached 0.75, after which the ratio value increased significantly, eventually becoming quite large at higher substitution levels. These results suggest three discussion points: (i) the cause of the sudden jump in the R_{gb} values of the Ca-LLZNb series, (ii) the reason for the gradual increase in the R_{gb} values of the Sr-LLZNb series upon adding more Sr and Nb, and (iii) the reason why the R_{gb} values of the Sr-LLZNb series are higher than those of the Ca-LLZNb. In order to clarify the above points, we investigated the morphologies at grain boundaries and the relationship between these morphologies and composition, using EPMA.

Figure 2A shows the Ca, Al, La, Zr, and Nb EPMA maps obtained from a 0.75Ca-LLZNb sample. Al was not a constituent element but was identified as a contaminant in the sample, likely originating from the Al_2O_3 crucible during the sintering process, and so the Al distribution was assessed. The linear analysis results obtained along the red line in the back scattered electron (BSE) map in **Figure 2A** are provided in **Figure 2B**. The distributions of the elements and the linear analysis results indicate that Al and Ca were located at the grain boundary triple points in the sample. In contrast, no segregation and no elemental concentration gradients were observed in garnet grains, suggesting that the composition of the garnet grains was uniform. In order to clarify the composition of the impurity in the Ca-LLZNb series, XRD patterns were acquired, as shown in **Figure 2C**. The Ca-LLZNb for which $X < 0.5$ generated only diffraction peaks corresponding to the garnet phase, indicating that a small amount of Ca can be substituted at the La sites. Conversely, impurity peaks assigned to CaO, $LiAlO_2$, and $CaAl_2O_4$ were produced by heavily Ca-substituted samples ($X \geq 0.5$). Thus, the solid solubility limit of Ca in LLZNb appears to coincide with an X value of ~0.5, after

TABLE 1 | R_{bulk} and R_{gb} values of the A-LLZNb (A = Sr or Ca, X = 0.25, 0.5, 0.75, or 0.1) series.

	$Li_7(La_{3-x}Ca_X)(Zr_{2-x}Nb_X)_{12}$		$Li_7(La_{3-x}Sr_X)(Zr_{2-x}Nb_X)_{12}$	
	R_{bulk} (Ω)	R_{gb} (Ω)	R_{bulk} (Ω)	R_{gb} (Ω)
$X = 0.25$	2.1×10^3	2.7×10^2	8.5×10^3	1.4×10^3
$X = 0.5$	3.4×10^3	1.8×10^2	5.2×10^3	1.9×10^3
$X = 0.75$	5.8×10^3	2.5×10^2	5.3×10^3	4.6×10^3
$X = 0.1$	2.7×10^4	2.7×10^5	7.5×10^3	1.0×10^4

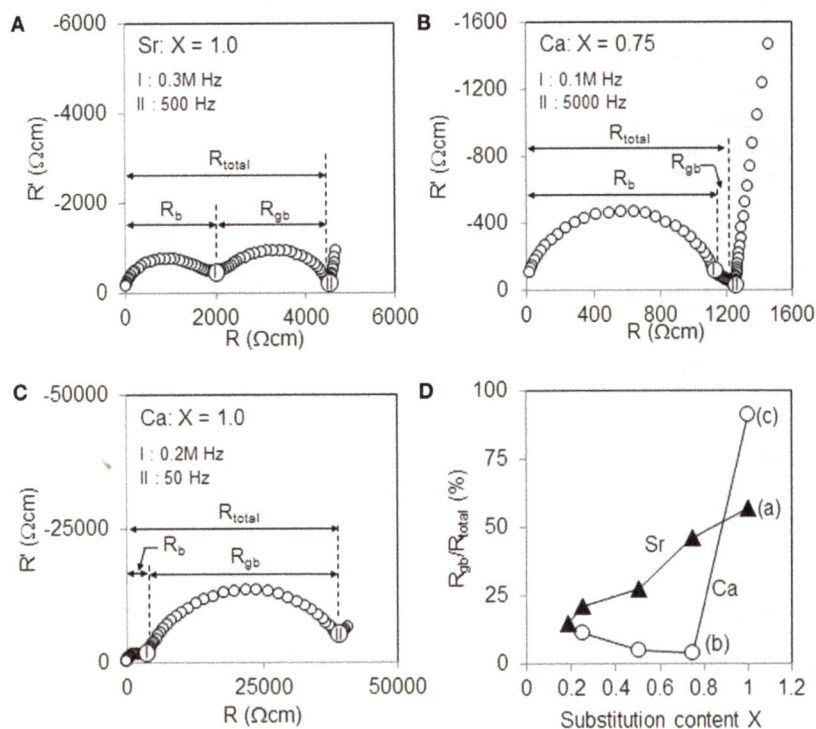

FIGURE 1 | Nyquist plots for Li_{7+X-Y} $(La_{3-X}A_X)(Zr_{2-Y}Nb_Y)O_{12}$ (A-LLZNb, A = Sr or Ca) in which (A) A = Sr, X = 1.0, (B) A = Ca, X = 0.75, (C) A = Ca, X = 1.0, and (D) the ratios of resistance due to grain boundaries to total resistance (R_{gb}/R_{total}) as functions of X. The terminal frequencies of both resistances are shown in the graph.

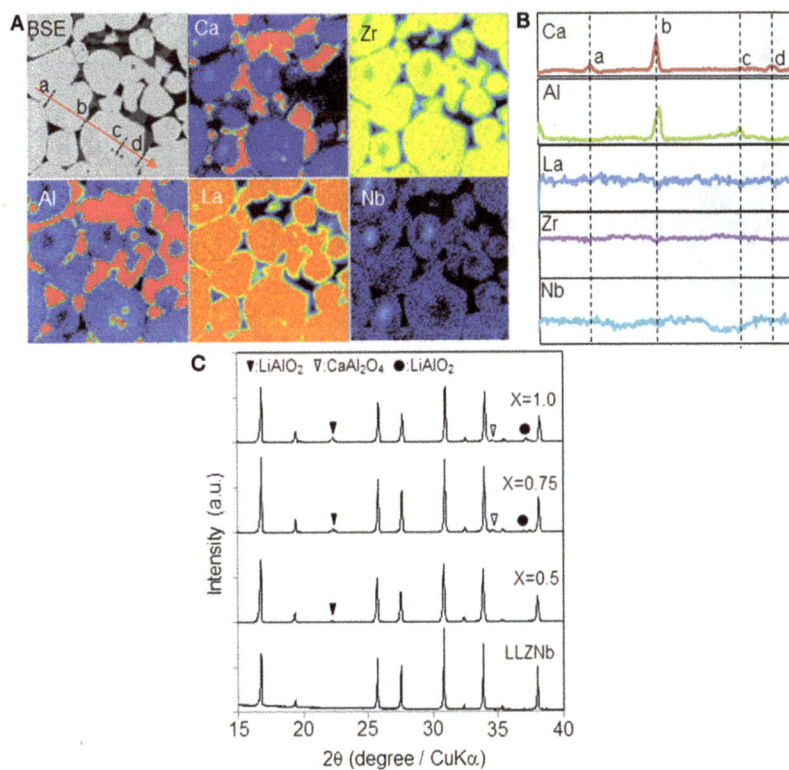

FIGURE 2 | (A) Cross-sectional EPMA mapping images of Ca-LLZNb, (B) linear analysis of Ca, Al, La, Zr, and Nb, and (C) XRD pattern obtained from Ca-LLZNb.

which the excess Ca accumulates at the grain boundaries. Based on this finding, an Al introduction mechanism can be proposed. In this mechanism, disproportionation of the garnet phase is induced upon the addition of Ca in excess of its solubility limit, after which excess Ca and Li accumulate at the grain boundaries. The excess Li reacts with the Al_2O_3 crucible such that Al is introduced into the sample, and $LiAlO_2$ and $CaAl_2O_4$ are subsequently formed during the sintering process.

Figure 3A presents the Sr, Al, La, Zr, and Nb EPMA maps generated by the 1.0Sr-LLZNb sample, while the linear analysis results obtained along the red line in the BSE map in **Figure 3A** are shown in **Figure 3B**. The garnet grains were determined to have a core–shell structure. The line scans show that, within the shell region and close to the grain boundaries at points a, b, c, and d in the BSE map in **Figure 3A**, there were higher concentrations of Sr and Nb and lower concentrations of La and Zr compared to the core region. Although the segregation of both Sr and Zr was observed in this field of view, $SrZrO_3$ diffraction peaks were not detected by XRD (**Figure 3C**). This result would suggest that the concentration of $SrZrO_3$ as an impurity was very low, and so we believe that the solubility limit of Sr in LLZNb is high compared to that of Ca. Al appears to have diffused into the grain boundaries of the sample from the Al_2O_3 crucible during sintering just as in the case of the Ca-LLZNb. A mechanism for this Al introduction can be proposed based on elemental distribution analysis, since this analysis found that there was an elemental concentration gradient inside the Sr-LLZNb grains. It thus appears that $LiAlO_2$

formation was induced by Li release due to structural disorder, just as is thought to have occurred with the Ca-LLZNb.

The elemental distribution and linear analysis results clearly show that the state of the grains and grain boundaries result from the effects of Sr or Ca substitution in the LLZNb. **Figure 4** presents schematic images of the grains and grain boundaries of Ca- or Sr-substituted LLZNb that appear at different R_{gb} ratios. Based on these schematics, we propose a reason for the gradual increase in the R_{gb} value of Sr-LLZNb with increasing Sr content. As greater amounts of Sr are added, the Sr will be forced out toward the boundaries to form a Sr-rich layer. The ionic radius of Sr^{2+} (1.44 Å) is greater than that of La^{3+} (1.36 Å), and their valence numbers are also different, and so the appearance of the Sr-rich layer must be balanced through an adjustment of both the structure and valency to form a garnet structure. Interestingly, Nb^{5+} (0.64 Å), which has a smaller ionic radius than Zr^{4+} (0.72 Å), and whose valence number is larger rather than smaller (as is the case with Sr^{2+} and La^{3+}), allows a balance to be maintained such that the Sr-rich garnet structure is unchanged (**Table 2**). Hence, although the solid solubility limit of Sr in LLZNb is higher than that of Ca, the gradual increase in the R_{gb} ratio in the Sr-LLZNb series is attributed to a Sr,Nb-rich layer that becomes increasingly thick as the amount of Sr added is increased (**Figures 4A,B**).

It is also possible to explain the lower solid solubility limit of Ca in LLZNb compared to that of Sr. The ionic radii of both Ca^{2+} (1.34 Å) and Nb^{5+} (0.64 Å) are smaller than those of La^{3+} (1.36 Å) and Zr^{4+} (0.72 Å), so the garnet structure cannot be

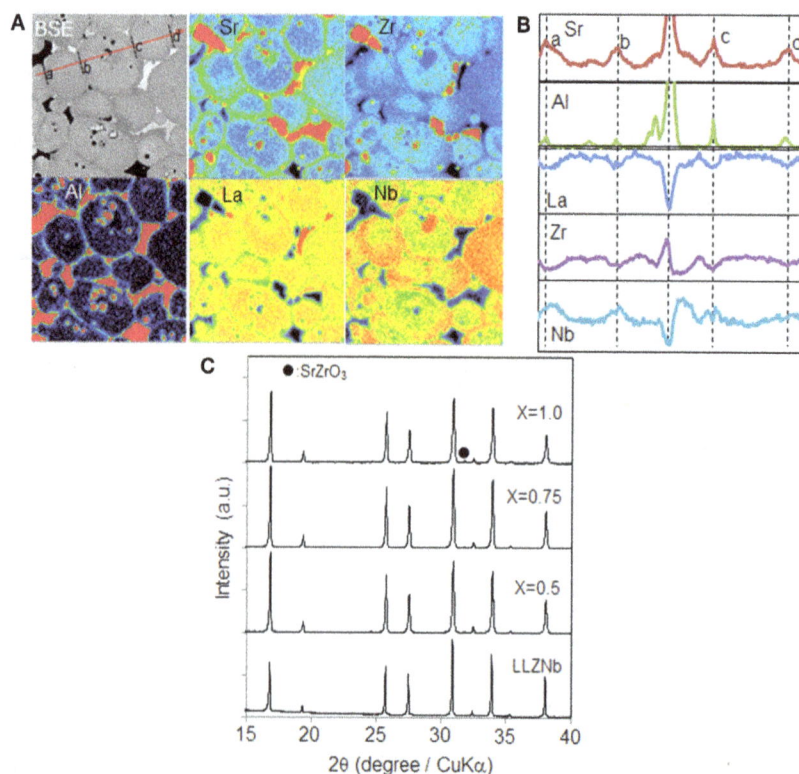

FIGURE 3 | (A) Cross-sectional EPMA mapping images of Sr-LLZNb, **(B)** linear analysis of Sr, Al, La, Zr, and Nb, and **(C)** XRD pattern obtained from Sr-LLZNb.

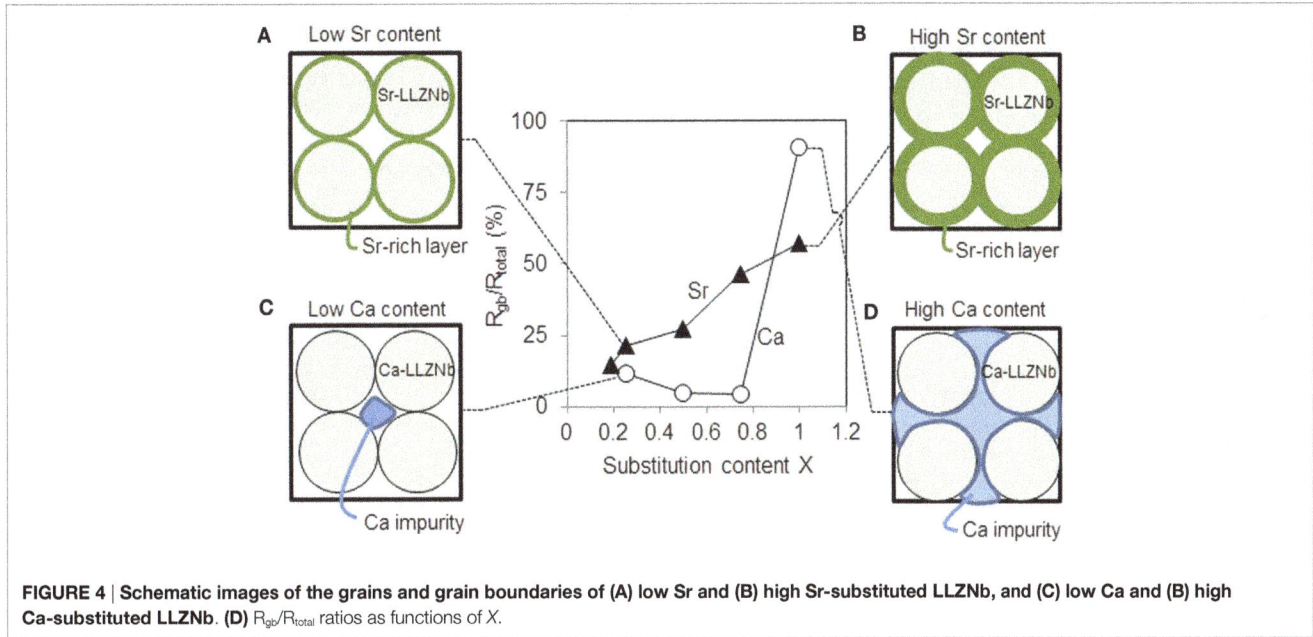

FIGURE 4 | Schematic images of the grains and grain boundaries of (A) low Sr and (B) high Sr-substituted LLZNb, and (C) low Ca and (B) high Ca-substituted LLZNb. (D) R_{gb}/R_{total} ratios as functions of X.

TABLE 2 | Valence numbers and ionic radii of elements substituted into the A-LLZNb (A = Sr or Ca) series.

| | Sr-LLZN | | | | | Ca-LLZN | | | | |
| | La site | | ZR site | | | La site | | ZR site | | |
	La		Sr	Zr		Nb	La		Ca	Zr		Nb
Charge	3+	→down	2+	4+	→up	5+	3+	→down	2+	4+	→up	5+
Radii	1.36	→up	1.44	0.72	→down	0.64	1.36	→down	1.34	0.72	→down	0.64

adjusted, although the relationship of their valance number is same as is possible in the case of Sr-LLZNb (**Table 2**). When the Ca substitution level is close to the solid solubility limit, although small amounts of Ca and Al impurities are accumulated at the grain boundary triple points, a Li conduction pathway can form due to sufficient contact between grains, such that the R_{gb} of the LLZNb remains low (**Figure 4C**). When the Ca substitution level is above the solid solubility limit, significant concentrations of Ca and Al impurities are accumulated at the grain boundaries, leading to insufficient contact between grains (**Figure 4D**). This is the reason why the R_{gb} of the LLZNb suddenly increases when high concentrations of Ca are substituted.

CONCLUSION

We successfully fabricated Li_{7+X-Y} $(La_{3-X}A_X)(Zr_{2-Y}Nb_Y)O_{12}$ (A-LLZNb; A = Ca, Sr) by solid-state reaction. The observed variations in the R_{gb} ratio between Sr-LLZNb and Ca-LLZNb are considered to result from the different substitution elements. Three issues were addressed in this work: (i) the sudden jump in the R_{gb} values of the Ca-LLZNb series, (ii) the higher R_{gb} values of

the Sr-LLZNb series, and (iii) the gradual increase in the R_{gb} values of the Sr-LLZNb upon adding increasing amounts of Sr and Nb.

With regard to issue (i), EPMA analysis found Ca and Al at the grain boundaries of the Ca-LLZNb at $X = 1.0$, explaining the very high R_{gb} ratio. Point (ii) was elucidated by results showing a Sr,Nb-rich layer on the surfaces of garnet grains. The effect in point (iii) is thought to result from this Sr,Nb-rich layer, which would be balanced by variations in ionic size and valence. In the case of Sr-Nb or Ca-Nb combinations, the larger Sr^{2+} ions at the La^{3+} sites would be balanced by smaller Nb^{5+} ions at Zr^{4+} sites to form a Sr,Nb-rich layer, while Ca^{2+} will not form a Ca-rich layer because of its mismatch in ionic size and valence with Nb^{5+}. Thus, the gradual increase in the R_{gb} ratio in the Sr-LLZNb series with increasing Sr content is considered to result from the Sr,Nb-rich layer that becomes increasingly thick as the amount of Sr rises. These results also demonstrate that the solid solubility limit of Sr in LLZNb is apparently higher than that of Ca. In the case of Ca, the elemental concentration gradient seen with Sr was not observed, and the Ca and other elements were distributed evenly inside the grains. Even though Ca and Ca–Zr accumulations were found at grain boundaries, the grains were connected with one another, which might have led to the low R_{gb} ratios in the Ca-LLZNb.

AUTHOR CONTRIBUTIONS

SO: conception and design of the study; YK: collection and assembly of data; TA: approval of this study.

REFERENCES

Allen, J. L., Wolfenstine, J., Rangasamy, E., and Sakamoto, J. (2012). Effect of substitution (Ta, Al, Ga) on the conductivity of Li7La3Zr2O12. *J. Power Sources* 206, 315–319. doi:10.1016/j.jpowsour.2012.01.131

Gao, X., Fisher, C. A. J., Kimura, T., Ikuhara, Y. H., Kuwabara, A., Moriwake, H., et al. (2014). Domain boundary structures in lanthanum lithium titanates. *J. Mater. Chem. A.* 2, 843–852. doi:10.1039/c3ta13726k

Inaguma, Y., Chen, L. Q., Itoh, M., Nakamura, T., Uchida, T., Ikuta, H., et al. (1993). High ionic-conductivity in lithium lanthanum titanate. *Solid State Commun.* 86, 689–693. doi:10.1016/0038-1098(93)90841-A

Jalem, R., Yamamoto, Y., Shiiba, H., Nakayama, M., Munakata, H., Kasuga, T., et al. (2013). Concerted migration mechanism in the Li ion dynamics of garnet-type Li7La3Zr2O12. *Chem. Mater.* 25, 425–430. doi:10.1021/cm303542x

Kihira, Y., Ohta, S., Imagawa, H., and Asaoka, T. (2013). Effect of simultaneous substitution of alkali earth metals and Nb in Li7La3Zr2O12 on lithium-ion conductivity. *ECS Electrochem. Lett.* 2, A56–A59. doi:10.1149/2.001307eel

Li, Y. T., Han, J. T., Wang, C. A., Xie, H., and Goodenough, J. B. (2012). Optimizing Li+ conductivity in a garnet framework. *J. Mater. Chem.* 22, 15357–15361. doi:10.1039/c2jm31413d

Logeat, A., Koohler, T., Eisele, U., Stiaszny, B., Harzer, A., Tovar, M., et al. (2012). From order to disorder: the structure of lithium-conducting garnets Li7-xLa3TaxZr2-xO12 (x=0-2). *Solid State Ionics.* 206, 33–38. doi:10.1016/j.ssi.2011.10.023

Moriwake, H., Gao, X., Kuwabara, A., Fisher, C. A. J., Kimura, T., Ikuhara, Y. H., et al. (2015). Domain boundaries and their influence on Li migration in solid-state electrolyte (La,Li)TiO3. *J. Power Sources* 276, 203–207. doi:10.1016/j.jpowsour.2014.11.139

Murugan, R., Thangadurai, V., and Weppner, W. (2007). Fast lithium ion conduction in garnet-type Li7La3Zr2O12. *Angew. Chem. Int. Ed.* 46, 7778–7781. doi:10.1002/anie.200701144

Nyman, M., Alam, T. M., McIntyre, S. K., Bleier, G. C., and Ingersoll, D. (2010). Alternative approach to increasing Li mobility in Li-La-Nb/Ta garnet electrolytes. *Chem. Mater.* 22, 5401–5410. doi:10.1021/cm101438x

ACKNOWLEDGMENTS

This study was partly supported by the project "Applied and Practical LiB Development for Automobile and Multiple Application" of the New Energy and Industrial Technology Development Organization (NEDO), Japan and Toyota Motor Corp.

Ohta, S., Kobayashi, T., and Asaoka, T. (2011). High lithium ionic conductivity in the garnet-type oxide Li7-X La-3(Zr2-X, Nb-X)O-12 (X=0-2). *J. Power Sources* 196, 3342–3345. doi:10.1016/j.jpowsour.2010.11.089

Ohtomo, T., Hayashi, A., Tatsumisago, M., and Kawamoto, K. (2013a). Suppression of H2S gas generation from the 75Li(2)S center dot 25P(2)S(5) glass electrolyte by additives. *J. Mater. Sci.* 48, 4137–4142. doi:10.1007/s10853-013-7226-8

Ohtomo, T., Hayashi, A., Tatsumisago, M., and Kawamoto, K. (2013b). Glass electrolytes with high ion conductivity and high chemical stability in the system LiI-Li2O-Li2S-P2S5. *Electrochemistery* 81, 428–431. doi:10.5796/electrochemistry.81.428

Rangasamy, E., Wolfenstine, J., Allen, J., and Sakamoto, J. (2013). The effect of 24c-site (A) cation substitution on the tetragonal-cubic phase transition in Li(7-x)La(3-x)A(x)Zr(2)O(12) garnet-based ceramic electrolyte. *J. Power Sources* 230, 261–266. doi:10.1016/j.jpowsour.2012.12.076

Shannon, R. D. (1976). Revised effective ionic radii and systematic studies of interatomic distances in halides and chalcogenides. *Acta Crystallograph. Sect. A* A32, 751–767. doi:10.1107/S0567739476001551

Stramare, S., Thangadurai, V., and Weppner, W. (2003). Lithium lanthanum titanates: are view. *Chem. Mater.* 15, 3974–3990. doi:10.1088/0953-8984/22/40/404203

Thompson, T., Wolfenstine, J., Allen, J. L., Johannes, M., Huq, A., David, I. N., et al. (2014). Tetragonal vs. cubic phase stability in Al – free Ta doped Li7La3Zr2O12 (LLZO). *J. Mater. Chem. A* 2, 13431–13436. doi:10.1039/c4ta02099e

Yashima, M., Itoh, M., Inaguma, Y., and Morii, Y. (2005). Crystal structure and diffusion path in the fast lithium-ion conductor La0.62Li0.16TiO3. *J. Am. Chem. Soc.* 127, 3491–3495. doi:10.1021/ja0449224

Conflict of Interest Statement: The authors declare that the research was conducted in the absence of any commercial or financial relationships that could be construed as a potential conflict of interest.

10

Aqueous Stability of Alkali Superionic Conductors from First-Principles Calculations

Balachandran Radhakrishnan and Shyue Ping Ong*

Department of NanoEngineering, University of California San Diego, La Jolla, CA, USA

Ceramic alkali superionic conductor solid electrolytes (SICEs) play a prominent role in the development of rechargeable alkali-ion batteries, ranging from replacement of organic electrolytes to being used as separators in aqueous batteries. The aqueous stability of SICEs is an important property in determining their applicability in various roles. In this work, we analyze the aqueous stability of twelve well-known Li-ion and Na-ion SICEs using Pourbaix diagrams constructed from first-principles calculations. We also introduce a quantitative free-energy measure to compare the aqueous stability of SICEs under different environments. Our results show that though oxides are, in general, more stable in aqueous environments than sulfides and halide-containing chemistries, the cations present play a crucial role in determining whether solid phases are formed within the voltage and pH ranges of interest.

Keywords: superionic conductors, Pourbaix diagrams, aqueous stability, passivation, corrosion

Edited by:
Guoxiu Wang,
University of Technology Sydney,
Australia

Reviewed by:
Li Wang,
Tsinghua University, China
Branimir Nikola Grgur,
University of Belgrade, Serbia

*Correspondence:
Shyue Ping Ong
ongsp@eng.ucsd.edu

Specialty section:
This article was submitted to
Energy Storage,
a section of the journal
Frontiers in Energy Research

Citation:
Radhakrishnan B and Ong SP (2016)
Aqueous Stability of Alkali Superionic
Conductors from First-Principles
Calculations.
Front. Energy Res. 4:16.

1. INTRODUCTION

Ceramic alkali superionic conductor solid electrolytes (SICEs) are key enablers to new rechargeable alkali-ion battery architectures that can significantly outperform today's Li-ion batteries (Goodenough et al., 1976; Aono et al., 1990; Kanno et al., 2000; Mizuno et al., 2005; Murugan et al., 2007; Kamaya et al., 2011; Rao and Adams, 2011; Rangasamy et al., 2012). SICEs are non-flammable and may potentially support wider electrochemical windows, leading to enhanced safety and higher energy densities compared to traditional liquid organic solvent electrolytes (Xu, 2014). A good SICE must satisfy many properties, namely, excellent alkali ionic conductivity, a wide electrochemical window and interfacial stability, mechanical compatibility with the electrodes, and phase and aqueous stability. Though properties other than ionic conductivity have not received as much attention in the past, there have been several recent efforts aimed at addressing other properties such as elastic properties (Deng et al., 2016) and electrochemical stability (Zhu et al., 2015; Richards et al., 2016) that have arguably become a more critical bottleneck in SICE design than ionic conductivity.

The aqueous stability of SICEs is also important in all-solid state batteries. In the event of mechanical abuse (Doughty and Pesaran, 2012), the SICE is exposed to environmental elements. On such exposures, the SICEs can react with moisture in the air to either form passivating layers that could alter their ionic conductivity or form hazardous materials. SICEs can also be used as separators between electrodes and electrolytes in aqueous batteries (Luo et al., 2010; Ma et al., 2015). In particular, separators are necessary when Li metal is used as an anode, for example, in Li-air systems (Visco and Chu, 2000; Liu et al., 2015), as Li reduces most electrolytes on contact. Separators in such applications must be stable against Li metal as well as aqueous electrolytes of varying pH values. In particular, the NAtrium SuperIonic CONductor (NASICON), such as $Li_{1.4}Al_{0.4}Ti_{1.4}Ge_{0.2}(PO_4)_3$ and garnet-type SICEs, have shown great promise in such roles (Zhang et al., 2008; Imanishi et al., 2014).

Determination of the aqueous stability of solid electrolytes has thus far been predominantly based on experimental observations. Such experiments take anywhere between a week to a month (Fuentes et al., 2001), where the electrolyte is immersed in an aqueous solution, sometimes with LiOH, LiCl, etc. (Hasegawa et al., 2009; Imanishi et al., 2014), and tested for changes in surface morphology, chemistry, and ionic conductivity. These experiments are time-consuming as well as limited in the set of environmental elements that can be tested. Also, typical experiments are performed at 0 V vs. the standard hydrogen electrode (SHE) and do not reflect the environment that the electrolyte is exposed during battery operation.

In this work, we apply and extend the Pourbaix diagram formalism developed by Persson et al. (2012) to develop a quantitative measure of the aqueous stability of solid electrolytes based on the free-energy change and the phase of the products formed. We then apply this methodology to twelve well-known SICEs that are of current interest. Unsurprisingly, we find that oxide SICEs are significantly more stable than sulfide and halide-containing

SICEs. However, we also find wide variations in aqueous stability even within the same anion chemistry. We will discuss these findings in context of the various potential applications of SICEs in energy storage.

2. MATERIALS AND METHODS

To keep the discussion self-contained, we provide a brief description of the equations used in plotting the aqueous stability plots here, and interested readers are referred to the previous work of Persson et al. (2012) for more details. For a given material, operating under an externally applied potential, V_{ext}, the chemical reaction with an aqueous medium can be represented by the following equation:

$$[Mat] + H_2O \overset{V_{ext}}{\Leftrightarrow} [Prod] + mH^+ + ze^-$$ (1)

From the Nernst equation, we can then derive the free-energy change as follows:

$$\Delta G_{rxn} = \Delta G_{rxn}^o + 2.303 \times RT \times \log \frac{[Prod]}{[Mat]}$$
$$-2.303 \times RT \times m \times pH - zV_{ext},$$ (2)

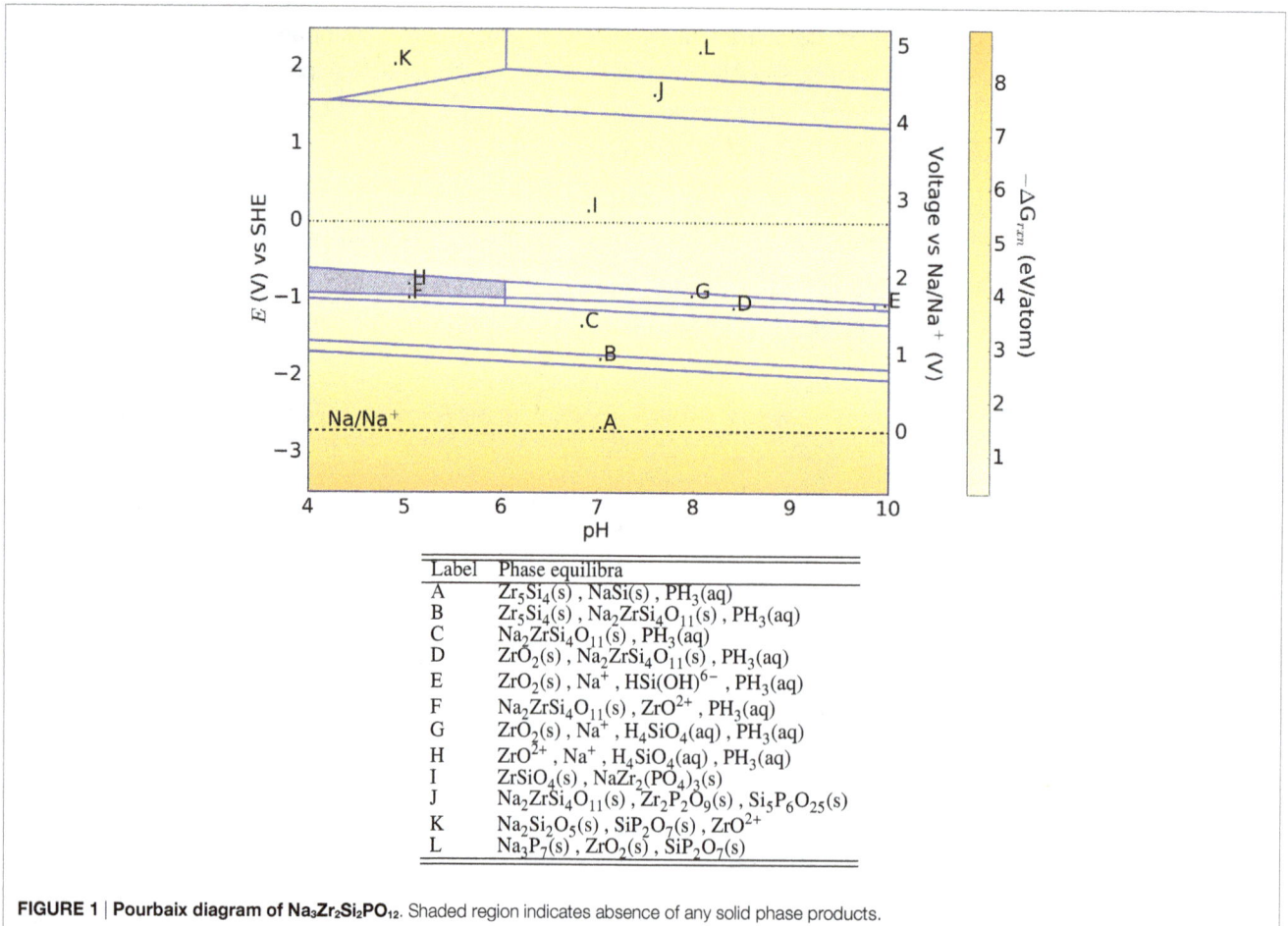

Label	Phase equilibra
A	$Zr_5Si_4(s)$, $NaSi(s)$, $PH_3(aq)$
B	$Zr_5Si_4(s)$, $Na_2ZrSi_4O_{11}(s)$, $PH_3(aq)$
C	$Na_2ZrSi_4O_{11}(s)$, $PH_3(aq)$
D	$ZrO_2(s)$, $Na_2ZrSi_4O_{11}(s)$, $PH_3(aq)$
E	$ZrO_2(s)$, Na^+ , $HSi(OH)^{6-}$, $PH_3(aq)$
F	$Na_2ZrSi_4O_{11}(s)$, ZrO^{2+} , $PH_3(aq)$
G	$ZrO_2(s)$, Na^+ , $H_4SiO_4(aq)$, $PH_3(aq)$
H	ZrO^{2+} , Na^+ , $H_4SiO_4(aq)$, $PH_3(aq)$
I	$ZrSiO_4(s)$, $NaZr_2(PO_4)_3(s)$
J	$Na_2ZrSi_4O_{11}(s)$, $Zr_2P_2O_9(s)$, $Si_5P_6O_{25}(s)$
K	$Na_2Si_2O_5(s)$, $SiP_2O_7(s)$, ZrO^{2+}
L	$Na_3P_7(s)$, $ZrO_2(s)$, $SiP_2O_7(s)$

FIGURE 1 | Pourbaix diagram of $Na_3Zr_2Si_2PO_{12}$. Shaded region indicates absence of any solid phase products.

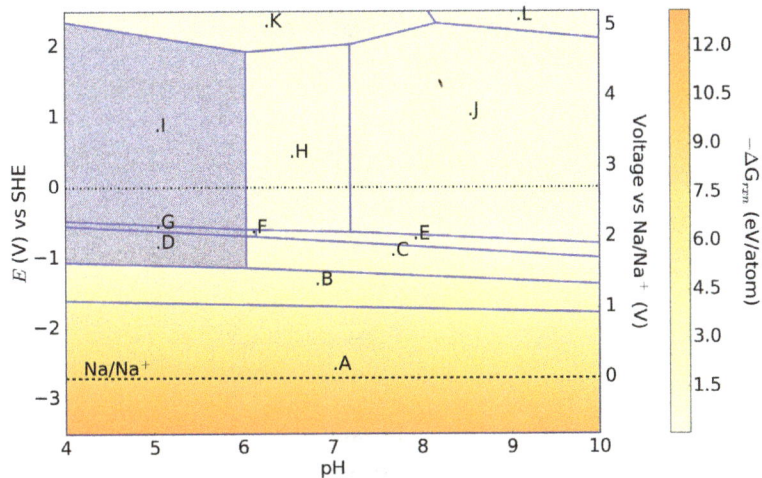

Label	Phase equilibra
A	$ZrH_2(s)$, $NaH(s)$, $PH_3(aq)$
B	$ZrH_2(s)$, Na^+ , $PH_3(aq)$
C	$ZrO_2(s)$, Na^+ , $PH_3(aq)$
D	ZrO^{2+} , Na^+ , $PH_3(aq)$
E	$ZrO_2(s)$, Na^+ , HPO_3^{2-}
F	$ZrO_2(s)$, Na^+ , $H_2PO_3^-$
G	ZrO^{2+} , Na^+ , $H_2PO_3^-$
H	$ZrO_2(s)$, Na^+ , $H_2PO_4^-$
I	ZrO^{2+} , Na^+ , $H_2PO_4^-$
J	$ZrO_2(s)$, Na^+ , HPO_4^{2-}
K	$Na_2P_3HO_9(s)$, $Zr_2P_2O_9(s)$
L	$ZrO_2(s)$, $Na_2O_2(s)$, HPO_4^{2-}

FIGURE 2 | Pourbaix diagram of NaZr₂(PO₄)₃. Shaded region indicates absence of any solid phase products.

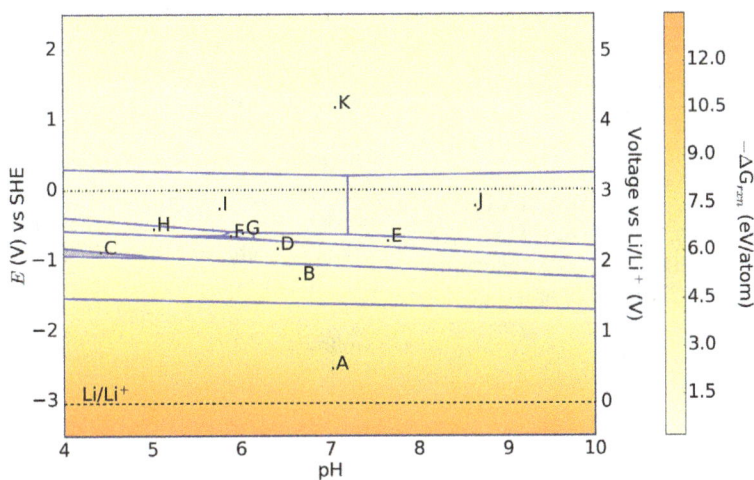

Label	Phase equilibra
A	$TiH_2(s)$, $LiH(s)$, $PH_3(aq)$
B	$TiH_2(s)$, Li^+ , $PH_3(aq)$
C	Li^+ , Ti^{2+} , $PH_3(aq)$
D	$TiO_2(s)$, Li^+ , $PH_3(aq)$
E	$TiO_2(s)$, Li^+ , HPO_3^{2-}
F	$TiO_2(s)$, $Li_3PO_4(s)$
G	$TiO_2(s)$, Li^+ , $H_2PO_3^-$
H	$Ti_5(PO_5)_4(s)$, Li^+
I	$TiO_2(s)$, Li^+ , $H_2PO_4^-$
J	$TiO_2(s)$, Li^+ , HPO_4^{2-}
K	$TiP_2O_7(s)$, $LiOH(s)$

FIGURE 3 | Pourbaix diagram of LiTi₂(PO₄)₃. Shaded region indicates absence of any solid phase products.

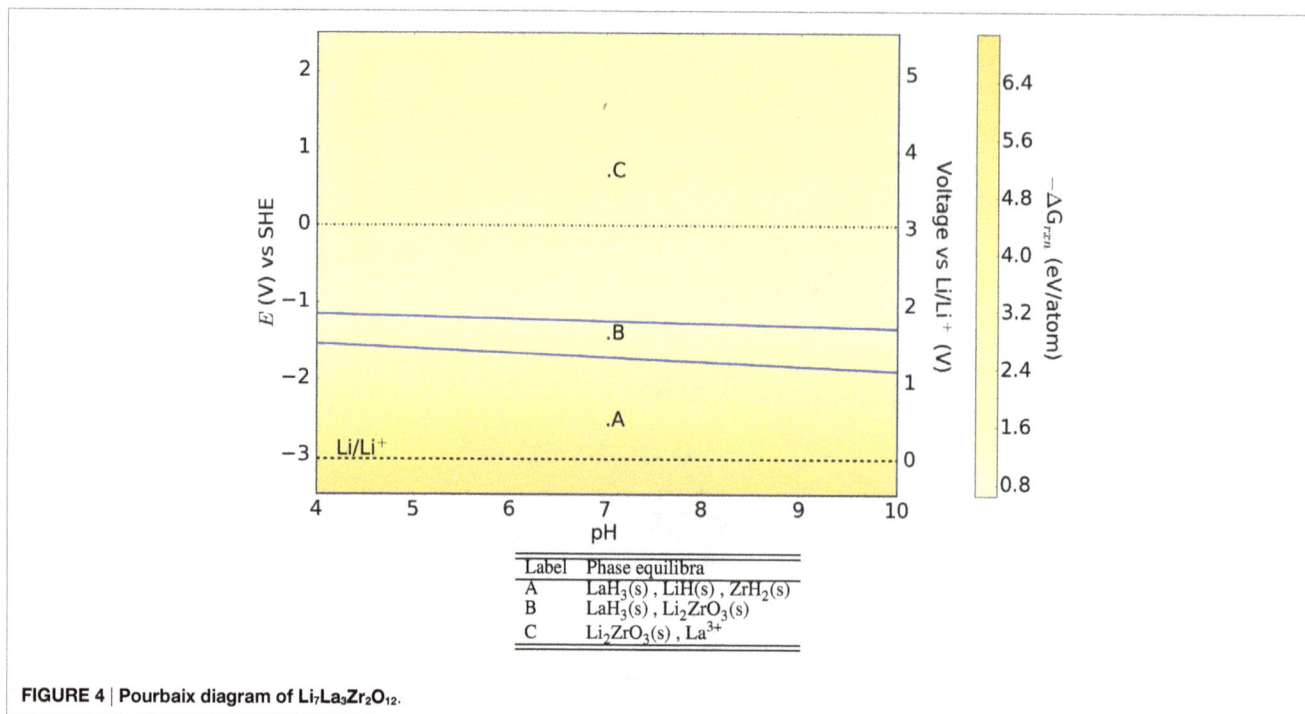

FIGURE 4 | Pourbaix diagram of Li₇La₃Zr₂O₁₂.

Label	Phase equilibra
A	$LaH_3(s)$, $LiH(s)$, $ZrH_2(s)$
B	$LaH_3(s)$, $Li_2ZrO_3(s)$
C	$Li_2ZrO_3(s)$, La^{3+}

where, $\Delta G^o_{rxn} = G^o_{Prod} - G^o_{Mat}$ is the free-energy change in the reaction under ideal environments (1 molar concentrations and $V_{ext} = 0$ V), R is the Universal gas constant, T is the temperature(=298 K), pH (=$-\log[H^+]$) is the measure of acidity of the aqueous medium, and [Mat] and [Prod] represent the concentration of the chemical species involved in the reaction. The activity of H_2O is assumed to be 1. In our analysis, we use $-\Delta G_{rxn}$ as a quantitative measure of aqueous stability of the material. The larger the value of this free-energy change, the greater is the thermodynamic driving force for the reaction to occur.

All first principles calculated energies are obtained with the Vienna *Ab initio* Simulation Package (VASP) (Kresse and Furthmüller, 1996) implementation of density functional theory (DFT), using the Perdew–Burke–Ernzerhof (PBE) generalized gradient approximation (Perdew et al., 1996) description of the exchange-correlation energy. The calculation parameters are similar to those used in the Materials Project (Jain et al., 2013), and all analyses were performed using the Python Materials Genomics package (Ong et al., 2013).

As per the formalism proposed by Persson et al. (2012), the reference energies for O_2 gas and H_2O are fitted to experimental values (Kubaschewski et al., 1993), and the H_2 energy is fixed to reproduce the H_2O formation energy. The free energies of aqueous ions were calculated from the dissolution energy of solids as given in Johnson et al. (1992) and Pourbaix (1966). The Pourbaix diagrams of SICEs are constructed by fixing the relative ratios of all chemical species except H and O. For example, to generate the Pourbaix diagram of the $Na_3Zr_2Si_2PO_{12}$ NASICON, the Na:Zr:Si:P ratio is fixed at 3:2:2:1, and the Pourbaix diagram at that composition is plotted as a function of pH and the applied voltage.

3. CHEMISTRIES INVESTIGATED

Twelve well-known Li-ion and Na-ion SICEs in a wide range of cation and anion chemistries are studied in this work. For each chemistry, we selected the most stable polymorph for analysis. While many metastable polymorphs are of greater interest due to their higher ionic conductivities, we argue that our analysis provides an upper limit to the free energy of decomposition in aqueous media, i.e., the metastable polymorph will be less stable (in a thermodynamic sense) than the specific phases investigated in this work. Furthermore, the relative energy differences between different polymorphs are on the order of ~10 meV, about 2–3 orders magnitude smaller than the free-energy change of solvation. Our analysis is also limited to the undoped structures as we do not expect small dopant concentrations to alter the aqueous stability significantly. The SICEs studied in this work are as follows:

1. *Oxides*: The NAtrium SuperIonic CONductor (NASICON) and garnet families are two of the most widely explored oxide SICEs. In this work, we have studied the $Na_3Zr_2Si_2PO_{12}$ (Goodenough et al., 1976), $NaZr_2(PO_4)_3$, and $LiTi_2(PO_4)_3$ (Subramanian et al., 1986) members of the NASICON family, which allows us to study the effect of alkali ion and cation chemistry on aqueous stability. We have also selected the tetragonal $Li_7La_3Zr_2O_{12}$ (LLZO) (Murugan et al., 2007) garnet, which is the stable polymorph of the cubic garnet phase that is of interest because of its good ionic conductivity of 0.4 mS/cm and excellent stability against Li metal anodes.

2. *Sulfides*: Sulfide SICEs have received wide attention recently due to their significantly higher ionic conductivities compared

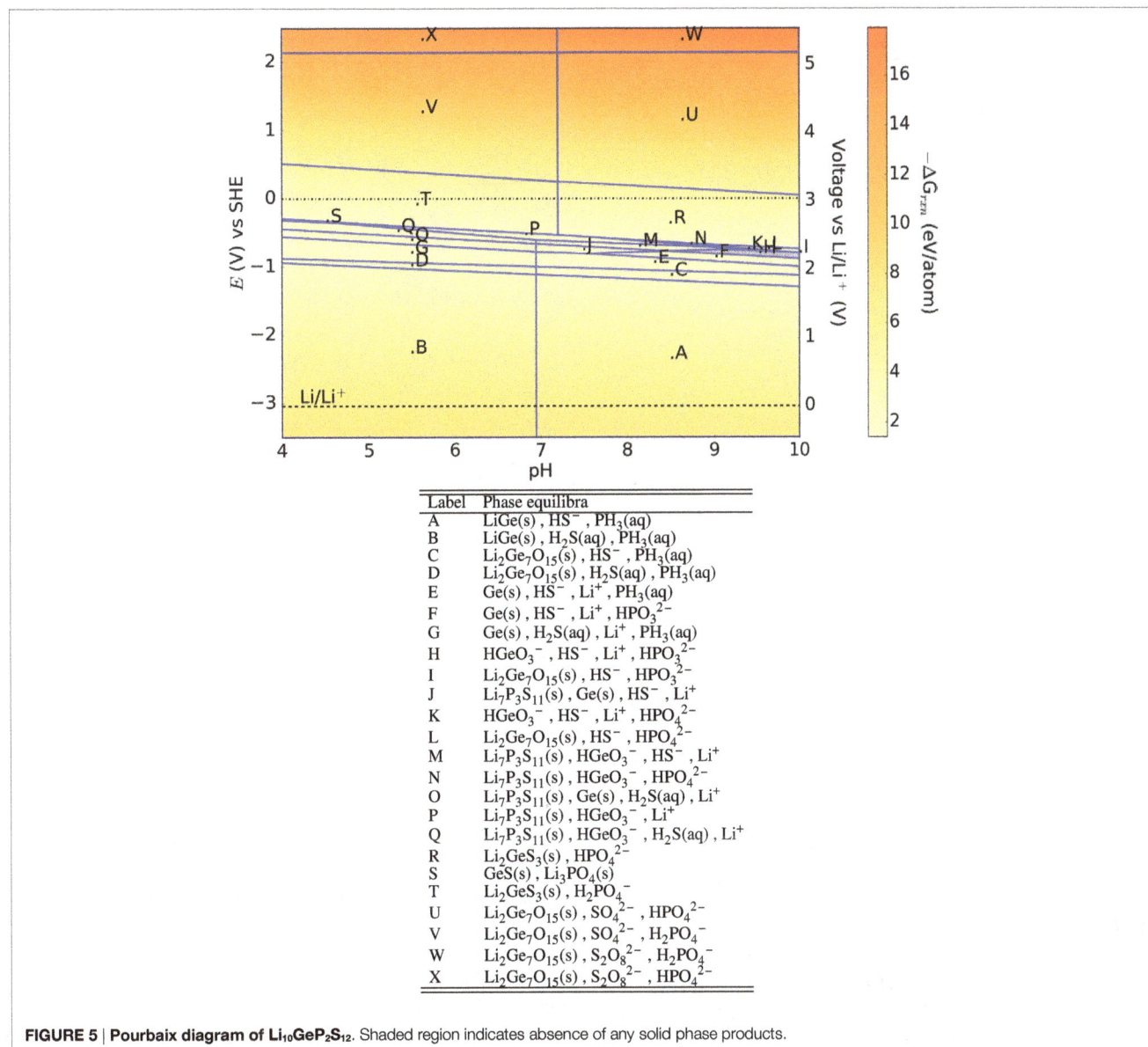

Label	Phase equilibra
A	$LiGe(s)$, HS^-, $PH_3(aq)$
B	$LiGe(s)$, $H_2S(aq)$, $PH_3(aq)$
C	$Li_2Ge_7O_{15}(s)$, HS^-, $PH_3(aq)$
D	$Li_2Ge_7O_{15}(s)$, $H_2S(aq)$, $PH_3(aq)$
E	$Ge(s)$, HS^-, Li^+, $PH_3(aq)$
F	$Ge(s)$, HS^-, Li^+, HPO_3^{2-}
G	$Ge(s)$, $H_2S(aq)$, Li^+, $PH_3(aq)$
H	$HGeO_3^-$, HS^-, Li^+, HPO_3^{2-}
I	$Li_2Ge_7O_{15}(s)$, HS^-, HPO_3^{2-}
J	$Li_7P_3S_{11}(s)$, $Ge(s)$, HS^-, Li^+
K	$HGeO_3^-$, HS^-, Li^+, HPO_4^{2-}
L	$Li_2Ge_7O_{15}(s)$, HS^-, HPO_4^{2-}
M	$Li_7P_3S_{11}(s)$, $HGeO_3^-$, HS^-, Li^+
N	$Li_7P_3S_{11}(s)$, $HGeO_3^-$, HPO_4^{2-}
O	$Li_7P_3S_{11}(s)$, $Ge(s)$, $H_2S(aq)$, Li^+
P	$Li_7P_3S_{11}(s)$, $HGeO_3^-$, Li^+
Q	$Li_7P_3S_{11}(s)$, $HGeO_3^-$, $H_2S(aq)$, Li^+
R	$Li_2GeS_3(s)$, HPO_4^{2-}
S	$GeS(s)$, $Li_3PO_4(s)$
T	$Li_2GeS_3(s)$, $H_2PO_4^-$
U	$Li_2Ge_7O_{15}(s)$, SO_4^{2-}, HPO_4^{2-}
V	$Li_2Ge_7O_{15}(s)$, SO_4^{2-}, $H_2PO_4^-$
W	$Li_2Ge_7O_{15}(s)$, $S_2O_8^{2-}$, $H_2PO_4^-$
X	$Li_2Ge_7O_{15}(s)$, $S_2O_8^{2-}$, HPO_4^{2-}

FIGURE 5 | Pourbaix diagram of $Li_{10}GeP_2S_{12}$. Shaded region indicates absence of any solid phase products.

to oxides. Among these sulfides, two of the most promising candidates are $Li_{10}GeP_2S_{12}$ (LGPS), which was reported by Kamaya et al. (2011) to have an ionic conductivity of 12 mS/cm, and $Li_7P_3S_{11}$ (Minami et al., 2007), which has the highest reported conductivity of 17 mS/cm thus far. We also included γ-Li_3PS_4 (Liu et al., 2013), which is one of the parent structures of the widely used LiPON solid electrolyte, and tetragonal Na_3PS_4 (t-Na_3PS_4), which is the stable polymorph of the cubic Na_3PS_4 SICE (Hayashi et al., 2012).

3. *Halide-containing chemistries*: Several halide-containing compounds have recently emerged as potential SICEs. The lithium-rich anti-perovskites (LRAP) (Zhao and Daemen, 2012) are a recently discovered class of SICEs, and we have included the Li_3OCl and Li_3OBr compounds in our analysis. We have also included both the oxide Li_6PO_5Cl and

sulfide Li_6PS_5Cl members of the argyrodite family (Kong et al., 2010).

4. RESULTS

Using the above methodology, we have analyzed the aqueous stability of twelve well-known SICEs in a pH range of 4.0–10.0, i.e., from acidic to basic environments. The externally applied voltage is varied between −3.5 V and 2.5 V (versus SHE), which encompasses an equivalent voltage range of −0.46 to 5.54 V versus Li/Li^+ (−3.04 V vs SHE) and −0.79 to 5.21 V versus Na/Na^+ (−2.71 V vs SHE) (Vanysek, 2011). In all Pourbaix diagrams, the corresponding alkali anode half-cell voltage is indicated for reference, and all aqueous ions other than H^+ are fixed at a molality of 10^{-6} mol/L.

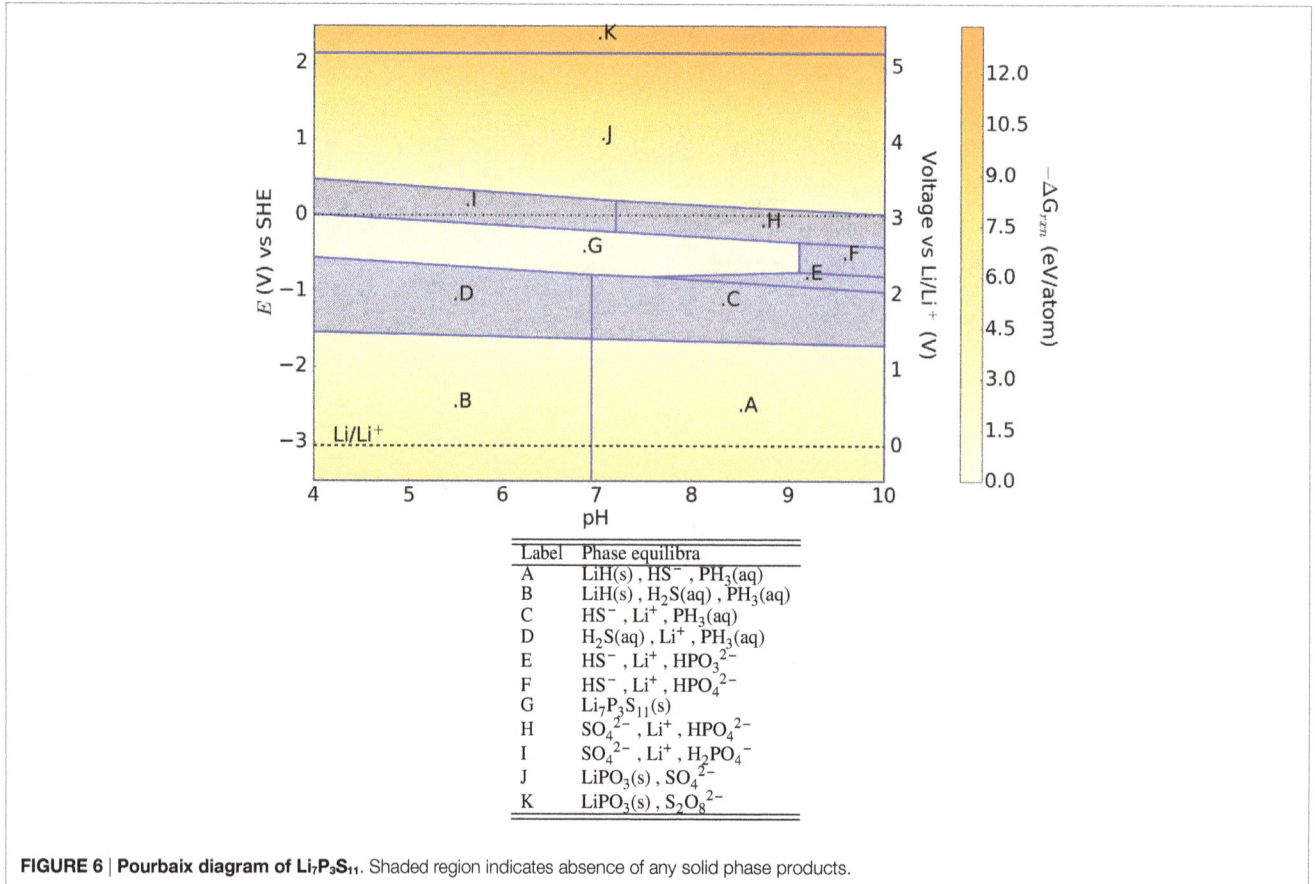

Label	Phase equilibra
A	$LiH(s)$, HS^- , $PH_3(aq)$
B	$LiH(s)$, $H_2S(aq)$, $PH_3(aq)$
C	HS^- , Li^+ , $PH_3(aq)$
D	$H_2S(aq)$, Li^+ , $PH_3(aq)$
E	HS^- , Li^+ , HPO_3^{2-}
F	HS^- , Li^+ , HPO_4^{2-}
G	$Li_7P_3S_{11}(s)$
H	SO_4^{2-} , Li^+ , HPO_4^{2-}
I	SO_4^{2-} , Li^+ , $H_2PO_4^-$
J	$LiPO_3(s)$, SO_4^{2-}
K	$LiPO_3(s)$, $S_2O_8^{2-}$

FIGURE 6 | Pourbaix diagram of $Li_7P_3S_{11}$. Shaded region indicates absence of any solid phase products.

4.1. Oxides

4.1.1. NASICON $Na_3Zr_2Si_2PO_{12}$, $NaZr_2(PO_4)_3$, and $LiTi_2(PO_4)_3$

The Pourbaix diagram of the $Na_3Zr_2Si_2PO_{12}$ NASICON is shown in **Figure 1**. On exposure to neutral environments, the non-silicon version of NASICON:$NaZr_2(PO_4)_3$ is formed along with $ZrSiO_4$ at 0 V, materials known to be less ionically conductive compared to NASICON. This is qualitatively in good agreement with the experiments explaining the change in the surface morphology as well as the conductivity (Ahmad et al., 1987; Mauvy et al., 1999; Fuentes et al., 2001). Fuentes et al. (2001) attributes the change in surface morphology to the formation of hydronium NASICON, but positive identification has been found lacking. In the voltage range −3.5 to −1 V (SHE), exposure to aqueous environments leads to formation of phosphine, which in gaseous form is known to be toxic and highly flammable. Otherwise, the Pourbaix diagram predicts that the NASICON forms solid insulating phases at high voltages. At low voltages, however, ZrH_2, a metallic alloy (Bickel and Berlincourt, 1970), is predicted to be the only solid product formed, which may not protect the SICE against further reaction.

The free-energy change shows that NASICON is more stable at reducing environments than oxidizing environments, especially at neutral environments with $-\Delta G_{rxn} \lesssim 1$ eV/atom. In comparison, we observe that $NaZr_2(PO_4)_3$ (**Figure 2**) is more stable at neutral environments than $Na_3Zr_2Si_2PO_{12}$ (lower free-energy change), but corrodes in acidic environments.

$LiTi_2(PO_4)_3$ (LTP), which also crystallizes in the NASICON structure, is one of the most stable electrolytes at 0 V across the pH range with a $-\Delta G_{rxn} \leq 0.2$ eV/atom, as shown in **Figure 3**. When exposed to aqueous environments at anodic voltages, it forms LiH, TiH_2, and phosphine gas. At voltages >0 V (SHE), the solid products formed, such as TiO_2 and Li_2O_2, can potentially provide a stable passivating layer. Acidity of the aqueous medium does not alter the composition of the passivating layers. The passivity predicted in the Pourbaix diagram explains the experimental observations that LTP is generally quite stable in aqueous environments (Zhang et al., 2008; Hasegawa et al., 2009). The same experiments also reported that exposure to LiCl and LiOH, which are used in aqueous batteries, alters only the surface chemistry while the electrolyte material itself was stable. At anodic voltages, the tendency for Ti^{4+} to be reduced to Ti^{2+}-forming TiH_2 may account for the larger driving force toward decomposition. It should be noted that TiH_2 (Ito et al., 2006) is metallic.

4.1.2. Garnet $Li_7La_3Zr_2O_{12}$

The Pourbaix diagram of garnet $Li_7La_3Zr_2O_{12}$ (**Figure 4**) shows that it is relatively stable in aqueous environments, with solid phases formed throughout the voltage range of interest. Li_2ZrO_3, formed at 0 V and oxidizing environments, is known to be a stable solid with low ionic conductivity. At anodic voltages <-1 V (SHE), hydrides of Li, Zr, and La are formed.

Like LTAP, garnet SICEs have been explored for the role of a separator in aqueous electrolytes with Li metal anodes (Imanishi et al., 2014). Similar to the NASICONs, we find that the acidity of the aqueous medium does not significantly alter the products formed. Also, the low free-energy change (≤ 1 eV/atom at 0 V) throughout the pH range shows that the garnet is relatively stable in aqueous environments, consistent with previous experimental observations (Murugan et al., 2007; Shimonishi et al., 2011).

4.2. Sulfides

4.2.1. $Li_{10}GeP_2S_{12}$ (LGPS)

From **Figure 5**, we may observe that the $Li_{10}GeP_2S_{12}$ (LGPS) (Kamaya et al., 2011) superionic conductor is predicted to form solid phases throughout the entire voltage range. At 0 V (SHE) and oxidizing environments, $Li_2Ge_7O_{15}$ is one of the products formed, which can potentially form a good passivation layer even though it has much lower ionic conductivity compared to LGPS. However, either LiGe or Ge are formed at voltages ≤ -1 V (SHE), which being poor electronic insulators, may not passivate the SICE against further reaction. Between voltages ± 0.5 V (SHE), Li_2GeS_3 is formed which has a reasonable ionic conductivity of

0.1 mS/cm (Seo and Martin, 2011). As with other phosphorous containing compounds, phosphine gas is one of the predicted products at voltages ≤ 0 V (SHE).

Though solid phases are generally formed for LGPS throughout the pH and voltage range of interest, we find that the driving force for the formation of these products is generally higher than for the oxide phases. Unlike the NASICON and garnet, LGPS shows greater driving force toward oxidation rather than reduction.

4.2.2. $Li_7P_3S_{11}$

In contrast to $Li_{10}GeP_2S_{12}$, the $Li_7P_3S_{11}$ (Minami et al., 2007) glass–ceramic is predicted to be significantly less stable in aqueous environments. There are large E–pH regions where no solid phases are stable (shaded regions in **Figure 6**). The potentially passivating $LiPO_3$ phase is formed at cathodic voltages of 5 V (vs SHE), while hydrides and phosphine are formed at anodic voltages ≤ -1.5 V (SHE), similar to other SICEs.

4.2.3. t-Na_3PS_4 and γ-Li_3PS_4

M_3PS_4 (M = Na and Li) also belongs to the M_2S–P_2S_5 glass–ceramic system. The qualitative features of their Pourbaix diagrams (**Figures 7** and **8**) are very similar to that of $Li_7P_3S_{11}$, with large E–pH regions where no solid phases are stable. Phosphates and

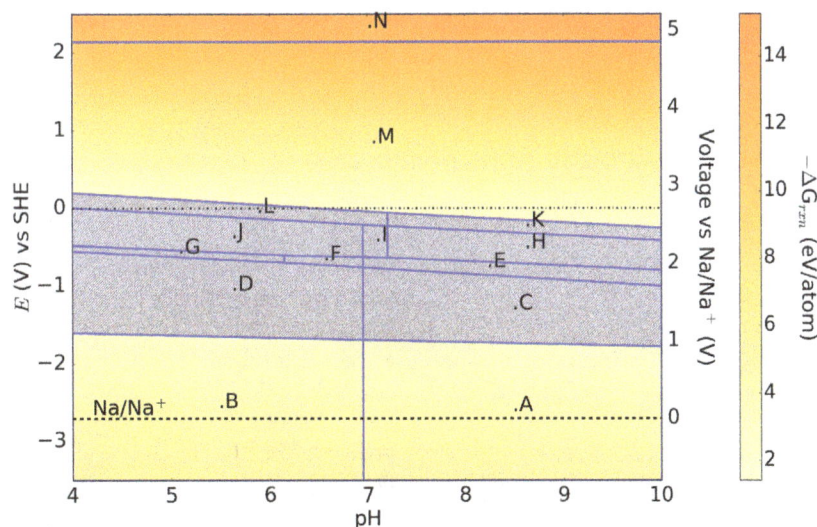

Label	Phase equilibra
A	$NaH(s)$, HS^-, $PH_3(aq)$
B	$NaH(s)$, $H_2S(aq)$, $PH_3(aq)$
C	HS^-, Na^+, $PH_3(aq)$
D	$H_2S(aq)$, Na^+, $PH_3(aq)$
E	HS^-, Na^+, HPO_3^{2-}
F	$H_2S(aq)$, Na^+, HPO_3^{2-}
G	$H_2S(aq)$, Na^+, $H_2PO_3^-$
H	HS^-, Na^+, HPO_4^{2-}
I	HS^-, Na^+, $H_2PO_4^-$
J	$H_2S(aq)$, Na^+, $H_2PO_4^-$
K	SO_4^{2-}, Na^+, HPO_4^{2-}
L	SO_4^{2-}, Na^+, $H_2PO_4^-$
M	$Na_2P_3HO_9(s)$, SO_4^{2-}
N	$Na_2P_3HO_9(s)$, $S_2O_8^{2-}$

FIGURE 7 | Pourbaix diagram of t-Na_3PS_4. Shaded region indicates absence of any solid phase products.

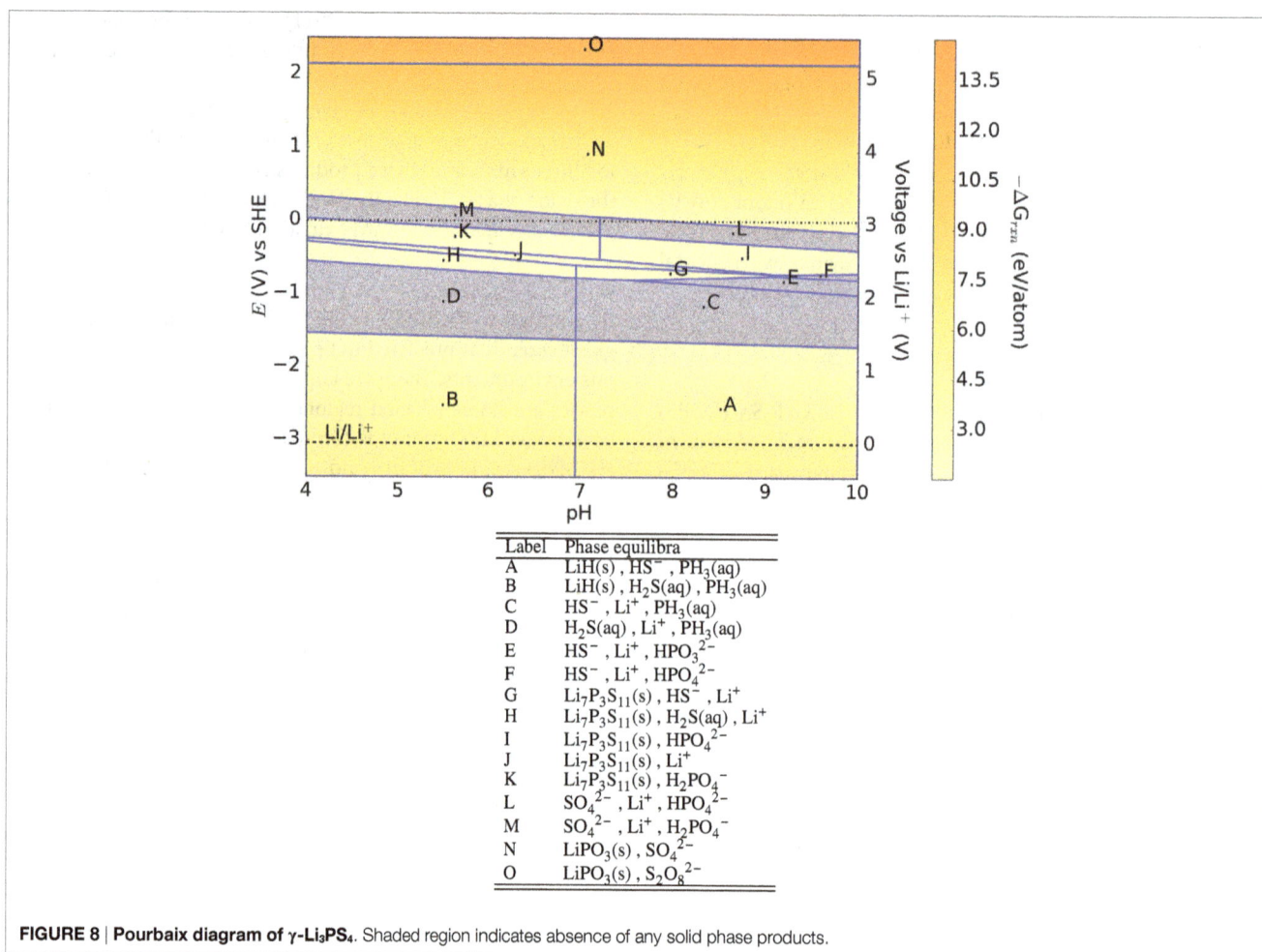

Label	Phase equilibra
A	$LiH(s)$, HS^-, $PH_3(aq)$
B	$LiH(s)$, $H_2S(aq)$, $PH_3(aq)$
C	HS^-, Li^+, $PH_3(aq)$
D	$H_2S(aq)$, Li^+, $PH_3(aq)$
E	HS^-, Li^+, HPO_3^{2-}
F	HS^-, Li^+, HPO_4^{2-}
G	$Li_7P_3S_{11}(s)$, HS^-, Li^+
H	$Li_7P_3S_{11}(s)$, $H_2S(aq)$, Li^+
I	$Li_7P_3S_{11}(s)$, HPO_4^{2-}
J	$Li_7P_3S_{11}(s)$, Li^+
K	$Li_7P_3S_{11}(s)$, $H_2PO_4^-$
L	SO_4^{2-}, Li^+, HPO_4^{2-}
M	SO_4^{2-}, Li^+, $H_2PO_4^-$
N	$LiPO_3(s)$, SO_4^{2-}
O	$LiPO_3(s)$, $S_2O_8^{2-}$

FIGURE 8 | Pourbaix diagram of γ-Li₃PS₄. Shaded region indicates absence of any solid phase products.

hydrated phosphates are generally formed at cathodic voltages, while hydrides are formed at anodic voltages.

4.3. Halide-Containing Chemistries
4.3.1. Lithium-Rich Anti-Perovskites (LRAP) Li₃OCl and Li₃OBr

Figures 9 and **10** show the Pourbaix diagrams of the Li₃OCl and Li₃OBr anti-perovskites, respectively. In general, the anti-perovskites are predicted to be unstable in aqueous environments with large regions where no solid phases are formed. Lithium peroxide Li₂O₂ is predicted to form at voltages ≥1 V (SHE), while LiH is predicted to form at voltages ≤−1.5 V (SHE). Between the two anion chemistries, the key difference is in the free-energy change of reaction. The chloride is predicted to have a higher free-energy change, i.e., larger driving force for decomposition, compared to the bromide. Nevertheless, these free-energy changes are much smaller in magnitude compared to the sulfides, which suggest that the LRAPs are less reactive compared to the sulfides in aqueous media.

4.3.2. Argyrodites Li₆PO₅Cl and Li₆PS₅Cl

The Pourbaix diagram of the oxide-argyrodite Li₆PO₅Cl (**Figure 11**) has generally similar features as that of the LRAPs,

with large E–pH regions containing no stable solid phases. Somewhat surprisingly, we find that the sulfur-argyrodite Li₆PS₅Cl may form a passivating layer of Li₇P₃S₁₁ at 0 V (SHE) as shown in **Figure 12**. A computational study on the sulfur-argyrodites (Li₆PS₅X) by Chen et al. (2015) hypothesized the formation of LiX, LiOH, H₂S, and Li₃PO₄ on hydrolysis with predicted decomposition energies of the order of 0.3 eV/atom. Our results predict decomposition energies ranging 1.5–12 eV/atom with a range of both solid and aqueous products formed across the E–pH range.

5. DISCUSSION

In this work, we investigated the aqueous stability of twelve well-known SICE chemistries using a Pourbaix diagram formalism. None of the SICEs investigated are found to be completely non-reactive on exposure to moisture. However, the decomposition of an electrolyte itself is not necessarily fatal to the electrochemical performance. After all, the organic solvents, such as ethylene carbonate and dimethylcarbonate, are unstable with respect to reduction by Li metal. The key is whether a good passivation layer can be formed that will protect the SICE against further

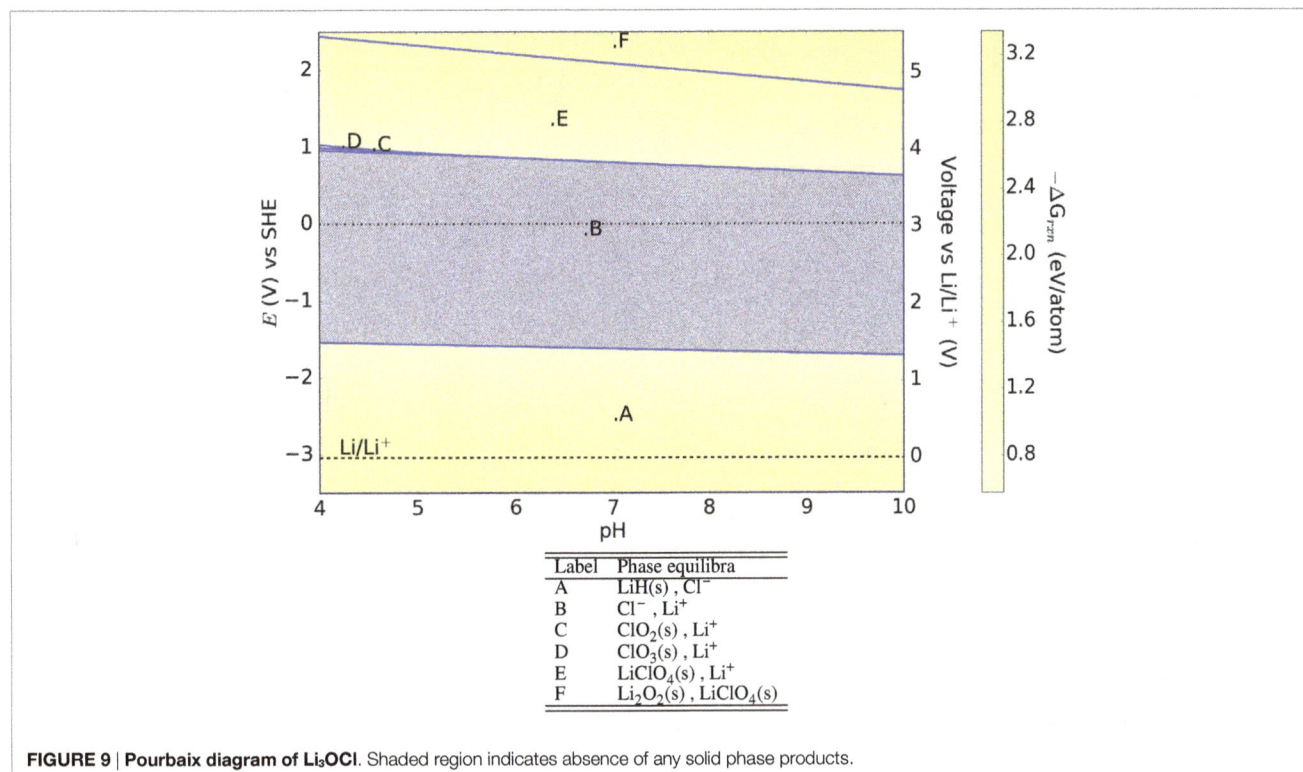

Label	Phase equilibra
A	$LiH(s)$, Cl^-
B	Cl^- , Li^+
C	$ClO_2(s)$, Li^+
D	$ClO_3(s)$, Li^+
E	$LiClO_4(s)$, Li^+
F	$Li_2O_2(s)$, $LiClO_4(s)$

FIGURE 9 | Pourbaix diagram of Li₃OCl. Shaded region indicates absence of any solid phase products.

reaction and still provide a reasonable pathway for alkali-ion transport.

Our broad findings are in general agreement with previous experimental studies and expected behavior of chemistries, i.e., that sulfide and halide-containing SICEs are less stable under aqueous media than oxide SICEs, with larger driving forces to oxidize at cathodic (high) voltages vs SHE. However, we find that there are significant variations in stability even within the same anion chemistry. For instance, comparing the $Na_3Zr_2Si_2PO_{12}$ and $NaZr_2(PO_4)_3$ NASICONs, we find that the former is more stable, particularly under acidic environments at high voltages, due to the presence of Si which promotes formation of solid silicate phases. Indeed, this observation suggests that the choice of cation (other than the alkali ion) may be a possible means to tune the phases formed at various E–pH. Another example can be seen by comparing the Pourbaix diagrams of LGPS, $Li_7P_3S_{11}$, and γ-Li_3PS_4. Although all the three materials are thiophosphates, the presence of Ge in LGPS results in the formation of stable solid phases across most of the investigated E–pH range.

In rechargeable alkali batteries, an ultimate goal is the use of a metal anode, which would yield significant increases in energy density over traditional carbon-based (graphitic for Li and hard carbons for Na) anodes. However, the inherent reactivity of the alkali metal, as well as the potential for dendrite formation during plating (for Li), has thus far pose a great challenge in the real-world applications. One potential strategy of addressing both issues is to use a separator to protect the alkali metal, as well as act as a barrier against dendrite formation. Indeed, the two common SICEs explored for this purpose, $LiTi_2(PO_4)_3$ and garnet LLZO (Zhang et al., 2008; Imanishi et al., 2014), are predicted to have relatively good stability in aqueous media. The garnet LLZO, in particular, is remarkable for its relative lack of reactivity across wide voltage and pH ranges, and low driving forces for decomposition.

It should be noted that the analysis presented is a purely thermodynamic one, and no kinetic factors are taken into account. Furthermore, while the lack of any stable solid phase would certainly preclude the possibility of passivation in a particular E–pH region, the existence of stable solid phases does not necessarily mean that a stable passivation layer will be formed. Ultimately, whether a solid phase can form a useful passivation layer depends on its morphology upon formation, its electronic conductivity, and its ionic conductivity. Nevertheless, we believe this work provides useful predictions of the phase equilibria at various E–pH and a quantification of the thermodynamic driving forces for reaction. These predictions can, and hopefully would be, validated against comprehensive experimental studies of SICEs in aqueous media.

6. CONCLUSION

To conclude, we have analyzed and compared the aqueous stability of twelve SICEs using a Pourbaix diagram formalism. In general, the predicted relative stabilities of the SICEs are in line with previous experimental results, where available, and expected behavior of chemistries. We find that the anion chemistry is a primary determinant of aqueous stability, though

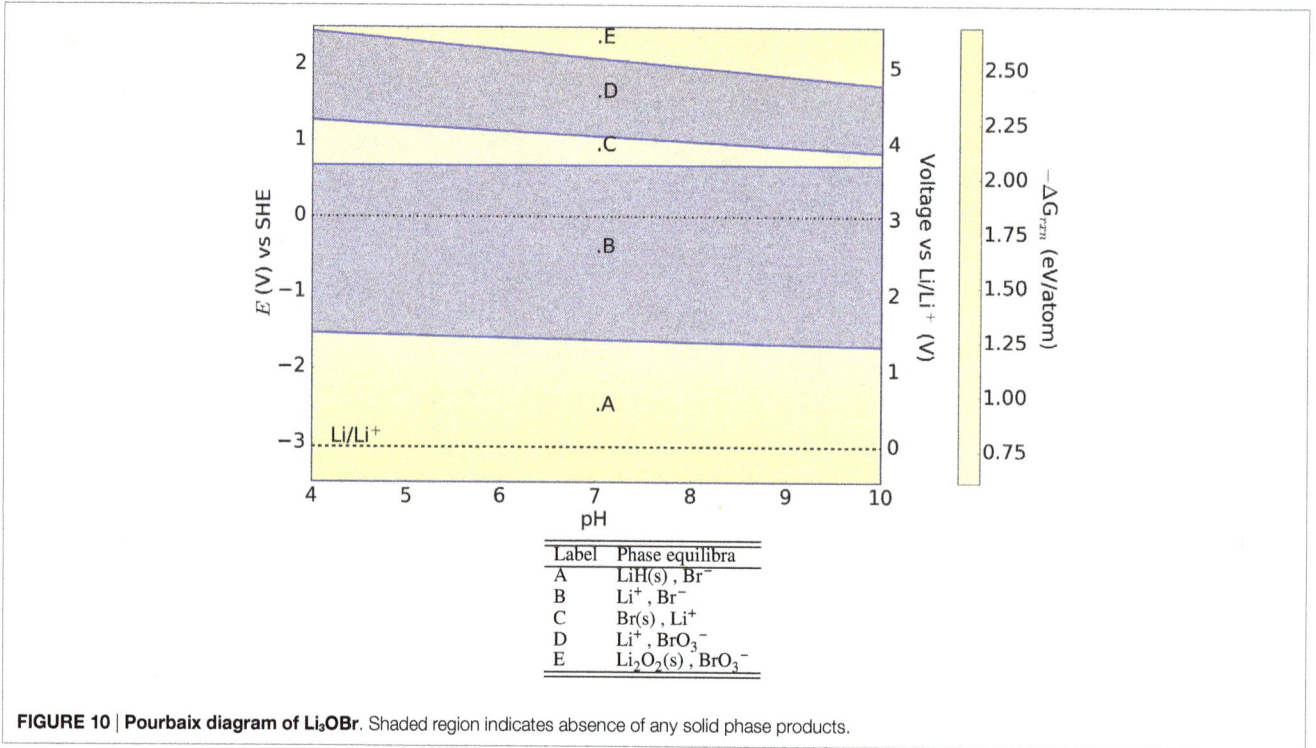

Label	Phase equilibra
A	$LiH(s)$, Br^-
B	Li^+ , Br^-
C	$Br(s)$, Li^+
D	Li^+ , BrO_3^-
E	$Li_2O_2(s)$, BrO_3^-

FIGURE 10 | Pourbaix diagram of Li₃OBr. Shaded region indicates absence of any solid phase products.

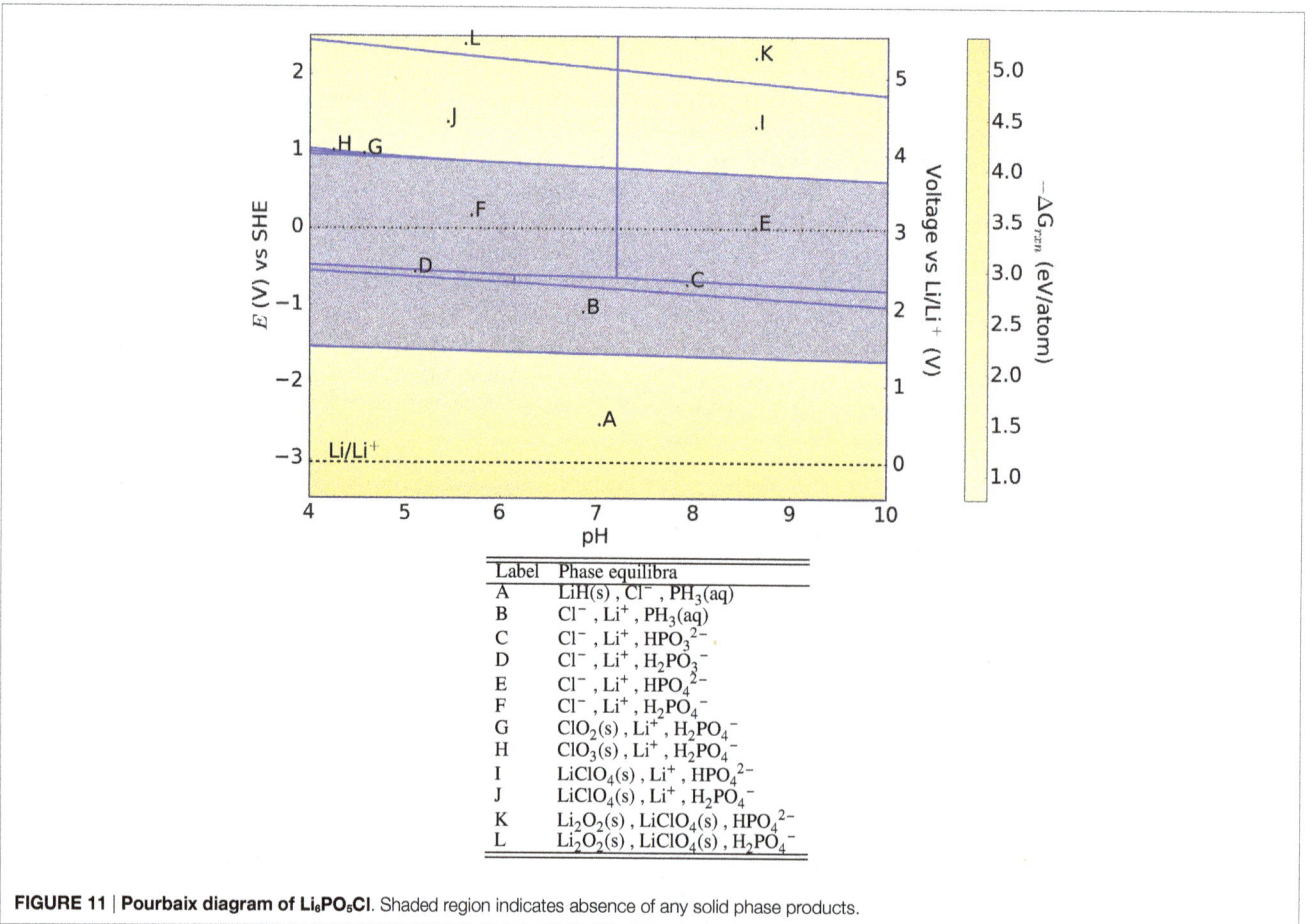

Label	Phase equilibra
A	$LiH(s)$, Cl^- , $PH_3(aq)$
B	Cl^- , Li^+ , $PH_3(aq)$
C	Cl^- , Li^+ , HPO_3^{2-}
D	Cl^- , Li^+ , $H_2PO_3^-$
E	Cl^- , Li^+ , HPO_4^{2-}
F	Cl^- , Li^+ , $H_2PO_4^-$
G	$ClO_2(s)$, Li^+ , $H_2PO_4^-$
H	$ClO_3(s)$, Li^+ , $H_2PO_4^-$
I	$LiClO_4(s)$, Li^+ , HPO_4^{2-}
J	$LiClO_4(s)$, Li^+ , $H_2PO_4^-$
K	$Li_2O_2(s)$, $LiClO_4(s)$, HPO_4^{2-}
L	$Li_2O_2(s)$, $LiClO_4(s)$, $H_2PO_4^-$

FIGURE 11 | Pourbaix diagram of Li₆PO₅Cl. Shaded region indicates absence of any solid phase products.

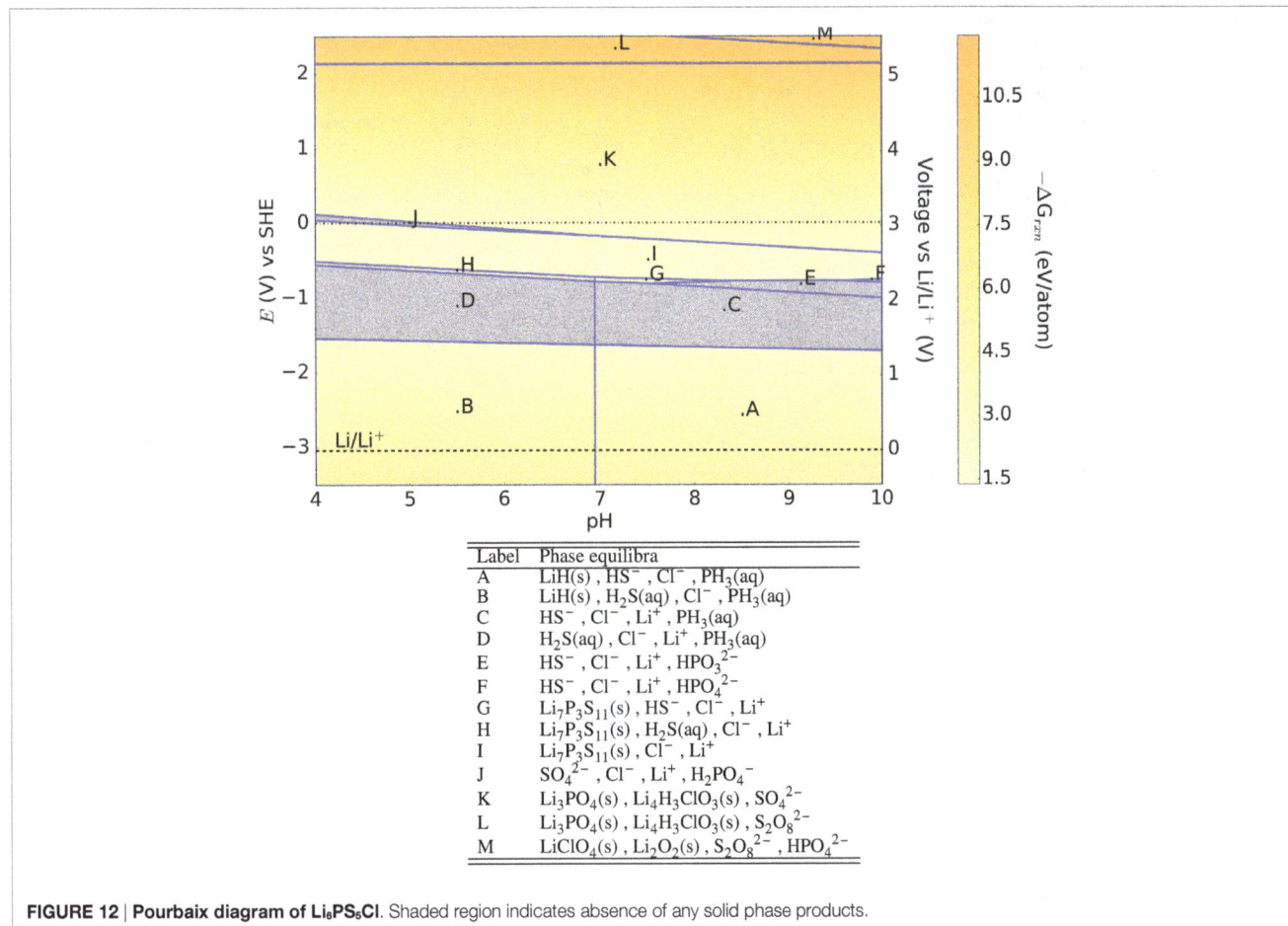

FIGURE 12 | Pourbaix diagram of Li₆PS₅Cl. Shaded region indicates absence of any solid phase products.

Label	Phase equilibra
A	$LiH(s)$, HS^- , Cl^- , $PH_3(aq)$
B	$LiH(s)$, $H_2S(aq)$, Cl^- , $PH_3(aq)$
C	HS^- , Cl^- , Li^+ , $PH_3(aq)$
D	$H_2S(aq)$, Cl^- , Li^+ , $PH_3(aq)$
E	HS^- , Cl^- , Li^+ , HPO_3^{2-}
F	HS^- , Cl^- , Li^+ , HPO_4^{2-}
G	$Li_7P_3S_{11}(s)$, HS^- , Cl^- , Li^+
H	$Li_7P_3S_{11}(s)$, $H_2S(aq)$, Cl^- , Li^+
I	$Li_7P_3S_{11}(s)$, Cl^- , Li^+
J	SO_4^{2-} , Cl^- , Li^+ , $H_2PO_4^-$
K	$Li_3PO_4(s)$, $Li_4H_3ClO_3(s)$, SO_4^{2-}
L	$Li_3PO_4(s)$, $Li_4H_3ClO_3(s)$, $S_2O_8^{2-}$
M	$LiClO_4(s)$, $Li_2O_2(s)$, $S_2O_8^{2-}$, HPO_4^{2-}

careful tuning of the cation chemistry can alter the solid phases formed and is a potential means of tuning SICEs for specific applications. Beyond SICEs, the first-principles approach presented here can be similarly applied to other technological applications, where aqueous stability of a material is a design concern.

AUTHOR CONTRIBUTIONS

BR designed and implemented the study on aqueous stability of alkali superionic conductors. SO is the principal investigator of this project and supervised the conceptualization and implementation. Both authors made equal contribution toward analysis of results and drafting the manuscript.

FUNDING

This work was supported by the National Science Foundation's Designing Materials to Revolutionize and Engineer our Future (DMREF) program under Grant No. 1436976. Some of the computations in this work were performed using the Extreme Science and Engineering Discovery Environment (XSEDE), which is supported by National Science Foundation grant number ACI-1053575.

REFERENCES

Ahmad, A., Wheat, T., Kuriakose, A., Canaday, J., and McDonald, A. (1987). Dependence of the properties of NASICONS on their composition and processing. *Solid State Ionics* 24, 89–97. doi:10.1016/0167-2738(87)90070-1

Aono, H., Sugimoto, E., Sadaoka, Y., Manaka, N., and Adachi, G. (1990). Ionic conductivity of solid electrolytes based on lithium titanium phosphate. *J. Electrochem. Soc.* 137, 1023–1027. doi:10.1149/1.2086597

Bickel, P. W., and Berlincourt, T. G. (1970). Electrical properties of hydrides and deuterides of zirconium. *Phys. Rev. B* 2, 4807–4813. doi:10.1103/PhysRevB.2.4807

Chen, H. M., Maohua, C., and Adams, S. (2015). Stability and ionic mobility in argyrodite-related lithium-ion solid electrolytes. *Phys. Chem. Chem. Phys.* 17, 16494–16506. doi:10.1039/c5cp01841b

Deng, Z., Wang, Z., Chu, I.-H., Luo, J., and Ong, S. P. (2016). Elastic properties of alkali superionic conductor electrolytes from first principles calculations. *J. Electrochem. Soc.* 163, A67–A74. doi:10.1149/2.0061602jes

Doughty, D. H., and Pesaran, A. A. (2012). *Vehicle Battery Safety Roadmap Guidance.* Golden: National Renewable Energy Laboratory.

Fuentes, R. O., Figueiredo, F., Marques, F. M. B., and Franco, J. I. (2001). Reaction of NASICON with water. *Solid State Ionics.* 139, 309–314. doi:10.1016/S0167-2738(01)00683-X

Goodenough, J. B., Hong, H. Y.-P., and Kafalas, J. A. (1976). Fast Na/+/-ion transport in skeleton structures. *Mater. Res. Bull.* 11, 203–220. doi:10.1016/0025-5408(76)90077-5

Hasegawa, S., Imanishi, N., Zhang, T., Xie, J., Hirano, A., Takeda, Y., et al. (2009). Study on lithium/air secondary batteries-stability of NASICON-type lithium ion conducting glass-ceramics with water. *J. Power Sources* 189, 371–377. doi:10.1016/j.jpowsour.2008.08.009

Hayashi, A., Noi, K., Sakuda, A., and Tatsumisago, M. (2012). Superionic glass-ceramic electrolytes for room-temperature rechargeable sodium batteries. *Nat. Commun.* 3, 856. doi:10.1038/ncomms1843

Imanishi, N., Matsui, M., Takeda, Y., and Yamamoto, O. (2014). Lithium ion conducting solid electrolytes for aqueous lithium-air batteries. *Electrochemistry* 82, 938–945. doi:10.5796/electrochemistry.82.938

Ito, M., Setoyama, D., Matsunaga, J., Muta, H., Kurosaki, K., Uno, M., et al. (2006). Electrical and thermal properties of titanium hydrides. *J. Alloys Comp.* 420, 25–28. doi:10.1016/j.jallcom.2005.10.032

Jain, A., Ong, S. P., Hautier, G., Chen, W., Richards, W. D., Dacek, S., et al. (2013). Commentary: the materials project: a materials genome approach to accelerating materials innovation. *APL Mater.* 1, 011002. doi:10.1063/1.4812323

Johnson, W. J., Oelkers, H. E., and Helgeson, C. H. (1992). SUPCRT92: a software package for calculating the standard molal thermodynamic properties of minerals, gases, aqueous species, and reactions from 1 to 5000 bar and 0 to 1000C. *Comput. Geosci.* 18, 899–947. doi:10.1016/0098-3004(92)90029-Q

Kamaya, N., Homma, K., Yamakawa, Y., Hirayama, M., Kanno, R., Yonemura, M., et al. (2011). A lithium superionic conductor. *Nat. Mater.* 10, 682–686. doi:10.1038/nmat3066

Kanno, R., Hata, T., Kawamoto, Y., and Irie, M. (2000). Synthesis of a new lithium ionic conductor, thio-LISICON – lithium germanium sulfide system. *Solid State Ionics.* 130, 97–104. doi:10.1016/S0167-2738(00)00277-0

Kong, S. T., Deiseroth, H. J., Maier, J., Nickel, V., Weichert, K., and Reiner, C. (2010). Li_6PO_5Br and Li_6PO_5Cl: the first lithium-oxide-argyrodites. *Z. Anorg. Allg. Chem.* 636, 1920–1924. doi:10.1002/zaac.201000121

Kresse, G., and Furthmüller, J. (1996). Efficient iterative schemes for ab initio total-energy calculations using a plane-wave basis set. *Phys. Rev. B Condens. Matter* 54, 11169–11186. doi:10.1103/PhysRevB.54.11169

Kubaschewski, O., Alcock, C. B., and Spencer, P. J. (1993). *Materials Thermochemistry*. New York: Pergamon Press.

Liu, T., Leskes, M., Yu, W., Moore, A. J., Zhou, L., Bayley, P. M., et al. (2015). Cycling $Li-O_2$ batteries via LiOH formation and decomposition. *Science* 350, 530–533. doi:10.1126/science.aac7730

Liu, Z., Fu, W., Payzant, E. A., Yu, X., Wu, Z., Dudney, N. J., et al. (2013). Anomalous high ionic conductivity of nanoporous beta-Li_3PS_4. *J. Am. Chem. Soc.* 135, 975–978. doi:10.1021/ja3110895

Luo, J.-Y., Cui, W.-J., He, P., and Xia, Y.-Y. (2010). Raising the cycling stability of aqueous lithium-ion batteries by eliminating oxygen in the electrolyte. *Nat. Chem.* 2, 760–765. doi:10.1038/nchem.763

Ma, C., Rangasamy, E., Liang, C., Sakamoto, J., More, K. L., and Chi, M. (2015). Excellent stability of a lithium-ion-conducting solid electrolyte upon reversible $Li+/H+$ exchange in aqueous solutions. *Angew. Chem. Int. Ed. Engl.* 127, 131–135. doi:10.1002/ange.201500056

Mauvy, F., Siebert, E., and Fabry, P. (1999). Reactivity of NASICON with water and interpretation of the detection limit of a NASICON based Na+ ion selective electrode. *Talanta* 48, 293–303. doi:10.1016/S0039-9140(98)00234-3

Minami, K., Mizuno, F., Hayashi, A., and Tatsumisago, M. (2007). Lithium ion conductivity of the $Li_2SP_2S_5$ glass-based electrolytes prepared by the melt quenching method. *Solid State Ionics* 178, 837–841. doi:10.1016/j.ssi.2007.03.001

Mizuno, F., Hayashi, A., Tadanaga, K., and Tatsumisago, M. (2005). New, highly ion-conductive crystals precipitated from $Li_2S-P_2S_5$ glasses. *Adv. Mater. Weinheim* 17, 918–921. doi:10.1002/adma.200401286

Murugan, R., Thangadurai, V., and Weppner, W. (2007). Fast lithium ion conduction in garnet-type $Li(7)La(3)Zr(2)O(12)$. *Angew. Chem. Int. Ed. Engl.* 46, 7778–7781. doi:10.1002/anie.200701144

Ong, S. P., Richards, W. D., Jain, A., Hautier, G., Kocher, M., Cholia, S., et al. (2013). Python Materials Genomics (pymatgen): a robust, open-source python library for materials analysis. *Comput. Mater. Sci.* 68, 314–319. doi:10.1016/j.commatsci.2012.10.028

Perdew, J., Burke, K., and Ernzerhof, M. (1996). Generalized gradient approximation made simple. *Phys. Rev. Lett.* 77, 3865–3868. doi:10.1103/PhysRevLett.77.3865

Persson, K. A., Waldwick, B., Lazic, P., and Ceder, G. (2012). Prediction of solid-aqueous equilibria: scheme to combine first-principles calculations of solids with experimental aqueous states. *Phys. Rev. B* 85, 235438. doi:10.1103/PhysRevB.85.235438

Pourbaix, M. (1966). *Atlas of Electrochemical Equilibria in Aqueous Solutions.* Oxford: Pergamon Press.

Rangasamy, E., Wolfenstine, J., and Sakamoto, J. (2012). The role of Al and Li concentration on the formation of cubic garnet solid electrolyte of nominal composition $Li_7La_3Zr_2O_{12}$. *Solid State Ionics* 206, 28–32. doi:10.1016/j.ssi.2011.10.022

Rao, R. P., and Adams, S. (2011). Studies of lithium argyrodite solid electrolytes for all-solid-state batteries. *Phys Status Solidi A Appl. Mater.* 208, 1804–1807. doi:10.1002/pssa.201001117

Richards, W. D., Miara, L. J., Wang, Y., Kim, J. C., and Ceder, G. (2016). Interface stability in solid-state batteries. *Chem. Mater.* 28, 266–273. doi:10.1021/acs.chemmater.5b04082

Seo, I., and Martin, S. W. (2011). Preparation and characterization of fast ion conducting lithium thio-germanate thin films grown by RF magnetron sputtering. *J. Electrochem. Soc.* 158, A465. doi:10.1149/1.3552927

Shimonishi, Y., Toda, A., Zhang, T., Hirano, A., Imanishi, N., Yamamoto, O., et al. (2011). Synthesis of garnet-type $Li_{7-x}La_3Zr_2O_{12-1/2x}$ and its stability in aqueous solutions. *Solid State Ionics* 183, 48–53. doi:10.1016/j.ssi.2010.12.010

Subramanian, M., Subramanian, R., and Clearfield, A. (1986). Lithium ion conductors in the system $AB(IV)2(PO4)3$ (B = Ti, Zr and Hf). *Solid State Ionics* 18-19, 562–569. doi:10.1016/0167-2738(86)90179-7

Vanysek, P. (2011). "Electrochemical series," in *Handbook of Chemistry and Physics*, ed. W. M. Haynes, 92nd Edn (London: CRC Press), 5-80–5-89.

Visco, S. J., and Chu, M.-Y. (2000). *Protective Coatings for Negative Electrodes.* Patent No: US 6025094 A

Xu, K. (2014). Electrolytes and interphases in Li-ion batteries and beyond. *Chem. Rev.* 114, 11503–11618. doi:10.1021/cr500003w

Zhang, T., Imanishi, N., Hasegawa, S., Hirano, A., Xie, J., Takeda, Y., et al. (2008). Li polymer electrolyte water stable lithium-conducting glass ceramics composite for lithium-air secondary batteries with an aqueous electrolyte. *J. Electrochem. Soc.* 155, A965. doi:10.1149/1.2990717

Zhao, Y., and Daemen, L. L. (2012). Superionic conductivity in lithium-rich anti-perovskites. *J. Am. Chem. Soc.* 134, 15042–15047. doi:10.1021/ja305709z

Zhu, Y., He, X., and Mo, Y. (2016). First principles study on electrochemical and chemical stability of the solid electrolyte-electrode interfaces in all-solid-state Li-ion batteries. *J. Mater. Chem. A* 4, 3253–3266. doi:10.1039/C5TA08574H

Conflict of Interest Statement: The authors declare that the research was conducted in the absence of any commercial or financial relationships that could be construed as a potential conflict of interest.

11

Radioactive Cs in the severely contaminated soils near the Fukushima Daiichi nuclear power plant

Makoto Kaneko[1], Hajime Iwata[1], Hiroyuki Shiotsu[1], Shota Masaki[1], Yuji Kawamoto[1], Shinya Yamasaki[1], Yuki Nakamatsu[1], Junpei Imoto[1], Genki Furuki[1], Asumi Ochiai[1], Kenji Nanba[2], Toshihiko Ohnuki[3], Rodney C. Ewing[4] and Satoshi Utsunomiya[1]*

[1] Department of Chemistry, Kyushu University, Fukuoka, Japan, [2] Department of Environmental Management, Faculty of Symbiotic System Science, Fukushima University, Fukushima, Japan, [3] Advanced Science Research Center Japan Atomic Energy Agency, Tokai, Japan, [4] Department of Geological Sciences, Center for International Security and Cooperation, Stanford University, Stanford, CA, USA

Edited by:
Muhammad Zubair,
University of Engineering and
Technology Taxila, Pakistan

Reviewed by:
Anca Melintescu,
Horia Hulubei National Institute of
Physics and Nuclear Engineering,
Romania
Giuseppe Vella,
University of Palermo, Italy
Qazi Muhammad Nouman Amjad,
University of Engineering and
Technology Taxila, Pakistan

*Correspondence:
Satoshi Utsunomiya,
Department of Chemistry, Kyushu
University, Hakozaki 6-10-1,
Higashi-ku, Fukuoka, Japan
utsunomiya.satoshi.998@
m.kyushu-u.ac.jp

Specialty section:
This article was submitted to Nuclear
Energy, a section of the journal
Frontiers in Energy Research

Citation:
Kaneko M, Iwata H, Shiotsu H,
Masaki S, Kawamoto Y, Yamasaki S,
Nakamatsu Y, Imoto J, Furuki G,
Ochiai A, Nanba K, Ohnuki T,
Ewing RC and Utsunomiya S (2015)
Radioactive Cs in the severely
contaminated soils near the
Fukushima Daiichi nuclear
power plant.
Front. Energy Res. 3:37.

Radioactive Cs isotopes (^{137}Cs, $t_{1/2} = 30.07$ years and ^{134}Cs, $t_{1/2} = 2.062$ years) occur in severely contaminated soils within a few kilometer of the Fukushima Daiichi nuclear power plant at concentrations that range from 4×10^5 to 5×10^7 Bq/kg. In order to understand the mobility of Cs in these soils, both bulk and submicron-sized particles elutriated from four surface soils have been investigated using a variety of analytical techniques, including powder X-ray diffraction analysis, scanning electron microscopy (SEM), transmission electron microscopy (TEM), and analysis of the amount of radioactivity in sequential chemical extractions. Major minerals in bulk soil samples were quartz, feldspar, and minor clays. The submicron-sized particles elutriated from the same soil consist mainly of mica, vermiculite, and smectite and occasional gibbsite. Autoradiography in conjunction with SEM analysis confirmed the association of radioactive Cs mainly with the submicron-sized particles. Up to ~3 MBq/kg of ^{137}Cs are associated with the colloidal size fraction (<1 μm), which accounts for ~78% of the total radioactivity. Sequential extraction of the bulk sample revealed that most Cs was retained in the residual fraction, confirming the high binding affinity of Cs to clays, aluminosilicate sheet structures. The chemistry of the fraction containing submicron-sized particles from the same bulk sample showed a similar distribution to that of the bulk sample, again confirming that the Cs is predominantly adsorbed onto submicron-sized sheet aluminosilicates, even in the bulk soil samples. Despite the very small particle size, aggregation of the particles prevents migration in the vertical direction, resulting in the retention of >98% of Cs within top ~5 cm of the soil. These results suggest that the mobility of the aggregates of submicron-sized sheet aluminosilicate in the surface environment is a key factor controlling the current Cs migration in Fukushima.

Keywords: cesium, colloid, electron microscopy, Fukushima Daiichi nuclear disaster, clay minerals

Introduction

Four years have passed since the Fukushima Daiichi nuclear power plant (FDNPP) event released ~520 PBq of radionuclides (Steinhauser et al., 2014) in the vicinity of FDNPP. An elongated area some ~27 miles in length and ~7 miles wide to the northwest of FDNPP is still contaminated with radioactivity, mainly ^{134}Cs and ^{137}Cs, with half-lives of 2.06 and 30.07 years, respectively. A recent study reported an ~40% decrease of radioactivity in this area (MEXT, 2013), which is greater than the decrease (~21%) that is expected from the decay constants for Cs, especially ^{134}Cs. A number of studies have reported geochemical behavior of Cs in the contaminated area [e.g., Koarashi et al. (2012) and Yoshida and Takahashi (2012)] in order to understand the distribution and migration of radioactive Cs in the future. After the Cs release to atmosphere, most Cs aerosols were deposited by precipitation on March 15 (wet deposition), which continued through March 16th (Kinoshita et al., 2011; Kaneyasu et al., 2012; Morino et al., 2013) based on the chemical transport model, which were in good agreement with the actual Cs distribution. In the contaminated Fukushima soils, the Cs remains within the top ~5 cm of the soil profile, strongly bound to clay minerals, probably vermiculite, and illite (Tanaka et al., 2012a). On the other hand, another study reported that other mineral component were also associated with irreversibly bound radioactive Cs, based on results obtained by chemical extraction and subsequent X-ray diffraction (XRD) analysis (Kozai et al., 2012; Ohnuki and Kozai, 2013). In addition, the presence of organic matter in soils may prevent the Cs adsorption onto mineral surfaces (Matsunaga et al., 2013).

Mukai et al. (2014) recently conducted transmission electron microscopy (TEM) analysis of Fukushima soils and showed that most particles associated with radioactive Cs are aggregates of clays and other phases, such as organic matter. This aggregation makes it difficult to characterize the submicron phases associated with Cs or to estimate their contribution to the total Cs radioactivity. It is well-known that submicron-sized particles behave as colloids in surface and subsurface aquifers (McCarthy and Zachara, 1989); thus, these colloids can potentially control the migration of trace radionuclides. Hence, we have completed a detailed analysis of the colloid sized and smaller fraction of the soils. Although the previous studies reported a high affinity of Cs with clay minerals in the Fukushima surface environments (Tanaka et al., 2012a,b; Matsunaga et al., 2013; Nakao et al., 2014), the role of fine particles <1 μm has not been quantitatively determined, which is presumably the size fraction dominated by clay minerals. Finally, the severely contaminated area close to FDNPP is not easily accessible, and thus, fewer data have been reported for the soil from this area as compared with data recently reported for the samples more than several kilometers from the FDNPP. The present study investigated that the soil particles collected within a few kilometer of FDNPP utilizing a variety of analytical techniques including conventional sequential extraction methods and electron microscopy in order to describe the mineralogy, identify the Cs speciation, and determine the dose contribution from Cs sorbed onto submicron-sized particles.

Materials and Methods

Sample Description

Samples were collected on March 15–16, 2012. The localities of the sample sites are shown in **Figure 1**. Four soil samples were collected in the highly contaminated area in Ohkuma within 5 km of the FDNPP: two samples from the east and west zones of the Ottozawa district and the other two samples from east and west zones of the Koirino district. Hereafter, these samples are identified as OTO-E, OTO-W, KOI-E, and KOI-W, respectively. The collected samples were dried and sieved with a 2-mm polyethylene mesh filter to remove pebbles and plant material, such as grass.

A soil sample core (30 cm in length × 55 mm in diameter) was also collected from KOI-E, using a DAIKI core sampler, DIK-110C. The core was initially used to obtain the vertical cross-sectional autoradiography in order to image the distribution of radionuclides as a function of depth. Then, the outermost part of this cylindrical core was carefully shaved off to exclude the contamination from the core boring process. The depth profile of ^{134}Cs and ^{137}Cs radioactivities was made at 1 cm intervals for

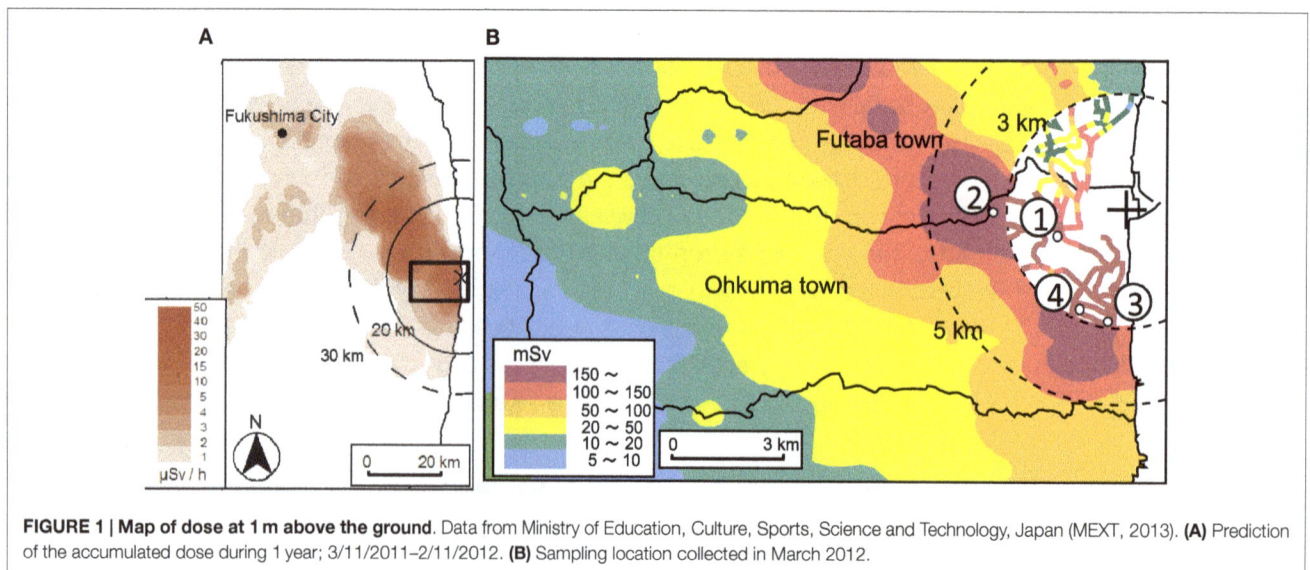

FIGURE 1 | Map of dose at 1 m above the ground. Data from Ministry of Education, Culture, Sports, Science and Technology, Japan (MEXT, 2013). **(A)** Prediction of the accumulated dose during 1 year; 3/11/2011–2/11/2012. **(B)** Sampling location collected in March 2012.

the core. The core sample was divided into 12 fractions by the depth: 0–1, 1–2, 2–3, 3–4, 4–5, 5–6, 6–7, 7–8, 8–9, 9–10, 10–13, and 13–16 cm.

Analytical Methods

Size Fractionation

The size-dependent ^{134,137}Cs activity was determined for three samples: OTO-E, OTO-W, and KOI-E. The size separation was performed with a sequence of size fractions: >597, 114–597, 48–114, 1–48, and <1 μm. The soil was suspended in ultra-pure water (milli-Q) and ultrasonicated for 15 min and filtered first with a 597 μm mesh. The residue was re-suspended in pure water, ultrasonicated, and filtered using a 114-μm mesh. The size ranges, >597, 114–597 and 48–114 μm, were obtained in the same manner. The soil particles <1 μm were separated from the <48 μm fraction by a sedimentation method, in which the hydrodynamic size was calculated based on the Stokes' law. The separated soils were air dried and weighed prior to further analysis. For the KOI-E sample, the size-fractionated samples for the >597, 114–597, and 48–114 μm fractions were subsequently ultrasonicated twice for 30 min in order to separate fine particles that were attached to

large-sized particles and to quantify the actual contribution of the fine particles to the total radioactivity.

Sequential Extraction of Radioactive Cs from Soils

Sequential extraction was performed for OTO-E, OTO-W, and KOI-E samples. The extraction method was based on Hou et al. (2003), which is a modified procedure after Tessier et al. (1979). Using this method, the chemical form of the Cs associated with soils is separated into six chemical categories: water soluble (F1), exchangeable (F2), bound to carbonate (F3), bound to metal oxide (F4), bound to organic matter (F5), and the residue (F6). Although the details of the extraction procedure were described in Hou et al. (2003), the procedure performed in the present study is briefly summarized as follows; first, pure water was added to 5 or 10 g soil and the suspension was continuously agitated for 24 h at room temperature. Then, the solution was centrifuged at 5000 rpm for 10 min, and the supernatant was filtered with 0.025 μm membrane filter. The filtrate was classified as water-soluble fraction (F1). The residue (residue-1) obtained by the filtering was re-suspended in 1.0M NH_4OAc (pH 8.0) and agitated for 24 h at room temperature. After the treatment, the suspension was centrifuged and filtered. The residue (residue-2) was rinsed with pure water and the rinsing water was added to the leachate. The filtrate was labeled as exchangeable (F2). Subsequently, the residue-2 was re-suspended in 1.0M NH_4OAc (pH 5.0) and agitated for 24 h at room temperature. The suspension was centrifuged and filtered. The residue (residue-3) was rinsed with pure water. The filtrate was labeled as the form bound to carbonate (F3). For the form bound to metal oxides (F4), 0.04M $NH_2OH\cdot HCl$ in 25% (v/v) HOAc at final pH of 2.0 was added to the residue-3, and the suspension was agitated for 24 h at $95 \pm 5°C$. After cooling to

TABLE 1 | ^{134}Cs and ^{137}Cs radioactivities of the four bulk soils.

	Radioactivity (Bq/kg)	
	^{134}Cs	^{137}Cs
① OTO-E	2.71×10^5	3.80×10^5
② OTO-W	8.37×10^5	1.11×10^6
③ KOI-E	1.85×10^5	2.50×10^5
④ KOI-W	2.03×10^7	2.66×10^7

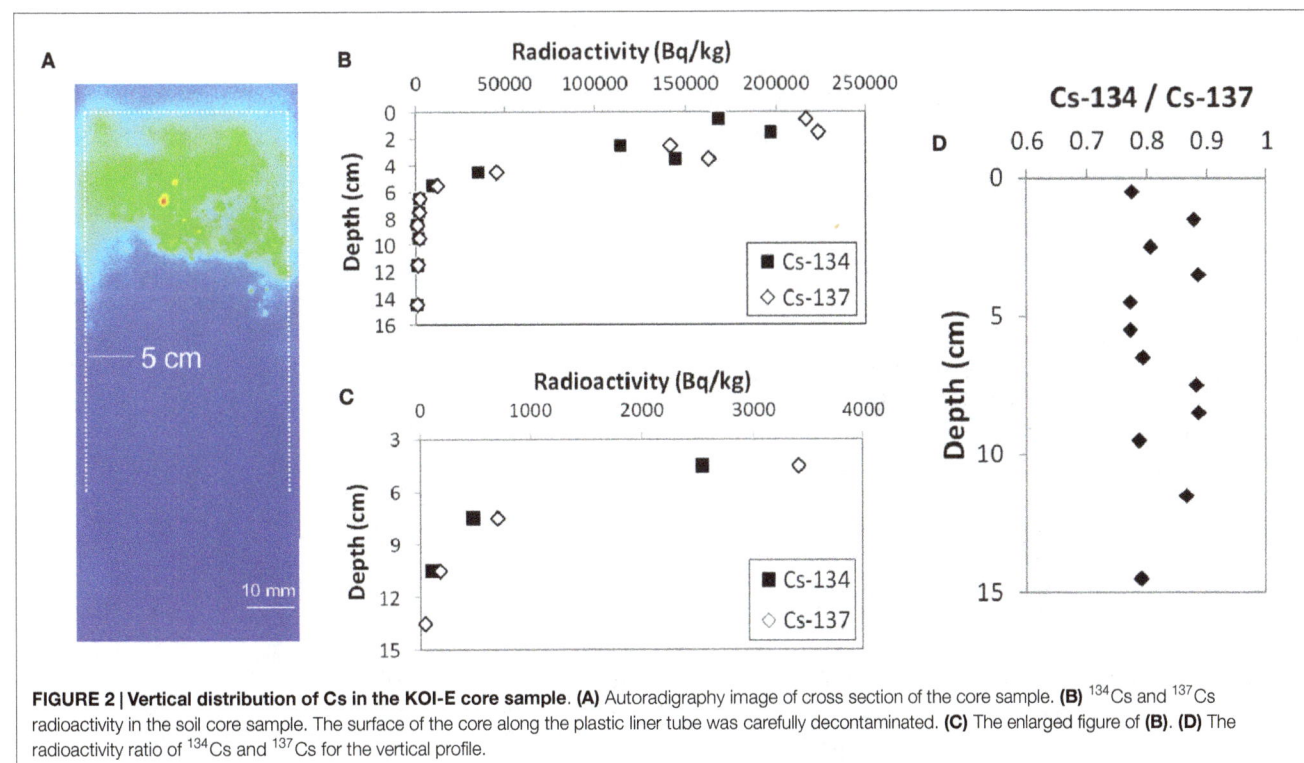

FIGURE 2 | Vertical distribution of Cs in the KOI-E core sample. (A) Autoradigraphy image of cross section of the core sample. **(B)** ^{134}Cs and ^{137}Cs radioactivity in the soil core sample. The surface of the core along the plastic liner tube was carefully decontaminated. **(C)** The enlarged figure of **(B)**. **(D)** The radioactivity ratio of ^{134}Cs and ^{137}Cs for the vertical profile.

room temperature, the suspension was centrifuged and filtered. The residue (residue-4) was rinsed with pure water. For the form bound to organic matter (F5), 30% H_2O_2–HNO_3 at pH 2.0 was added to the residue-4, and the suspension was agitated for 2 h at $95 \pm 5°C$. After cooling to room temperature, the suspension was centrifuged and filtered. This extraction procedure with peroxide and nitric acid was repeated twice. Subsequently, 3.2M NH_4OAc–20% HNO_3 at pH of 2.0 was added to the residue (residue-5a), and the suspension was agitated for 1 h at room temperature. The suspension was centrifuged and filtered. The residue (residue-5b) was rinsed with pure water. Insoluble residue (F6) was calculated by subtracting the sum of the radioactivities of F1–F5 from the total radioactivity.

Powder X-Ray Diffraction Analysis

Major mineral phases in the soils were determined by powder XRD (Rigaku MultiFlex) with Cu target and reflected beam monochromator. Measurement was conducted for the scan range of 3°–63° with the scanning speed of 1°/min of 2θ and the step angle of 0.02° for bulk soil sample and elutriated soil sample (<1 μm). In order to obtain the detailed structural information, the elutriated samples were measured for the focused angle ranging 3°–13° at the scan rate of 0.125° or 0.25°/min of 2θ with the step angle of 0.02°. In addition, ethylene glycol treatment was also tested for the elutriated sample for the range of 3°−13° at the scan rate of 0.125° or 0.25° min^{-1} of 2θ with the step angle of 0.02°.

Micro-Scale Analysis

Individual particle analysis of the soil samples was performed using a scanning electron microscopy (SEM, SHIMADZU SS-550) equipped with energy dispersive X-ray spectroscopy (EDX). Secondary electron (SE) imaging was conducted at 5 kV of the acceleration voltage, and the EDX point analysis and mapping were completed at 25 kV. SEM wave-length dispersive X-ray spectroscopy (SEM-WDX, Hitachi High-Technologies) was also employed for searching the Cs peak in soil samples. The operating condition was 20 kV of the acceleration voltage and 37.6 nA of the beam current. The measurement was conducted in the range of 4213–4359 eV at the scan speed of 1 or 2 eV/ch. All samples were coated with carbon using a carbon coater (SANYU SC-701C) to make conductivity.

High-resolution TEM (HRTEM), with EDX, and high-angle annular dark-field scanning electron microscopy (HAADF-STEM) were performed using JEOL ARM200F with the acceleration voltage of 200 kV. The JEOL Analysis Station software was used to control the STEM-EDX mapping. To minimize the effect of sample drift, a drift-correction mode was used in the acquisition of the elemental map. The STEM probe size was ~1.0 nm. The collection angle of the HAADF detector was ~50–110 mrad. The condenser aperture was 20 μm in diameter. TEM specimens were prepared by dispersing the sample on the holey-carbon thin film supported by a Cu mesh grid.

Cs Radioactivity Measurement

The $^{134,137}Cs$ radioactivities were measured using a gamma spectrometer equipped with a low background type Ge detector (EG&G ORTEC Ltd GMX, relative efficiency 55.4%) housed in the radioisotope center of Kyushu University. A spectrum analyzing software (SEIKO EG&G Ltd. Gamma Studio) was used to determine specific isotope contributions to the radioactivity.

Results and Discussion

Distribution of Cs Radioactivity in the Soil Profile

Radioactivities of ^{134}Cs and ^{137}Cs in the soil samples are summarized in **Table 1**. All samples revealed both ^{134}Cs and ^{137}Cs activities >100,000 Bq/kg, confirming the severe contamination. In particular, 2.66×10^7 Bq/kg of ^{137}Cs was detected in KOI-W sample due to the Cs accumulation beneath a rainwater pipe. Radioactivities of the other soil samples are of a similar order of magnitude, for samples collected from the open fields.

Figure 2A shows the cross-sectional autoradiograph image of KOI-E sample. It is evident that most of the radioactive Cs is concentrated in the top ~4 cm. Slightly elevated contrast along the outer-edge facing toward the plastic tube is a result of contamination from the surface soil during the coring process. There are some hot particles (appearing as red colored spots) near surface, indicating that some particles possesses relatively higher radioactivity among the other soil components.

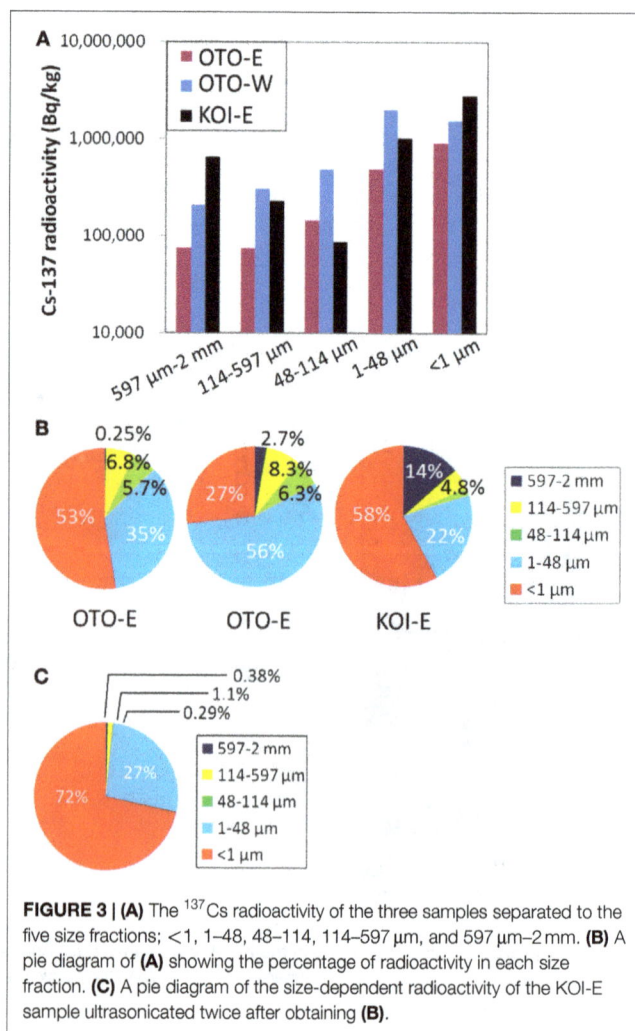

FIGURE 3 | **(A)** The ^{137}Cs radioactivity of the three samples separated to the five size fractions; <1, 1–48, 48–114, 114–597 μm, and 597 μm–2 mm. **(B)** A pie diagram of **(A)** showing the percentage of radioactivity in each size fraction. **(C)** A pie diagram of the size-dependent radioactivity of the KOI-E sample ultrasonicated twice after obtaining **(B)**.

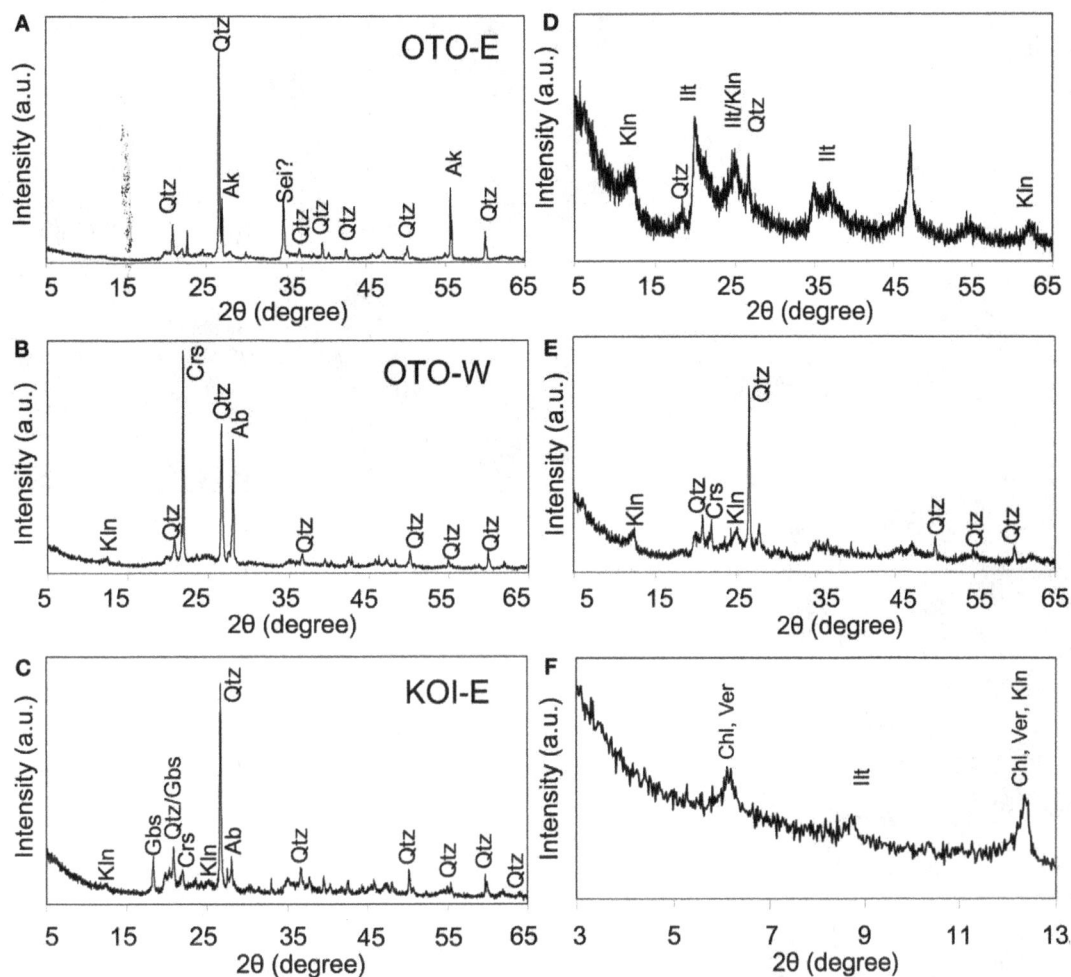

FIGURE 4 | X-ray diffraction spectrum of three soil samples: OTO-E, OTO-W, and KOI-E. (A–C) The patterns obtained from the bulk samples. **(D–F)** The patterns obtained from submicron-sized particle after elutriation of the samples **(A–C)**, respectively. Qtz, quartz; Kln, kaolinite; Crs, cristobalite; Ab, albite; Ilt, mica and illite; Gbs, gibbsite; Crs, cristobalite; Ab, albite; Ver, vermiculite; Chl, chlorite.

Figures 2B–D shows the Cs radioactivities in the KOI-E core sample and the activity ratio of $^{134}Cs/^{137}Cs$ as a function of depth from the soil surface. The Cs contamination on the side surface of the core derived from the coring process was removed. A large fraction of the radioactive Cs is retained at the top of the profile (**Figure 2A**); ^{137}Cs radioactivity in the 0–5 cm interval accounts for more than 98% of the total ^{137}Cs radioactivity. These data suggest that Cs has not migrated to deeper parts of the soil profile, even 1 year after the accident. Retention of Cs in upper soil is consistent with previous studies reporting results for other Fukushima soils collected 1 month after the accident (Kato et al., 2012; Tanaka et al., 2012a). The $^{134}Cs/^{137}Cs$ activity ratios in the depth profile are almost constant, ranging 0.8–0.9 (**Figure 2D**), which is also consistent with the results of Kinoshita et al. (2011).

Size Fraction of Cs Radioactivity in the Soils

In order to investigate the particle size dependence of Cs radioactivity, three soils (OTO-E, OTO-W, and KOI-E) were separated into five fractions, 597 µm–2 mm, 114–597, 48–114, 1–48, and <1 µm by a filtering or sedimentation method. The size-fractionated ^{137}Cs radioactivity revealed that relatively high ^{137}Cs is present in the smaller size fraction; <1 and <48 µm, while about 10^5 Bq/kg of ^{137}Cs still exist in the large particle fractions: 114–597 µm and 597 µm–2 mm (**Figure 3A**). The highest radioactivity was detected in the <1 µm fraction in OTO-E and KOI-E (**Figure 3B**). Specifically, the more than 3.0×10^6 Bq/kg of ^{137}Cs in this fraction from KOI-E accounts for 58% of the total radioactivity, confirming that the greatest proportion of the radioactivity is to be found in the submicron-sized particles. Because small particles bearing ^{137}Cs can aggregate and form a larger particle, a certain portion of radioactivity derived from submicron-sized particles must also occur in the large size fractions. In order to quantify the actual Cs radioactivity of the large size fractions, the same specimen of KOI-E was sieved for the size fraction, 597 µm–2 mm, 114–597, and 48–114 µm in **Figure 3** was ultrasonicated, and the radioactivities were measured again. As shown in **Figure 3C**, the Cs radioactivity of the three larger size fractions decreased dramatically by an order of magnitude. This significant

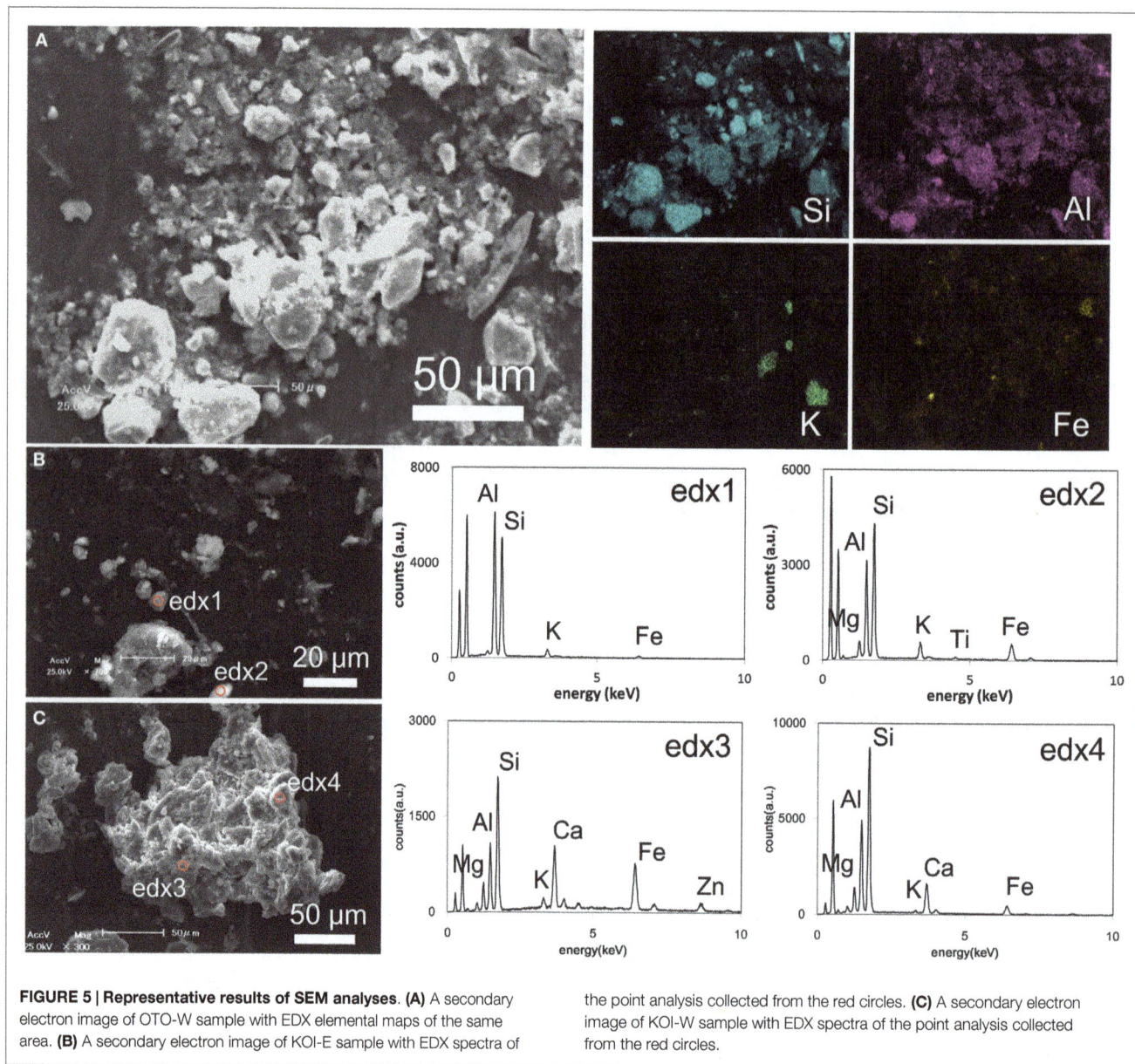

FIGURE 5 | Representative results of SEM analyses. (A) A secondary electron image of OTO-W sample with EDX elemental maps of the same area. (B) A secondary electron image of KOI-E sample with EDX spectra of the point analysis collected from the red circles. (C) A secondary electron image of KOI-W sample with EDX spectra of the point analysis collected from the red circles.

decrease of radioactivity by the ultrasonication confirms of the extensive aggregation and attachment of submicron-sized Cs particles to larger particles. The specific surface area, which is typically proportional to the number of surface adsorption site for ions, increases as the particle size decreases. However, in this case, the preferential distribution of Cs in the submicron-sized particle cannot be simply explained by the increased number of adsorption site for Cs. The cation exchange capacity, as determined for Fukushima soils is on the order of 1 ~ 5 cmol/kg (Iwata et al., 2012), which is eight orders of magnitude greater than the Cs concentrations. If 10^6 Bq/kg of ^{137}Cs occur in a soil, this corresponds to only 3×10^{-2} ppb. Thus, despite the great abundance of adsorption sites in the small particle fraction, only a small number of these sites are occupied by Cs. In the case of OTO-W, the ^{137}Cs radioactivities in the bulk and submicron fraction are 1.11×10^6 and 1.53×10^6 Bq/kg, respectively. This difference is not consistent with the difference in the surface areas that are simply estimated based on the particle size. Thus, it appears that the very high-surface area of the small particle fraction is not entirely responsible for the high level of radioactivity in very small particle size fractions.

Characterization of Soil Minerals

Major mineral phases in the three soils, OTO-E, OTO-W, and KOI-E, were determined by XRD analysis (**Figures 4A–C**). Quartz was dominant in the three samples. Other mineral phases vary depending on the soils; possible akaganeite in OTO-E, kaolinite, cristobalite, and albite in OTO-W, kaolinite, gibbsite, cristobalite, and albite in KOI-E. The XRD patterns of the elutriated samples containing only submicron-sized particles in OTO-W and KOI-E revealed that the intensity of the diffraction peaks of quartz and albite decreased, while gibbsite and 7 Å peaks were enhanced (**Figures 4D,E**). Slow-scan analysis of the elutriated sample of KOI-E at the low angles showed the characteristic diffraction peaks of clay minerals corresponding to 14, 10, and 7 Å (**Figure 4F**). The possible phases based on these diffraction maxima are vermiculite, smectite, and chlorite for 14 Å, micas for 10 Å, and kaolinite for 7 Å.

A SEM–EDX elemental map of OTO-W sample revealed that the presence of various minerals, such as quartz, feldspar, Fe-oxide, and aluminosilicate (**Figure 5A**). It was difficult to detect Cs peak because the low concentration adsorbed on the soil minerals was below the detection limit of EDX, ~0.1 wt%. Aluminosilicates were further analyzed for their composition (**Figures 5B,C**). The particle of edx1 may be kaolinite or zeolite. The edx2 represents biotite composition. The edx 3 and 4 likely indicate a mixture of biotite–vermiculite and possible Ca-carbonate. Wave-length dispersive X-ray analysis was also attempted on the sheet silicate minerals; however, Cs was not detected (data not shown).

A droplet of the elutriated suspension of OTO-W that contains submicron-sized particle was dried on a silicon wafer (**Figure 6**). The particles concentrated at the edge of droplet while drying due to the surface tension and arrayed as ring shape on the substrate. The autoradiograph of this sample revealed that the radioactive Cs is also distributed in the same manner as the submicron-sized particles that were confirmed in the SEM image (**Figures 6A,B**). The average chemical composition of these particles was Si, Al, Fe, and little K (**Figure 6C**). The SEM analysis in conjunction with the autoradiography indicates that radioactive Cs was mainly associated with the submicron-sized aluminosilicates.

The submicron-sized particles elutriated from OTO-E were further characterized by TEM. **Figures 7A–C** show HAADF–STEM images of an aggregate of the submicron-sized particles. A wide area view (**Figure 7A**) shows numerous flaky-shaped particles in the aggregate, which correspond to the texture of clay minerals like vermiculite. The flaky-shape particle appears

to consist of a few domains of layered particles with different contrast (**Figure 7B**). Single layered particles were analyzed by EDX point analysis (edx1 and 2), revealing that the major composition is consistent with that of vermiculite (**Figure 7C**). Unfortunately, the Cs concentration was too low to detect by EDX. **Figure 7D** is the elemental maps of **Figure 7A** revealing the dominant occurrence of aluminosilicate with some Fe and occasional little K. The Fe-rich and Ti-rich particles are likely Fe-oxide and Ti-oxides, respectively. Bright-field TEM image of the clay particles clearly shows the shape of the clay aggregates as large as a few hundred nanometer in length and several 10 nm in width (**Figure 7E**). A HRTEM image of individual particles revealed that tetrahedral–octahedral–tetrahedral layers continue only <10 layers with spacing of ~1.0 nm **Figure 7F**. These clays are most likely anhydrated vermiculite in conjunction with EDX analysis.

Irreversible Sorption of Cs on the Surface Soils

Figure 8A shows the results of sequential extraction for three samples: OTO-E, OTO-W, and KOI-E revealing that 3.95–21.4% of ^{137}Cs was leached by 1M NH_4OAc (pH 8) solution and more than 63% of Cs was retained in the insoluble fraction. This result is consistent with the previous studies reporting that a small fraction of radiocesium in the interlayer of phyllosilicate minerals is exchangeable with NH_4^+ and a large fraction is fixed in the structure, probably along frayed edges (Evans et al., 1983; Cremer et al., 1988; McKinley et al., 2004). Because the Cs in the aerosols emitted just after the accident were predominantly water soluble, 50–90% (Tanaka et al., 2012b), the soluble Cs was delivered by rain water and fixed in the structure of phyllosilicate minerals. Comans and Hockley (1992) studied the irreversibility of Cs sorbed onto the clay minerals and concluded that the highly selective Cs sorption site was at frayed edge sites (FES) that develop along the weathered edges of micas. Their immediate weathering products are hydrous-mica and illite. At the FES, wedge zones form, and

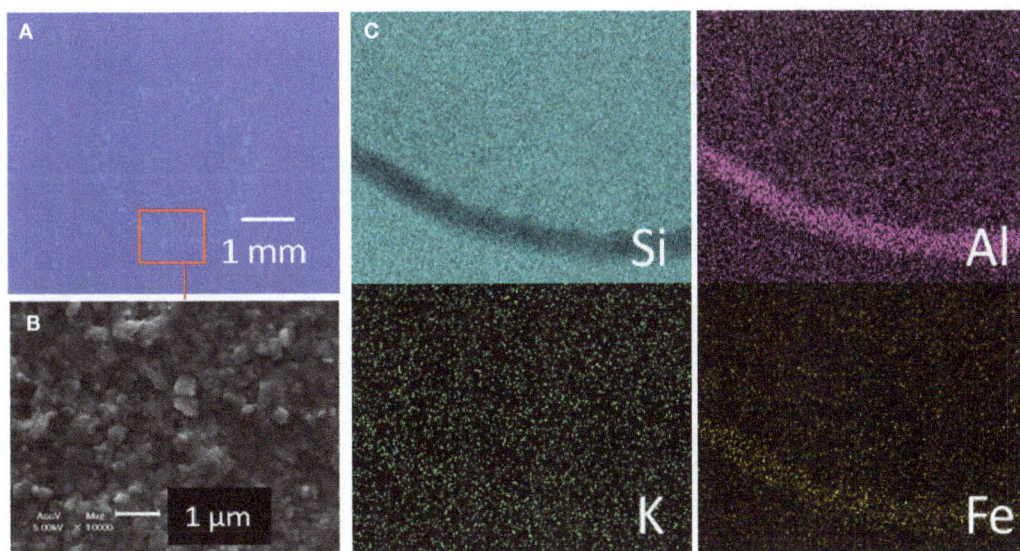

FIGURE 6 | (A) Autoradiography of a droplet of the suspended submicron-sized particles elutriated from OTO-W sample, which was dried on a smooth Si-metal plate. **(B)** A secondary electron image of a part of the area indicated by the red square in **(A)**. **(C)** A SEM–EDX elemental map obtained from the area indicated by the red square in **(A)**.

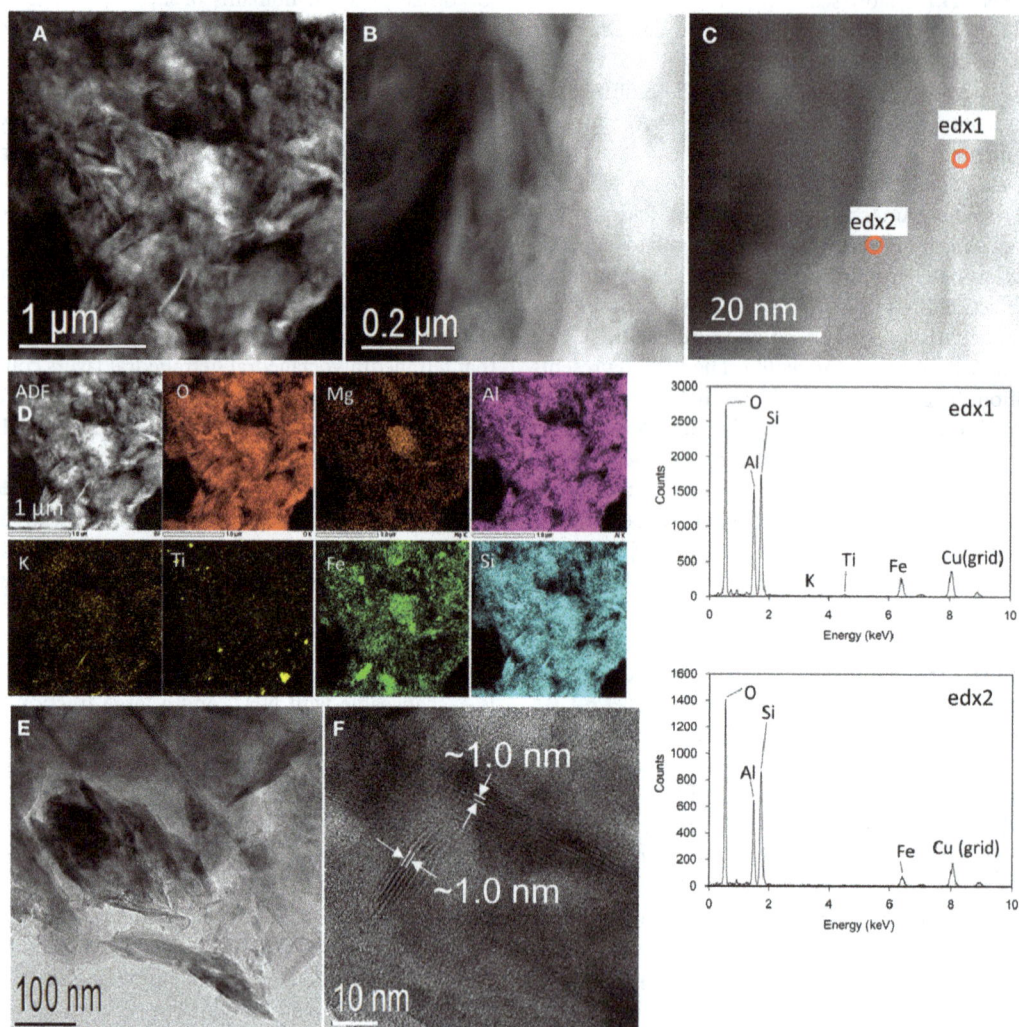

FIGURE 7 | Results of TEM analyses of the submicron-sized particles elutriated from OTO-E. (A–C) HAADF-STEM images of an aggregate of the submicron-sized particles. **(A)** A wide view, **(B)** an enlarged image of **(A)** showing flaky-shaped particles, and **(C)** a further enlarged image indicating the position of EDX point analyses. **(D)** Elemental maps of the area imaged in **(A)**. **(E)** Bright-field TEM image of the clay particles. **(F)** A HRTEM image of individual particles showing the lattice spacing of ~1.0 nm.

the 1.0 nm mica sheet opens to 1.4 nm vermiculite, providing a stereoselective environment for Cs^+ sorption (Jackson, 1963). The selectivity coefficients of Cs relative to the other cations (Na^+, K^+, and Ca^{2+}) expressed as $logK_c$, ($K_c = [A^{u+}][E_{CsX}]^u/[Cs^+]^u[E_{AXu}]$, where E_{CsX} and E_{AXu} represent adsorption density divided by CEC for Cs and cation A) estimated to be up to ~6.76 based on the ion exchange models (Bradbury and Baeyens, 2000; Zachara et al., 2002), stating the high affinity of Cs with sheet aluminosilicate. Hence, in the soils at Fukushima, similar Cs-binding mechanisms may play a role in fixing Cs onto aluminosilicate sheet structures; hence, the release of radioactive Cs from soil minerals should be minimal. A limited amount of Cs can be released by an ion exchange process, but it is expected that the released soluble Cs will be immediately adsorbed onto aluminosilicate minerals with a sheet structure, that is clays.

Sequential extraction was also conducted for the submicron-sized particle elutriated from OTO-W sample (**Figure 8B**).

The chemical form of Cs in the submicron-sized particle fraction is similar to that in the bulk, although the major mineral constituents were different from that of the bulk. Quartz and feldspar were major minerals in bulk; whereas, clays and mica-like minerals were dominant in the submicron-sized fraction. The results of sequential extraction indicate that Cs is predominantly adsorbed onto the clays in the bulk sample, as well as in the submicron-sized fraction, because the number of adsorption site on the sheet structure of the aluminosilicate clays is much greater than the amount of Cs available. The sequential extraction for Cs in Fukushima soils probably represents the chemical form of Cs adsorbed on the submicron-sized sheet alumino silicate.

Role of Submicron-Sized Particles as Cs Carrier

In general, submicron-sized particles in aquifer behave as colloids, which have unique properties in the subsurface. Colloid-facilitated transport of toxic elements is typically ascribed to

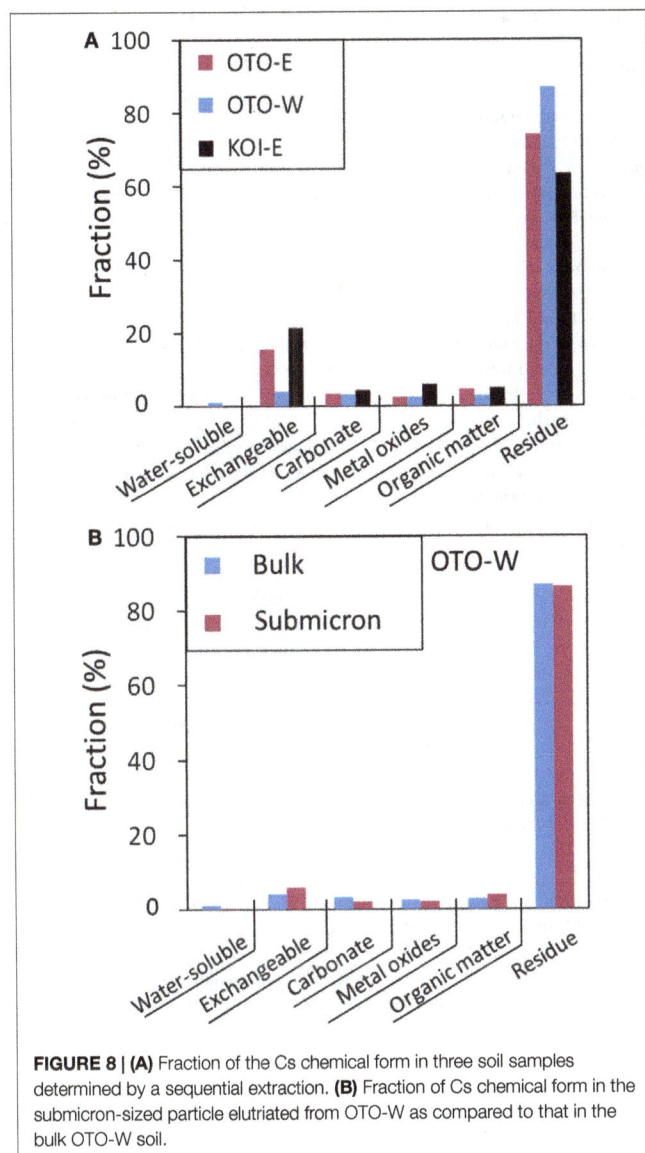

FIGURE 8 | (A) Fraction of the Cs chemical form in three soil samples determined by a sequential extraction. **(B)** Fraction of Cs chemical form in the submicron-sized particle elutriated from OTO-W as compared to that in the bulk OTO-W soil.

did not occur for the period from March 2011 to March 2012. As shown in the results above, a large fraction of Cs was adsorbed onto the submicron-sized aluminosilicates with sheet structures, that is clays, and although they are potentially mobile, their movement has been limited because the small particles have aggregated due to the fact that the pH of the ground water is near neutral and the point of zero charge (PZC) of sheet aluminosilicate is ~6–7 [based on the PZC of biotite as reported in Goldberg et al. (2012)]. Formation of aggregates has prevented the Cs particles from being transported vertically through the soils due to a filtering effect. Rather, the aggregates of small Cs particles currently migrate through surface water. Indeed, 90–98% of the radioactive Cs is transported as suspended fine particle from hydrographic basin to rivers, which is supposed to be a major dose contributor of the Cs flux to the ocean: 200–500 GBq/year for 5 years after the accident (JAEA, 2013). As a consequence, transport of Cs particle aggregates through surface water should have been responsible for the removal of Cs radioactivity of surface environment to the lower level than that simply calculated by radioactive decay of ^{134}Cs.

Conclusion

The submicron-sized particles in the contaminated soils near Fukushima are mainly composed of aluminosilicates with sheet structures, clays. These clays, mica, vermiculite, chlorite, and smectite, are responsible for adsorbing more than 70% of the total Cs in the soil profile. The adsorbed Cs does not desorb easily, probably because of strong binding at FES of the clays (Iwata et al., 2012). Aggregation of the very small clay particles into larger particles has further limited their migration downward through the soil profile; hence, most of the radioactivity remains within the top 5 cm of the soil profile. Currently, the mobility of the aggregates of submicron-sized sheet aluminosilicate in the surface environment is a key factor controlling the Cs migration in Fukushima. Still, the aggregates can be potentially transported as suspended colloids in case of deflocculation in a specific geochemical condition of ground water, and this should be considered in transport models that predict future migration of Cs in these highly contaminated regions.

Acknowledgments

This study is partially supported by JST Initiatives for Atomic Energy Basic and Generic Strategic Research and by the Science Grant of the Ministry of Education, Science and Culture. SU is also supported by ESPEC Foundation for Global Environment Research and Technology (Charitable Trust) (ESPEC Prize for the Encouragement of Environmental Studies). SU is also grateful to Dr. Watanabe for her assistance on SEM and XRD analyses at the Center of Advanced Instrumental Analysis, Kyushu University.

hydrodynamic chromatographic effects (McCarthy and Zachara, 1989; Kim, 1991). In the heavily contaminated soils of the present study, the ^{137}Cs radioactivity was ~3,000,000 Bq/kg on the colloids and these submicron-sized particles can potentially contribute to the migration of radioactive Cs as pseudo-colloids. Short range migration of Cs particles might occur only within top of the profile as reported in Ohnuki and Tanaka (1989) and Tanaka and Ohnuki (1996). However, the particles that have sorbed the radioactive Cs remain within the top ~5 cm of vertical profile (**Figure 2**), and the amount of radioactivity in groundwater collected from a well at Iitate village was below the detection limit (data not shown). This suggests that Cs migration through vadose zone to groundwater

References

Bradbury, M., and Baeyens, B. (2000). A generalised sorption model for the concentration dependent uptake of caesium by argillaceous rocks. *J. Contam. Hydrol.* 42, 141–163. doi:10.1016/S0169-7722(99)00094-7

Comans, R., and Hockley, D. (1992). Kinetics of cesium sorption on illite. *Geochim. Cosmochim. Acta* 56, 1157–1164. doi:10.1016/0016-7037(92)90053-L

Cremer, A., Elsen, A., De Preter, P., and Maes, A. (1988). Quantitative analysis of radiocaesium retention in soils. *Nature* 335, 247–249. doi:10.1038/335247a0

Evans, D., Alberts, J., and Clark, R. (1983). Reversible ion-exchange fixation of cesium-137 leading to mobilization from reservoir sediments. *Geochim. Cosmochim. Acta* 47, 1041–1049. doi:10.1016/0016-7037(83)90234-X

Goldberg, S. R., Lebron, I., Seaman, J. C., and Suarez, D. L. (2012). "Chapter 15: soil colloidal behavior," in *Handbook of Soil Sciences Properties and Processes*, 2nd Edn, eds P. M. Huang, Y. Li, and M. E. Sumner (Boca Raton, FL: CRC Press, Taylor and Francis Group), 15-1–15-39.

Hou, X., Fogh, C., Kucera, J., Andersson, K., Dahlgaard, H., and Nielsen, S. (2003). Iodine-129 and caesium-137 in chernobyl contaminated soil and their chemical fractionation. *Sci. Total Environ.* 308, 97–109. doi:10.1016/S0048-9697(02)00546-6

Iwata, H., Shiotsu, H., Kaneko, M., and Utsunomiya, S. (2012). "Nuclear accidents in Fukushima, Japan and exploration of effective decontaminant for the ^{137}Cs-contaminated soils," in *Advances in Nuclear Fuel*, ed. T. R. Shripad (Rijeka: Intech), 123–142.

Jackson, M. L. (1963). Interlayering of expansible layer silicates in soils by chemical weathering. *Clays Clay Miner.* 11, 29–46. doi:10.1346/CCMN.1962.0110104

JAEA. (2013). Available at: http://fukushima.jaea.go.jp/initiatives/cat03/entry06.html

Kaneyasu, N., Ohashi, H., Suzuki, F., Okuda, T., and Ikemori, F. (2012). Sulfate aerosol as a potential transport medium of radiocesium from the Fukushima nuclear accident. *Environ. Sci. Technol.* 46, 5720–5726. doi:10.1021/es204667h

Kato, H., Onda, Y., and Teramage, M. (2012). Depth distribution of Cs-137, Cs-134, and I-131 in soil profile after Fukushima Dai-Ichi nuclear power plant accident. *J. Environ. Radioact.* 111, 59–64. doi:10.1016/j.jenvrad.2011.10.003

Kim, J. I. (1991). Actinide colloid generation in groundwater. *Radiochim. Acta* 52/53, 71–81.

Kinoshita, N., Sueki, K., Sasa, K., Kitagawa, J., Ikarashi, S., Nishimura, T., et al. (2011). Assessment of individual radionuclide distributions from the Fukushima nuclear accident covering central-east Japan. *Proc. Natl. Acad. Sci. U.S.A.* 108, 19526–19529. doi:10.1073/pnas.1111724108

Koarashi, J., Atarashi-Andoh, M., Matsunaga, T., Sato, T., Nagao, S., and Nagai, H. (2012). Factors affecting vertical distribution of Fukushima accident-derived radiocesium in soil under different land-use conditions. *Sci. Total Environ.* 431, 392–401. doi:10.1016/j.scitotenv.2012.05.041

Kozai, N., Ohnuki, T., Arisaka, M., Watanabe, M., Sakamoto, F., Yamasaki, S., et al. (2012). Chemical states of fallout radioactive Cs in the soils deposited at Fukushima Daiichi nuclear power plant accident. *J. Nucl. Sci. Technol.* 49, 473–478. doi:10.1080/00223131.2012.677131

Matsunaga, T., Koarashi, J., Atarashi-Andoh, M., Nagao, S., Sato, T., and Nagai, H. (2013). Comparison of the vertical distributions of Fukushima nuclear accident radiocesium in soil before and after the first rainy season, with physicochemical and mineralogical interpretations. *Sci. Total Environ.* 447, 301–314. doi:10.1016/j.scitotenv.2012.12.087

McCarthy, J., and Zachara, J. (1989). Subsurface transport of contaminants – mobile colloids in the subsurface environment may alter the transport of contaminants. *Environ. Sci. Technol.* 23, 496–502. doi:10.1021/es00063a602

McKinley, J., Zachara, J., Heald, S., Dohnalkova, A., Newville, M., and Sutton, S. (2004). Microscale distribution of cesium sorbed to biotite and muscovite. *Environ. Sci. Technol.* 38, 1017–1023. doi:10.1021/es034569m

MEXT. (2013). Available at: http://radioactivity.nsr.go.jp/ja/contents/7000/6749/24/191_258_0301_18.pdf

Morino, Y., Ohara, T., Watanabe, M., Hayashi, S., and Nishizawa, M. (2013). Episode analysis of deposition of radiocesium from the Fukushima Daiichi nuclear

power plant accident. *Environ. Sci. Technol.* 47, 2314–2322. doi:10.1021/es304620x

Mukai, H., Hatta, T., Kitazawa, H., Yamada, H., Yaita, T., and Kogure, T. (2014). Speciation of radioactive soil particles in the Fukushima contaminated area by IP autoradiography and microanalyses. *Environ. Sci. Technol.* 48, 13053–13059. doi:10.1021/es502849e

Nakao, A., Ogasawara, S., Sano, O., Ito, T., and Yanai, J. (2014). Radiocesium sorption in relation to clay mineralogy of paddy soils in Fukushima, Japan. *Sci. Total Environ.* 46, 523–529. doi:10.1016/j.scitotenv.2013.08.062

Ohnuki, T., and Kozai, N. (2013). Adsorption behavior of radioactive cesium by non-mica minerals. *J. Nucl. Sci. Technol.* 50, 369–375. doi:10.1080/00223131.2013.773164

Ohnuki, T., and Tanaka, T. (1989). Migration of radionuclides controlled by several different migration mechanisms through a sandy soil layer. *Health Phys.* 56, 47–53. doi:10.1097/00004032-198901000-00004

Steinhauser, G., Brandl, A., and Johnson, T. E. (2014). Comparison of the chernobyl and Fukushima nuclear accidents: a review of the environmental impacts. *Sci. Total Environ.* 47, 800–817. doi:10.1016/j.scitotenv.2013.10.029

Tanaka, K., Takahashi, Y., Sakaguchi, A., Umeo, M., Hayakawa, S., Tanida, H., et al. (2012a). Vertical profiles of iodine-131 and cesium-137 in soils in Fukushima prefecture related to the Fukushima Daiichi nuclear power station accident. *Geochem. J.* 46, 73–76. doi:10.2343/geochemj.1.0137

Tanaka, K., Sakaguchi, A., Kanai, Y., Tsuruta, H., Shinohara, A., and Takahashi, Y. (2012b). Heterogeneous distribution of radiocesium in aerosols, soil and particulate matters emitted by the Fukushima Daiichi nuclear power plant accident: retention of micro-scale heterogeneity during the migration of radiocesium from the air into ground and river systems. *J. Radioanal. Nucl. Chem.* 295, 1927–1937. doi:10.1007/s10967-012-2160-9

Tanaka, T., and Ohnuki, T. (1996). Colloidal migration behavior of radionuclides sorbed on mobile fine soil particles through a sand layer. *J. Nucl. Sci. Technol.* 33, 62–68. doi:10.1080/18811248.1996.9731862

Tessier, A., Campbell, P. G. C., and Bisson, M. (1979). Sequential extraction procedure for the speciation of particulate trace metals. *Anal. Chem.* 51, 844–851. doi:10.1021/ac50043a017

Yoshida, N., and Takahashi, Y. (2012). Land-surface contamination by radionuclides from the Fukushima Daiichi nuclear power plant accident. *Elements* 8, 201–206. doi:10.2113/gselements.8.3.201

Zachara, J., Smith, S., Liu, C., McKinley, J., Serne, R., and Gassman, P. (2002). Sorption of Cs^+ to micaceous subsurface sediments from the Hanford Site, USA. *Geochim. Cosmochim. Acta* 66, 193–211. doi:10.1016/S0016-7037(01)00759-1

Conflict of Interest Statement: The authors declare that the research was conducted in the absence of any commercial or financial relationships that could be construed as a potential conflict of interest.

12

Business Model Innovation for Local Energy Management: A Perspective from Swiss Utilities

Emanuele Facchinetti[1]*, Cherrelle Eid[2], Andrew Bollinger[3] and Sabine Sulzer[1]

[1] Lucerne Competence Center for Energy Research, Lucerne University of Applied Science and Arts, Horw, Switzerland, [2] Faculty of Technology, Policy and Management, Delft University of Technology, Delft, Netherlands, [3] Urban Energy Systems Laboratory, EMPA, Dübendorf, Switzerland

Edited by:
Sgouris Sgouridis,
Masdar Institute of Science and
Technology, United Arab Emirates

Reviewed by:
Xu Tang,
China University of Petroleum, China
Dimitrios Angelopoulos,
National Technical University
of Athens, Greece
Haris Doukas,
National Technical University
of Athens, Greece

*Correspondence:
Emanuele Facchinetti
emanuele.facchinetti@hslu.ch

Specialty section:
This article was submitted
to Energy Systems and Policy,
a section of the journal
Frontiers in Energy Research

Citation:
Facchinetti E, Eid C, Bollinger A and
Sulzer S (2016) Business Model
Innovation for Local
Energy Management:
A Perspective from Swiss Utilities.
Front. Energy Res. 4:31.

The successful deployment of the energy transition relies on a deep reorganization of the energy market. Business model innovation is recognized as a key driver of this process. This work contributes to this topic by providing to potential local energy management (LEM) stakeholders and policy makers a conceptual framework guiding the LEM business model innovation. The main determinants characterizing LEM concepts and impacting its business model innovation are identified through literature reviews on distributed generation typologies and customer/investor preferences related to new business opportunities emerging with the energy transition. Afterwards, the relation between the identified determinants and the LEM business model solution space is analyzed based on semi-structured interviews with managers of Swiss utilities companies. The collected managers' preferences serve as explorative indicators supporting the business model innovation process and provide insights into policy makers on challenges and opportunities related to LEM.

Keywords: local energy management, energy hub, business models, business innovation, distributed generation, renewable energy

INTRODUCTION

The European ambitions for attaining sustainability targets are visible in the policies and measures deployed by the European Commission to achieve its 2020 and 2030 objectives for emissions reduction, energy efficiency, and increase in share of renewable energy (European Commission, 2012, 2014) and in the recent adoption of the Energy Union strategy (European Commission, 2015). However, in force EU policies will be insufficient to achieve the long-term target defined in the European Commission' Energy Road Map 2050 (European Commission, 2011). Countries leading the energy transition, such as Germany and Switzerland, adopted policies even more ambitious setting clear and long-term targets for 2050, including substantial reduction in primary energy consumption and carbon emission, increase of renewable energy share, and the phase out of nuclear power (BMWI, 2010; SFOE, 2013; Markard et al., 2015).

In this respect, a range of supportive energy policies, e.g., feed-in-tariffs and subsidies, favoring bottom-up investments in renewable energy and energy efficiency measures has been put in place (Anaya and Pollitt, 2015). Virtuous examples are the penetration of solar photovoltaic in

---

Abbreviations: LEM, local energy management.

Germany (EPIA, 2014) and combined heat and power in many other European countries (Lund and Münster, 2006; Toke and Fragaki, 2008; Fragaki and Andersen, 2011). As a result of such policies, the increasing market penetration of distributed generation – based on renewables or favoring energy efficiency in fossil-based energy systems – is observed in many industrialized countries (IEA, 2014; Anaya and Pollitt, 2015) and is expected to significantly transform the energy supply value chain (Schleicher-Tappeser, 2012).

The intrinsic technological distinctness between distributed generation and traditional centralized generation is reflected on both the economic and organizational perspectives. As discussed by Schleicher-Tappeser (2012), distributed generation challenges traditional utility business models opening up new business opportunities to horizontal integrate diverse energy services – i.e., including electricity supply, cooling, heating, and additional energy-related services – and increasing autonomy and flexibility of customers. Such customer-oriented multi-services approaches require to be addressed by appropriate innovative business models. The capability, on the one hand, of policy makers to facilitate the transition with effective regulations and, on the other hand, of market players to develop successful business models will substantially affect the speed and effectiveness of the energy transition (Schleicher-Tappeser, 2012).

The role of business model innovation in supporting the required fundamental change of value proposition and value creation logic to generally promote the energy transition has been acknowledged by a number of recent scientific works. Loock (2012) reported the results of choice experiments with investment managers for renewable energy aiming to identify which business models could succeed in the market. The study showed that business models focusing on customers and proposing high-quality services are considered more attractive than business models oriented to low price and best technologies. Richter (2012) reviewed the existing business model approaches adopted by utilities with regard to renewable energy. The results showed that even though business models focusing on large projects insure a better risk–benefit compromise, utilities should urge to invest in business model innovation to take advantage of the forthcoming business opportunities related to smaller distributed generation projects. The same author (Richter, 2013b) analyzed the attitude of German utilities with respect to photovoltaic-based distributed generation and showed that utilities tended to fail perceiving photovoltaic generation as a new business opportunity and identified the causes behind this fact. Furthermore, it has been found that creating separated units within the company to address new businesses and to emphasize external partnerships can effectively foster the business model innovation process. The establishment of collaboration between distributed generation firms has been acknowledged as a key driver fostering business model innovation also by Hellström et al. (2015). In addition, the authors concluded that business models for distributed energy systems are not only the outcomes of a decision-making process across a certain number of options dictated by internal and external conditions but instead also a continuously active process aiming to keep the local business ecosystem profitable.

As presented in this concise literature review, the key role of business model innovation in fostering the energy transition has been described from different perspectives in the available scientific literature. Nevertheless clear gaps remain in the identification of specific business model patterns applicable in the energy transition context. Research addressing this topic has been initiated in a previous work presenting a heuristic methodology easing the identification of business model patterns best suited for local energy management (LEM) – the management of energy supply, demand and storage within a given geographical area (Facchinetti and Sulzer, 2016). Building upon the previously developed conceptual framework, the present contribution aims to specifically identify and explore the impact of the main determinants (or factors) that should be considered by stakeholders undertaking a business model innovation process for an intended LEM concept. First, the main determinants have been selected through the analysis of the existing scientific literature on distributed generation typologies and related customer/investor preferences. In a second step, the impacts of the identified determinants on the business model innovation process have been investigated via the implementation of semi-structured interviews with utility managers.

Utility companies, the focus of attention for this work, are on the edge of the energy transition and face the difficult challenge of innovating their business model in accordance with a very changing environment (Sühlsen and Hisschemöller, 2014). In particular, Swiss utilities, which have been involved in the present investigation, are currently exposed to a dual challenge. On the one hand, Switzerland is one of the countries leading the energy transition with its Energy Strategy 2050 (SFOE, 2013). On the other hand, Switzerland is still in the process of achieving full market liberalization. At the moment, locally established (city or canton level) utilities control the Swiss energy retail market. However, they are being prepared to face the upcoming full market liberalization.

The outcome of this study is a set of Swiss utility manager's preference indications providing an orientation toward the most appropriate business model pattern(s) to select for different LEM concepts. The present work represents a starting point toward the characterization of aspects driving the business model innovation process for LEM concepts.

The paper is organized as follows. In Section "Methodology," the applied conceptual framework and research methodology are described. In Section "Results," the results of the main determinants selection and interviews with utility managers are presented. The results are discussed in Section "Discussion." Finally, in Section "Conclusion," conclusions and policy implications are outlined.

METHODOLOGY

The applied methodological framework is depicted in **Figure 1**. In Section "The Conceptual Business Model Solution Space," the conceptual business model solution space for LEM developed in a previous work is summarized. In Section "The Selection of the Determinants," the methodology applied for the selection of the determinants is described. Finally, in Section "The Utility

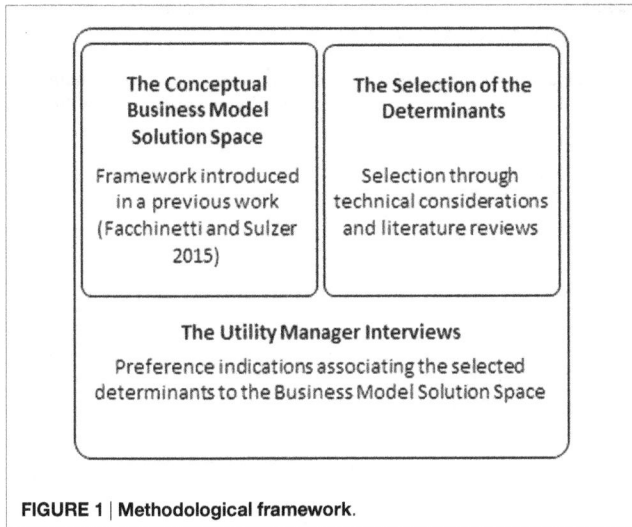

FIGURE 1 | Methodological framework.

The Conceptual Business Model Solution Space

Framework introduced in a previous work (Facchinetti and Sulzer 2015)

The Selection of the Determinants

Selection through technical considerations and literature reviews

The Utility Manager Interviews

Preference indications associating the selected determinants to the Business Model Solution Space

TABLE 1 | The business model reference patterns within the business model solution space (Facchinetti and Sulzer, 2016).

		Delivery of energy services		
		Basic services (no frills)	Tailored services (user designed)	High-quality comprehensive services (experience selling)
Procurement of infrastructure/operation and control	Leasing to customers (rent instead of buying)			Pattern III
	Shared ownership (fractional ownership)		Pattern II	
	Customer ownership (orchestrator)	Pattern I		

Manager Interviews," the research approach applied for the manager's interviews is outlined.

Prior to describing the methodological framework, the considered definition of business model and business model innovation are here introduced. In this work, the term business model refers to the definition proposed by Osterwalder and Pigneur (2010): "the rationale of how an organization creates, delivers, and captures value." This definition is widely accepted within the literature and in particular in the energy sector (Okkonen and Suhonen, 2010; Richter, 2012; Facchinetti and Sulzer, 2016). Osterwalder and Pigneur fully characterize a business model based on four elements, namely the value proposition, describing the offer; the customers, characterizing the customer targets; the infrastructures, including all means required to deliver the value proposition; and, finally, the financial viability, explaining how profit is generated.

Business model innovation is defined as a discipline supporting the change of value proposition to customers (Bocken et al., 2014), involving the change of the way a business is run (Zott and Amit, 2013), and considering a large number of stakeholders and a broad value-network going well beyond the existing firm perspective (Beattie and Smith, 2013; Zott and Amit, 2013).

The Conceptual Business Model Solution Space

In a previous publication (Facchinetti and Sulzer, 2016), the authors developed a conceptual framework characterizing the LEM business model solution space. The conceptual framework supports LEM business model innovation offering a structured and comprehensive overview on available business model patterns. The business model pattern is defined as a portion of the solution space and it is characterized by a number of potentially applicable business model ideas organized with respect to the different steps of the LEM value chain. The business model ideas comprised in the selected business model solution space portion can be used to develop specific business models suitable to the intended LEM concept. In order to put into context and describe the defined solution space,

three reference patterns spanning across the solution space are defined and compared.

The business model solution space and the location of the three reference patterns are depicted in **Table 1**. The solution space is defined with respect to the available options on the procurement of infrastructures and on the delivery of energy services sides. On the procurement of infrastructures side, from the LEM perspective, the available options span from the customers owning infrastructures (*Orchestrator*), shared ownership between LEM and customers (*Fractional Ownership*), to leasing the LEM owned infrastructures to the customers (*Rent instead of Buying*). On the delivery of energy services side, the alternatives range from offering essential services (*No Frills*), offering customized services (*User Designed*), to offering comprehensive high-quality services (*Experience Selling*).

Following the business model definition of Osterwalder and Pigneur (2010), the key features characterizing the three reference business model patterns are presented in **Table 2**. The bottom line of Pattern I is to focus on the operation and control of third party owned infrastructures and to offer basic quality services to customers. In Pattern II, the LEM shares infrastructure ownership with the customers and offers them personalized solutions compatible with their own infrastructures and needs. Within Pattern III, the LEM offers to the customer the possibility to lease all-inclusive turnkey solutions, including high-quality services going beyond mere energy supply. For further details on the business model patterns and the business model ideas, refer to Facchinetti and Sulzer (2016).

The Selection of the Determinants

The first objective of the present work is the selection of the most relevant boundary conditions characterizing the business model innovation process. Such selected boundary conditions are defined as determinants. The determinants have been organized into three categories: (1) related to the LEM typology, including infrastructures and building characteristics; (2) related to customers, including socio-demographic aspects; and (3) related to the external determinants. Within each

TABLE 2 | Main features of the identified reference business model patterns (Facchinetti and Sulzer, 2016).

Reference business model patterns	Pattern I (orchestrator-no frills)	Pattern II (fractional ownership – user designed)	Pattern III (rent instead of buying – experience selling)
Value proposition	Multiside platform connecting consumers, prosumers, and energy market	Tailored energy services adaptable and complementary to customer infrastructures	Comprehensive turnkey solutions going beyond energy services
Customers	Cost-sensitive customers Prosumers owning the infrastructures	Customers participating to the infrastructure investments	Customers inclined to pay higher prices to get the best service quality
Infrastructures	No investment on infrastructures Strong partnerships	Infrastructure ownership shared with customers	Owned infrastructures leased to customers
Financial viability	Revenues from energy trading only	Revenues from energy trading and service on infrastructures	Revenues from energy trading, leasing and additional services

TABLE 3 | List of Swiss utilities participating to the interviews.

Utility company	Main area of operation	Revenues 2014 (MCHF)
BKW	Bern Canton	2844
SIG	Geneva Canton	1022
EWZ	Zurich city	791
Groupe E	Fribourg Canton	645
Romande Energie	Vaud Canton	583
IBAARAU	Aarau city	147
SHPOWER	Schaffhausen Canton	112
EnergieThun	Thun city	78

TABLE 4 | Overview on selected main determinants and manager's preference indications.

Determinant categories	Determinants	Literature review references	Manager's preference indications
LEM typology	Project typology	Chicco and Mancarella (2009) and Mancarella (2014)	**Table 5**
	Density of buildings	Mancarella (2014) and Lund et al. (2014)	**Table 6**
	Buildings typology	Mancarella (2014) and Lund et al. (2014)	**Table 7**
	Energy conversion infrastructures	Chicco and Mancarella (2009), Orehounig et al. (2015), and Fazlollahi et al. (2015)	**Table 8**
	Self-sufficiency level	Chicco and Mancarella (2009), Orehounig et al. (2015), and Lasseter (2011)	**Table 9**
Customer socio-demographic	Willingness to pay	Kaufmann et al. (2013) and Sagebiel et al. (2014)	**Table 10**
	Customer awareness	Curtius et al. (2012), Kaufmann et al. (2013), and Tabi et al. (2014)	**Table 11**
	Building ownership	Sagebiel et al. (2014) and Ebers and Wüstenhagen (2015)	**Table 12**
External determinants	Energy policy	Provance et al. (2011), Yildiz et al. (2015), Bürer and Wüstenhagen (2009), and Wüstenhagen and Menichetti (2012)	**Table 13**
	Macro-economy	Hofman and Huisman (2012), Masini and Menichetti (2012), and Masini and Menichetti (2013)	**Table 14**

category a number of determinants have been selected. The LEM typology determinants have been derived from technical considerations based on the available scientific literature on distributed generation typologies aiming to generally cover all possible LEM typologies. The review of the existing literature on customer and investor preferences related to new business opportunities emerging with the energy transition have supported the selection of, respectively, customer-related and external determinants.

The Utility Manager Interviews

In order to investigate the influence of the selected determinants on the business model innovation process, managers of utility companies have been involved in an explorative qualitative research approach. Explorative qualitative approaches are suitable to early stage research (Silvermann, 2009), such as the one on business models for the energy transition (Richter, 2013a).

Ten managers from eight among the largest Swiss utilities were involved. The eight utilities cover more than one-third of the Swiss energy retail market. The represented utility

companies are listed in **Table 3** with their main area of operation and revenues. The managers were identified through their collaboration with the Swiss Competence Centers for Energy Research (Swiss Commision for Technology and Innovation, 2014), the Swiss national energy research program under which this project has been carried out. The participants included asset, business development, marketing, and product managers or directors[1]. The variety of the participant's business function and the differences in size and operational region of the represented companies ensured a representative and consistent sample suiting the explorative purposes of this qualitative analysis.

The managers were collectively invited to attend a workshop structured in two sessions. In a first common session, the managers were informed about the research project objectives and research methodology applied. In a second session, they were split into two groups and received a semi-structured questionnaire, including around 20 close-ended questions referring to the identified determinants. Going through the questionnaire, the moderators of each group introduced each question and confirmed in a preliminary discussion the understanding of the query. Afterwards each manager was asked to independently

---

[1] Two managers belonged to the upper management, seven to the middle management, and one to the lower management.

provide his/her answer(s). The answers were first collected and then discussed to reveal the reasons behind the decisions[2].

The data collected during the workshop were elaborated in three steps. The first step focused on the collection of the answers/reasons per question across the two groups. In the second step, the answers for each question were compared and general trends were identified. In the third step, making use of the collected answer explanations, the identified trends were put into context and substantiated to derive preference indications specific to each determinant. The applied explorative qualitative research approach does not allow for drawing statistically relevant conclusions. The results of such data processing are presented in the next section.

RESULTS

This section explores the outcomes of the selection of the main determinants and of the utility managers interviews. The determinants are organized into three categories: related to the LEM typology, related to the socio-demographic characterization of customers, and related to external aspects. The identified determinants are summarized in **Table 4**.

Three sections focusing on each determinant category are presented hereafter. Within these sections, first, the selection of the specific determinants is discussed and each determinant characterized. Afterwards, the association of the determinants to the business model solution space, derived from the manager interviews, is outlined in form of preference indication (**Tables 5–14**).

LEM Typology Determinants
Determinants Selection
This determinant category comprises the LEM typology-related features with a potential impact on the business model innovation process. In this respect, based on basic technical considerations and considering the available literature referring to distributed generation typologies (Chicco and Mancarella, 2009), including general concepts as smart grids (Mancarella, 2014), energy hubs (Orehounig et al., 2015), micro grids (Lasseter, 2011), and district energy systems (Lund et al., 2014;

---

[2]Answers and explanations are kept anonymous by request of the participants.

Fazlollahi et al., 2015), the LEM typologies are classified with respect to the following characteristics: *project type* (Chicco and Mancarella, 2009; Mancarella, 2014), *density of buildings* (Lund et al., 2014; Mancarella, 2014), *building typologies* (Lund et al.,

TABLE 6 | Density of buildings.

			Delivery of energy services		
			Basic services	Tailored services	High-quality comprehensive services
Procurement of infrastructure/ operation and control		Leasing to customers		High density of buildings	
		Shared ownership			
		Customer ownership		Low density of buildings	

TABLE 7 | Buildings typology.

			Delivery of energy services		
			Basic services	Tailored services	High-quality comprehensive services
Procurement of infrastructure/ operation and control		Leasing to customers			Heterogeneous buildings typology
		Shared ownership	Homogeneous buildings typology		
		Customer ownership			

TABLE 8 | Energy conversion infrastructures.

			Delivery of energy services		
			Basic services	Tailored services	High-quality comprehensive services
Procurement of infrastructure/ operation and control		Leasing to customers			Multi energy carriers
		Shared ownership	Single energy carrier		
		Customer ownership			

TABLE 5 | Project typology.

			Delivery of energy services		
			Basic services	Tailored services	High-quality comprehensive services
Procurement of infrastructure/ operation and control		Leasing to customers			New development projects
		Shared ownership		Retrofit projects	
		Customer ownership			

TABLE 9 | Self-sufficiency.

			Delivery of energy services		
			Basic services	Tailored services	High-quality comprehensive services
Procurement of infrastructure/ operation and control		Leasing to customers		High self-sufficiency	
		Shared ownership			
		Customer ownership			

TABLE 10 | Willingness to pay.

		Delivery of energy services		
		Basic services	Tailored services	High-quality comprehensive services
Procurement of infrastructure/ operation and control	Leasing to customers	Price sensitive customers		Higher willingness to pay customers
	Shared ownership			
	Customer ownership			

TABLE 11 | Customer awareness.

		Delivery of energy services		
		Basic services	Tailored services	High-quality comprehensive services
Procurement of infrastructure/ operation and control	Leasing to customers	Skeptical customers		
	Shared ownership		Sustainability-oriented customers	
	Customer ownership	Skeptical customers		

TABLE 12 | Building ownership.

		Delivery of energy services		
		Basic services	Tailored services	High-quality comprehensive services
Procurement of infrastructure/ operation and control	Leasing to customers		Real estates ownership	
	Shared ownership	Cooperatives ownership		
	Customer ownership		House owners	

TABLE 13 | Energy Policy.

		Delivery of energy services		
		Basic services	Tailored services	High-quality comprehensive services
Procurement of infrastructure/ operation and control	Leasing to customers			Supporting policies
	Shared ownership			
	Customer ownership	Not supporting policies		

TABLE 14 | Macro economy.

		Delivery of energy services		
		Basic services	Tailored services	High-quality comprehensive services
Procurement of infrastructure/ operation and control	Leasing to customers			
	Shared ownership	Low economic growth		Stable economic growth
	Customer ownership			

paragraphs, the relevance of these features and the related derived preference indications on the business model pattern selection are outlined.

Determinants Exploration

Project Typology

Local energy management can be integrated and operate either in a completely new district project or in a retrofit project of an existing district. Although in both cases the implementation objectives are the same - to maximize the energy efficiency, sustainability, and profitability of the energy system - the challenges faced are clearly different. Developing the energy concept from scratch allows considering a larger amount of options, whilst adapting to the existing situation bears more constraints and uncertainties. New LEM projects can benefit from a potentially higher energy efficiency performance and flexibility levels that can lead to higher profitability for the same investment compared to retrofit projects. The latter must be developed selecting complementary infrastructures complying with existing (not optimized) technical infrastructures heterogeneously operated and owned by different customer typologies.

The managers generally agreed that tailored business models specifically addressing the particular requirements of the various involved stakeholders are required to overcome the additional barriers typically characterizing retrofit projects. For this reason, they expect retrofit projects to take advantage of business model approaches involving the customers in the infrastructure procurements and/or operation (Preference indication: *Retrofit projects*, **Table 5**). Conversely, manager's opinion is that, since new projects can be addressed more flexibly, they potentially appear as more appropriate and financially attractive also to external large investors completely or partially taking over the procurement and operation of the infrastructures. Due to the lack of pre-existing technical and legally binding agreements with the customers, they substantially agreed that the valorization of comprehensive high-quality solutions is easier in new projects (Preference indication: *New development projects*, **Table 5**).

Density of Buildings

The density of buildings within the LEM operation area represents an influential aspect driving the development especially of LEM concepts dealing with thermal and chemical energy networks. Infrastructure costs and energy losses related to the energy

2014; Mancarella, 2014), *energy conversion devices* (Chicco and Mancarella, 2009; Fazlollahi et al., 2015; Orehounig et al., 2015), *and self-sufficiency level devices* (Chicco and Mancarella, 2009; Lasseter, 2011; Orehounig et al., 2015). In the following

services distribution increase significantly when the customers' and partners' density decrease. Districts can be categorized with three typical density levels: high density, typical of city's districts; moderate density, typical of suburban districts; and low density, typical of rural districts.

High-density districts can profit from reduced energy losses and need for distribution infrastructures. For these reasons, the managers argued that, on the one hand, high-density districts can offer higher profitability potential to the LEM investors and so be suitable to leasing solutions. On the other hand, they also offer opportunity for centralized infrastructures (e.g., district heating), characterized by long term, large investments and, thus, more appropriate to business models based on shared ownership (Preference indication: *High density of buildings*, **Table 6**). The fact that moderate and low-density districts are penalized by distribution costs and attract smaller investments in distributed heat, cooling, and storage infrastructures explains the managers inclination toward business models favoring customer ownership for such districts (Preference indication: *Low density of buildings*, **Table 6**). No preference has been highlighted on the quality level of energy services to be provided.

Buildings Typology

The LEM buildings typology has a significant impact on both technical and business model developments. Buildings typologies generally include residential buildings, commercial buildings, industry, and farms. Energy demand and generation profiles of these typologies are very heterogeneous in time, quantity, and flexibility. The development of business models tailored to the customer energy requirements and flexibility is the main driver to maximize the LEM profitability. Within cities a high share of residential and commercial buildings is more likely. In suburban and rural areas, the relevance of industry and farms rises substantially. Heterogeneous customers and partners enable the LEM to operate across various energy demand/load profiles and, thus, potentially to have more internal flexibility opportunities to valorize. Consequently, a LEM operating across more diverse building typologies is expected to require smaller investments in infrastructure to achieve the same levels of profitability and energy independence.

The managers converged on the idea that business models offering tailored service solutions on both the ownership and quality of services perspectives should address very heterogeneous customer portfolios. In particular, they argued that leasing solutions and high-quality comprehensive services should be considered for this scenario characterized by higher complexity and profitability potential (Preference indication: *Heterogeneous buildings typology*, **Table 7**). Evaluating the homogeneous customer portfolio scenario, the managers expressed a preference toward basic services and multiple infrastructure ownership options. The idea behind this choice is to provide a limited number of standardized services to a large number of similar customers, while leaving all options open on the infrastructure ownership side (Preference indication: *Homogeneous buildings typology*, **Table 7**).

Energy Conversion Infrastructures

A LEM could generally operate energy conversion and storage infrastructures across multiple energy carriers, including electrical, thermal, and chemical carriers. Distributed generation implies a mix of customer-sited energy conversion devices (e.g., photovoltaic panels, boilers) and centralized plants (e.g., combined heat and power plants, power to gas plants). Different combinations of devices enable different combinations of energy consumption, generation, and storage. Furthermore, each infrastructure can be used, operated, and owned by different entities. The increased complexity of the technical architecture with respect to traditional decoupled single energy carrier energy systems is also reflected on the business model structure. LEM business models should capitalize on the economic value of the flexibility offered by operating across different energy carriers, while maintaining the level of clarity and transparency associated with traditional single energy carrier business models.

The managers were asked to provide their opinion on LEM scenarios, including single or multiple energy carriers. Evaluating the multiple energy carriers' case, the managers argued that the more complex the LEM infrastructure, the larger the required investments and the higher should be the profitability. For this reason, the majority of the managers favored business models focusing on procurement of infrastructures and on the creation of added value through offering high-quality comprehensive and tailored services. They considered these business model approaches to be the most suitable to address the higher investment risks and related higher profitability characterizing multiple energy carrier scenarios (Preference indication: *Multi energy carriers*: **Table 8**). Conversely, they evaluated simpler LEM scenarios, including only a single energy carrier as conveniently associable with business models providing basic services and attracting a larger number of customers through the offering of different ownership solutions (Preference indication: *Single energy carrier*: **Table 8**).

Self-Sufficiency Level

The LEM self-sufficiency level defines the LEM dependency on the external energy market. This feature is closely related to the energy conversion infrastructures available: a high self-sufficiency level is likely to require investments in infrastructures complementary to the ones contributed by the customers. On the other hand, self-sufficiency represents an additional value proposition to the customer, a self-sufficient LEM could offer: reduced exposure to the external determinants influencing the prices on the wholesale energy market; and certified local origin and quality (i.e., renewable share) of the provided energy services.

The managers generally appeared rather skeptical of the possibility of self-sufficient LEM due to techno-economic constraints. However, following the same logic applied when evaluating multi-energy carriers scenarios, they agreed that LEM characterized by high self-sufficiency levels should be favorably addressed by business models targeting customers willing to pay the extra price of the provided added values and optionally willing to participate to the infrastructure investment (Preference indication: *Self-sufficiency level*, **Table 9**).

Customer Socio-Demographic Determinants

Determinants Selection

Many recent publications have addressed the characterization of customer's segments with respect to new business opportunities emerging with the energy transition. A concise literature review is presented hereafter. The outlined findings enable identification of the most important determinants related to socio-demographic aspects to be considered in the business model innovation process.

Curtius et al. (2012) explored the customer segmentation for smart grids on the basis of a European study and derived a number of generic business models best suited to address the different customer segments. This work highlighted that no single business model can guarantee the successful penetration of smart grids. Instead various business models characterized by optimized value propositions matching the heterogeneous customer values perceptions should be developed. Three different customer segments have been identified: *The Supporters*, including customers across all ages strongly supporting smart grids; *The Ambiguous*, characterized mainly by young customers willing to support smart grid diffusion but also concerned by data security issues; and *The Skeptical*, including mainly older customers not particularly concerned by environmental problems and expecting small benefits from smart grid.

Kaufmann et al. (2013) investigated the preference for smart metering of private consumers in Switzerland. Using conjoint analysis they assessed, on the one hand, the overall willingness of customers to pay for smart meters and, on the other hand, the existence of four customer segments with significantly different value perception: the *risk averse*, including customers not convinced on the benefit from smart meters; the *technology minded*, including customers perceiving a high value from smart metering; the *price sensitive*, including customers strongly concerned by the price and interested to reduce their costs with smart meters; and the *safety oriented*, including customers attracted by the values offered by smart meters associated with home security. As an outcome, the heterogeneity of customers in terms of value perception encourages the offering of different tailored value propositions rather than a standard offer for the whole market.

Tabi et al. (2014) investigated the differences between customers adopting renewable energy and potential adopters. Through conjoint analysis, a different customer segmentation has been identified based on customers' preferences concerning electricity products. Although a large majority of customers demonstrated clear preferences for renewable energy, only a small fraction purchased it. The study showed that demographic variables play a minor role, while a significantly higher education level appears as key driver. Psychographic and behavioral factors have a strong impact on the choice of adopting renewable energy. The identified aspects that should influence potential adopters are: the preference for locally produced electricity, the sensitivity to increases in the cost of electricity, and the reduced awareness on environmental issues.

Sagebiel et al. (2014) showed how in Germany customers have a slightly higher willingness to pay for renewable electricity produced within cooperatives. In particular, customers appears to be mainly attracted by the fact that electricity from a cooperative is produced from renewable resources and to a smaller extent they value the facts that this electricity is produced locally, democratically, and transparently.

Ebers and Wüstenhagen (2015) investigated the investment decision-making with respect to different financing options for renewable energy projects of consumers in Switzerland. The result of the analysis highlighted the existing market potential for new financial products, such as community finance projects and retirement investment funds. Furthermore, the analysis showed how homeowners would preferably rely on their own funds to finance a renewable energy installation or to make use of mortgage and only to a minor extent they consider loan and leasing solutions.

Customer's socio-demographic aspects appear to play a major and complex role in determining their preferences. The strong correlation between these variables is reflected in very heterogeneous customer's preferences and perceptions. Effective business models should cope with the consequent large customer segmentation offering more flexible and tailored solutions than in traditional energy-related business models. Based on the presented literature review, three main determinants related to customer socio-demographic aspects are selected: willingness to pay (Kaufmann et al., 2013; Sagebiel et al., 2014); customer awareness (Curtius et al., 2012; Kaufmann et al., 2013; Tabi et al., 2014); and building ownership (Sagebiel et al., 2014; Ebers and Wüstenhagen, 2015). These three determinants are explored hereafter.

Determinants Exploration

Willingness to Pay

In traditional energy-related business models, little or no attention is paid to differentiating the offers addressing customers with different willingnesses to pay. The required new business models, focusing on services and potentially including the procurement of infrastructures value chain activity, should flexibly target customer segments characterized by all levels of investment potential and price sensitivity.

In this regard, the managers generally agreed on the need to develop flexible business models offering adequate financial instruments and basic services to be offered to price-sensitive customers and to customers with reduced possibility to invest in owned infrastructures (Preference indication: *Price sensitive customers*, **Table 10**). Conversely, customers with higher willingness to pay should be attracted by high-quality service packages and different options of investments opportunities (Preference indication: *Higher willingness to pay customers*, **Table 10**).

Customer Awareness

The awareness and consciousness of environmental issues is expected to play a significant role on the energy transition process and is a key aspect to be considered while developing LEM business models.

The managers suggested that customers particularly concerned by energy transition and environmental issues are more prone to engage in investments in the required infrastructures.

Moreover, they agreed that such customers should be addressed with offers, including more than basic services (Preference indication: *Sustainability oriented customers*, **Table 11**). At the opposite end, conservative-minded customers, skeptical of the energy turnaround and of new technologies, are expected to require only the basic traditional services and to favor either the traditional approach of owning the required infrastructures or the option to lease it (Preference indication: *Skeptical customers*, **Table 11**).

Building Ownership

Local energy management could operate across privately owned, real estates and cooperative owned buildings. The building ownership is an important aspect to consider while developing successful business models. Real estate investors, private house owners, and cooperatives represent highly diversified customer segments.

The managers presumed that private owners would be interested in investment opportunities in owned infrastructure and in its potential valorization within the energy market. For this reason, they agreed on suggesting business models focusing on customer ownership. No preferences were indicated on the quality level of energy services to be provided (Preference indication: *House owners*, **Table 12**). With respect to cooperatives, the managers agreed on the fact that business models based on shared ownership and including basic or tailored solutions are likely to be the most appropriate. Real estates are expected to favor a range of different quality level leasing solutions, which could be offered to their tenants (Preference indication: *Real estates ownership*, **Table 12**).

External Determinants

Determinants Selection

Innovative entrepreneurial approaches, such as the one required by the energy transition, strongly rely on external investors. Understanding investor preferences and the drivers behind their decision-making process represents an additional key driver for the development of appealing business models with a higher probability to succeed. A number of recent studies investigating investor preferences on the renewable energy sectors are available in the literature.

Studying the microgeneration sector, Provance et al. (2011) highlighted that in strongly institutionalized markets, such as energy, business models innovation is driven not only by resolute decision-making based on available internal values but also on local external determinants, such as politico-institutional and socio-institutional aspects. Yildiz (2014) investigated the business models fostering financial citizen participation in investments in renewable energy infrastructures in Germany. The study highlights the importance of coordinating the development of new business models and new policies in fostering private contribution to renewables investment.

Bürer and Wüstenhagen (2009) analyzed the policy preferences of private investors in clean tech companies in Europe and North America. By interviewing 60 venture capital and private equity investors, they assessed the overall preference for feed-in tariffs-based policies. Feed-in tariffs appeared to be more effective in reducing investment risks than trading mechanism policies, such as green certificates. An additional outcome was that investors agreed on the need for a mix of consistent policies to stimulate interest in investment in clean technology. Hofman and Huisman (2012) repeated part of the same survey 3 years later in order to analyze the impact of the economic crisis on investor preferences. The study highlighted that generally the popularity of energy policies has decreased especially on policies involving subsidies and trade schemes and on European investors. However, feed-in tariffs remained the preferred option.

Masini and Menichetti (2012, 2013) investigated the decision-making process of investors in renewable energy technologies in the context of the current global economic crisis. Investors' preferences on policy instruments and technological risks were analyzed. The results revealed how investors value more the proven technical reliability of a technology than its market efficiency, since market efficiency can be influenced by policy measures. Furthermore, the analyses identified a segment of investors strongly preferring short-term policies providing high financial incentives than long-time policies characterized by lower financial support. Wüstenhagen and Menichetti (2012) discussed the relation between the strategic decision process for renewable energy investments and energy policies. The authors outlined the need for a segmentation of energy policies to support and promote the heterogeneous segmentation of investors characterizing the renewable energy market.

Existing literature highlighted the strong influence of external determinants on investors preferences and, thus, on business model development. Based on the literature analysis, energy policy (Bürer and Wüstenhagen, 2009; Provance et al., 2011; Wüstenhagen and Menichetti, 2012; Yildiz et al., 2015) and macro-economy (Hofman and Huisman, 2012; Masini and Menichetti, 2012, 2013) are selected as the main external determinants to be considered. The description of these two determinants and the related preference indications derived from the utility manager opinions are outlined in the following paragraphs.

Determinants Exploration

Energy Policy

Policy makers regulate the energy transition's evolving pace through market regulations, subsidies (on fossil fuel and/or renewable energies), as well as taxes (e.g., carbon tax). Time length, typology, and amplitude of the financial incentives together with the level of clarity and stability of the political strategy are the main factors influencing potential investor decisions.

Supporting energy policies frameworks characterized by short-term high financial incentives are the most attractive for small to large investors. The managers expected this scenario to be suitable to business models open to all infrastructures ownership options and focusing on high-quality comprehensive services (Preference indication: *Supporting policies*, **Table 13**). Considering an uncertain energy policy framework, such as those mostly charactering the current phase of the energy transition, the managers indicated that large investments are discouraged and, thus, business models favoring customer ownership and relying on basic or tailored energy services are the most preferred (Preference indication: *Not supporting policies*, **Table 13**).

Macro-Economy

The macro-economic situation, including growth or recession cycles, interest rates, and inflation levels, has a significant impact on investors' behavior in any market. The typically large and long-term investments required in the energy sector accentuate the influence of this determinant.

In uncertain low growth macro-economic scenarios characterized by low inflation and low interest rates, such as the current global economic situation, business models oriented to customer or shared ownerships were indicated as most suitable due to their compatibility toward both small private and large public investments (Preference indication: *Low economic growth*, **Table 14**). Conversely, the managers expected that large investors would become protagonists in more favorable economic conjunctures, characterized by stable growth and high interest rates. For this case, they suggested business models providing high-quality services and characterized by higher investments and potential profitability (Preference indication: *Stable economic growth*, **Table 14**).

DISCUSSION

For most of the investigated determinants, the managers provided agreeing or at least compatible answers. The reasons behind their choices were sometimes more heterogeneous, showing how the different perspectives related to their different job functions and experiences played an important role. Combining answers and explanations enabled the recognition of leading general trends pointing to more or less specific portions of the business model solution space.

Especially at the beginning of the semi-structured interviews, the managers struggled to cope with the proposed approach of investigating each determinant independently. They highlighted the fact that – the identified determinants being to a certain extent correlated to each other – it is challenging to generally judge the impact of each one separately. Nevertheless, once the managers became familiar with the proposed approach they acknowledged the advantage of focusing on a single aspect at time. This reduces the unsuitable attitude of associating existing combination of determinants to traditional business model approaches. Instead it forces broad re-examination of each determinant to identify unexplored options.

The analysis of possible correlations/dominances between different determinants is an interesting aspect to be addressed in future investigations. A better understanding of these dependencies would further strengthen the ability of the conceptual framework to support business model innovation.

Also, during the discussion, the managers highlighted the difficulty that they are currently experiencing in switching from the continuous adaptation of their traditional long established business models, which focus on infrastructures and feature little or no consideration of customer preferences, toward the creation of innovative customer-oriented business models. The related challenge is threefold. First, it entails the identification of economically viable new business opportunities in a changing market and policy framework. Second, it requires the organization of the new indispensable activities of collecting, monitoring,

and understanding customer preferences. Third, it demands the development from scratch of the business models required to exploit the broader variety of emerging business opportunities and customer segments. With respect to the last point, the managers praised the possibility offered by the proposed conceptual framework to provide a comprehensive and organized solution space favoring a systematic approach to initiate the business model innovation process. Furthermore, they valued the proposed research work aiming to provide an orientation within said conceptual framework through the identification and characterization of major determinants.

Going generally across the provided preference indications, it appears that when the business conditions are favorable (e.g., growing economy, high willingness to pay of customers, high density of buildings), the managers generally orient their preferences toward business models focusing on comprehensive high-quality services. Conversely, when conditions are not encouraging, the preference goes toward basic or tailored services often in combination with customer ownership of infrastructures. This confirms that the current economic and political situation favors business models oriented to customer ownership and basic or tailored services (Facchinetti and Sulzer, 2016). Assuming that the next steps of the energy transition are characterized by a more suitable and less uncertain regulatory framework, business models offering customer-oriented high-quality services and infrastructure leasing solutions are expected to attract more and more attention from utilities. Improved macro-economic conditions would accelerate this trend. From this general perspective, the role of shared ownership approaches appears controversial. The managers rarely indicated individual preference for these approaches; instead they mostly suggested them as an alternative/complementary option to LEM ownership approaches. This evidence could be explained by the fact that these business approaches have proven their market success in the context of the energy transition in public or cooperatives businesses (Viardot, 2013; Yildiz et al., 2015). However, so far they have not been at the core of the utilities' market strategy (Richter, 2013a).

Clearly, the present work bears limitations that should be considered. The fact that the analysis has been restricted to a very specific empirical context and to a limited sample of managers makes the outcomes difficult to generalize. The applied explorative qualitative research approach is not meant to derive exhaustive and general conclusions, but instead to provide a first contribution to understanding the link between business model innovation and stakeholder preferences at this stage of the energy transition. We expect to address in follow-up works the further consolidation, refinement, and extension of the proposed preference indications through a deeper analysis involving a larger number and variety of stakeholders.

CONCLUSION

The effective accomplishment of the energy turnaround relies on a deep reorganization of the energy market. Business model innovation is recognized as a key driver of this process. The present paper contributes to this topic by providing to stakeholders and policy makers an orientation within the vast business model solution space of LEM concepts – defined as the management of energy supply, demand, and storage within a given geographical area. The contribution of the present work to the existing literature on business model innovation for the energy transition is twofold. First, based on comprehensive literature reviews, it provides a structured selection of main factors to be considered by stakeholders undertaking a business model innovation process for an intended LEM concept. Second, based on interviews with utility managers it provides indications for the identification of specific business model patterns most appropriate to valorize LEM.

Building upon a recent previous work (Facchinetti and Sulzer, 2016) in which the LEM business model solution space was characterized through the definition of a conceptual framework, the present work goes one step further. The main determinants to be considered within the business model innovation process have been selected through reviews on distributed generation typologies and customer/investor preferences related to new business opportunities emerging with the energy transition. The identified determinants have been organized into three categories: related to the LEM typology, related to customer characterization, and related to external aspects.

The impact of the identified determinants on the business model innovation process has been investigated through an explorative qualitative research approach collecting the opinions of Swiss utility managers through semi-structured interviews. The outcome of this study is a set of preference indications associating the identified determinants with the most suitable region of the LEM business model solution space (**Tables 5–14**).

The interviews confirmed existing literatures (Richter, 2013b; Helms, 2016) and shed light on the challenge that the utility managers are currently facing: discontinuing the adaptation of traditional long established business models, focusing on infrastructures and with little or no customer preferences consideration, in favor of pursuing new business models built around customers' needs and preferences. From this perspective, the managers acknowledged the support offered by the proposed systematic and comprehensive conceptual framework for LEM business model innovation.

The provided conceptual framework offers to potential LEM stakeholders a structured guide through the business model innovation process. Characterizing the intended LEM with regard to the determinants identified in this work, and then following the corresponding preference indications, would lead stakeholders to the most appropriate portion(s) of the business model solution space in accordance with the utility manager's preferences. Each solution space portion is characterized by clusters of generic business model ideas referring to the different LEM value chain steps (Facchinetti and Sulzer, 2016). Through combination and adaptation of the identified business model ideas, stakeholders can develop suitable business models tailored to the intended LEM concept.

Policy makers could take advantage of the presented conceptual framework by using it to understand and predict a LEM stakeholder's decision-making process in a comprehensive variety of scenarios. Analyzing the impact factors influencing the business model innovation process should ease the determination of outdated policies hindering the penetration of innovative business approaches, and accelerate the development of a more segmented and flexible regulatory framework appropriate to favoring LEM market penetration.

The following main outcomes have been derived from the managers' preference indications. The utility managers appeared generally inclined to favor business models focusing on high-quality comprehensive services when business conditions are favorable. The mentioned favorable business conditions include supportive policies scenarios, a growing economy, new development projects involving multiple energy carriers in areas characterized by a high building density, and target customers with a high willingness to pay. When conditions are more adverse, the manager's preferences are generally oriented toward business models relying on basic or tailored services, often in combination with customer ownership of infrastructures. This evidence suggests that in the current political and economic situation business models focusing on customer ownership and basic or tailored services are preferred. This conclusion confirms previous findings based on a market analysis (Facchinetti and Sulzer, 2016). In other ways, assuming movement toward a more suitable and less uncertain regulatory framework, business models focusing on customer-oriented high-quality services and infrastructure leasing solutions should attract more and more attention from the utilities. From this perspective, in order to increase the appeal of LEM to large investors, policy makers should develop suitable regulatory frameworks guaranteeing consistent and stable policies across a time period compatible with the long-term investment required. Finally, the analysis has shown that, in the view of utility managers, the role of business models characterized by shared ownership approaches is controversial. Indeed, the managers rarely indicated individual specific preference for such approaches. Considering the intrinsic suitability of shared ownership approaches to LEM concepts, policy makers should encourage such approaches so far only to a minor extent addressed by the legislation and applied in the energy market.

AUTHOR CONTRIBUTIONS

EF: main author, main contribution to the conception of the work. CE, AB, and SS: significant contribution to the conception and critical revision of the work.

ACKNOWLEDGMENTS

This work has been accomplished in the frame of the Swiss Competence Center for Energy Research on Future Energy Efficient Buildings & Districts SCCER FEEB&D, funded by The Commission for Technology and Innovation of the Swiss Confederation. The authors are particularly grateful to the utility companies that kindly accepted to participate to this study.

REFERENCES

Anaya, K. L., and Pollitt, M. G. (2015). Integrating distributed generation: regulation and trends in three leading countries. *Energy Policy* 85, 1–12. doi:10.1016/j.enpol.2015.04.017

Beattie, V., and Smith, S. J. (2013). Value creation and business models: refocusing the intellectual capital debate. *Br. Account. Rev.* 45, 243–254. doi:10.1016/j.bar.2013.06.001

Bundesministerium für Wirtschaft und Technologie (BMWi). (2010). Energiekonzept. Available at: https://www.bmwi.de/BMWi/Redaktion/PDF/E/energiekonzept-2010,property=pdf,bereich=bmwi2012,sprache=de,rwb=true.pdf

Bocken, N. M. P., Short, S. W., Rana, P., and Evans, S. (2014). A literature and practice review to develop sustainable business model archetypes. *J. Clean. Prod.* 65, 42–56. doi:10.1016/j.jclepro.2013.11.039

Bürer, M. J., and Wüstenhagen, R. (2009). Which renewable energy policy is a venture capitalist's best friend? Empirical evidence from a survey of international cleantech investors. *Energy Policy* 37, 4997–5006. doi:10.1016/j.enpol.2009.06.071

Chicco, G., and Mancarella, P. (2009). Distributed multi-generation: a comprehensive view. *Renew. Sustain. Energy Rev.* 13, 535–551. doi:10.1016/j.rser.2007.11.014

Curtius, H. C., Künzel, K., and Loock, M. (2012). Generic customer segments and business models for smart grids. *der markt* 51, 63–74. doi:10.1007/s12642-012-0076-0

Ebers, A., and Wüstenhagen, R. (2015). 5th Consumer Barometer of Renewable Energy in Cooperation with Raiffeisen. Available at: http://www.iwoe.unisg.ch/~/media/internet/content/dateien/instituteundcenters/iwoe/news/150522_kundenbarrometer2015_e_web.pdf?fl=en

EPIA. (2014). *Global Market Outlook for Photovoltaics 2014–2018*. Brussels. Available at: http://www.cleanenergybusinesscouncil.com/site/resources/files/reports/EPIA_Global_Market_Outlook_for_Photovoltaics_2014-2018_-_Medium_Res.pdf

European Commission. (2011). *COM/2011/0885 Energy Roadmap 2050*. Brussels

European Commission. (2012). *Directive 2012/27/EU of the European Parliament and of the Council of 25 October 2012 on Energy Efficiency*. Brussels

European Commission. (2014). *COM/2014/015 A Policy Framework for Climate and Energy in the Period from 2020 to 2030*. Brussels

European Commission. (2015). *COM/2015/080 A Framework Strategy for a Resilient Energy Union with a Forward-Looking Climate Change Policy*. Brussels

Facchinetti, E., and Sulzer, S. (2016). General business model patterns for local energy management concepts. *Front. Energy Res* 4, 7. doi:10.3389/fenrg.2016.00007

Fazlollahi, S., Becker, G., Ashouri, A., and Maréchal, F. (2015). Multi-objective, multi-period optimization of district energy systems: IV – A case study. *Energy* 84, 365–381. doi:10.1016/j.energy.2015.03.003

Fragaki, A., and Andersen, A. N. (2011). Conditions for aggregation of CHP plants in the UK electricity market and exploration of plant size. *Appl. Energy* 88, 3930–3940. doi:10.1016/j.apenergy.2011.04.004

Hellström, M., Tsvetkova, A., Gustafsson, M., and Wikström, K. (2015). Collaboration mechanisms for business models in distributed energy ecosystems. *J. Clean. Prod.* 102, 226–236. doi:10.1016/j.jclepro.2015.04.128

Helms, T. (2016). Asset transformation and the challenges to servitize a utility business model. *Energy Policy* 91, 98–112. doi:10.1016/j.enpol.2015.12.046

Hofman, D. M., and Huisman, R. (2012). Did the financial crisis lead to changes in private equity investor preferences regarding renewable energy and climate policies? *Energy Policy* 47, 111–116. doi:10.1016/j.enpol.2012.04.029

IEA. (2014). *World Energy Outlook 2014*.

Kaufmann, S., Künzel, K., and Loock, M. (2013). Customer value of smart metering: explorative evidence from a choice-based conjoint study in Switzerland. *Energy Policy* 53, 229–239. doi:10.1016/j.enpol.2012.10.072

Lasseter, R. H. (2011). Smart distribution: coupled microgrids. *Proc. IEEE* 99, 1074–1082. doi:10.1109/JPROC.2011.2114630

Loock, M. (2012). Going beyond best technology and lowest price: on renewable energy investors' preference for service-driven business models. *Energy Policy* 40, 21–27. doi:10.1016/j.enpol.2010.06.059

Lund, H., and Münster, E. (2006). Integrated energy systems and local energy markets. *Energy Policy* 34, 1152–1160. doi:10.1016/j.enpol.2004.10.004

Lund, H., Werner, S., Wiltshire, R., Svendsen, S., Thorsen, J. E., Hvelplund, F., et al. (2014). 4th Generation district heating (4GDH). *Energy* 68, 1–11. doi:10.1016/j.energy.2014.02.089

Mancarella, P. (2014). MES (multi-energy systems): an overview of concepts and evaluation models. *Energy* 65, 1–17. doi:10.1016/j.energy.2013.10.041

Markard, J., Suter, M., and Ingold, K. (2015). Socio-technical transitions and policy change – Advocacy coalitions in Swiss energy policy. *Environ. Innov. Soc. Transit* 18, 215–237. doi:10.1016/j.eist.2015.05.003

Masini, A., and Menichetti, E. (2012). The impact of behavioural factors in the renewable energy investment decision making process: conceptual framework and empirical findings. *Energy Policy* 40, 28–38. doi:10.1016/j.enpol.2010.06.062

Masini, A., and Menichetti, E. (2013). Investment decisions in the renewable energy sector: an analysis of non-financial drivers. *Technol. Forecast. Soc. Change* 80, 510–524. doi:10.1016/j.techfore.2012.08.003

Okkonen, L., and Suhonen, N. (2010). Business models of heat entrepreneurship in Finland. *Energy Policy* 38, 3443–3452. doi:10.1016/j.enpol.2010.02.018

Orehounig, K., Evins, R., and Dorer, V. (2015). Integration of decentralized energy systems in neighbourhoods using the energy hub approach. *Appl. Energy* 154, 277–289. doi:10.1016/j.apenergy.2015.04.114

Osterwalder, A., and Pigneur, Y. (2010). *Business Model Generation: A Handbook for Visionaries, Game Changers, and Challengers*. Hoboken, NJ: Wiley. Available at: http://www.amazon.com/Business-Model-Generation-Visionaries-Challengers/dp/0470876417

Provance, M., Donnelly, R. G., and Carayannis, E. G. (2011). Institutional influences on business model choice by new ventures in the microgenerated energy industry. *Energy Policy* 39, 5630–5637. doi:10.1016/j.enpol.2011.04.031

Richter, M. (2012). Utilities' business models for renewable energy: a review. *Renew. Sustain. Energy Rev.* 16, 2483–2493. doi:10.1016/j.rser.2012.01.072

Richter, M. (2013a). Business model innovation for sustainable energy: German utilities and renewable energy. *Energy Policy* 62, 1226–1237. doi:10.1016/j.enpol.2013.05.038

Richter, M. (2013b). German utilities and distributed PV: how to overcome barriers to business model innovation. *Renew. Energy* 55, 456–466. doi:10.1016/j.renene.2012.12.052

Sagebiel, J., Müller, J. R., and Rommel, J. (2014). Are consumers willing to pay more for electricity from cooperatives? Results from an online Choice Experiment in Germany. *Energy Res. Soc. Sci.* 2, 90–101. doi:10.1016/j.erss.2014.04.003

Schleicher-Tappeser, R. (2012). How renewables will change electricity markets in the next five years. *Energy Policy* 48, 64–75. doi:10.1016/j.enpol.2012.04.042

SFOE. (2013). *Swiss Energy Strategy 2050*. Available at: www.energystrategy2050.ch (accessed November 11, 2015)

Silvermann, D. (2009). *Doing Qualitative Research*. London: SAGE.

Sühlsen, K., and Hisschemöller, M. (2014). Lobbying the "Energiewende". Assessing the effectiveness of strategies to promote the renewable energy business in Germany. *Energy Policy* 69, 316–325. doi:10.1016/j.enpol.2014.02.018

Swiss Commision for Technology and Innovation. (2014). *Swiss Competence Centers for Energy Research*. Available at: https://www.kti.admin.ch/kti/en/home/ueber-uns/foerderbereiche/foerderprogramm-energie.html

Tabi, A., Hille, S. L., and Wüstenhagen, R. (2014). What makes people seal the green power deal? – customer segmentation based on choice experiment in Germany. *Ecol. Econ.* 107, 206–215. doi:10.1016/j.ecolecon.2014.09.004

Toke, D., and Fragaki, A. (2008). Do liberalised electricity markets help or hinder CHP and district heating? The case of the UK. *Energy Policy* 36, 1448–1456. doi:10.1016/j.enpol.2007.12.021

Viardot, E. (2013). The role of cooperatives in overcoming the barriers to adoption of renewable energy. *Energy Policy* 63, 756–764. doi:10.1016/j.enpol. 2013.08.034

Wüstenhagen, R., and Menichetti, E. (2012). Strategic choices for renewable energy investment: conceptual framework and opportunities for further research. *Energy Policy* 40, 1–10. doi:10.1016/j.enpol.2011.06.050

Yildiz, Ö (2014). Financing renewable energy infrastructures via financial citizen participation – the case of Germany. *Renew. Energy* 68, 677–685. doi:10.1016/j. renene.2014.02.038

Yildiz, Ö, Rommel, J., Debor, S., Holstenkamp, L., Mey, F., Müller, J. R., et al. (2015). Renewable energy cooperatives as gatekeepers or facilitators? Recent developments in Germany and a multidisciplinary research agenda. *Energy Res. Soc. Sci.* 6, 59–73. doi:10.1016/j.erss.2014.12.001

Zott, C., and Amit, R. (2013). The business model: a theoretically anchored robust construct for strategic analysis. *Strateg. Organ.* 11, 403–411. doi:10.1177/ 1476127013510466

Conflict of Interest Statement: The authors declare that the research was conducted in the absence of any commercial or financial relationships that could be construed as a potential conflict of interest.

13

Progress in Electrolyte-Free Fuel Cells

Yuzheng Lu[1], Bin Zhu[2,3]*, Yixiao Cai[4], Jung-Sik Kim[5], Baoyuan Wang[2,3], Jun Wang[1]*, Yaoming Zhang[1] and Junjiao Li[6]

[1] Jiangsu Provincial Key Laboratory of Solar Energy Science and Technology, School of Energy and Environment, Southeast University, Nanjing, China, [2] Faculty of Physics and Electronic Technology, Hubei Collaborative Innovation Center for Advanced Organic Materials, Hubei University, Wuhan, China, [3] Department of Energy Technology, Royal Institute of Technology KTH, Stockholm, Sweden, [4] Ångström Laboratory, Department of Engineering Sciences, Uppsala University, Uppsala, Sweden, [5] Department of Aeronautical and Automotive Engineering, Loughborough University, Loughborough, UK, [6] Nanjing Yunna Nano Technology Co., Ltd., Nanjing, China

Edited by:
Baibiao Huang,
Shandong University, China

Reviewed by:
Weifeng Yao,
Shanghai University of
Electric Power, China
Kesong Yang,
University of California
San Diego, USA

***Correspondence:**
Bin Zhu
binzhu@kth.se;
Jun Wang
wj-jw@seu.edu.cn

Specialty section:
This article was submitted to
Nanoenergy Technologies
and Materials, a section of the
journal Frontiers in Energy Research

Citation:
Lu Y, Zhu B, Cai Y, Kim J-S, Wang B,
Wang J, Zhang Y and Li J (2016)
Progress in Electrolyte-Free
Fuel Cells.
Front. Energy Res. 4:17.

Solid oxide fuel cell (SOFC) represents a clean electrochemical energy conversion technology with characteristics of high conversion efficiency and low emissions. It is one of the most important new energy technologies in the future. However, the manufacture of SOFCs based on the structure of anode/electrolyte/cathode is complicated and time-consuming. Thus, the cost for the entire fabrication and technology is too high to be affordable, and challenges still hinder commercialization. Recently, a novel type of electrolyte-free fuel cell (EFFC) with single component was invented, which could be the potential candidate for the next generation of advanced fuel cells. This paper briefly introduces the EFFC, working principle, performance, and advantages with updated research progress. A number of key R&D issues about EFFCs have been addressed, and future opportunities and challenges are discussed.

Keywords: fuel cell, single component/layer fuel cell, electrolyte (layer)-free, research progress, solid oxide fuel cell

INTRODUCTION

Energy crisis and environmental pollution continue to challenge all countries around the world in terms of both the global economy and the planetary environment (Zhang and Cooke, 2010). Fossil energy, which can meet the demand of human energy consumption, will release SO_2, NOx, CO, CO_2, and other toxic substances. All these emissions will seriously harm the environment. Meanwhile, the utilized efficiency of fossil energy is not high enough, and the fossil resources themselves are limited and not renewable. However, fuel cells (FCs), as one of the most important energy technologies, can convert chemical energy from fossil fuels to electricity by electrochemical reaction. It increases efficiency up to 40–80%, which is very promising without being limited by the Carnot cycle (Dufour, 1998). The characteristics of cleanliness, quietness, and high efficiency, delivered by FCs, make it possible to be one of the most effective methods to resolve the environmental pollution from fossil fuel combustion. Among all types of FCs, SOFC, and proton exchange membrane fuel cells (PEMFC) are considered to be the most promising technologies in stationary and transportation applications (Yi, 1998).

Solid oxide fuel cell (SOFC) is a kind of FC that can operate at high temperatures between 500 and 1000°C. The advantages of SOFC include high efficiency, multi-fuel flexibility, environment-friendly, lower material cost, long service life, and so on. The core part is the three-layered structure of anode, electrolyte [yttrium-stabilized zirconia (YSZ)], and cathode as shown in **Figure 1A**. Today, the YSZ electrolyte SOFCs still face commercialization challenges due to high costs. Removal of the electrolyte layer could provide a completely new technology that would be simple and most cost-effective. Recently, an electrolyte (layer)-free fuel cell (EFFC) was invented. It exhibits a new energy conversion technology (Zhu et al., 2011a,b). The novel structure of EFFC is shown in **Figure 1B**. Of primary importance is that EFFC maintains the core function of traditional SOFC to convert chemical energy from fuel to electricity. However, the single layer structure of EFFC is significantly different from the conventional three-layer structure.

ELECTROLYTE (LAYER)-FREE FUEL CELL

The Working Principle of EFFC

Traditional FC is constructed by three layers – a typical anode–electrolyte–cathode structure. The porous anode and cathode are separated by an electrolyte that is composed of dense solid oxide

YSZ. The electrolyte is employed here to separate oxidant (oxygen) from reductant (fuel) and facilitate the transfer of ions (O^{2-}), as shown in **Figure 2A**. The electrolyte is the key component of this structure. Unlike the SOFC structure, there is no macroscopic electrolyte in EFFCs. The structure of EFFC is a homomorphous layer that is constituted by a mixture of semiconductor and oxygen ion conductor. Its working temperature ranges from 300 to 600°C. The different working principle of EFFC has been presented by Zhu et al. (2011c, 2012).

It can be seen from **Figure 2B** that the EFFC fuel is oxidized and electrons are released at the anode (fuel side). At the cathode (air side), oxygen (oxidant) is reduced to oxygen ions (O^{2-}), which combine with electrons from external circuit to generate electricity. Hydrogen is employed here to explain this electrochemical process of the EFFC.

In proton conduction case:

At the hydrogen-contacting side:

$$H_2 \rightarrow 2H^+ + 2e^-$$

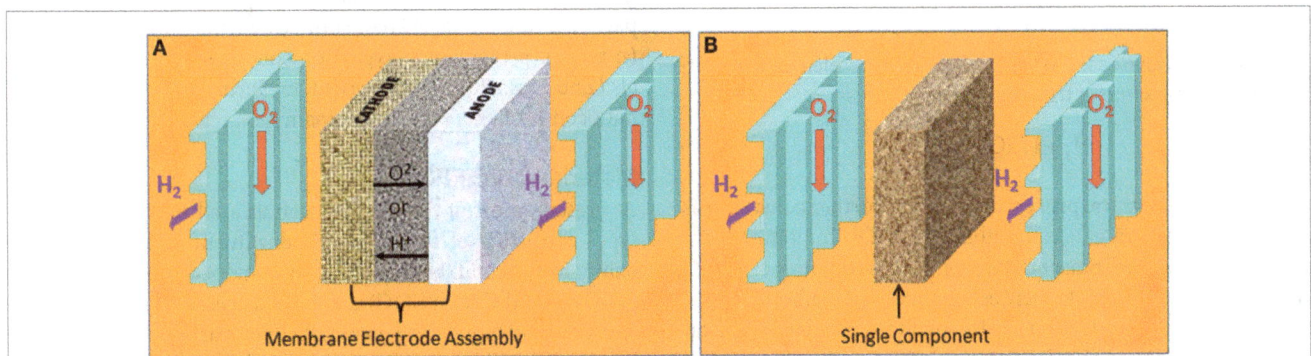

FIGURE 1 | Structures between (A) SOFC and (B) EFFC, or called single component.

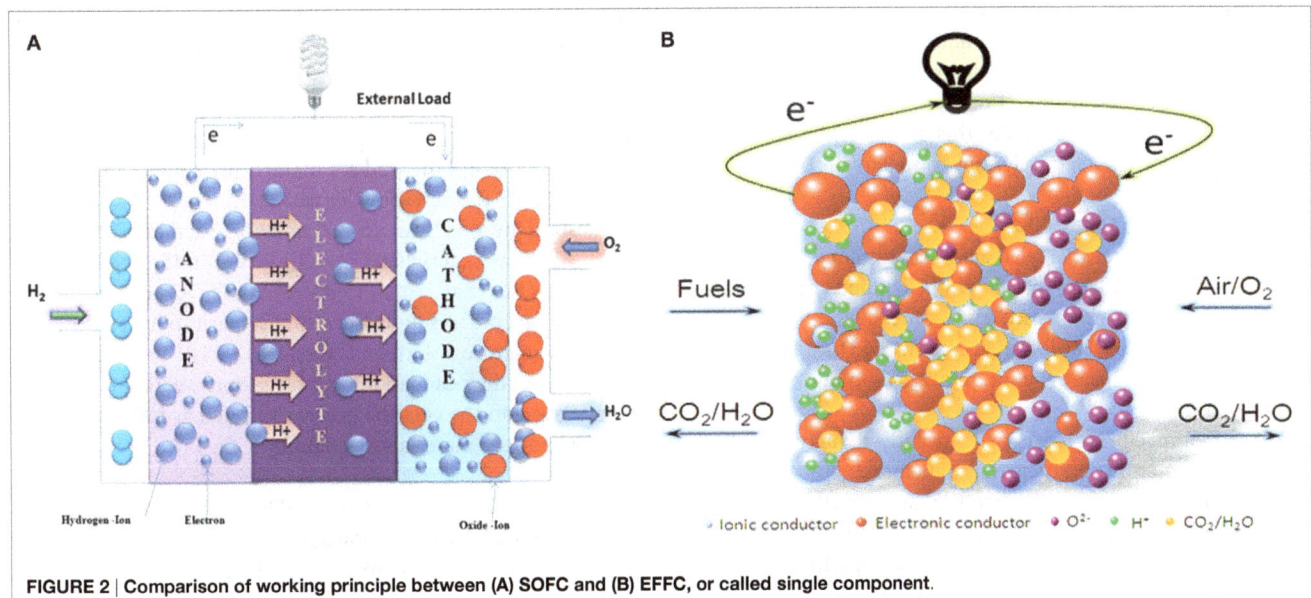

FIGURE 2 | Comparison of working principle between (A) SOFC and (B) EFFC, or called single component.

At the air (O_2)-contacting side:

$$1/2O_2 + 2H^+ + 2e^- \rightarrow H_2O$$

In oxide ion conducting case:

At the hydrogen-contacting side:

$$H_2 + O^{2-} \rightarrow H_2O$$

At the air (O_2)-contacting side:

$$1/2O_2 + 2e^- \rightarrow O^{2-}$$

It is also common that the single layer may conduct both proton and oxide ions, in this case:

At the hydrogen-contacting side:

$$H_2 \rightarrow 2H^+ + 2e^-$$

At the air (O_2)-contacting side:

$$1/2O_2 + 2e^- \rightarrow O^{2-}$$

To all the above cases, the overall electrochemical reaction in the EFFC is the same as the common FC process:

$$H_2 + 1/2O_2 \rightarrow H_2O$$

The electrochemical process of FC can be further construed as H^+ transfering from anode to cathode or as O^{2-} migrating from cathode to anode. Transportation of O^{2-} in traditional SOFC occurs at the electrolyte between anode and cathode. More specifically, ionic conduction can be enhanced and conducted on the particles' surface (Veldsink et al., 1995; Suzuki et al., 2005), which is used in the EFFC device with co-ionic H^+/O^{2-} transport.

This device is named as EFFC according to the new principle. The preparation technology of EFFC is very simple. Only one component is required, which can be fabricated by the mixture of electrode (anode and cathode) and electrolyte, the so-called "Three in one," highlighted by Nature Nanotechnology (Zhu, 2011). Different from the traditional SOFC constructed by three-layer anode–electrolyte–cathode, this new device with single component/layer can effectively convert fuel to electricity.

Advantages of EFFCs

Electrolyte (layer)-free fuel cell is a new energy conversion device showing many advantages that cannot be obtained by traditional SOFC since their operation mechanisms are different (Zhu, 2011, Zhu et al., 2011d, 2013a; Fan et al., 2012, 2013). In brief, some special merits of EFFC can be summarized as follows:

(i) The manufacturing cost is significantly reduced due to its simple structure and preparation method.
(ii) The interfaces of the electrolyte/the anode and the electrolyte/the cathode contribute to major polarization losses in the traditional three-layer structure. In addition, the three-layer device requests strict thermal compatibility and chemical

stability among the anode, electrolyte, and cathode, which have different material components. These present a serious SOFC technology challenge resulting in high cost as well. However, there is only one layer in the EFFC. Hence, losses or thermal stress problem from the interface can be avoided here. This ensures the long-term stability of the EFFCs.
(iii) Electrolyte (layer)-free fuel cells fabricated by multi-functional nanocomposite materials include oxygen ion conductor and semiconductor. Ionic conductivity of the cell materials is enhanced greatly for oxygen ion conducts on the surface of the cell and in itself, simultaneously. Meanwhile, there exists a synergistic effect established among the ions, electrons, holes between n-type or p-type semiconductor, and oxygen ion conductor. It helps to prevent short-circuit current and improve ionic conductivity of the single cell. This directly reduces the working temperature of the cell and provides a chance for introducing a wide range of materials.

Technical Developments on EFFCs

"Three in one" single layer materials are the key to realize the EFFCs with semiconductor-ion conductivities, which can integrate all anode, electrolyte, and cathode functions into one. In particular, a single layer is constituted by semiconductor materials (n-type or p-type) and oxygen ion conductor composite materials (Zhu et al., 2011e). At present, exploring new multi-function nanocomposites are one of the main research activities of EFFCs.

Materials for EFFCs

At present, most of the EFFC materials are based on composite types consisting of two types of constituent materials, i.e., (1) oxide ionic conductors, e.g., Sm_2O_3-doped CeO_2 (SDC) or Gd_2O_3-doped CeO_2 (GDC) and various ceria-based composites and (2) semiconducting materials, e.g., various transition element metal oxides, e.g., Ni, Cu, Fe, Zn, etc., and their complex or composite types. Recent research shows that the electrochemical performance strongly depends on the properties of the constituent materials, stoichiometric proportions, compositions between the constituent ion and semiconductor, their morphology and microstructure as well as experimental conditions.

Oxide Ion Conductors

Sm2O3-Doped CeO2. At present, chemical co-precipitating method and sol–gel method are mainly carried out in the preparation of SDC. The SDC powder materials within nanoscale were synthesized by one step co-precipitation method (Zhu et al., 2011b). The raw materials of $Ce(NO_3)_3 \cdot 6H_2O$ and $Sm(NO_3)_3 \cdot 6H_2O$ were used (Xia et al., 2012) to prepare the SDC powder materials. In sol–gel process, both ceria and samarium nitrate hydrates formed a mixture solution of $Ce(NO_3)_3 \cdot 6H_2O$ and Sm_2O_3, the solid citric acid was added into with vigorous stirring at 60–70°C until turning to the gel. The gel was sintered at 800°C to obtain the SDC powder materials by grinding.

Gd2O3-Doped CeO2. Apart from SDC, GDC is another material for EFFC. The GDC powder materials were obtained by using carbonate co-precipitation method (Zhu et al., 2011a). $Ce(NO_3)_3 \cdot 6H_2O$ and $Gd(NO_3)_3 \cdot 6H_2O$ ($Ce^{3+}:Gd^{3+} = 9:1$) were first dissolved in deionized water to form a mixed solution. Then, Na_2CO_3 was added into this solution by vigorous stirring to form white precipitate at 120°C. The precipitating precursor was further subjected to washing and drying. The GDC powder materials can be obtained after sintering it at 800°C.

Modified and Doped Nanomaterials Based on SDC and GDC. Hu et al. (2014) had prepared MgZn–SDC materials by chemical co-precipitation method. $Mg(NO_3)_2 \cdot 6H_2O$ and $Zn(NO_3)_2 \cdot 6H_2O$ were used in the co-precipitation process for nanometer material preparation. The cost of this material has been further reduced by introducing Mg and Zn contents. Zhu et al. (2014) had prepared GDC–KAlZn (KAZ) materials by two-step co-precipitating method. They first dissolved $Ce(NO_3)_3 \cdot 6H_2O$ and $Gd(NO_3)_3 \cdot 6H_2O(Ce^{3+}:Gd^{3+} = 4:1)$ in deionized water to get the properly distributed mixture in solution. A 0.5 M Na_2CO_3 solution was gradually added into the mixed solution while it was stirring at 120°C. The white precipitate was formed. Following filtrating, washing, and drying processes, a GDC precursor was obtained. In parallel, KAlZn composite was also prepared by co-precipitation method. $Al(NO_3)_3 \cdot 9H_2O$ and $Zn(NO_3)_2 \cdot 6H_2O$ ($Al^{3+}:Zn^{2+} = 4:3$) were dissolved in deionized water, and K_2CO_3 was selected as a precipitant agent to form the co-precipitation. The GDC precursor was added and then dried at 150°C. The GDC–KAlZn (KAZ) precursor was thus obtained. After washing, filtrating, drying, and then sintering at 800°C for 4 h, the GDC–KAlZn (KAZ) nanocomposite materials were obtained.

Semiconducting Materials Up to now, semiconductor materials (n-type or p-type) for EFFCs are selected among those transition metal oxides, e.g., NiO, CuO, FeOx, ZnO, and CoOx-doped $LiNiO_2$ or further complex or composite metal oxide materials. The first semiconductor material, $LiNiO_2$, was reported (Zhu et al., 2011a) by using LiOH and $Ni(NO_3)_2 \cdot 6H_2O$. Then, the composite was prepared by mixing semiconductor materials $LiNiO_2$ with GDC to fabricate the EFFC device. The maximum power output had reached 450 mW/cm² at 550°C. Next, LiNiZn-oxide was processed and adopted by sintering Li_2CO_3, $NiCO_3 \cdot 2Ni(OH)_2 \cdot 6H_2O$,

$Zn(NO_3)_2 \cdot 6H_2O$ *via* a solid-state reaction. Then, the semiconductor material LiNiZn-oxide was mixed with GDC to fabricate the EFFC. The maximum power outputs reached 300–600 mW/cm² at 450–550°C. The maximum output power could reach 700 mW/cm², when it was further doped by the oxidation–reduction catalyst Fe (Zhu et al., 2011b). The $LiNiZnO_{2-\delta}$–SDC composite materials achieved the maximum power output of the EFFC device with up to 600 mW/cm² at 550°C (Zhu et al., 2011c). The total conductivity of $LiNiZnO_{2-\delta}$–SDC-mixed materials of ion conductor and semiconductor is up to 0.1–1 S/cm.

The ceria–carbonate materials were also used to get semiconductor-ion single layer materials. The maximum power output of EFFCs could reach 700 mW/cm² at 550°C (Zhu et al., 2011e) by mixing LiNiCuZnFe-oxide and Na_2CO_3–SDC (NSDC) composite. The maximum power output of EFFC obtained by mixing LiNiCuSr-oxide and MgZnDC, according to a certain mass ratio, can reach 600 mW/cm² at 550°C. Its open circuit voltage (OCV) is up to 1.02 V at the same temperature (Hu et al., 2014). Xia et al. (2011) has developed $Sr_2Fe_{1.5}Mo_{0.5}O_{6-\delta}$ (SFM) and NSDC in a certain mass ratio as the single layer material for the EFFC. They reported that to achieve the best device performance the electronic and ionic conductivities must be closely matched, so 30% mass of SFM was able to achieve the best performance of the maximum power output of 360 mW/cm² at 750°C. This result is in agreement to the earlier studies as reported by Xia et al. for the SDC–LiNiZn-oxide single layer conductivities by adjusting the manipulation of concentration rations of ionic SDC to electronic conductors LiNiZn-oxide. They found that the single layer containing 30% mass of LiNiZn-oxide exhibits an almost uniform distribution of the two constituent components to establish a balance between the ionic and electronic conductors (Xia et al., 2012). Hei et al. (2014) investigated a composite of a perovskite oxide proton conductor ($BaCe_{0.7}Zr_{0.1}Y_{0.2}O_{3-\delta}$ and BCZ10Y20) and alkali carbonates ($2Li_2CO_3:1Na_2CO_3$ and LNC). The cell shows a maximum power density of 957 mW/cm² at 600°C with hydrogen as the fuel and oxygen as the oxidant. The summary of the typical performance demonstrated by various EFFCs are presented in **Table 1**.

TABLE 1 | Overview of EFFC performances.

No.	Performance	Temperature (°C)	Article	Key materials	Journal	Year
1	360 mW/cm² (maximum power output)	750	Single layer fuel cell based on a composite of $Ce_{0.8}Sm_{0.2}O_{2-\delta}$–$Na_2CO_3$ and a mixed ionic and electronic conductor $Sr_2Fe_{1.5}Mo_{0.5}O_{6-\delta}$	$Sr_2Fe_{1.5}Mo_{0.5}O_{6-\delta}$ (SFM) and NSDC	Journal of Power Sources	2014
2	600 mW/cm² (maximum power output)	550	Fabrication of electrolyte-free fuel cell with $Mg_{0.4}Zn_{0.6}O$/$Ce_{0.8}Sm_{0.2}O_{2-\delta}$–$Li_{0.3}Ni_{0.6}Cu_{0.07}Sr_{0.03}$–$O_{2-\delta}$ layer	LiNiCuSr-oxide and MgZnDC	Journal of Power Sources	2014
3	10×10^{-2} S/cm (total conductivity)	600	Electrical conductivity optimization in electrolyte-free fuel cells by single component $Ce_{0.8}Sm_{0.2}O_{2-\delta}$–$Li_{0.3}Ni_{0.45}Zn_{0.4}$ layer	SDC–LNZ oxides	RSC Advances	2012
4	450 mW/cm² (maximum power output)	550	A fuel cell with a single component functioning simultaneously as the electrodes and electrolyte	LiNiZn-oxide was mixed with GDC	Electrochemistry Communications	2011
5	700 mW/cm² (maximum power output)	550	A single component fuel cell reactor	Doped the oxidation–reduction catalyst Fe	International Journal of Hydrogen Energy	2011
6	600 mW/cm² (maximum power output)	550	An electrolyte-free fuel cell constructed from one homogenous layer with mixed conductivity	$LiNiZnO_{2-\delta}$-SDC	Advanced Functional Materials	2011
7	700 mW/cm² (maximum power output)	550	Single component and three-component fuel cells	LiNiCuZnFe-oxide and Na_2CO_3–SDC (NSDC)	Journal of Power Sources	2011

CRITICAL ISSUES ON EFFC

The first critical issue is to address whether penetration of the fuel (H_2) and oxidant (air) through the single layer material causes any electrochemical leakage due to H_2/O_2 thermal combustion. In this case, first, the security of the EFFC is of concern; the second is the electronic short-circuiting problem, by using the homogenously mixed semiconductor-ionic single layer material instead of the purely ionic-conducting electrolyte layer between the anode and cathode.

Non-Potential Safety Hazard

A separate layer of dense electrolyte is crucial for traditional FC. However, the dense electrolyte is not a must as long as it is gas tight, which can prevent gas penetration through it. Here, a major question is concerned: can oxygen and hydrogen pass through the porous single layer device and react directly to cause electrochemical leakage and potentially an explosion? The feasibility of eliminating the potential of explosion is a must for EFFC commercialization. Security level is determined by operating conditions, porosity, catalytic activity, and other features of EFFC. Liu et al. (2012) developed a dynamic model of anode and cathode reactions based on electrochemical impedance spectroscopy (EIS) to assess the performance of EFFCs. The security issue had been analyzed by using the single layer EFFCs by supplying the oxidant and fuel under the situation of open circuit conditions and operating conditions, respectively. The experimental and theoretical results show that the safety of EFFC is guaranteed. Assuming that porosity is equal to 0.5, define the torsion resistance τ is $\epsilon/\tau = \epsilon 1.5$, under a standard atmospheric pressure. The reaction depth of anode is 5.3–1.0×10^{-5} m and cathode is 1.4–0.25×10^{-6} m when the current is changed from 100 to 2000 mA·cm^{-2}, as shown in **Figure 3**.

The effect of porosity on the reaction depth is given in **Figure 4**. The reaction depth of O_2 is in the range of 0.6–1.4×1.0^{-6} m with the porosity of 0.2–0.5. For H_2, it is in the range of 2.1–5.3×1.0^{-5} m. It is shown that the reaction depth, both of H_2 and O_2, increase with porosity. However, all of them are far below the thickness of EFFC, which is in the microscale/10^{-3} m level. It is decisively proven that there is no risk of explosion under the given operating conditions.

Therefore, the EFFC security level is very high according to the calculations. In addition, there is a quenching distance so that the explosion cannot occur if the distance between the oxygen and hydrogen molecule is smaller than a certain separation distance. According to Janicke et al. (2000), this quenching distance is

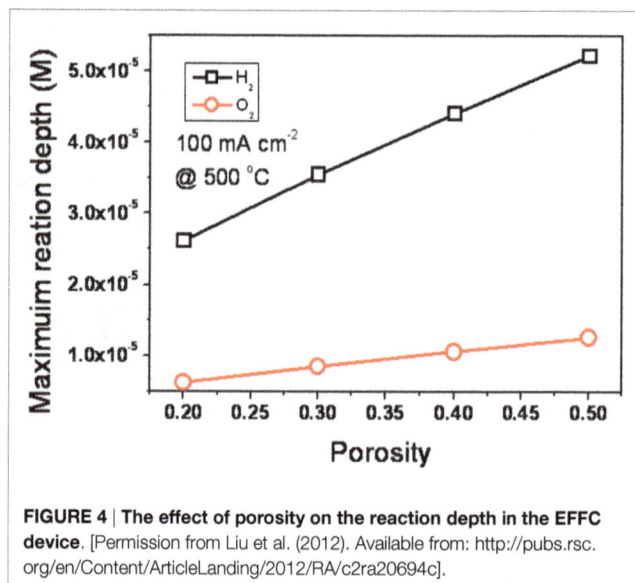

FIGURE 4 | The effect of porosity on the reaction depth in the EFFC device. [Permission from Liu et al. (2012). Available from: http://pubs.rsc.org/en/Content/ArticleLanding/2012/RA/c2ra20694c].

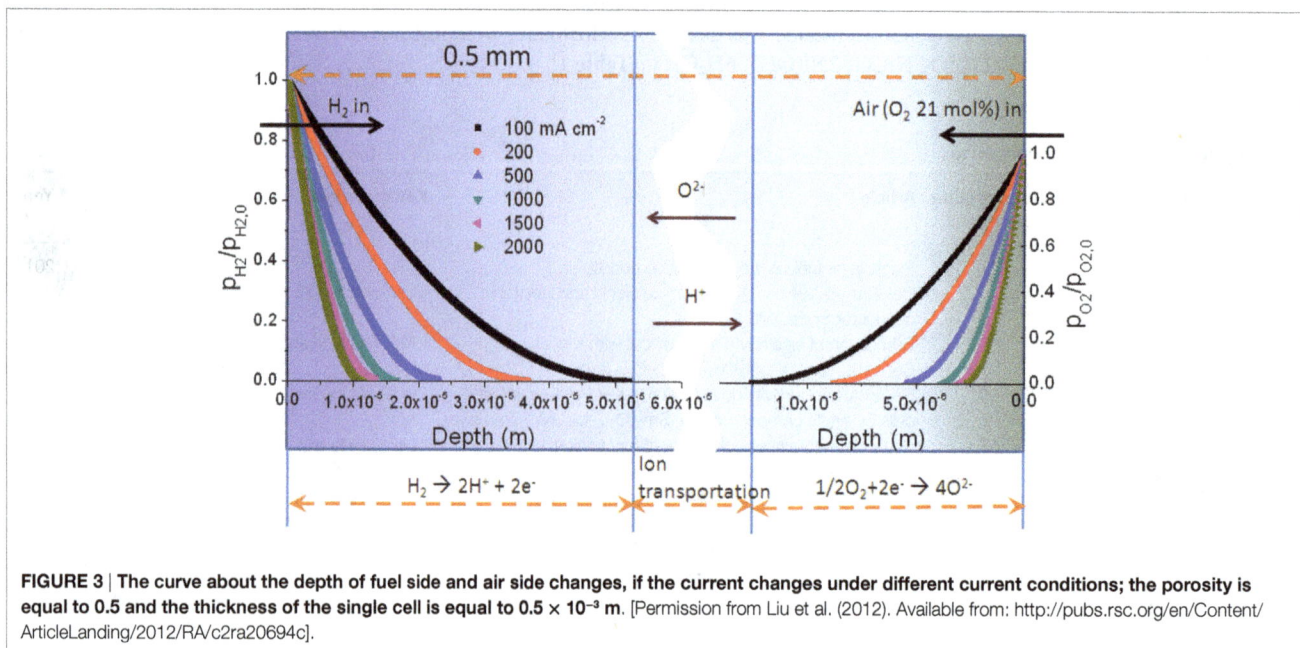

FIGURE 3 | The curve about the depth of fuel side and air side changes, if the current changes under different current conditions; the porosity is equal to 0.5 and the thickness of the single cell is equal to 0.5×10^{-3} m. [Permission from Liu et al. (2012). Available from: http://pubs.rsc.org/en/Content/ArticleLanding/2012/RA/c2ra20694c].

1 mm. However, EFFC is a device constructed in basis of nano-structure. Its pore diameter is up to dozens of nanometer. The maximum micro-distance between the oxygen and hydrogen is far less than the quenching distance in a millimeter level. Hence, it is impossible to explode, even though they have met inside the EFFC device.

Short-circuit Issue and Rectification Effect

Electrolyte (layer)-free fuel cell worked as electron-blocking layer. Does electron produced from fuel side pass through the EFFC single layer device and arrive to the air side? Hence, is there a short-circuit or electrochemical leakage current? As reported in literature, OCVs of the EFFC are higher than that using the pure ionic ceria, e.g., SDC electrolyte FC. These results do not support the existence of the short-circuit current.

The rectification effect for the junction was observed for the EFFC device, which is clearly seen in the measured $I-V$ curve when applying a bias voltage within –5 to +5 V range, as shown in **Figure 5**. All measurements were taken under the device OCV (open circuit condition) by supplying H_2/air. The rectification effect is decreased with increasing the ionic component of NSDC. These results provide evidence that the SJFC single layer device shows the junction barrier in the presence of the constituent semiconducting material phase in the single layer device.

The single layer is constituted by semiconductor-ionic material acting as an electron-blocking layer. Actually, under the FC conditions, supplying the H_2 and air, respectively, from both sides of the single layer device, there will form an n-type conducting zone in H_2-contacting side, and p-type conducting zone in the air side because the semiconducting transition metal oxides possess amphoteric properties, i.e., the transition metal oxide can behave with n-type conductivity in reducing atmosphere and as p-type conductor in oxidant environment due to their defect properties

(Singh et al., 2013). In this situation, a spatial n–p junction could be established, after the fuel and oxidant were supplied over the single layer device. Hydrogen is decomposed into protons and electrons on the surface of n-type semiconductor when inserting the H_2 and O_2 into the device. Meanwhile, oxygen receives electrons and forms oxygen ion at the surface of p-type semiconductor. More removable negative and positive charges are generated in areas of n-type and p-type semiconductor nanoparticles, respectively. A dynamic space-charge region may be re-formed between them and then internal electric field is generated similar to the traditional p–n junction. The space-charge region has two functions. The first is maintaining the electro dynamic potential of EFFC to ensure the output of electricity. The second is generating p–n junction, which can prevent electron produced at fuel side migrating to air side. It can also prevent the short-circuit current. Thus, the scientific principle is similar to solar cell.

There is still something unique for this single layer FC when compared with solar cell. Also, p–n junction is dynamic in EFFC, but static in solar cell. Spatial distribution of p–n junction is dependent on the composition of the atmosphere. We will give a more complete description about EFFC, including synergy problems of electron, ion, hole and proton, and the dynamic distribution of p–n junction, by updating the latest development in these regards.

Latest Progress on Scientific Studies of EFFCs

The invention of EFFC initiated a new scientific research frontier. However, the understanding of the electrochemistry in the existing FCs is not explicit, e.g., a key issue is how to prevent the electron passing through the device without using the electrolyte separator. The second issue is how to avoid short-circuit and electrochemical leakage when the mixing semiconductor and ionic-conducting layer are both introduced into the single layer. It is known that the mixed ionic and electronic conductors (MIECs) developed for SOFCs cannot replace the electrolyte. Otherwise, the short-circuiting problem will cause serious device OCV and power output losses (Eguchi et al., 1992; Riess et al., 1996; Shen et al., 2014). The true situation in the EFFCs is very different from the MIEC behavior, because the EFFC does not show OCV loss in a proper composition range of doped ceria ionic materials and semiconductor compared to the pure ionic-doped ceria electrolyte device, which usually exhibits rather lower OCV of 0.85–0.9 V due to the ceria-based electrolyte reduced electronic conduction. It was overcome in EFFCs using the doped ceria mixed with semiconductor materials. In our latest development, we have used the Sm-Ca co-doped ceria (SCDC), which has a high O^{2-} conductivity of 0.12 S/cm at 700°C (Banerjee et al., 2007) and perovskite $La_{0.6}Sr_{0.4}Co_{0.2}Fe_{0.8}O_{3-\delta}$ (LSCF), a p-type semiconductor that is one of the most promising SOFC cathodes due to its decent electrocatalytic properties for redox reactions (Esquirol et al., 2004), high electrical (p-type) conductivity, e.g., 230 S/cm at 900°C (Jiang, 2002) as well as good oxygen ion conductivity (0.1 S/cm at 800°C) (Kostgloudis and Ftikos, 1999).

Figure 6 shows typical $I-V$ and $I-P$ characteristics for this device compared to the pure ionic-doped ceria electrolyte device, which was fabricated using conventional FC anode/electrolyte/

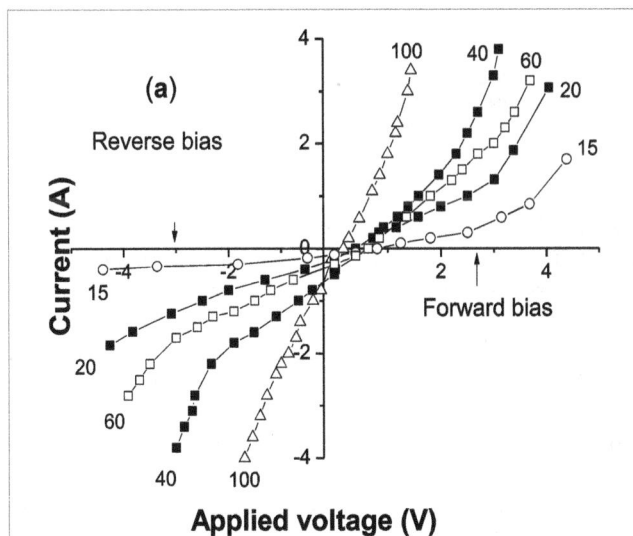

FIGURE 5 | The device I–V measured for the SJFC devices with various compositions under applied bias voltages. wt% represents the percentage of NSDC in NSDC–LCN composite, wt = 16, 20, 40, 60, 100. [Permission from Zhu et al. (2015). Available from: http://onlinelibrary.wiley.com/wol1/doi/10.1002/aenm.201401895/full].

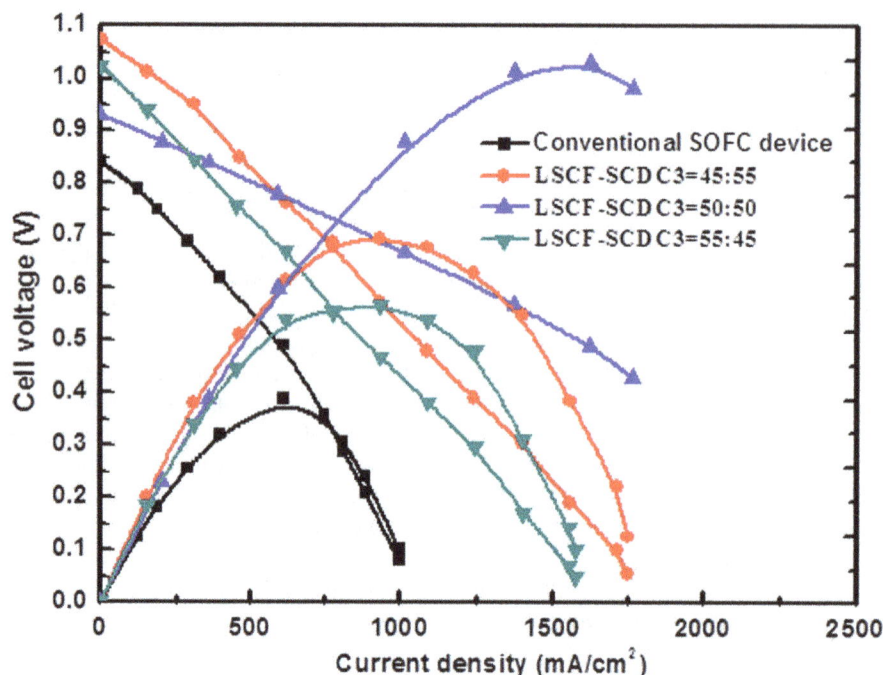

FIGURE 6 | Typical *I–V* and *I–P* characteristics for LSCF–SCDC fuel cells in comparison to the fuel cell using the ionic SCDC electrolyte.

cathode configuration. It can be seen from **Figure 6** that the OCV increased from pure ionic conductor SCDC electrolyte device 0.85 V up to 1.1 V at 45% LSCF:55% SCDC. Increasing the weight ratio of LSCF does not cause the OCV and power loss but increased significantly from pure SCDC at 300 mW/cm^2 up to 496 mW/cm^2 at 40% LSCF:55% SCDC. Though the OCV is somewhat lower than 0.94 V, it is still higher than that of the SCDC fuel cell 0.85 V, and the power output of the device with 55% LSCF: 45% SCDC raises to 798 mW/cm^2, see **Figure 6**. These facts, excluding the SCDC case, are in strong disagreement to the MIEC, which would have acted as a membrane instead of the electrolyte, and would cause significant losses in both voltage and power of the assembled device (Eguchi et al., 1992; Riess et al., 1996; Shen et al., 2014). This has been also proved by using the SCDC electrolyte due to its MIEC behavior in FC environment. But in the LSCF–SCDC membrane, we can see clearly that the incorporation of electronic (hole) conduction into the SCDC to form the semiconductor-ion membrane resulted in much better performance. It indicates a completely new phenomenon because of the unique LSCF–SCDC materials, which may be defined as a new type of functional semiconductor-ionic material. To the best of our knowledge, the semiconductor-ionic material has not been reported in literature so far.

Understanding the scientific principle, the underlying processes have been recently investigated. The device is activated by fuel. Charge carrier (e$^-$) is activated by fuel. When a proton (H$^+$) is formed, an electron is produced at the same time. Negative and positive charges are generated on the surface of air side and fuel side, respectively. The junction between them becomes depleted of charges/carriers, i.e., non-conduction. Consequently, a cell potential is generated, and electric energy can be taken out the

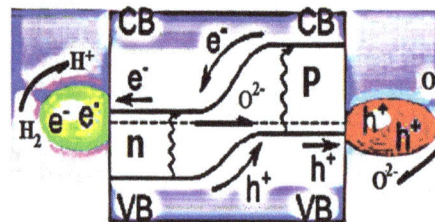

FIGURE 7 | EFFC built on the BHJ structure and working principle. [Permission from Zhu et al. (2013a). Available from: http://www.sciencedirect.com/science/article/pii/S2211285513000827].

device. The device functions based on nano-redox processes, as described by the formation of bulk heterojunction structures (BHJ). The energy band difference between the n and p semiconductors allows the charge separation at particles to prevent the electron crossing over internally to avoid the short-circuit problem (**Figure 7**) (Zhu et al., 2013b).

Also, Schottky junction has been discovered for the EFFC device. A potential or barrier can be built up at the interface between a metal and an n- or p-type semiconductor, known as the Schottky junction device (Mönch, 1994). Such device is preferably built only on p-type of semiconductor. In this case, it is a compatible anode metal, e.g., Ni or its alloy reduced from FC operation at H$_2$ side, on the semiconductor surface, typically, p-type semiconducting oxide, e.g., LiNiO-based oxide, as reported (Zhu et al., 2015) in a Schottky type contact/junction. Such junction can directly prevent the electrons crossing over the junction to avoid the short-circuiting problem due to its built-in

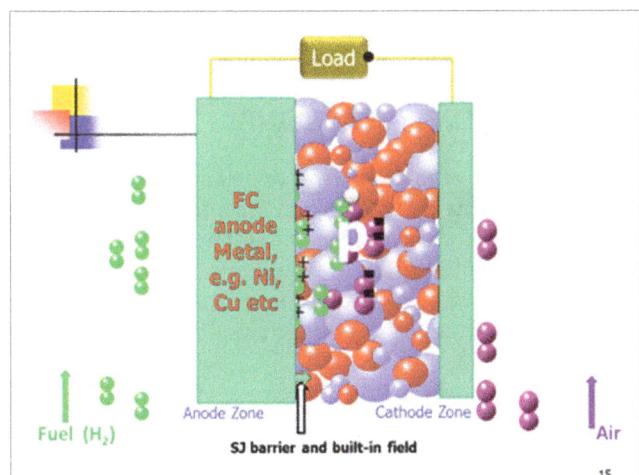

FIGURE 8 | The EFFC with the Shottky junction barrier built on p-type semiconductor and working principle (black ball: e$^-$ charge; white ball: hole charge; green ball$^+$: H$^+$; purple ball$^=$: O$^{2=}$; blue and red balls simply semiconductor-ionic material).

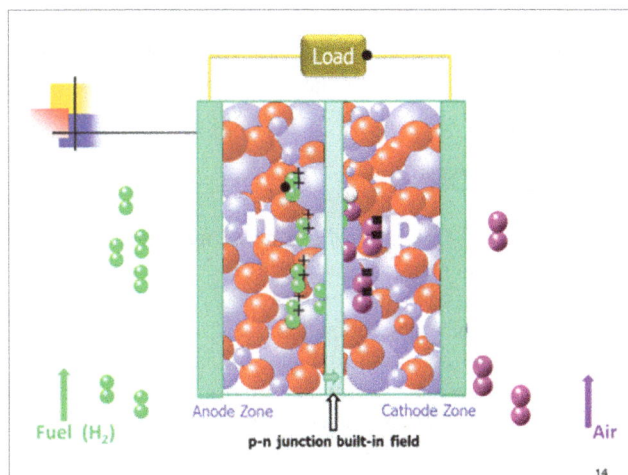

FIGURE 9 | p–n junction type EFFC working principle (black ball: e$^-$ charge; white ball: hole charge; green ball$^+$: H$^+$; purple ball$^=$: O$^{2=}$; blue and red balls simply semiconductor-ionic material).

field or Schottky junction barrier. The working principle for such type EFFC device is presented in **Figure 8**.

The bilayer EFFC device was discovered before Zhu et al. (2011e). At that time, no knowledge about the semiconductor junction was possessed, though it was constructed by using LiNiCuZn (Fe)-oxide material mixed with the electrolyte, $Sm_{0.2}Ce_{0.8}O_{2-x}$ (SDC), and perovskite cathode $Ba_{0.6}Sr_{0.4}Co_{0.85}Fe_{0.15}O$ (BSCF) (p-type) mixed with SDC were employed without using the electrolyte layer. More than 500 mW/cm^2 power output was achieved at 550°C, which is comparable with the performance achieved for conventional anode LiNiCuZn (Fe)/electrolyte (SDC)/cathode (BSCF) electrolyte-based FC. Now, it is recognized in retrospect as EFFC that is based on p–n junction principle, where LiNiCuZn (Fe) reduced by H_2 in FC operation forms the n-conducting layer, and thus the bilayer is actually the n–p device configuration. It is well known that p–n junction forms the depletion zone, which can prevent the electronic conduction only in one direction transport, i.e., diode effect (Constantinescu et al., 1973). The working principle for such type EFFC device is presented in **Figure 9**.

The scientific principle and underlying processes can be further understood by combining semiconducting and energy band theories. Using the LSCF–SCDC device case as the example, the energy band alignment between the LSCF and NCAL can block the electron in anode injecting into LSCF. Such well-aligned band level enables the creation of a new mechanism to prevent the semiconductor-ionic membrane device from short-circuiting and promote charge flows, and the ion transport and charge transfer more directly rather than the use of the electrolyte separator in the FC. On the other hand, the NCAL can be reduced at the anode side, to form Ni–Co metal surface; thus, a Schottky junction may be established between the metal layer and semiconductor material LSCF. The built-in field directs from the metal to LSCF (p-type), which can prevent the electrons passing through the interface between metal/LSCF–SCDC and hence establishes the

device OCV. Moreover, the built-in field can actually promote the O_2^- across over the junction, i.e., at the anode zone. Differing from the conventional FCs, the working principle here is unique in combining semiconductor physics and electrochemistry.

OPPORTUNITIES AND CHALLENGES FOR EFFC

Reviewing the history of the FC for over 170 years, basic technology and breakthrough are not enough to conquer the technical challenges, hence delaying the commercialization. Eventual barrier is caused or limited by the materials and the three-layer device structure. The appearance of EFFC could be not only a breakthrough but also a revolution in the FC sector. First, we overcome the inherent model that ionic conduction only happens in the electrolyte. The nanocomposite porous super ionic conduction material is built where ions can conduct both in the internal and on the surface to make it fast in transport and charge transfer to realize effectively the redox reaction for power generation. Second, the latest EFFC developments have realized much better power output that that from traditional structure SOFCs.

Electrolyte (layer)-free fuel cell was invented in 2010 and provides a new direction for FCs and SOFCs with new commercialization opportunity. It was selected as research highlight on Nature Nanotechnology in 2011 and named as "three in one" (Zhu, 2011). Introducing FC with high efficiency, zero/low emission and low noise into applications early and in large scale is of great importance. It is not only the demand of energy crisis but also for addressing environmental and low carbon emission issues. New science and energy technologies are urgently needed. EFFC semiconductor-ionic materials that can integrate FC all anode, electrolyte, and cathode functions have been invented and developed successfully. It has great potential in the competition of market issues. It is estimated that a cost of 100$/kW is feasible. The EFFC science and technology are expected to have wide application, e.g., other fuel cell technologies, not only for SOFCs, but also for other electrochemical devices, Fe–air, Al–air, Li–air batteries etc., super-capacitor, electrolysis, photoelectrochemical devices etc. It will exert great influence on the science, technology, and economy.

CONCLUSION

Electrolyte (layer)-free fuel cell is a new energy device and the overall function of generating electricity is done in an analogous way to a FC. It can bring high power efficiency as theoretical calculation and further expectation. On the other hand, it may also function for the electrolysis with high performance to electrolyze water. EFFC can remove the bottleneck from electrolyte, which used to be a barrier for commercialization.

This design of new device aims to conquer the conventional FC drawbacks, such as high cost, complex construction, and so on. The EFFC may rapidly drive FC industrialization and commercialization. The marketization of EFFCs largely depends on not only the materials but also the improvement of technology and the performance of the devices. There still remains a lot of future work that needs to be carried out. Initiating from the long-term development demanding of science and technology, research activities about EFFC is expanding.

AUTHOR CONTRIBUTIONS

YL's contribution of this work was to find and read the related literature and write the manuscript. Main idea of this work is from Dr. BZ, who also guided all the writing process. YC, J-SK, and BW's contribution to this work were to modify the details. JW and YZ mainly focused on investigating the manuscript. JL's work about this review was to draw the pictures.

FUNDING

This work was supported by the Natural Science Foundation of Hubei Province, major project (Grant No. 2015CFA120), the Swedish Research Council (Grant No. 621-2011-4983), the European Commission FP7 TriSOFC-project (Grant No. 303454), and the Swedish Agency for Innovation Systems (VINNOVA). One of the lead authors would also like to thank the Hubei Provincial 100-Talent Distinguished Professor grant.

REFERENCES

Banerjee, S., Devi, P. S., Topwal, D., Mandal, S., and Menonr, K. (2007). Enhanced ionic conductivity in $Ce_{0.8}Sm_{0.2}O_{1.9}$: unique effect of calcium co-doping. *Adv. Funct. Mater.* 17, 2847–2854. doi:10.1002/adfm.200600890

Constantinescu, C., Goldenblum, A., and Sostarich, M. (1973). Photovoltaic effects in laterally illuminated p n junctions. *Int. J. Electron.* 35, 65–72. doi:10.1080/00207217308938517

Dufour, A. U. (1998). Fuel cells: a new contributor to stationary power. *J. Power Sources* 71, 19–25. doi:10.1016/S0378-7753(97)02732-8

Eguchi, K., Setoguchi, T., Inoue, T., and Arai, H. (1992). Electrical properties of ceria-based oxides and their application to solid oxide fuel cells. *Solid State Ionics* 52, 165. doi:10.1016/0167-2738(92)90102-U

Esquirol, A., Brandon, N. P., Kilner, J. A., and Mogensen, M. (2004). Electrochemical characterization of $La_{0.6}Sr_{0.4}Co_{0.2}Fe_{0.8}O_3$ cathodes for intermediate-temperature SOFCs. *J. Electrochem. Soc.* 151, A1847–A1855. doi:10.1149/1.1799391

Fan, L. D., Wang, C. Y., Chen, M. M., and Zhu, B. (2013). Recent development of ceria-based (nano)composite materials for low temperature ceramic fuel cells and electrolyte-free fuel cells. *J. Power Sources* 234, 154–174. doi:10.1016/j.jpowsour.2013.01.138

Fan, L. D., Wang, C. Y., Osamudiamen, O., Raza, R., Singh, M., and Zhu, B. (2012). Mixed ion and electron conductive composites for single component fuel cells: I. Effects of composition and pellet thickness. *J. Power Sources* 217, 164–169. doi:10.1016/j.jpowsour.2012.05.045

Hei, Y. F., Huang, J. B., Wang, C., and Mao, Z. Q. (2014). Novel doped barium cerate-carbonate composite electrolyte material for low temperature solid oxide fuel cells. *Int. J. Hydrogen Energy* 39, 14328–14333. doi:10.1016/j.ijhydene.2014.04.031

Hu, H. Q., Lin, Q. Z., Zhu, Z. G., Zhu, B., and Liu, X. R. (2014). Fabrication of electrolyte-free fuel cell with $Mg_{0.4}Zn_{0.6}O/Ce_{0.8}Sm_{0.2}O_{2-\delta}-Li_{0.3}Ni_{0.6}Cu_{0.07}Sr_{0.03}-O_{2-\delta}$ layer. *J. Power Sources* 248, 577–3581. doi:10.1016/j.jpowsour.2013.09.095

Janicke, M. T., Kestenbaum, H., Hagendorf, U., Schuth, F., Fichtner, M., and Schubert, K. (2000). The controlled oxidation of hydrogen from an explosive mixture of gases using a microstructured reactor/heat exchanger and Pt/Al_2O_3 catalyst. *J. Catal.* 194, 282–293. doi:10.1006/jcat.2000.2819

Jiang, S. P. (2002). A comparison of O-2 reduction reactions on porous $(La,Sr)MnO_3$ and $(La,Sr)(Co,Fe)O$-3 electrodes. *Solid State Ionics* 146, 1–22. doi:10.1016/S0167-2738(01)00997-3

Kostogloudis, G. C., and Ftikos, C. (1999). Properties of A-site-deficient $La_{0.6}Sr_{0.4}Co_{0.2}Fe_{0.8}O_3-\delta$-based perovskite oxides. *Solid State Ionics* 126, 143–151. doi:10.1016/S0167-2738(99)00230-1

Liu, Q. H., Qin, H. Y., Raza, R., Fan, L. D., Li, Y. D., and Zhu, B. (2012). Advanced electrolyte-free fuel cells based on functional nanocomposites of a single porous component: analysis, modeling and validation. *RSC Adv.* 2, 8036–8040. doi:10.1039/c2ra20694c

Mönch, W. (1994). Metal-semiconductor contacts: electronic properties. *Surf. Sci.* 299-30, 928–944. doi:10.1016/0039-6028(94)90707-2

Riess, I., Gödickemeier, M., and Gauckler, L. J. (1996). Characterization of solid oxide fuel cells based on solid electrolytes or mixed ionic electronic conductors. *Solid State Ionics* 90, 91–104. doi:10.1016/S0167-2738(96)00355-4

Shen, S. L., Yang, Y. P., Guo, L. J., and Liu, H. T. (2014). A polarization model for a solid oxide fuel cell with a mixed ionic and electronic conductor as electrolyte. *J. Power Sources* 256, 43–51. doi:10.1016/j.jpowsour.2014.01.041

Singh, K., Nowotny, J., and Thangadurait, V. (2013). Amphoteric oxide semiconductors for energy conversion devices: a tutorial review. *Chem. Soc. Rev.* 42, 19611. doi:10.1039/c2cs35393h

Suzuki, T., Jasinski, P., Petrovsky, V., Anderson, H. U., and Dogan, F. (2005). Impact of anode microsturcture on solid oxid fuel cells. *J. Electrochem. Soc.* 152, A527–A531. doi:10.1149/1.1858811

Veldsink, J. W., van Damme, R. M. J., Versteeg, G. F., and van Swaajj, W. P. M. (1995). The use of the dusty-gas model for the description of mass transport with chemical reaction in porous media. *Chem. Eng. J.* 57, 115–125. doi:10.1016/0923-0467(94)02929-6

Xia, D., Li, T., Jiang, L., Zhao, Y. C., Tian, Y., and Li, Y. D. (2011). Single layer fuel cell based on a composite of $Ce_{0.8}Sm_{0.2}O_{2-\delta}-Na_2CO_3$ and amixed ionic and electronic conductor $Sr_2Fe_{1.5}Mo_{0.5}O_{6-\delta}$. *J. Power Sources* 249, 270–276. doi:10.1016/j.jpowsour.2013.10.045

Xia, Y. J., Liu, X. J., Bai, Y. J., Li, H. P., Deng, X. L., Niu, X. D., et al. (2012). Electrical conductivity optimization in electrolyte-free fuel cells by single-component $Ce_{0.8}Sm_{0.2}O_{2-\delta}-Li_{0.15}Ni_{0.45}Zn_{0.4}$ layer. *RSC Adv.* 2, 3828–3834. doi:10.1039/c2ra01213h

Yi, B. L. (1998). Status and future of fuel cell. *Chin. J. Power Sources* 22, 2–6.

Zhang, F. Z., and Cooke, P. (2010). Hydrogen and fuel cell development in China: a review. *Eur. Plan. Stud.* 18, 1153–1165. doi:10.1080/09654311003791366

Zhu, B. (2011). Nature nanotechnology research highlights. Three in one. *Nat. Nanotechnol.* 6, 330–330.

Zhu, B., Fan, L. D., and Lund, P. (2013a). Breakthrough fuel cell technology using ceria-based multi-functional nanocomposites. *Appl. Energy* 106, 163–175. doi:10.1016/j.apenergy.2013.01.014

Zhu, B., Lund, P., Raza, R., Patakangas, J., Huang, Q. A., Fan, L. D., et al. (2013b). A new energy conversion technology based on nano-redox and nano-device processes. *Nano Energy* 2, 1179–1185. doi:10.1016/j.nanoen.2013.05.001

Zhu, B., Fan, L. D., Zhao, Y. F., Tan, W. Y., Xiong, D. B., and Wang, H. (2014). Functional semiconductor – ionic composite GDC-KZnAl/LiNiCuZnOx for single-component fuel cell. *RSC Adv.* 4, 9920–9925. doi:10.1039/c3ra47783e

Zhu, B., Lund, P., Raza, R., Ma, Y., Fan, L. D., Afzal, M., et al. (2015). Schottky junction effect on high performance fuel cells based on nanocomposite materials. *Adv. Energy Mater.* 5, 1401895. doi:10.1002/aenm.201401895

Zhu, B., Ma, L., Wang, X. D., Raza, R., Qin, H. Y., and Fan, L. D. (2011a). A fuel cell with a single component functioning simultaneously as the electrodes and electrolyte. *Electrochem. Commun.* 13, 225–227. doi:10.1016/j.elecom.2010.12.019

Zhu, B., Qin, H. Y., Raza, R., Liu, Q. H., Fan, L. D., Patakangas, J., et al. (2011b). A single-component fuel cell reactor. *Int. J. Hydrogen Energy* 36, 8536–8541. doi:10.1016/j.ijhydene.2011.04.082

Zhu, B., Raza, R., Abbas, G., and Singh, M. (2011c). An electrolyte-free fuel cell constructed from one homogenous layer with mixed conductivity. *Adv. Funct. Mater.* 21, 2465–2469. doi:10.1002/adfm.201002471

Zhu, B., Raza, R., Qin, H. Y., and Fan, L. D. (2011d). Fuel cells based on electrolyte and non-electrolyte separators. *Energy Environ. Sci.* 4, 2986–2992. doi:10.1039/c1ee01202a

Zhu, B., Raza, R., Qin, H. Y., and Fan, L. D. (2011e). Single-component and three-component fuel cells. *J. Power Sources* 196, 6362–6365. doi:10.1016/j.jpowsour.2011.03.078

Zhu, B., Raza, R., Liu, Q. H., Qin, H. Y., Zhu, Z. G., Fan, L. D., et al. (2012). A new energy conversion technology joining electrochemical and physical principles. *RSC Adv.* 2, 5066–5070. doi:10.1039/c2ra01234k

Conflict of Interest Statement: The authors declare that the research was conducted in the absence of any commercial or financial relationships that could be construed as a potential conflict of interest.

14

Value of Lost Load: An Efficient Economic Indicator for Power Supply Security?

Thomas Schröder* and Wilhelm Kuckshinrichs

Institute of Energy and Climate Research – Systems Analysis and Technology Evaluation (IEK-STE), Forschungszentrum Jülich GmbH, Jülich, Germany

Security of electricity supply has become a fundamental requirement for well-functioning modern societies. Because of its central position in all sections of society, the present paper considers the economic consequences of a power supply interruption. The value of lost load (VoLL) is a monetary indicator expressing the costs associated with an interruption of electricity supply. This paper reviews different methods for calculating VoLL, provides an overview of recently published studies, and presents suggestions to increase the explanatory power and international comparability of VoLL.

Keywords: power outage, energy security, value of lost load, energy systems, qualitative analysis

Edited by:
Luís Alexandre Duque Moreira De Sousa,
Public Research Centre Henri Tudor, Luxembourg

Reviewed by:
Behnam Mohammadi-ivatloo,
University of Tabriz, Iran
Payman Dehghanian,
Texas A&M University, USA

*Correspondence:
Thomas Schröder
t.schroeder@fz-juelich.de

Specialty section:
This article was submitted to Energy Systems and Policy, a section of the journal Frontiers in Energy Research

Citation:
Schröder T and Kuckshinrichs W (2015) Value of Lost Load: An Efficient Economic Indicator for Power Supply Security? A Literature Review.
Front. Energy Res. 3:55.

INTRODUCTION

Power blackouts or interruptions of supply all over the world demonstrate the potential of severe socioeconomic disruptions and economic losses. A selection of events during the last 20 years includes blackouts, such as those of 26 April 1995 (USA), 8 June 1995 (Israel), 20 June 1998 (Bangladesh), 21 January 2003 (Brazil), and 14 March 2005 (Australia). The latest blackout in March 2015 plunged Turkey into darkness (Reevell, 2015). Therefore, it is obviously of great importance to analyze blackout events, identify technical options, and develop strategies and instruments to avoid blackouts or to deal successfully with such events (Makarov et al., 2005; Barkans and Zalostiba, 2009) Typically, blackouts are not caused by a single event but by a combination of several malfunctions, such as unforeseen simultaneous interruptions of several power plants, sudden simultaneous high power demand, breakdown of electrical equipment, human errors during maintenance work, switching operations, or power line collapse. Beside this reason, an increasing international interconnection and interdependence of networks may lead to situations in which even failures of a small fraction of nodes in one network can lead to the complete fragmentation of a system of several networks (Buldyrev et al., 2010). Such events are called cascading events. Impressive examples comprise the blackouts in the European Power System on 28 September 2003 and 4 November 2006 (Bundesnetzagentur, 2007; Barkans and Zalostiba, 2009; Buldyrev et al., 2010).

Generally, the power supply industry identified *Liberalization and Privatization* (which mainly took place in the 1990s) and *Expansion of Renewable Energy Production Capacities* (which forms an essential option for sustainable energy systems) as the two major trends in the last 10 to 20 years that increase the risk of power blackouts (Aichinger et al., 2011). For industrialized countries aiming at energy sustainability through the increased use of renewable energies for power supply, additional efforts are necessary to preserve the level of power supply security, such as grid adaptations, as Pesch et al. (2014) have shown for Germany. All these options involve increasing costs, which must be considered if power supply security is to be maintained.

On the other hand, efforts to maintain or increase the level of power supply security should be balanced against the damage as a consequence of blackouts, because it is obvious that blackouts involve far-reaching consequences for the entire socioeconomic system (Petermann et al., 2011). Obviously, (nearly) every economic process is highly dependent on a safe and reliable supply of electricity. Technical indices, such as SAIFI, SAIDI, and CAIDI, statistically reflect the security of the system focusing on average power interruption frequency, duration, and intensity. From a socioeconomic perspective, the *value of lost load (VoLL)*[1] is an important indicator addressing the economic consequences of power blackouts and the monetary evaluation of uninterruptedness of power supply. It has a long history and current studies present quantifications.

The present paper takes up the issue of economically evaluating the security of the power supply using the indicator VoLL. However, first of all, the nature of an interruption of the power supply will be described in detail. Consideration will be given to the various factors influencing a blackout (see Characteristics of Power Interruptions). The cost aspects of a blackout will then be discussed and various methods for determining VoLL will be qualitatively evaluated (see Costs and VoLL Measurement Methods). A structured overview will be used to analyze the informative value of a range of various VoLL studies from the past 10 years (see Current VoLL Studies). This will be followed by a qualitative evaluation of the VoLL approach as an economic indicator of power supply security. Furthermore, a framework for VoLL will be presented that will improve the temporal and international comparability of the results (see Suggestions to Increase the Explanatory Power of VoLL). The text concludes with a summary and conclusions (see Summary).

CHARACTERISTICS OF POWER INTERRUPTIONS

Technical and Systemic Characteristics

A power outage occurs when electricity customers (industry, state, individuals) are supplied with less electricity than they require from the electricity system (Ajodhia, 2006). A blackout describes a situation when no electricity is supplied at all. It can have many causes, such as malfunctions or overloading of the various levels of the electricity system, malfunctions in the generation, transmission, or distribution structure, or as a consequence of a lack of raw materials (Ajodhia, 2006). In supply systems with a high and possibly increasingly proportion of renewable energy sources (RES), which cannot be regulated easily and are not suitable for maintaining the base load, there is a growing danger of interruptions at the level of the transmission and distribution grids, as shown by the example of the increasing number of interventions by transmission system operators in Germany to regulate supply.[2] However, this is largely of no significance from the perspective of

electricity customers. The consequences for electricity customers (material damage, costs) are not usually affected by the cause of the interruption and are contingent on how much they depend on electricity (Sanghvi, 1982) as well as how long they are being interrupted, which will be thoroughly explored later in the paper. The consequences are affected by the factors influencing the outage, which are inherent to each individual case. The character of the individual factors and their combination determine the extent of the consequences. Each outage, therefore, represents a unique event that affects electricity customers to different extents.

In order to represent the multidimensionality of a blackout, the different factors characterizing an interruption of the electricity supply can be broken down into various subcategories. Based on Ratha et al. (2013), the factors influencing the blackout are divided into the subcategories of "technical factors," "load-side factors," and "social factors" (**Table 1**).

The technical factors describe the framework conditions constraining the interruption, the characteristics of which are decisive for the consequences of a blackout. The load-side factors concern the effects that exacerbate the damage arising as a consequence of the structure of the electricity customer affected. In this respect, the customers' pattern of electricity use is also decisive (Caves et al., 1990). The load-side factors are naturally determined by the technical factors. Finally, the social factors describe the influences that affect the consequences of the blackouts but which are difficult to assess objectively. These are mainly culturally related differences in the economic and social structures of different regions, which lead to differences in power supply security. According to Ratha et al. (2013), it is particularly the cultural factors that cannot be modeled appropriately.

Time Characteristics

In addition to the multifaceted parameters of a blackout described in Section "Technical and Systemic Characteristics," the time course of a blackout must also be considered in a differentiated manner. The duration of an interruption of the power supply is an essential influencing factor and requires closer consideration. Three basic phases can be identified, each following the other. The first phase concerns preparation for an interruption (if the interruption is planned and announced), such as the modification of working procedures. This requires the workforce and resources to be used for restructuring and preparatory work preventing them from fulfilling their usual duties or functions. The second phase describes the period of the actual interruption of the power supply.

---

[1]In electric industry and in the literature following terms "Value of Customer Reliability," "Cost of Unserved Energy," "Cost of Power Interruption," or "Cost of Electricity Outages," are used synonymously to address the phenomenon of VoLL.
[2]In 2003, the transmission system operator TenneT had to intervene twice, whereas this increased to 387 times in 2007 and in 2012 the number of interventions rose to 1213 (Falthauser and Geiß, 2012; Barth, 2013).

TABLE 1 | Factors influencing power interruptions.

Technical factors	Load-side factors	Social factors
• Duration	• Type of electricity customer	Special cultural and social features
• Region	• Number of customers affected and level of dependence on electricity	
• Frequency		
• Time		
• Dimension	• Degree to which process steps can be substituted	
• Advance warning		
• Accustomed level of supply security	• Existence of standby power supply	

Source: adapted from Ratha et al. (2013).

The third and final phase is taken to be the interval before the usual production processes are up and running again (Rose et al., 2004). During this phase, the opportunities increase once again to intervene and guide events, although the duration and characteristics of the final phase strongly depend on the crisis management abilities of the executives in a company (Caves et al., 1992).

This breakdown into phases can be applied to all affected electricity customers. If advance notice is given, then, for example, the first phase also begins with preparatory measures for private electricity customers, such as data backup or controlled shutdown of electrical appliances. During the power interruption, all electricity-dependent activities are affected (both housework and leisure activities). The phase of restoring usual activities and rectifying any damage starts with the end of the power interruption.

The different phases may vary in their duration and characteristics so that, for instance, the first step of making preparations may be inapplicable if there is no advance warning. In this case, the two subsequent phases are more extreme.

Technical Indicators for Characterizing Power Interruptions

Following on from Section "Time Characteristics," a power interruption is a very complex phenomenon that is influenced by a large number of stochastic factors. The Institute of Electrical and Electronics Engineers (IEEE) has compiled standardized technical indices in order to measure, evaluate, and compare the reliability and quality of the power supply. The most important indices for supply security are System Average Interruption Frequency Index (SAIFI), System Average Interruption Duration Index (SAIDI), and Customer Average Interruption Duration Index (CAIDI). They refer to the low-and medium-voltage grids (IEEE, 2004). The determination of the technical indices is subject to clearly defined regulations, for example, only supply interruptions lasting longer than 3 min are taken into consideration. These technical indices form the framework for the regulatory authorities' process of monitoring supply security. The uniform procedure to be applied in collecting data ensures that values can be compared internationally and also over time. Nevertheless, the informative value of the indices has been criticized in various ways. For example, industry is not satisfied with the length of time during which data on power interruptions are collected. Since industrial plants are often liable to brief power failures, they frequently shut down automatically if the power supply is interrupted for more than about 0.2 s (Schlandt, 2012). Furthermore, only the duration of the interruption itself is considered. Periods of advance warning and restart times are not covered according to the definitions of the technical indices. The indices merely indicate that an interruption has occurred. They do not refer to the measures employed or to the efforts made to avoid an interruption.

COSTS AND VoLL MEASUREMENT METHODS

Over and above the acquisition of these purely technical indices, the question arises of the resulting damage and the macroeconomic costs of a power interruption. This requires an economic consideration of power interruptions. To this end, the structure of the different cost categories will be examined in the next section.

Cost Categories for Damage and Mitigation

It seems appropriate to present the various cost types and categories according to the different end users (roughly: industry and commercial users, individuals).

First of all, a distinction can be made between two types of costs. On the one hand, there are costs that can be termed damage costs. On the other hand, end users incur costs that can be better described as mitigation costs. Damage costs can be broken down into direct and indirect costs. Direct damage costs are taken to mean those that are incurred directly by the company or the individual affected. For example, loss of production can be regarded as direct damage for the manufacturer. This loss of production then makes itself felt as indirect damage for other companies in the form of delayed deliveries. Mitigation costs are understood, for example, as costs for the procurement and operation of standby generators. **Table 2** gives an overview of the

TABLE 2 | Structure of damage and mitigation costs.

Economy (industry, commercial users)			Private individuals		
Damage costs		Mitigation costs	Damage costs		Mitigation costs
Direct	Indirect		Direct	Indirect	
(a) Opportunity costs of idle resources • Labor • Country • Capital • Profits (b) Production holdups and restart times (c) Adverse effects and damage to capital goods, data loss (d) Health and safely aspects	(a) Delayed deliveries along the value chain (b) Damage for consumers if the company produces an end product (c) Costs/benefits for some manufacturers (d) Health and safely aspects	Procurement of standby generators, batteries, etc. Investments in grid construction via charges (network tariffs)	(a) Restrictions on activities, lost leisure, stress (b) Financial costs • Damage to premises and real estate • Food spoilage • Data loss (c) Health and safely aspects	Restrictions on acquisition of goods Costs for other private individuals and companies	Procurement of standby generators, batteries, etc. Investments in grid construction via charges (network tariffs)

Source: adapted from Munasinghe and Sanghvi (1988).

types of costs structured according to end user and according to whether they are damage or mitigation costs.

However, Rose et al. (2004) object that in the course of a supply interruption not only do costs arise, but there are also some market participants who profit from the interruption such as companies commissioned to undertake repair and restoration work as a result of the interruption. At the same time, this means follow-up costs for the companies placing the orders. Even if it does not seem particularly desirable for the companies affected in this context, a power interruption also means savings in electricity costs (Caves et al., 1992), although they generally tend to be small in comparison to the costs of production downtimes (except in the case of energy-intensive manufacturing sectors).

Cost Optimum for Power Supply Security

In order to assess the damage costs of a power interruption, an important approach is the VoLL. VoLL can be understood as an economic indicator for power supply security. VoLL is determined by relating the monetary damage arising from a power outage due to the loss of economic activities to the level of kWh that were not supplied during an interruption (van der Welle and van der Zwaan, 2007). In addition to plotting monetary units against kWh, costs can also be plotted in relation to time. However, a representation in monetary units/kWh is more commonly used (Ajodhia, 2006). Since the VoLL is an economic indicator, the cause of the power interruption is of no interest (Frontier Economics, 2008).

In the optimum case, the level of supply security should be defined in such a way that the marginal damage costs, expressed by VoLL, are equal to the marginal costs for ensuring uninterrupted electricity supply (Röpke, 2013) (see **Figure 1**).

Accordingly, the calculation of the economic indicator VoLL represents, on the one hand, an opportunity to determine the level of damage caused by a power interruption, the result of which, on the other hand, describes the value of power supply security (van der Welle and van der Zwaan, 2007).

VoLL Measurement Approaches

Modern industrialized societies are extremely dependent on electricity so that electricity can be regarded both as an essential input factor for all economic processes and also as the basis for many forms of leisure activities. From the economic point of view, it can be argued that all economic activities cease when there is no electricity (Holmgren, 2007). A secure power supply is thus indispensable for electricity-based life in society. It is also regarded as an important locational advantage (von Roon, 2013) and has a macroeconomic value.

Even if VoLL offers the opportunity of expressing the value of power supply security in monetary terms, there is no market on which power interruptions can be traded, which is why VoLL cannot be directly derived as market performance. Consequently, VoLL must be determined by using scientific measuring techniques (van der Welle and van der Zwaan, 2007). The individual techniques are grouped differently in the literature [see Caves et al. (1990), Woo and Pupp (1992), Sullivan and Keane (1995), Lijesen and Vollaard (2004), Ajodhia (2006), de Nooij et al. (2007), and London Economics (2013a)]. A general distinction can be made between direct or survey methods and indirect methods (**Table 3**). Direct or survey methods obtain their information on the costs of power interruptions directly from end users, whereas indirect methods require other sources of information, such as statistical data (Ajodhia, 2006).

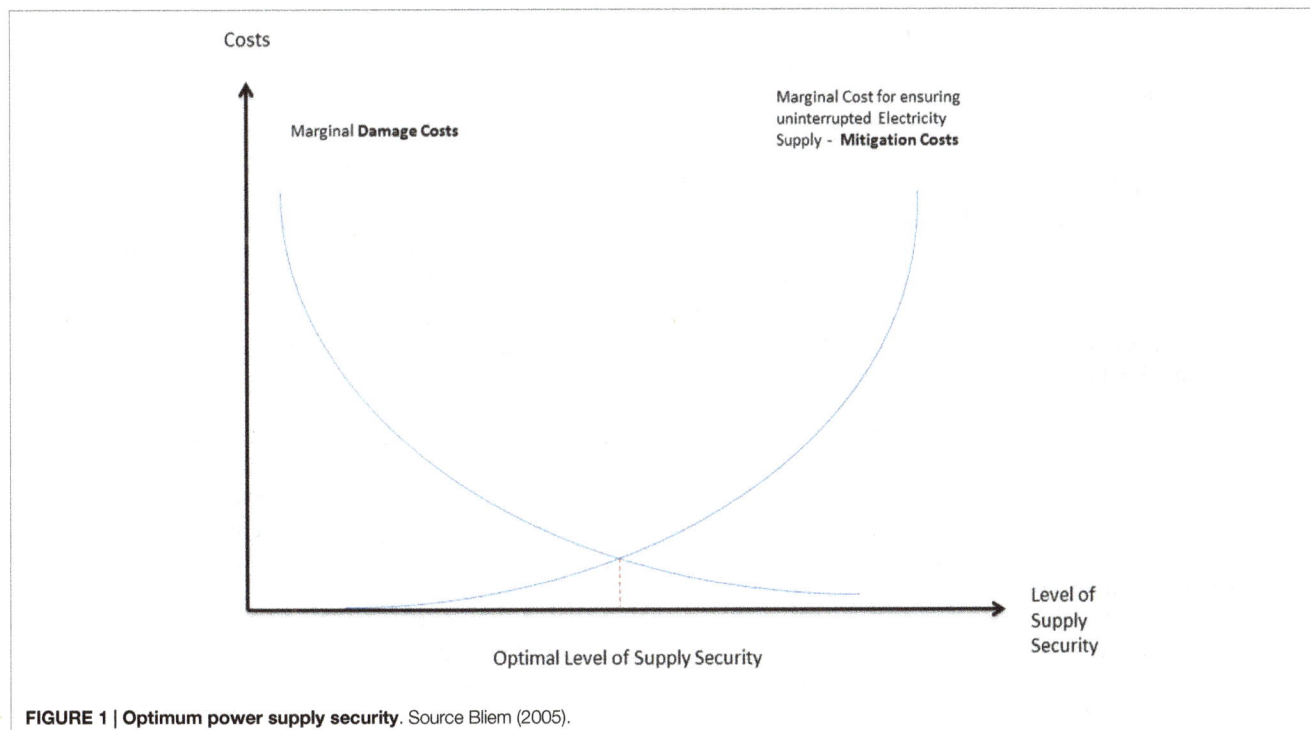

FIGURE 1 | Optimum power supply security. Source Bliem (2005).

TABLE 3 | Overview of different VoLL measurement techniques.

	Direct (Survey)		Indirect	
Blackout studies	Willingness to pay/avoid	Direct costs	Macroeconomic approaches	Revealed preference

In the literature, the VoLL values of different user groups are calculated separately. This particularly applies to industrial or commercial users and private users. Further subdivisions can also be made, e.g., on the basis of branches of industry. The reason for this is that different users are affected differently by the same power interruption. A subdivision also makes sense since no marketable output results from power consumption in private households (Ratha et al., 2013), which makes it more difficult to quantify the costs of the interruption (LaCommare and Eto, 2006). In the following sections, the techniques will be briefly presented and their advantages and disadvantages will be discussed (see **Tables 4** and **5**).

Direct Approaches

Blackout Studies

In this approach, the resulting damage costs of a real power interruption are recorded retrospectively. The interruption parameters are clearly defined. This method is mainly used for long-lasting and large-area interruptions (Billinton et al., 1993). A blackout study can also be used as a reference for verifying other VoLL methods. As part of the cost survey, blackout studies also frequently analyze both the performance of the emergency services as well as the impacts on the ecological system (Ajodhia, 2006).

The basis of the survey – a real power interruption – is both the strength and the weakness of this method. The advantage is that VoLL can be measured for a real event. The disadvantage is, however, that blackouts, at least in industrialized countries, are relatively rare and generally occur without warning so that researchers cannot adequately prepare themselves for the event and that information gathering is time-consuming and expensive.

Willingness to Pay/Avoid

A number of econometric methods are available for assessing power outages, of which the best known are contingent valuation and contingent ranking (or Choice) (Ajodhia, 2006). In the contingent valuation method (CVM), individuals are asked, either in questionnaires or direct interviews, to give a monetary value to certain unmarketable goods (in this case a blackout). A clearly defined scenario must be formulated in advance. Accordingly, all the payments and earnings determined by this method are of a hypothetical nature. The questions posed using this method can be generally subdivided into two approaches: willingness to pay (WTP) and willingness to accept (WTA). With respect to a power interruption, customers are asked WTP questions about how much they would pay to either avoid a blackout or to be guaranteed a higher level of supply security. The WTA approach represents the opposite strategy according to which questions are formulated about how much money consumers would have to be offered for them to accept a reduction in supply security or to

TABLE 4 | Pros and cons of direct (survey) measurement techniques.

Advantages	Disadvantages
Blackout studies	
• Great advantage is the investigation of real blackouts, no hypothetical basis (de Nooij et al., 2003)	• Very limited application (Frontier Economics, 2008)
Willingness to pay/avoid	
• Consideration of the interruption parameters using scenarios (Frontier Economics, 2008) • Possible to integrate periods of advance warning and restart times • Possible to consider making up for loss of production • Flexible with respect to the variables to be measured (Frontier Economics, 2008) (e.g., can be expanded to include socioeconomic aspects)	• Strong priority for retaining the *status quo*, which may lead to distortions in WTP vs. WTA • Large discrepancy between WTP and WTA values determined (WTA > WTP) → verifiability of the results? (Woo and Pupp, 1992); explainability by calculating the benefits and behavioral economics? (Schubert et al., 2013) • Subject of investigation unknown, problems in putting oneself in the hypothetical situation (Sanghvi, 1982) • Results dependent on type of survey (questionnaire, interview), possible distortion due to wording (formulation, emphasis) (Sanghvi, 1982) • Time-consuming (London Economics, 2013a)
Direct costs	
• Information direct from end users • Possible to evaluate various scenarios (Frontier Economics, 2008) • Above all, suitable for industrial and commercial users (Caves et al., 1990)	• Subject of investigation unknown • Less clear-cut costs are not satisfactorily covered; very difficult to assess work undertaken in households in terms of money (Caves et al., 1990; Billinton et al., 1993) • Problem of finding economic value for uncertainties, annoyance, and stress for private end users

retain the present level of security instead of being upgraded to a higher level (Caves et al., 1990).

Another technique is the contingent ranking method (CRM). In this case, individuals are asked to rank a number of options (here, interruption scenarios) (Kling et al., 2012). Each option is linked to a certain monetary value, i.e., compensation or cost. WTP and WTA values can be derived from the interviewees' answers in order to determine the consumers' preferences (Caves et al., 1990). In CRM, the interviewers set the prices (Ajodhia, 2006). WTP/WTA studies can also take the socioeconomic characteristics of the interviewees into consideration (Portney, 1994).

Direct Costs

In this method, interviewees are given a set of different blackout scenarios, for example, of different durations or starting times, to give them a feeling for the general issues involved in a power outage. For each scenario, the end users are asked about the damage costs that they would experience in each situation. In some studies, interviewees are asked to divide the damage costs into categories. This procedure is mainly applied for industrial and commercial users (Billinton et al., 1993), where different damage categories may result from the different operations in the companies (see **Table 2**). The identification of different cost categories has two

Value of Lost Load: An Efficient Economic Indicator for Power Supply Security?

131

TABLE 5 | Pros and cons of indirect measurement techniques.

Advantages	Disadvantages
Production function	
• Little data required, data available at low cost, simple application (Woo and Pupp, 1992) • Sectoral calculation, aggregation for the whole economy (de Nooij et al., 2007) • Simple linkage to input-output analysis (Chen and Vella, 1994; Praktiknjo, 2013)	• Greatly simplified assumptions (Frontier Economics, 2008) • Inadequate consideration of advance warning periods, system restarts, and restart times • No temporal differentiation, e.g., assumption of identical time-independent costs, no consideration of factors influencing the interruption (Sanghvi, 1982) • No consideration of making up for loss of production (London Economics, 2013a)
Household income	
• Average income taken as estimated value for costs of private individuals' lost leisure (Munasinghe and Gellerson, 1979)	• Great discrepancy between theory and practice of freely managing one's time, e.g., traditional 40-hour week, fixed working hours (Woo and Pupp, 1992)] • Optimum hourly rates of pay for a homemaker, pensioner, child, etc. (Sanghvi, 1982) • No temporal differentiation, e.g., assumption of time-independent costs, no consideration of factors influencing the interruption (Sanghvi, 1982) • Method only appropriate for electricity-dependent leisure activities (Woo and Pupp, 1992) • Poor availability of data (London Economics, 2013a)
Revealed preferences	
• Directly shows the willingness to pay and, thus, the value of supply security for electricity customers • Provides up-to-date data	• Hardly applicable in industrialized countries • Only relevant if investments are made in backup systems or interruptible contracts (London Economics, 2013a) • Does not seem possible to apply method efficiently for a whole economy

objectives. First, they help provide the interviewees, who may have little experience of blackouts, with an overview of possible types of damage and the consequences so that they can evaluate them. Second, they supply the interviewers with important information on the major cost components. Knowledge of the major cost categories can help to minimize the real damage costs of a blackout (Caves et al., 1990). Ajodhia et al. (2002) summarize these methods in three points:

1. identifying the cost categories,
2. weighting each category with an economic value, and
3. determining the interruption costs by adding up the individual damage costs.

Indirect Approaches
Macroeconomic Approaches
Macroeconomic approaches include the production function approach for calculating VoLL for industrial and commercial electricity customers and the determination of VoLL for private

customers by means of the household income as a special case of the production function.

Production Function
The production function approach is based on the understanding that electricity is an important input factor like work or capital for the production of goods and services (Munasinghe and Gellerson, 1979; Munasinghe and Sanghvi, 1988). If an essential input factor in the production process ceases to exist, then a drop in production inevitably results or even a complete production shutdown. The production function approach calculates the consequences of an interruption by relating production outages during the interruption to the kWh that have not been supplied. Essentially statistical data are required and evaluated in order to calculate the outage costs using a production function (de Nooij et al., 2007).

Table 5 compares the advantages and disadvantages. It can be seen that this approach has some weak points that should be noted. However, a great advantage is that the data basis required can usually be supplied by official statistical bureaus and can, thus, be obtained at relatively low cost. Furthermore, by integrating the production function into the input–output calculation, it is possible to determine the consequences of a power outage beyond regional and sectoral boundaries at different impact levels.

Household Income
The approach of determining the interruption costs using household income is based on the logic of evaluating leisure in terms of money. This approach based on Becker (1965) is summarized very succinctly by de Nooij et al. (2007). According to de Nooij et al. the essence of Becker's theory is that private individuals do not benefit from money or goods alone but from a combination of goods and time purchased with money. In this way, private individuals produce added value by using time and money as input factors. According to this logic, private individuals can also be regarded as production units. As an example, de Nooij et al. (2007) say that merely owning a television set does not in itself represent a benefit or added value for an individual, since the owner also needs time in order to watch television.

In general, it can be said that the value of the income from additional working hours drops the more hours a person works. At the same time, the value of leisure increases since longer working hours necessarily lead to less leisure. Correspondingly, each person has an optimum number of working hours. In this optimum state, the pay for the last hour of work is equivalent to the value of an additional hour of leisure so that the value of an hour of leisure corresponds to a person's hourly rate of pay.

Furthermore, it is assumed that housework interrupted by a blackout must be undertaken at a later time so that this time cannot then be used for recreation. This assumption leads to 1 hour of housework being equated to 1 hour of leisure (de Nooij et al., 2003, 2007).

A power outage, thus, limits the freedom of private electricity customers in managing their time and forces them to change their preferred habits even if many of their activities can be performed at a later time without great effort or financial expenditure. Overall, determining VoLL using the income of private households follows a clear theoretical derivation. Unfortunately, transferring this logic to reality causes certain restrictions, such as differentiating between a homemaker, pensioner, and child, or if temporal differentiation is not taken into consideration.

Revealed Preferences

Another approach to determining the costs of power outages is the derivation of VoLL from current market behavior. In this case, VoLL is either derived from the behavior of companies and households with respect to their investment activities, for example, standby generators or batteries, or from the conclusion of interruptible supply contracts. These expenditures can then be analyzed with respect to the willingness of electricity customers to pay for uninterruptible power supply. These investments are not damage costs but rather mitigation costs (**Table 2**). However, the question arises of whether investments in backup systems are voluntary or rather, as in the case of hospitals, governed by legislation (Röpke, 2013).

Assuming a very high level of supply security, as in most developed countries, this method is not applicable in practice since the investment activities of electricity customers are not accessible to analysis. At the same time, conditions, for instance for concluding interruptible supply contracts, are at least in Germany insufficient to provide comprehensive information on the WTP on the part of industrial and private electricity customers.

CURRENT VoLL STUDIES

Characteristics

Now that the economic indicator for evaluating the security of the power supply, VoLL, and the various survey and calculation methods have been presented in Section "VoLL Measurement Approaches," the following will give an overview of current studies on VoLL. The studies were published between 2004 and 2014.

Table 6 distinguishes the studies according to the country or region investigated and the base year analyzed. Consideration is also given to which interruption scenarios were assumed in the studies and what methods were used to calculate VoLL. Finally, the table shows the areas on which the respective studies focused.

Some trends can be perceived from **Table 6**:

- It is immediately apparent that studies on determining VoLL have been performed in a large number of countries. The determination of VoLL is, thus, an issue that has been considered and explored in an international context. However, a regional cluster is clearly visible. Of the 21 studies, 3 relate to the USA and 16 to member states of the European Union. Germany takes first place with six studies followed by Austria, the Netherlands, and Ireland with two each. From the studies relating to Germany, it can be concluded that the motivation for the analyses is the increasing integration of RES into the energy system.
- With respect to the past decade, the number of publications increased in the period 2011–2014. Fourteen studies were published during this period, but only seven from 2004 to 2010. This trend can also be explained by the significant increase in the expansion of RES in recent years (for example, all the studies on Germany were published after 2011), which

leads to power outages being increasingly regarded as a real danger.

- Furthermore, it can be seen from the table that almost without exception macroeconomic approaches and surveys on WTP are applied.[3] In this context, studies that make use of macroeconomic approaches tend to give more differentiated consideration to the industrial sectors in their calculations than studies that apply WTP approaches.
- The table also shows the boundary conditions that are of particular significance for the different authors. A comparison of the scenario frameworks shows that different degrees of consideration were given to the factors influencing a power outage (**Table 1**). This is particularly apparent in the assumed duration of the interruption. The periods analyzed range from a few seconds to 3 days.
- Moreover, it becomes apparent that the analyses focus on different end-user groups, whereby a distinction was generally made between industrial or commercial and private electricity customers. Considerable differences can be seen in the depth of the sectoral differentiation of the industrial sectors. For example, Tol (2007) differentiates 19 industrial sectors whereas Baarsma and Hop (2009) regard industry as one overall sector.

Quantitative Results

The following figures show the results of the VoLL studies from **Table 6**, broken down according to methodology applied and end-user group (industrial and commercial end users in **Figure 2** and private end users in **Figure 3**). Due to the different degrees of differentiation, the VoLL results of the studies are shown as ranges.[4]

As a whole, the following trends can be perceived:

- Heterogeneity of the VoLL level in and among the end-user groups: for the industrial and commercial sectors, the results range from a few €/kWh to more than € 250/kWh. This could be due, for example, to the different industrial structures in the individual countries. These ranges for the industrial and commercial sectors are therefore high. The large differences between the VoLL values for individual countries or groups of countries are also striking. The differences range from a few €/kWh for EU member states (Bliem, 2005; Tol, 2007; Lineares and Rey, 2012) to more than € 250/kWh for the USA and New Zealand (Sullivan et al., 2009; New Zealand Electriciy Authority, 2013). For private end users, the values range from a few €/kWh (Reichl et al., 2013) and up to about € 45/kWh (Tol, 2007). In this case as well, structural differences, such as country-specific industrial structures and differences in salary level, may provide an explanation. The VoLL for industrial and commercial end users tends to be considerably higher than that for private end users.

---

[3] Only Centolella et al. (2006) use the direct cost approach in a subsection of their study.

[4] All values in 2013 euros (Deutsche Bundesbank, 2014; Wirtschaftskammer Österreich, 2014).

TABLE 6 | VoLL studies 2004–2014 – an overview.

Study	State/Region	Base Year	Method/Scenario	Focus
Chowdhury et al. (2004)	USA – Midwest Region	2002	Willingness to pay: differentiation according to event, 2 s, 1 min, 20 min, 1 h, 4 h, 8 h	Industry, commercial users, private users, organizations
Bliem (2005)	Austria	2002	Macroeconomic approach: regional differentiation (federal states), consideration of different points in time (weekday/Sunday)	Industry (six sectors), private households
Centolella et al. (2006)	USA – Midwest Region	2005	Direct cost survey: differentiation into larger (>1 million kWh/a) and smaller (<1 million kWh/a) industrial and commercial users; determination for an interruption of 1 h, 2 h, 3 h Willingness to pay	Industry (nine sectors) Private households
Tol (2007)	Ireland	2005	Macroeconomic approach: for 2005, calculation differentiated according to 19 sectors; calculation of industrial VoLL from 1990 to 2005, but average values for industry broken down according to time of day/week/year	Industry (19 sectors), private households
de Nooij et al. (2007)	The Netherlands	2001	Macroeconomic approach: differentiation according to regions, broken down according to days of the week (weekday/Saturday/Sunday) and time of day (day/evening/night)	Industry (six sectors), government, private households
Baarsma and Hop (2009)	The Netherlands	2003-2004	Willingness to pay: differentiation according to event: 1 event/a lasting 0.5 h, 1 h, 4 h, 8 h, 24 h, and a 2-h event 1, 2, 4, 6, or 12 times/a	Industry, private households
Sullivan et al. (2009)	USA	2008	Willingness to pay: a metadatabase was compiled from 28 studies (surveys on willingness to pay between 1989 and 2005); differentiation into larger (>50,000 kWh/a) and smaller (<50,000 kWh/a) industrial and commercial users; differentiation according to length of event: short-term, 30 min, 1 h, 4 h, 8 h; calculation for different points in time (summer/winter; weekday/ weekend; mornings/daytime/evenings)	Industry (nine sectors), private households
Praktiknjo et al. (2011)	Germany	2002	Macroeconomic approach: combined with a Monte Carlo simulation	Industry (four sectors), private households
Leahy and Tol (2011)	Ireland and Northern Ireland	2008/2010	Macroeconomic approach: differentiated consideration of Ireland and Northern Ireland; period from 2000 to 2007, consideration of average values for industry broken down according to weekday/weekend; day/evening/night; spring/summer/autumn/winter	Industry, services, private households
Carlsson et al. (2011)	Sweden	2004	Willingness to pay: distinction between planned and unplanned; differentiation according to event 1 h, 4 h, 8 h, 24 h, consideration of the influence of socioeconomic factors; comparison before and after actual power interruption	Private households
Lineares and Rey (2012)	Spain	2008	Macroeconomic approach: for 2008, calculation differentiated according to 15 sectors; calculation of industrial VoLL from 2000 to 2008, but average values for industry for five sectors; differentiated according to Spanish regions for 2008	Industry (15 sectors), private households
Zachariadis and Poullikkas (2012)	Cyprus	2009	Macroeconomic approach: differentiated according to seasons; weekday/ weekend; time of day (hours); only industrial/commercial/private users are considered in the temporal differentiation	Industry (15 sectors), private households
Reichl et al. (2013)	Austria	2009	Macroeconomic approach: 12-h interruption in summer Willingness to pay: 12-h interruption in summer; consideration of the influence of socioeconomic factors	Industry (15 sectors) Private households
Growitsch et al. (2013)	Germany	2007	Macroeconomic approach: results differentiated according to federal state and sector; overall costs determined for a period of 1 h for the federal states	Industry (15 sectors), private households
Röpke (2013)	Germany	2008-2010	Macroeconomic approach	Industry (five sectors), private households
Piaszeck et al. (2013)	Germany	2010	Macroeconomic approach: regional subdivision on the level of local districts; breakdown into time of day/course of the week	Industry (six sectors), private households
New Zealand Electriciy Authority (2013)	New Zealand	2010	Willingness to pay: differentiation into small/medium-sized/large enterprises; regional differentiation; differentiation and event, 10 min, 1 h, 8 h; scenarios according to time of day and season	Industry, private households
Schubert et al. (2013)	Germany, Munich	2012	Willingness to pay: investigation of a blackout on 15 November 2012, duration 4 h	Private households
London Economics, (2013b)	UK	2011	Willingness to pay: differentiation into small and medium-sized enterprises/ industrial and commercial enterprises; scenarios according to season and working day/weekend	Small and medium-sized enterprises (SMEs), industrial and commercial enterprises I&C), private households
Praktiknjo (2014)	Germany	2011	Willingness to pay: combined with a Monte Carlo simulation; blackout scenarios lasting 15 min, 1 h, 4 h, 1 day, 4 days	Private households
Kim et al. (2014)	South Korea	2010	Willingness to pay: differentiation according to event (1 s, 3 s, 1 min, 20 min, 1 h, 2 h, 4 h, 8 h, 1 day, 3 days); at the same time socioeconomic factors also surveyed	Industry, private households

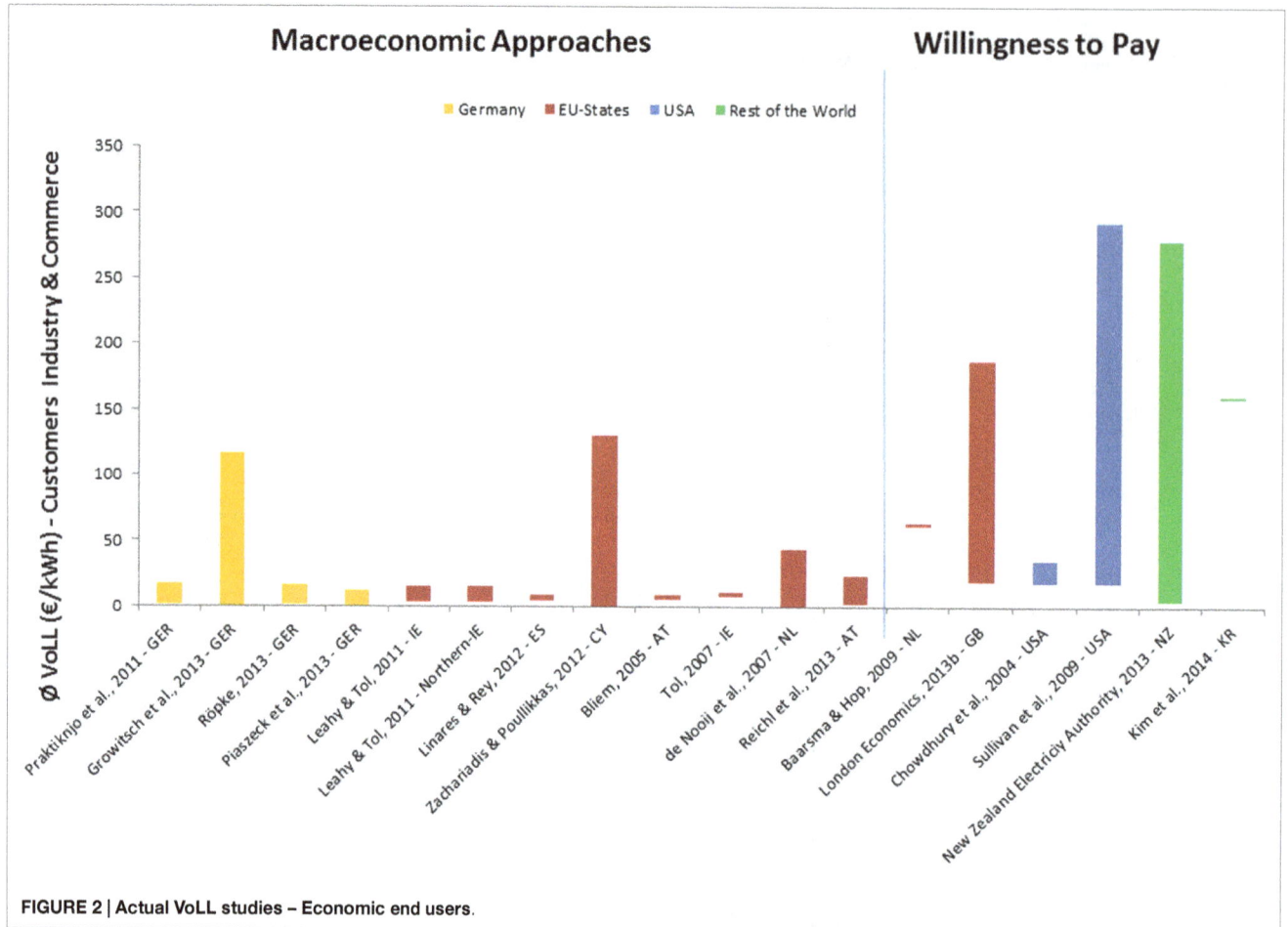

FIGURE 2 | Actual VoLL studies – Economic end users.

- Vice versa, the level of VoLL in the end-user groups depends on the methodological approach. In the studies examined here, there are great differences between the results obtained by the different methods. An obvious explanation is the fact that the results of the studies shown here were determined by two fundamentally different methodological approaches (macroeconomic vs. WTP). It is striking that the average VoLL values for private electricity customers in studies whose results are calculated using the macroeconomic approach are appreciably higher than those in studies based on the WTP approach. In the macroeconomic approaches, VoLL is generally in the range of ~€ 10 to € 25/kWh, whereas in surveys of WTP the maximum VoLL is generally ~€ 10/kWh. For industry and commerce, in contrast, the VoLL results based on the WTP method considerably exceed those obtained on the basis of macroeconomic approaches.

Nevertheless, these explanations can only justify some of the differences. Upon closer consideration, it becomes apparent that, for example, the studies by Praktiknjo et al. (2011), Growitsch et al. (2013), Piaszeck et al. (2013), and Röpke (2013) regard the whole of Germany as the area of study, but the damage costs differ greatly. Another explanation can, therefore, be found in the differences in the more detailed methodological structure of the studies. In this respect, two essential factors have an impact. First, the consideration and weighting of the technical factors from **Table 1** play a major role. If different assumptions are made in structuring the framework of the scenario, for instance about the duration or the regional location of the blackout [in the present case, Praktiknjo et al. (2011) and Röpke (2013) consider Germany as a whole, Growitsch et al. (2013) consider the level of the federal states, and Piaszeck et al. (2013) consider the regional districts], then this affects the results obtained. Second, the subdivision of the industrial sectors influences the resulting VoLL values. If the economy as a whole is considered, then VoLL represents a mean. The more extensively the economy is broken down, the more differentiated the VoLL values, and the ranges tend to increase.

In summary, apart from the typical structural features of the income and industrial structures of the countries, three fundamental influences can be identified:

- choice of method,
- structure of the scenario framework of the hypothetical power outage, and
- breakdown of the industrial sectors, as well as boundaries and level of differentiation.

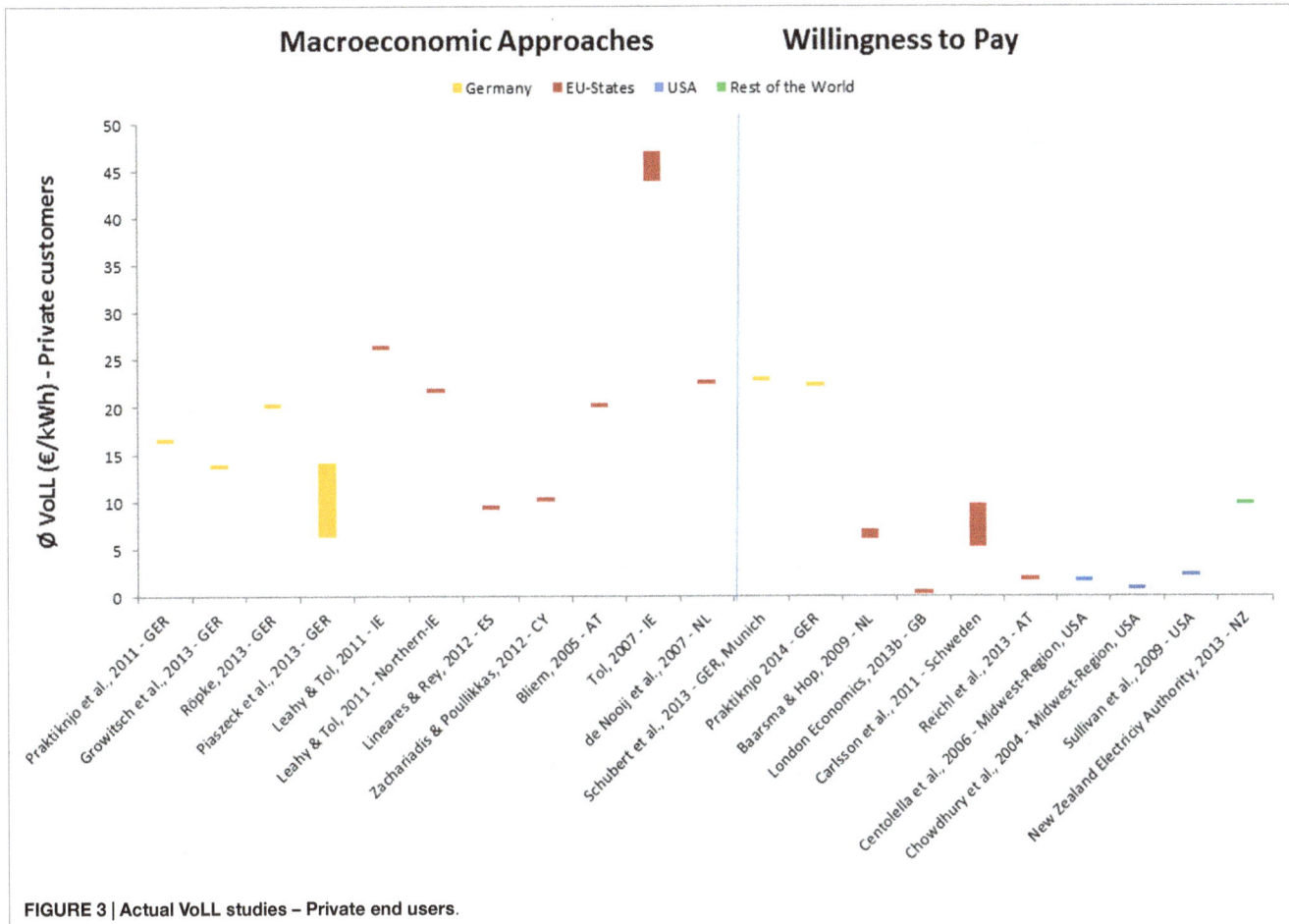

FIGURE 3 | Actual VoLL studies – Private end users.

SUGGESTIONS TO INCREASE THE EXPLANATORY POWER OF VoLL

According to definition, VoLL is determined by relating the monetary damage arising from a power interruption (due to the loss of economic activities) to the level of the kWh that are not supplied during an interruption.

On the basis of this definition, a number of methodological approaches and a wide variety of differently structured methods have been developed, as shown in Section "Introduction."

From the discussion in Section "Quantitative Results," it can be concluded that the framework definition of VoLL is so broad that the understanding of the concept and of the problem is not sufficiently harmonized and the indicator values cannot, therefore, be properly compared.[5] This is particularly problematic since it is claimed that the indicator should ensure recognized international comparability of the economic evaluation of power supply security. A general criticism is that the definition of VoLL is so all-embracing that it can be used to justify a large number of

different procedures. This reduces the informative value of VoLL considerably since the results cannot be integrated into a larger context by means of comparisons and, thus, it is only possible to represent individual cases. Consequently, a uniform framework as a basis for comparison makes an important contribution to improving this indicator enabling it to provide more informative value on an international level. This requires clearly defined specifications. Technical indices that map power supply security, such as SAIDI, are subject to clearly defined criteria and, thus, fulfill these conditions.

The following procedure is proposed in order to develop a uniform framework for determining VoLL. First of all, as an essential prerequisite, it must be ensured that one single method is employed (macroeconomic or WTP). In the next step, the framework of the outage must be clearly defined, i.e., the factors from **Table 1** must be given equal consideration. And in a third step, the breakdown of the industrial sectors must be coordinated both with respect to their delimitation and also the degree of differentiation.[6]

---

[5]CEER (2010) makes a proposal for a guideline of estimation of costs due to electricity interruption. However, this is based on a fragmentary consideration of VoLL methods, because macroeconomic approaches are not recognized (although approx. 50% of recently published Studies apply this method).

---

[6]In this context, attention must be paid that the data basis is harmonized on an international level. It is important not to aim for the greatest possible level of differentiation but rather to ensure international comparability.

The clarification of the general analytical framework ensures a uniform and harmonized procedure, thus, providing the basis for international comparability. The VoLL determined on this basis offers the opportunity to place the individual results in an international context and also to consider them over time.

The realistic nature of the VoLL determined in this way can be increased even further by integrating damage-aggravating factors (e.g., influence of preparatory and restart times on the duration of the outage) and damage-mitigating factors (e.g., proportion of internal electricity generation, stockkeeping, making up for production losses, restructuring process steps, advance warning). However, the integration of such factors leads to a reduction in international comparability since the inclusion of such factors requires a solid data basis that in many cases cannot be supplied (or only in part) by official statistical offices and is, thus, frequently not accessible.

This context reveals the problems associated with deciding on a focus, i.e., either as realistic as possible or with the greatest possible international comparability. Since VoLL is completely lacking in international comparability, a uniform analytical framework is urgently required. As soon as this common basis has been created, further steps should be taken to clarify which damage-aggravating and which damage-mitigating influences can be integrated – while maintaining international comparability. However, this still requires extensive discussions both with respect to the methodological approach and also to harmonizing the data basis.

SUMMARY

The review and analysis of 21 studies on the VoLL published in the past 10 years revealed four different aspects that have a fundamental influence on the calculated VoLL.

REFERENCES

Aichinger, M., Bruch, M., Münch, V., Kuhn, M., Weymann, M., and Schmid, G. (2011). Power Blackout Risks. Munich: Allianz Global Corporate & Specialty (AGCS) and The Chief Risk Officer Forum (CRO). Available at: http://www.agcs.allianz.com/insights/white-papers-and-case-studies/?c=&page=11

Ajodhia, V. (2006). *Regulating Beyond Price – Integrated Price-Quality Regulation for Electricity Distribution Networks*. Delf: Delft University.

Ajodhia, V., van Gemert, M., and Hakvoort, R. (2002). Electricity outage cost valuation: a survey. *Discussion Paper, DTe*, The Hague.

Baarsma, B. E., and Hop, J. P. (2009). Pricing power outages in the Netherlands. *Energy* 34, 1378–1386. doi:10.1016/j.energy.2009.06.016

Barkans, J., and Zalostiba, D. (2009). *Protection against Blackouts and Self-Restoration of Power Systems*. Riga: RTU Pulishing House.

Barth, T. (2013). "The German energiewende: shining or warning example for Europe? (Vortrag)," in *5th Conference ELECPOR*, 2013. Lisboa.

Becker, G. S. (1965). A theory of the allocation of time. *Econ. J.* 75, 493–517. doi:10.2307/2228949

Billinton, R., Tollefson, G., and Wacker, G. (1993). Assessment of electric service reliability worth. *Int. J. Electr. Power Energy Syst.* 15, 95–100. doi:10.1016/0142-0615(93)90042-L

Bliem, M. (2005). *Eine makroökonomische Bewertung zu den Kosten eines Stromausfalls im österreichischen Versorgungsnetz*. Kärnten: Institut für Höhere Studien (IHSK).

- country-specific features of the industrial and social structures,
- choice of method,
- structure of the scenario framework of the hypothetical power outage, and
- breakdown of the industrial sectors, as well as boundaries and level of differentiation.

Country-specific features are of particular significance in identifying the results but cannot be varied for the analysis. The three other factors that can affect the VoLL calculation are the choice of method, the structuring of the scenario framework, and the breakdown of the industrial structure by data processing. The different weighting of these aspects is responsible for the large range of VoLL in the results of the analyzed studies.

Overall, the analysis of recent VoLL studies has shown that, according to the present state of the art, VoLL is only capable of mapping one individual case as an economic evaluation index of power supply security, and the respective results must be considered and assessed against the background of the analytical framework. The informative value of these results is not sufficient for comparisons with the results of other studies.

If VoLL can be determined according to a uniformly defined procedure, it may become a decisive factor on which decisions for and against investments for grid optimization and expansion may be based. Furthermore, VoLL may also become extremely important for location decisions on the part of companies. Regions with high blackout probabilities and costs are less attractive for companies to retain existing or establish new operations. Moreover, VoLL could help to ensure an optimum distribution to end users of the remaining electricity in the case of a power outage, as far as this is possible with the technical options available. Further developing the VoLL approach as an economic index would, thus, effectively complement other technical indices.

Buldyrev, S. V., Parshani, R., Paul, G., Stanley, H. E., and Havlin, S. (2010). Catastrophic cascade of failures in interdependent networks. *Nat. Lett.* 464, 1025–1028. doi:10.1038/nature08932

Bundesnetzagentur. (2007). *Report by the Federal Network Agency for Electricity, Gas, Telecommunications, Post, and Railways on the Disturbance in the German and European Power System on the 4th of November 2006*. Available at: http://www.bundesnetzagentur.de/SharedDocs/Downloads/EN/BNetzA/Areas/ElectricityGas/Special%20Topics/Blackout2005/BerichtEnglischeVersionId9347pdf.pdf?__blob=publicationFile

Carlsson, F., Martinsson, P., and Akay, A. (2011). The effect of power outages and cheap talk on willingness to pay to reduce outages. *Energy Econ.* 33, 790–798. doi:10.1016/j.eneco.2011.01.004

Caves, D. W., Herriges, J. A., and Windle, R. J. (1990). Customer demand for service reliability in the electric power industry: a synthesis of the outage cost literature. *Bull. Econ. Res.* 42, 79–121. doi:10.1111/j.1467-8586.1990.tb00294.x

Caves, D. W., Herriges, J. A., and Windle, R. J. (1992). The cost of electric power interruptions in the industrial sector: estimates derived from interruptible service programs. *Land Econ.* 68, 49–61. doi:10.2307/3146742

CEER. (2010). *Guidelines of Good Practice on Estimation of Costs Due to Electricity Interruptions and Voltage Disturbances*.

Centolella, P., FARBER-Deanda, M., Greening, L. A., and Kim, T. (2006). *Estimates of the Value of Uninterrupted Service for the Mid-West Indipendent System Operator*. McLean: Science Applications International Corporation.

Chen, C.-Y., and Vella, A. (1994). Estimating the economic costs of electricity outages using input-output analysis – the case of Taiwan. *Appl. Econ.* 26, 1061–1069. doi:10.1080/00036849400000122

Chowdhury, A. A., Mielnik, T. C., Lawion, L. E., Sullivan, M. J., and Katz, A. (2004). "Reliability worth assessment in electric power delivery systems," in *Power Engineering Society General Meeting, 2004*, (Denver: IEEE), 654–660.

de Nooij, M., Bijvoet, C., and Koopmans, C. (2003). "The demand for supply security," in *Research Symposium European Electricity Markets*, (The Hague).

de Nooij, M., Koopmanns, C., and Bijvoet, C. (2007). The value of supply security. The costs of power interruptions: economic input for damage reduction and investment in networks. *Energy Econ.* 29, 277–295. doi:10.1016/j.eneco.2006.05.022

Deutsche Bundesbank. (2014). *Euro-Referenzkurse der Europäischen Zentralbank – Jahresendstände und -durchschnitte*. Available at: http://www.bundesbank.de/Redaktion/DE/Downloads/Statistiken/Aussenwirtschaft/Devisen_Euro_Referenzkurs/stat_eurorefj.pdf?__blob=publicationFile

Falthauser, M., and Geiß, A. (2012). *Zahlen und Fakten zur Stromversorgung in Deutschland*. Münch: Wirtschaftsbeirat Bayern.

Frontier Economics. (2008). *Kosten von Stromversorgungsunterbrechungen*. Essen: RWE AG.

Growitsch, C., Malischek, R., Nick, S., and Wetzel, H. (2013). *The Costs of Power Interruptions in Germany – An Assessment in the Light of the Energiewende*. Cologne: Institute of Energy Economics at the University of Cologne (EWI). Working Paper 13/07.

Holmgren, ÅJ. (2007). "A framework for vulnerability assessment of electric power systems," in *Critical Infrastructure Reliability and Vulnerability*, eds Murray A. and Grubesic T. (Berlin: Springer-Verlag), 31–55.

IEEE. (2004). *IEEE Guide for Electric Power Distribution Reliability Indices*. New York: IEEE, 1–50.

Kim, C.-S., Jo, M., and Koo, Y. (2014). Ex-ante evaluation of economic costs from power grid blackout in South Korea. *J. Electr. Eng. Technol.* 9, 796–802. doi:10.5370/JEET.2014.9.3.796

Kling, C. L., Phaneuf, D. J., and Zhao, J. (2012). From exxon to BP: has some number become better than no number? *J. Econ. Perspect.* 26, 3–26. doi:10.1257/jep.26.4.3

LaCommare, K. H., and Eto, J. H. (2006). Cost of power interruptions to electricity consumers in the United States (US). *Energy* 31, 1845–1855. doi:10.1016/j.energy.2006.02.008

Leahy, E., and Tol, R. S. J. (2011). An estimate of the value of lost load for Ireland. *Energy Policy* 39, 1514–1520. doi:10.1016/j.enpol.2010.12.025

Lijesen, M., and Vollaard, B. (2004). *Capacity Spare? A Cost-Benefit Approach to Optimal Spare in Electricity Production*. The Hague: CPB – Central Planning Bureau. CPB document No. 60.

Lineares, P., and Rey, L. (2012). *The Costs of Electricity Interruptions in Spain. Are We Sending the Right Signals?*. Vigo: Economics for Energy. WP FA5/2012.

London Economics. (2013a). *Estimating the Value of Lost Load – Briefing Paper Prepared for the Electric Reliability Council of Texas, Inc.* Boston: London Economics.

London Economics. (2013b). *The Value of Lost Load (VoLL) for Electricity in Great Britain*. London: London Economics.

Makarov, Y., Reshetov, V., Stroev, V., and Voropai, N. (2005). "Blackouts in North-America and Europe: analysis and generelization," in *IEEE St. Petersburg PowerTech*, (St. Petersburg: IEEE), 1–7.

Munasinghe, M., and Gellerson, M. (1979). Economic criteria for optimizing power system reliability levels. *Bell Econ. J.* 10, 353–365. doi:10.2307/3003337

Munasinghe, M., and Sanghvi, A. P. (1988). Reliability of electricity supply, outage costs and value of service: an overview. *Energy J.* 9, 1–18. doi:10.5547/ISSN0195-6574-EJ-Vol9-NoSI2-1

New Zealand Electriciy Authority. (2013). *Investigation into the Value of Lost Load in New Zealand – Report on Methodology and Key Findings*. Wellington: New Zealand Electriciy Authority.

Pesch, T., Allelein, H. J., and Hake, J. F. (2014). Impacts of the transformation of the German energy system on the transmission grid. *Eur. Phys. J. Spec. Top.* 223, 2561–2575. doi:10.1140/epjst/e2014-02214-y

Petermann, T., Bradke, H., Lüllmann, A., Paetzsch, M., and Riehm, U. (2011). *Was bei einem Blackout geschieht – Folgen eines langandauernden und großflächigen Stromausfalls*. Berlin: TAB – Technikfolgen-Abschätzung beim Deutschen Bundestag.

Piaszeck, S., Wenzel, L., and Wolf, A. (2013). *Regional Diversity in the Costs of Electricity Outages: Results for German Counties*. Hamburg: Hamburg Institute of International Economics (HWWI). Research Paper 142.

Portney, R. P. (1994). The contingent valuation debate – why economists should care. *J. Econ. Perspect.* 8, 3–17. doi:10.1257/jep.8.4.3

Praktiknjo, A. J. (2013). *Sicherheit der Elektrizitätsversorgung – Das Spannungsfeld von Wirtschaftlichkeit und Umweltverträglichkeit*. Berlin: Institut für Energietechnik, Fachgebiet Energiesysteme Berlin, Technische Universität Berlin.

Praktiknjo, A. J. (2014). Stated preferences based estimation of power interruption costs in private households: an example from Germany. *Energy* 76, 82–90. doi:10.1016/j.energy.2014.03.089

Praktiknjo, A. J., Hähnel, A., and Erdmann, G. (2011). Assessing energy supply security: outage costs in private households. *Energy Policy* 39, 7825–7833. doi:10.1016/j.enpol.2011.09.028

Ratha, A., Iggland, E., and Andersson, G. (2013). "Value of lost load: how much is supply security worth?," in *Power and Energy Society General Meeting (PES), 2013*, (Vancouver, BC: IEEE), 1–5.

Reevell, P. (2015). *Turkey blackout: massive power outage plunges country into darkness*. ABC News. Available at: http://abcnews.go.com/International/massive-blackout-plunges-turkey-darkness/story?id=30024749

Reichl, J., Schmidthaler, M., and Schneider, F. (2013). The value of supply security: the costs of power outages to Austrian households, firms and the public sector. *Energy Econ.* 36, 256–261. doi:10.1016/j.eneco.2012.08.044

Röpke, L. (2013). The development of renewable energies and supply security: a trade-off anaysis. *Energy Policy* 61, 1011–1021. doi:10.1016/j.enpol.2013.06.015

Rose, A., Oladosu, G., and Salvino, D. (2004). "Regional economic impacts of electricity outages in Los Angeles: a computable general equilibrium analysis," in *Obtaining the Best from Regulation and Competition*, eds Crew M. and Spiegel M. (Dordrecht: Kluwer), 179–210.

Sanghvi, A. P. (1982). Economic costs of electricity supply interruptions: US and foreign experience. *Energy Econ.* 4, 180–198. doi:10.1016/0140-9883(82)90017-2

Schlandt, J. (2012). *Die Angst vorm Dunkel*. Frankfurter Rundschau vom. Available at: http://www.fr-online.de/energie/stromausfaelle-die-angst-vorm-dunkel,1473634,19496664.html

Schubert, D. K. J., von Selasinsky, A., Meyer, T., Schmidt, A., THUß, S., Erdmann, N., et al. (2013). *Gefährden Stromausfälle die Energiewende? Einfluss auf Akzeptanz und Zahlungsbereitschaft*, Vol. 63. Munich: Energiewirtschaftliche Tagesfragen, 35–39.

Sullivan, M. J., and Keane, D. M. (1995). *Outage Cost Estimation Guidebook*. San Francisco, CA: Electric Power Research Institute.

Sullivan, M. J., Mercurio, M., Schellenberg, J., and Sullivan, F. (2009). *Estimated Value of Service Reliability for Electric Utility Customers in the United States*. LBNL Research Project Final Report.

Tol, R. S. J. (2007). The value of lost load. *ESRI Working Paper*, 214. Dublin.

van der Welle, A., and van der Zwaan, B. (2007). An overview of selected studies on the value of lost load. *Working Paper, Energy Research Centre of the Netherlands (ECN)*. Amsterdam, 1–23.

von Roon, S. (2013). *Versorgungsqualität und -zuverlässigkeit als Standortfaktor. Energieeffizienz und Erneuerbare Energien im Wettbewerb – der Schlüssel für eine Energiewende nach Maß*. Munich: Tagungsband zur FfE-Fachtagung FfE-Schriftenreihe – Band, 31.

Wirtschaftskammer Österreich. (2014). *Inflationsraten*. Available at: http://wko.at/statistik/eu/europa-inflationsraten.pdf

Woo, C.-K., and Pupp, R. L. (1992). Costs of service disruptions to electricity consumers. *Energy* 17, 109–126. doi:10.1016/0360-5442(92)90061-4

Zachariadis, T., and Poullikkas, A. (2012). The cost of power outages: a case study from Cyprus. *Energy Policy* 51, 630–641. doi:10.1016/j.enpol.2012.09.015

Conflict of Interest Statement: The authors declare that the research was conducted in the absence of any commercial or financial relationships that could be construed as a potential conflict of interest.

15

Development of Sulfide Solid Electrolytes and Interface Formation Processes for Bulk-Type All-Solid-State Li and Na Batteries

Akitoshi Hayashi[1]*, Atsushi Sakuda[1,2] and Masahiro Tatsumisago[1]

[1] *Department of Applied Chemistry, Graduate School of Engineering, Osaka Prefecture University, Sakai, Osaka, Japan,*
[2] *Department of Energy and Environment, Research Institute of Electrochemical Energy, National Institute of Advanced Industrial Science and Technology (AIST), Ikeda, Osaka, Japan*

Edited by:
Shyue Ping Ong,
University of California,
San Diego, USA

Reviewed by:
Liqiang Mai,
Wuhan University of
Technology, China
Xiao-Liang Wang,
Seeo Inc., USA

*Correspondence:
Akitoshi Hayashi
hayashi@chem.osakafu-u.ac.jp

Specialty section:
This article was submitted
to Energy Storage,
a section of the journal
Frontiers in Energy Research

Citation:
Hayashi A, Sakuda A and
Tatsumisago M (2016) Development
of Sulfide Solid Electrolytes and
Interface Formation Processes
for Bulk-Type All-Solid-State
Li and Na Batteries.
Front. Energy Res. 4:25.

All-solid-state batteries with inorganic solid electrolytes (SEs) are recognized as an ultimate goal of rechargeable batteries because of their high safety, versatile geometry, and good cycle life. Compared with thin-film batteries, increasing the reversible capacity of bulk-type all-solid-state batteries using electrode active material particles is difficult because contact areas at solid–solid interfaces between the electrode and electrolyte particles are limited. Sulfide SEs have several advantages of high conductivity, wide electrochemical window, and appropriate mechanical properties, such as formability, processability, and elastic modulus. Sulfide electrolyte with $Li_7P_3S_{11}$ crystal has a high Li^+ ion conductivity of 1.7×10^{-2} S cm^{-1} at 25°C. It is far beyond the Li^+ ion conductivity of conventional organic liquid electrolytes. The Na^+ ion conductivity of 7.4×10^{-4} S cm^{-1} is achieved for $Na_{3.06}P_{0.94}Si_{0.06}S_4$ with cubic structure. Moreover, formation of favorable solid–solid interfaces between electrode and electrolyte is important for realizing solid-state batteries. Sulfide electrolytes have better formability than oxide electrolytes. Consequently, a dense electrolyte separator and closely attached interfaces with active material particles are achieved *via* "room-temperature sintering" of sulfides merely by cold pressing without heat treatment. Elastic moduli for sulfide electrolytes are smaller than that of oxide electrolytes, and Na_2S–P_2S_5 glass electrolytes have smaller Young's modulus than Li_2S–P_2S_5 electrolytes. Cross-sectional SEM observations for a positive electrode layer reveal that sulfide electrolyte coating on active material particles increases interface areas even with a minimum volume of electrolyte, indicating that the energy density of bulk-type solid-state batteries is enhanced. Both surface coating of electrode particles and preparation of nanocomposite are effective for increasing the reversible capacity of the batteries. Our approaches to form solid–solid interfaces are demonstrated.

Keywords: all-solid-state battery, lithium battery, sodium battery, sulfide, solid electrolyte, electrode–electrolyte interface

INTRODUCTION

All-solid-state batteries using inorganic solid electrolytes (SEs), used in place of conventional organic liquid electrolytes, have been studied because of their high safety (non-flammability with no liquid leakage), long cycle life, and versatile geometries (Takada, 2013; Tatsumisago et al., 2013; Tatsumisago and Hayashi, 2014). These features are important for large rechargeable lithium batteries with high energy density for application to eco-cars, such as electric vehicles and plug-in hybrid vehicles. Rechargeable sodium batteries are also attractive for large-scale applications for stationary load-leveling because sodium is expected to be the next targeted element after lithium based on its atomic weight, standard potential, and natural abundance (Yabuchi et al., 2015b; Yamada, 2014).

A key material to realize all-solid-state rechargeable batteries is a superior SE. In lithium ion conductors, sulfide electrolytes of $Li_{10}GeP_2S_{12}$ and $Li_7P_3S_{11}$ have high room-temperature conductivity of more than 10^{-2} S cm^{-1}, which is as high as the conductivity of conventional organic liquid electrolytes (Kamaya et al., 2011; Seino et al., 2014). The lithium ion transport number of liquid electrolytes is below 0.5, whereas that of SE is 1. It is noteworthy that the conductivity of lithium ions of sulfide SEs already exceeds that of organic liquid electrolytes. Very recently, $Li_{9.54}Si_{1.74}P_{1.44}S_{11.7}Cl_{0.3}$ has been reported to show the highest conductivity of 2.5×10^{-2} S cm^{-1}. Using this superior electrolyte, high power competing with that of supercapacitors can be achieved in all-solid-state rechargeable lithium batteries (Kato et al., 2016b). Sulfide SEs have several important benefits of high conductivity, wide electrochemical window, and appropriate mechanical properties, such as formability and elastic modulus. A shortcoming of sulfide electrolytes is its low air stability. To realize bulk-type all-solid-state batteries, the formation of favorable solid–solid interfaces between electrode and electrolyte is important in addition to the development of superior sulfide electrolytes. Direct coating of sulfide electrolytes on active material particles instead of adding electrolyte particle is effective for forming a close solid–solid interface with large contact area. Insulative active materials, such as sulfur and Li_2S, should be blended not only with SE but also with carbon-conductive additive to form ion and electron conduction paths to active materials. Nanocomposites of the three components are useful by preparation with high-energy ball milling, which pulverizes and combines them. Transition metal sulfides (MS_x) with high conductivity have good compatibility with sulfide SEs having the same sulfide anions. Especially, sulfur-rich compounds, such as TiS_3, are attractive active materials with high reversible capacity. Lithium metal is a supremely negative electrode, but issues related to lithium dendrites prevent its commercialization. Combination with a SE is a promising solution. Interface modification between Li metal and the SE is important.

As described in this paper, the recent development of sulfide SEs and interface formation processes for bulk-type all-solid-state Li and Na batteries are reviewed. Significant properties as SEs of conductivity, chemical stability, and mechanical property for Li$^+$ or Na$^+$ ion conducting sulfides are reported. Procedures for preparing sulfide electrolytes, such as mechanical milling and liquid-phase synthesis, are also described. Several approaches to form favorable solid–solid interfaces developed by our research group are demonstrated. Processes for the coating of sulfide electrolytes via gas-phase or liquid-phase process on $LiCoO_2$ or graphite particles have been developed. Preparation of nanocomposite electrodes with sulfur, Li_2S, and Li_3PS_4 (as a bifunctional material of electrolyte and electrode) particles is described, along with the use of MS_x active materials. Formation of a solid–solid interface for using lithium metal negative electrodes is also discussed.

DEVELOPMENT OF SULFIDE SOLID ELECTROLYTES

Sulfide SEs with Li$^+$ or Na$^+$ ion conductivity have been developed during the past three decades. Recently, chemical stability and mechanical properties as well as conductivity for sulfide electrolytes have attracted much attention. Sulfide electrolytes are prepared using several techniques with solid-phase reaction, melt-quenching, mechanical milling, crystallization of mother glasses, and liquid-phase reaction. Detailed information related to electrolyte properties will be presented in the following sections of this report.

Conductivity

Inorganic sulfide SEs with high Li$^+$ or Na$^+$ ion conductivities have been developed. Sulfide electrolytes with high conductivities are presented in **Table 1**. In Li$^+$ ion conducting sulfide electrolytes, crystalline $Li_{10}GeP_2S_{12}$ (LGPS; Kamaya et al., 2011) and glass–ceramic $Li_7P_3S_{11}$ (Seino et al., 2014) have considerably high conductivity of more than 10^{-2} S cm^{-1} at 25°C, which is beyond the Li$^+$ ion conductivity of conventional organic liquid electrolytes. Very recently, $Li_{9.54}Si_{1.74}P_{1.44}S_{11.7}Cl_{0.3}$ with LGPS structure has been reported to show the highest conductivity of 2.5×10^{-2} S cm^{-1} (Kato et al., 2016b). Studies investigating new electrolytes with much higher conductivity are in progress. Several crystals, such as $Li_{10}SnP_2S_{12}$ (Boron et al., 2013) and Li_6PS_5Cl (Boulineau et al., 2012), have high conductivity of more than 10^{-3} S cm^{-1}. This conductivity is also achieved by the addition of lithium halides to sulfide glass and glass–ceramic electrolytes (Wada et al., 1983; Ujiie et al., 2014).

The Na$^+$ ion conductivity is lower than Li$^+$ ion conductivity in glassy electrolytes. Ionic conduction of Na$^+$ ion with ionic radius larger than Li$^+$ ion is unfavorable in glasses (Souquet et al., 1981). For sodium ion conductors, sulfides with high conductivity had not been found since this report described Na_3PS_4 glass–ceramic electrolytes in 2012. A cubic Na_3PS_4 phase is precipitated by crystallization of a mother Na_3PS_4 glass. The prepared glass–ceramic electrolyte shows Na$^+$ ion conductivity of greater than 10^{-4} S cm^{-1} at 25°C (Hayashi et al., 2012b). Furthermore, partial substitution of Si for P in Na_3PS_4 is useful for increasing conductivity (Tanibata et al., 2014). **Figure 1** presents the composition dependence of conductivities at 25°C for Na_3PS_4–Na_4SiS_4 glass–ceramic electrolytes. The replacement of 6 mol% Na_3PS_4 by Na_4SiS_4 increases the conductivity from 4.6×10^{-4} to 7.4×10^{-4} S cm^{-1}. The electron density distribution of the cubic Na_3PS_4 structure obtained using the maximum entropy method is shown in the inset of **Figure 1**. Na_3PS_4 has three-dimensional

TABLE 1 | Li⁺ ion and Na⁺ ion conductivities for sulfide solid electrolytes.

Composition	Conductivity at 25°C (S cm⁻¹)	Classification	Reference
$Li_{9.54}Si_{1.74}P_{1.44}S_{11.7}Cl_{0.3}$	2.5×10^{-2}	Crystal	(Kato et al. 2016b)
$Li_{10}GeP_2S_{12}$	1.2×10^{-2}	Crystal	Kamaya et al. (2011)
$Li_{10}SnP_2S_{12}$	4×10^{-3}	Crystal	Boron et al. (2013)
$Li_{3.833}Sn_{0.833}As_{0.166}S_4$	1.4×10^{-3}	Crystal	Sahu et al. (2014)
Li_6PS_5Cl	1.3×10^{-3}	Crystal	Boulineau et al. (2012)
$70Li_2S\cdot30P_2S_5$ ($Li_7P_3S_{11}$)	1.7×10^{-2}	Glass–ceramic	Seino et al. (2014)
$63Li_2S\cdot27P_2S_5\cdot10LiBr$	8.4×10^{-3}	Glass–ceramic	Ujiie et al. (2014)
$80Li_2S\cdot20P_2S_5$	1.3×10^{-3}	Glass–ceramic	Mizuno et al. (2006)
$30Li_2S\cdot26B_2S_3\cdot44LiI$	1.7×10^{-3}	Glass	Wada et al. (1983)
$50Li_2S\cdot17P_2S_5\cdot33LiBH_4$	1.6×10^{-3}	Glass	Yamauchi et al. (2013)
$63Li_2S\cdot36SiS_2\cdot1Li_3PO_4$	1.5×10^{-3}	Glass	Aotani et al. (1994)
$70Li_2S\cdot30P_2S_5$	1.6×10^{-4}	Glass	Zhang and Kennedy (1990)
Na_3PSe_4	1.2×10^{-3}	Crystal	Zhang et al. (2015)
Na_3PS_4 (tetragonal)	1×10^{-6}	Crystal	Jansen and Henseler (1992)
$94Na_3PS_4\cdot6Na_4SiS_4$	7.4×10^{-4}	Glass–ceramic	Tanibata et al. (2014)
Na_3PS_4 (cubic)	2×10^{-4}	Glass–ceramic	Hayashi et al., (2012b)
$60Na_2S\cdot40GeS_2$	7.3×10^{-6}	Glass	Souquet et al. (1981)

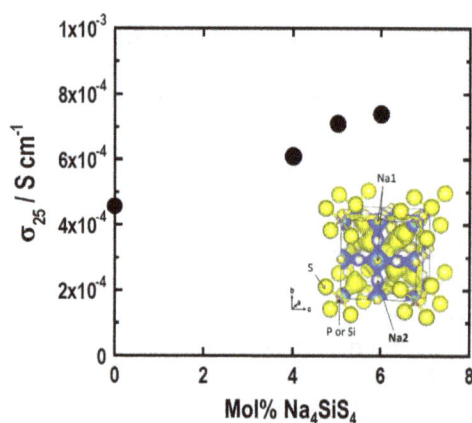

FIGURE 1 | Composition dependence of conductivity at 25°C for (100-x)Na₃PS₄·xNa₄SiS₄ (mol%) glass–ceramic electrolytes. The inset shows the electron density distribution of the cubic Na₃PS₄ structure.

Na⁺ ion conduction paths along Na1 and Na2 sites. Increased Na⁺ ion concentration is presumed as the reason for the enhancement of conductivity. Very recently, high conductivity of more than 10^{-3} S cm⁻¹ has been reported in crystalline Na_3PSe_4 (Zhang et al.,

2015). However, studies of new sulfide Na⁺ ion conductors are few. A first-principles calculation indicates that Sn-doped cubic Na_3PS_4 is predicted to have a higher Na⁺ ion conductivity of 10^{-2} S cm⁻¹ (Zhu et al., 2015). Synthesis of SEs with a favorable structure for Na⁺ ion conduction based on calculation results is important for finding new Na⁺ ion conducting SEs.

Chemical Stability

A shortcoming of sulfide electrolytes is their lower chemical stability in air atmosphere. Sulfides tend to be decomposed by hydrolysis, generating harmful H_2S. Suppression of hydrolysis of sulfides is an important task for developing sulfide electrolytes. Based on our early experiments, the selection of compositions in sulfide electrolytes gives moderate stability in air to sulfide electrolytes (Muramatsu et al., 2011). The chemical stability of sulfide glass electrolytes in the binary system $Li_2S–P_2S_5$ was examined by exposing them to air atmosphere. Amounts of generated H_2S from the sulfides depend on the composition of the glasses; H_2S generation is minimized at the composition $75Li_2S\cdot25P_2S_5$ (mol%). The glass comprises Li⁺ and PS_4^{3-} ions. An isolated anion PS_4^{3-} without bridging sulfurs is useful for high tolerance for hydrolysis. Li_3PS_4-based SEs with both good chemical stability and high conductivity have been prepared by the combination of oxides (Li_2O or P_2O_5) and iodides (LiI) (Ohtomo et al., 2013a,b). The addition of metal oxides, such as ZnO, which act as an absorbent for H_2S, is also effective for decreasing H_2S (Hayashi et al., 2013). It is noteworthy that the use of a favorable M_xO_y (M_xO_y: Fe_2O_3, ZnO, and Bi_2O_3) with a larger negative Gibbs energy change (ΔG) for the reaction with H_2S is effective for improving the chemical stability of sulfide electrolytes. Another approach is the use of sulfide compositions based on the hard and soft acids and bases theory (HSAB; Sahu et al., 2014). Lithium tin thiophosphate, Li_4SnS_4, has better air stability than that of Li_3PS_4. Actually, as-substituted Li_4SnS_4 has good features of both high conductivity of 10^{-3} S cm⁻¹ and high air stability.

The chemical stability of SEs tends to affect battery performance. The electrochemical performance of all-solid-state $C/LiCoO_2$ cells using Li_3PS_4 glass or $Li_7P_3S_{11}$ glass–ceramic as a SE is compared. The cell with Li_3PS_4 glass electrolytes exhibits better cycle performance, although Li_3PS_4 glass has lower conductivity than $Li_7P_3S_{11}$ electrolyte (Ohtomo et al., 2013c). High performance of the battery would be based on the higher chemical stability of Li_3PS_4 electrolytes. Chemical stability as well as conductivity is an important factor of SEs for developing superior solid-state batteries.

Mechanical Property

Adhesion of the solid–solid interface is a key to the utilization of electrode active materials in all-solid-state batteries. Formability or processability of SEs is examined by the molding pressure dependence of the relative density of compressed powder pellets.

Figure 2A shows the dependence of the relative density of $75Li_2S\cdot25P_2S_5$ and $75Na_2S\cdot25P_2S_5$ (mol%) glass electrolytes on molding pressure. The relative densities increase gradually with an increase in molding pressure in both glasses at the same alkali compositions of 75 mol% M_2S (M = Li or Na), whereas the relative density for the Na_2S system is higher than that for the Li_2S

system (Sakuda et al., 2013a,b; Nose et al., 2015). Cross-sectional SEM images of the 75 mol% M_2S pellets pressed at 360 MPa are shown in the inset. Grain boundaries and voids in the pellets are more decreased in the $75Na_2S \cdot 25P_2S_5$ glass compared with the $75Li_2S \cdot 25P_2S_5$ glass. Sulfide glasses are densified by cold pressing without heat treatment, and this densification phenomenon is called "room-temperature pressure sintering" (Sakuda et al., 2013a,b). It is noteworthy that the $75Na_2S \cdot 25P_2S_5$ glass has better formability than the $75Li_2S \cdot 25P_2S_5$ glass. Both the glasses comprise Li^+ or Na^+ ion and PS_4^{3-} ion, which are thought to diffuse at the particle boundaries on pressing at room temperature. Na^+ ion with a larger ionic radius than Li^+ ion has a weaker interaction with PS_4^{3-} ion. Therefore, both Na^+ and PS_4^{3-} ions would diffuse readily by cold press, leading to better densification.

Retaining solid–solid contacts between active materials and SEs during charge–discharge processes brings about long cycle lives of all-solid-state batteries. Young's moduli of SEs are important for keeping favorable contacts even at volume changes of active materials. Those for densified sulfide electrolytes prepared by hot-pressing are determined by an ultrasonic pulse-echo technique and the uniaxial compression tests (Sakuda et al., 2013a,b; Nose et al., 2015). Young's moduli of the sulfide glasses in the systems $Li_2S–P_2S_5$ and $Na_2S–P_2S_5$ are presented in **Figure 2B**. They are increased gradually with the increase in the alkali content in

both systems. The $Na_2S–P_2S_5$ glasses have smaller Young's moduli of 15–19 GPa than the $Li_2S–P_2S_5$ glasses (18–25 GPa). The difference on Young's modulus is understood based on the Coulomb force and the mean atomic volume of the glasses. These sulfide glasses have an intermediate Young's modulus between oxide glasses and organic polymers. Sulfide electrolytes deforming elastically are expected to act as a buffer in response to volume changes of active materials during charge–discharge processes. In fact, most all-solid-state batteries that use sulfide SEs exhibit good cycle performance.

Preparation Process

Sulfide SEs were prepared *via* various techniques. Crystalline electrolytes are prepared *via* solid-phase reaction, whereas glass electrolytes are obtained using the melt-quenching method. In general, sulfide starting materials are sealed in a carbon-coated quartz tube under vacuum and then heat-treated because of high vapor pressure of sulfides at high temperatures. Heating temperatures and cooling rates at preparation process affect precipitated crystals. A phase diagram in ternary system $Li_2S–GeS_2–P_2S_5$ is complicated (Hori et al., 2015), and crystalline $Li_{10}GeP_2S_{12}$ with a conductivity of 10^{-2} S cm^{-1} is prepared by selecting experimental conditions. Another preparation technique of sulfide electrolytes is mechanical milling using a high-energy planetary ball mill apparatus. This technique is fundamentally a room-temperature process. Therefore, sulfides are reacted at ordinary temperature and pressure. Electrolyte particles are obtained directly by milling. They are applicable to all-solid-state batteries without additional pulverization of electrolytes. Crystallization of glass electrolyte tends to precipitate a metastable phase, such as high-temperature phase, which generally has high ionic conductivity (Tatsumisago et al., 1991; Hayashi et al., 2003). Crystalline $Li_7P_3S_{11}$ (a high-temperature phase at the composition) is precipitated as a primary phase by heat treatment of a corresponding mother glass, and the obtained glass–ceramic electrolytes show conductivities of 10^{-3}–10^{-2} S cm^{-1}, which depend on the degree of crystallinity and the grain boundary of $Li_7P_3S_{11}$ (Mizuno et al., 2005, 2006; Seino et al., 2014). Cubic Na_3PS_4 is crystallized from the Na_3PS_4 glass prepared by mechanical milling. The prepared glass–ceramic electrolytes with Na_3PS_4 show Na^+ ion conductivity of 10^{-4} S cm^{-1} at 25°C (Hayashi et al., 2012b).

Electrolyte preparation *via* liquid phase is a suitable process for cost-effective quantity synthesis without using a special reaction apparatus. In general, this process has benefits of lowering the reaction temperature, shortening the reaction time, and controlling the particle morphology and size. Prepared electrolyte solutions are also useful for the coating of active material particles. Very recently, liquid-phase synthesis of sulfide SEs has been reported. The reaction processes for sulfides are divided into two categories: one uses suspension and the other uses a homogeneous solution. For the former synthesis, β-Li_3PS_4 synthesized in tetrahydrofuran (Liu et al., 2013) or dimethyl carbonate (Phuc et al., 2016) as reaction medium and $Li_7P_3S_{11}$ synthesized in 1,2-dimethoxyethane (Ito et al., 2014) are reported. In these reactions, precursors with precipitates are obtained. Compressed pellets of the heat-treated sulfide electrolytes show conductivity of greater than 10^{-4} S cm^{-1} at 25°C. As the latter one *via*

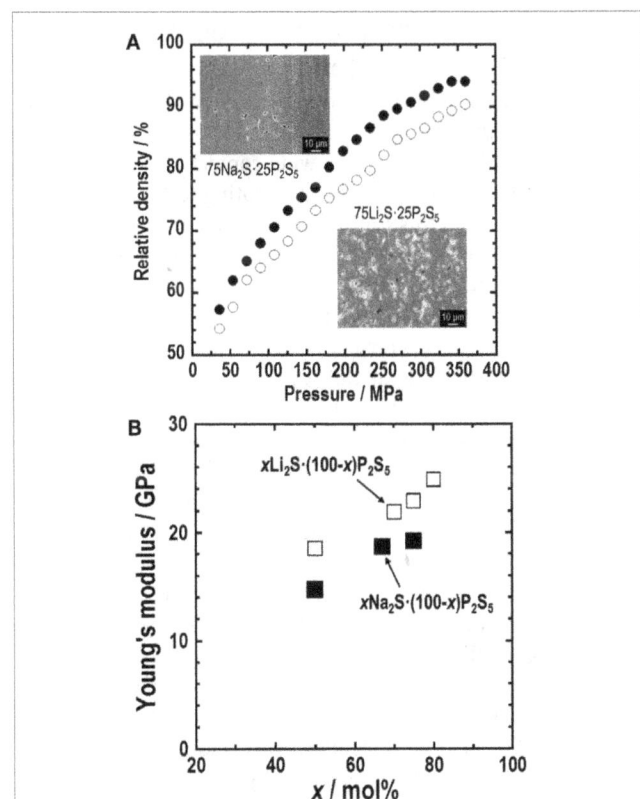

FIGURE 2 | (A) Molding pressure dependence of relative density of $75M_2S \cdot 25P_2S_5$ (M = Li or Na, mol%) glass electrolyte. Cross-sectional SEM images of these glass pellets prepared by pressing at 360 MPa are also shown in the inset. **(B)** Composition dependence of Young's modulus for the $xM_2S \cdot (100-x)P_2S_5$ glasses.

homogeneous liquid, Li_3PS_4 is synthesized from a mixture of Li_2S and P_2S_5 with N-methlyformamide (NMF) (Teragawa et al., 2014a). The Li_3PS_4 SEs can also be prepared using a dissolution–reprecipitation process in NMF from $80Li_2S\cdot20P_2S_5$ (mol%) glass prepared in advance using mechanical milling (Teragawa et al., 2014b). The prepared Li_3PS_4 electrolyte shows low conductivity of 10^{-6} S cm^{-1} at the present stage, but conductivity can be enhanced by selecting electrolyte compositions. Argyrodide-type Li_6PS_5Cl is dissolved into ethanol. Then, the argyrodite phase is reprecipitated by removing ethanol at 80°C under vacuum for 3 h (Yubuchi et al., 2015a,b). A pellet of the product shows conductivity of 10^{-5} S cm^{-1} at 25°C, which is somewhat lower than that of the original Li_6PS_5Cl. Grain boundary resistance, which is affected by surface structure and morphology, might be greater in the prepared Li_6PS_5Cl. Optimization of posttreatments for the prepared powders will enhance the Li_6PS_5Cl conductivity. Furthermore, the combination of sulfide SEs and ionic liquids produces pseudo-SEs (Minami et al., 2010; Oh et al., 2015). The prepared electrolytes give a new category of electrolytes having both high conductivity and good formability.

Sodium-ion conducting sulfide electrolytes with cubic Na_3PS_4 are also synthesized *via* a liquid-phase process from the mixture of Na_2S and P_2S_5 in NMF solvent (Yubuchi et al., 2015a). The room-temperature conductivity of the obtained electrolyte is 10^{-6} S cm^{-1}, which is lower than the conductivity of the electrolyte prepared by mechanical milling. Studies investigating sulfide SEs with Na^+ ion conductivity are extremely few at present. New electrolytes produced *via* a simple liquid-phase process will be researched widely.

PREPARATION OF SOLID–SOLID INTERFACE IN ALL-SOLID-STATE BATTERIES

Sulfide glasses are well-balanced SEs with high conductivity, good formability, appropriate Young's modulus, and moderate chemical stability. They are therefore highly promising SEs for use in all-solid-state batteries. A schematic diagram of bulk-type all-solid-state batteries is depicted in **Figure 3**. A positive electrode layer is composed not only of active material particles but also of SE ones. The Li^+ ion is supplied from SEs attached to active materials. Electrons are mobile through active materials. To enhance the rate of performance of the batteries, conductive additives of nanocarbons are added to the electrode layer. A lithium alloy or lithium metal is used as the negative electrode. A SE layer as a separator is sandwiched with the positive and negative electrode layers. Then, because of good formability of sulfide SEs, it can be pressed uniaxially at room temperature to fabricate bulk-type all-solid-state batteries.

An all-solid-state $In/LiCoO_2$ or $Li-In/Li_4Ti_5O_{12}$ cell with Li_2S–P_2S_5 glass–ceramic electrolytes exhibits good cycle performance for hundreds of times at 25°C (Tatsumisago and Hayashi, 2008; Tatsumisago et al., 2013). These cells operate at a high temperature of 100°C, where it is difficult for a liquid electrolyte cell to be used. The all-solid-state cell with $Li_4Ti_5O_{12}$ shows a discharge–charge capacity of about 140 mAh g^{-1}. It retains the capacity for 700

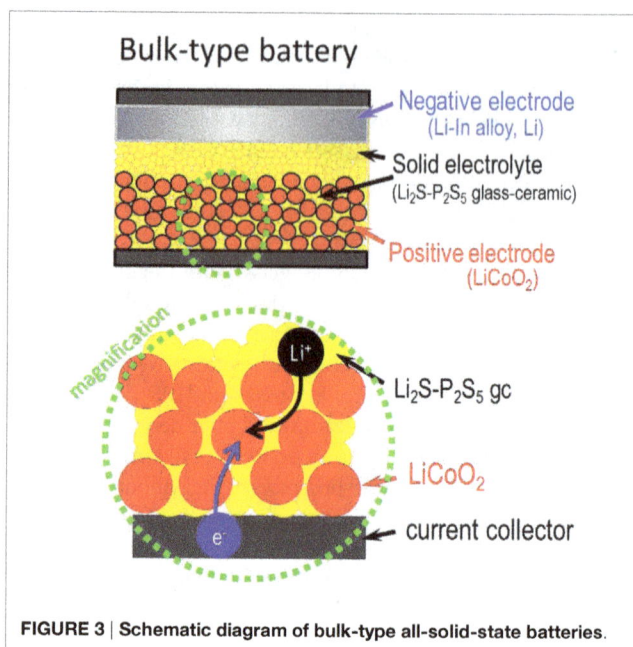

FIGURE 3 | Schematic diagram of bulk-type all-solid-state batteries.

cycles with no degradation under a high current density of more than 10 mA cm^{-2} (Minami et al., 2011).

Coating of Sulfide Electrolytes on $LiCoO_2$ and Graphite Particles

Direct coating of sulfide SEs on $LiCoO_2$ particles, instead of mixing electrolyte particles, is effective for forming good electrolyte–$LiCoO_2$ interfaces with a wide contact area. Pulsed laser deposition (PLD) was first used as a coating technique (Sakuda et al., 2010, 2011). In this study, $LiNbO_3$-coated $LiCoO_2$ particles were fluidized with a vibrator to ensure uniform coating of sulfide electrolytes on the particles.

Figure 4 presents cross-sectional SEM images of positive electrodes with (a) a conventional mixture of $LiCoO_2$ and Li_2S–P_2S_5 SE particles with the weight ratio of 70/30 and (b) SE-coated $LiCoO_2$ particles, where the weight ratio of $LiCoO_2$ and SE coatings was 90/10. **Figure 2A** shows that solid–solid contacts between $LiCoO_2$ and SE are formed by cold pressing of the mixture electrode because SE particles are deformed easily by pressing. **Figure 4A** shows that the aggregation of $LiCoO_2$ particles engenders less utilization of $LiCoO_2$ because many voids and less contact with SE are observed among $LiCoO_2$ particles. A close-packed electrode layer is formed using only SE-coated $LiCoO_2$ (**Figure 4B**). Its enlarged figure, which will be displayed in **Figure 7D** later, shows that a close solid–solid interface with fewer voids appears, and Li^+ ion conduction paths are therefore formed among $LiCoO_2$ particles. It is noteworthy that SE amounts in the electrode are decreased considerably by SE coating on $LiCoO_2$ particles. The use of minimum amounts of SE in an electrode layer contributes to enhanced energy density of all-solid-state batteries.

A cross-sectional high-angle annular dark field (HAADF)-STEM image and EDX mappings of O, P, S, Co, and Nb elements for the SE-coated $LiCoO_2$ electrodes are presented in **Figure 5**. A SE coating layer is observed between two $LiCoO_2$ particles,

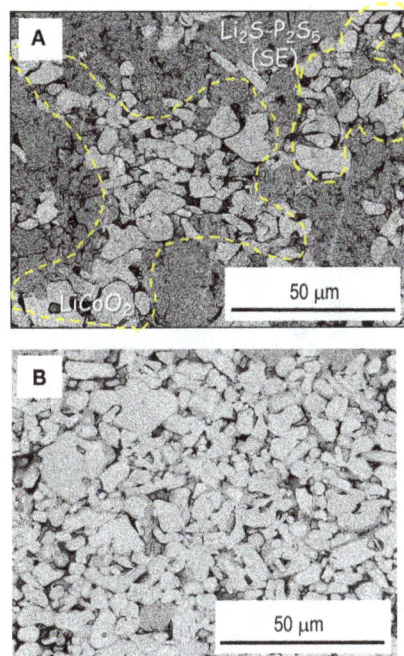

FIGURE 4 | Cross-sectional SEM images of LiCoO₂ positive electrode layers consisting of (A) mixture electrode with LiCoO₂ and Li₂S–P₂S₅ electrolyte particles and (B) electrolyte-coated LiCoO₂ particles.

where one particle has several cracks. The EDX mapping of Nb element reveals that a $LiNbO_3$ coating layer exists on the surface of $LiCoO_2$, but the layer is missing at the cracks. The EDX mappings of P and S elements indicate that SE penetrates into the cracks as a liquid electrolyte does. Favorable formability of sulfide SE is effective for forming good contacts with the surface of active materials and using active materials even at newly formed crystal faces with cracks.

Solid electrolyte coatings were also done for graphite particles using PLD. The mixture of SE-coated graphite and SE particles was used as a negative electrode. The weight ratio of graphite and SE was 90/10. All-solid-state cells with $Li_2S–P_2S_5$ SE as a separator layer were charged and discharged at a constant current density of 0.064 mA cm⁻² (ca. 0.05 C) for voltages of 2.8–4.3 or 4.6 V at room temperature (Sakuda et al., 2013b). **Figure 6** shows charge–discharge curves for all-solid-state SE-coated graphite/SE-coated $LiCoO_2$ batteries. The battery with a cutoff voltage of 4.6 V at the charge process shows a higher discharge capacity than that of 4.3 V. Battery has a discharge capacity of 133 mAh g⁻¹, as calculated from the total mass of the composite positive electrode.

Figure 7 shows cross-sectional SEM images of SE-coated $LiCoO_2$ positive electrodes; **Figure 7A** as-prepared, **Figure 7B** after the initial charge-discharge (cutoff voltage: 4.3 V), and **Figure 7C** after the initial charge–discharge (cutoff voltage: 4.6 V). **Figures 7D–F**, respectively, portray enlarged images of **Figures 7A–C**. Closely attached SE–$LiCoO_2$ interfaces with large contact area are achieved by PLD coating, as shown in **Figures 7A,D**. The electrode morphology (**Figure 7B**) does not change greatly after the initial charge–discharge process at the

cutoff of 4.3 V. SE coatings still attach on the $LiCoO_2$ particles (**Figure 7E**), and empty spaces among particles observed in panel (**Figure 7B**) are similar to those observed in the as-prepared electrode (**Figure 7A**). **Figure 7C** shows that voids among $LiCoO_2$ particles increase after the charge–discharge process at a higher cutoff voltage of 4.6 V. A $LiCoO_2$ particle charged to 4.6 V suffers from deterioration. Many cracks are formed in the particle, as shown in the magnified image of panel (**Figure 7F**). This morphological change is attributable to (1) reduction of mechanical strength and (2) excess strain at the solid–solid interfaces among $LiCoO_2$ particles during their large volume expansion and/or phase transition.

Figure 8 shows cross-sectional SEM images of SE-coated graphite negative electrodes; **Figure 8A** as-prepared and **Figure 8B** after the initial charge–discharge (cutoff voltage: 4.3 V). Graphite particles are close-packed in a negative electrode layer. SE coatings are observed around graphite particles. They form Li⁺ ion conduction paths through the electrode layer. After the initial charge–discharge, no obvious void is observed in the negative electrode. The close-packed electrode is present. Good solid–solid interfaces are also retained between the graphite layer and an electrolyte separator layer.

As described in Section "Preparation Process," liquid-phase synthesis of sulfide electrolytes is also useful for forming favorable electrode–electrolyte solid–solid interfaces, which can be achieved by removing solvents from electrolyte solutions. An all-solid-state cell using $LiCoO_2$ coated with Li_3PS_4 electrolyte via an NMF solution operates as rechargeable batteries without the addition of SE and carbon-conductive additive particles to the positive electrodes (Teragawa et al., 2014a,b). However, the cell capacities are lower than those of the cells using $LiCoO_2$ coated with Li_3PS_4 electrolytes by PLD, because the electrolytes synthesized via the liquid-phase has lower ionic conductivities than those prepared by PLD. The $LiCoO_2$ particles coated with Li_6PS_5Cl electrolytes via ethanol solution with a higher conductivity than Li_3PS_4 were therefore applied to all-solid-state cells. The weight ratio of $LiCoO_2$/SE layer was 92.5/7.5 in SE-coated $LiCoO_2$ particles. An all-solid-state cell using the electrolyte-coated $LiCoO_2$ shows an initial discharge capacity of 45mAh g⁻¹, which is greater than that of cells using Li_3PS_4-coated $LiCoO_2$ (Yubuchi et al., 2015b). SE coating via electrolyte liquids is a simple and cost-effective process. Increasing conductivity of SE at lower temperatures is necessary to improve battery performance.

Sulfur-Based Nanocomposite Positive Electrodes

Sulfur is a fascinating positive electrode with high energy density because it is an abundant resource with high theoretical capacity of 1672 mAh g⁻¹ and environmental friendliness. Lithium polysulfides (Li_2S_x), which are formed during discharge (lithiation process), are readily dissolved in organic liquid electrolytes, leading to lack of a sulfur positive electrode. Dissolution of lithium polysulfides is suppressed by absorbing sulfurs in nanocarbon pores. This approach has been studied extensively (Ji and Nazar, 2010). The use of inorganic SEs fundamentally resolves the problem.

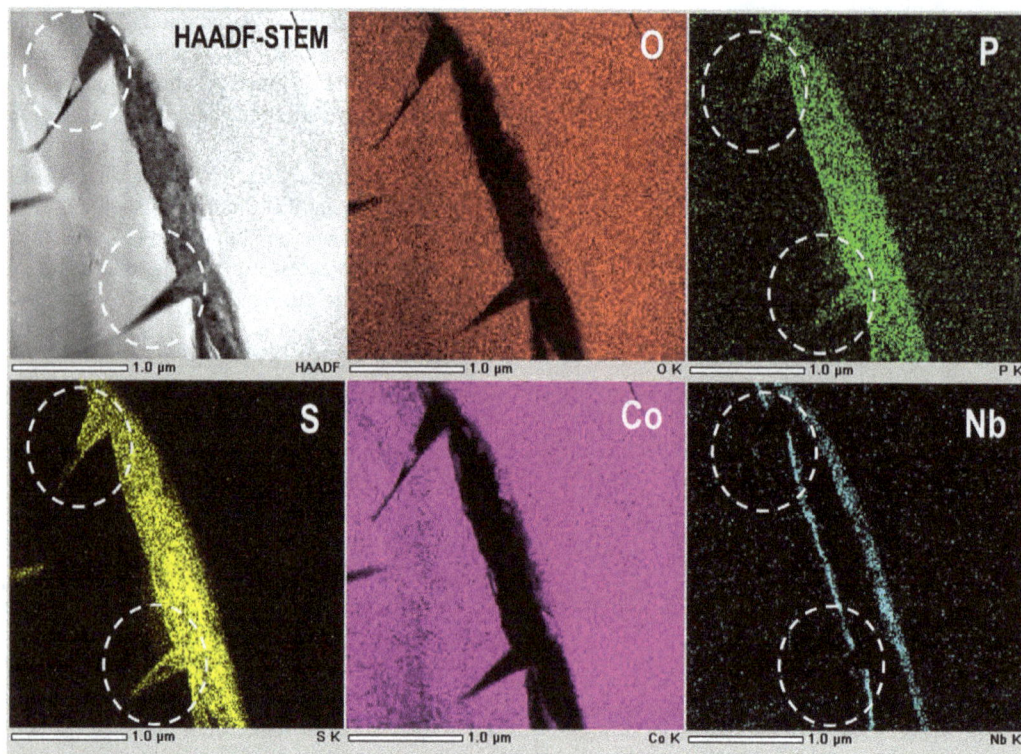

FIGURE 5 | Cross-sectional high-angle annular dark field (HAADF)-TEM image and EDX mappings of O, P, S, Co, and Nb elements for the SE-coated LiCoO₂ electrodes.

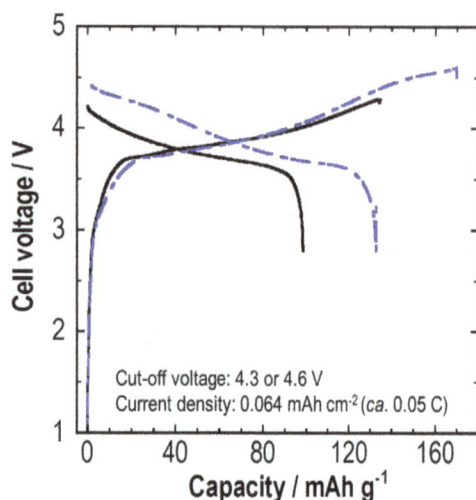

FIGURE 6 | Initial charge–discharge curves for an all-solid-state SE-coated graphite/SE-coated LiCoO₂ battery.

Composite sulfur electrodes consisting of S, acetylene black (AB), and $Li_2S-P_2S_5$ SE powders with a weight ratio of 25/25/50 were prepared using high-energy planetary ball milling to produce favorable contacts among the three components (Nagao et al., 2011). The Li-In/S cell exhibits a large reversible capacity of greater than 1500 mAh g⁻¹ of sulfur with average potential of ca. 2.1 V (vs. Li⁺/Li). The cell with sulfur electrode shows 15 times higher capacity than the cell with LiCoO₂, although the operating potential of the former cell is almost half that of the latter cell.

Application of Li_2S as a discharge product of sulfur active material offers the important benefits of high theoretical capacity of 1167 mAh g⁻¹ and versatility of negative electrode materials without lithium sources. Because of the insulative nature of Li_2S, composite electrodes, Li_2S mixed with conductive additives, such as nanocarbons and SEs, should be prepared for the use of Li_2S as an active material. Composite Li_2S electrodes were prepared by mechanical milling. A typical weight ratio of Li_2S/AB/$Li_2S-P_2S_5$ SE is 25/25/50. The prepared composite gives broad peaks attributable to Li_2S in X-ray diffraction patterns. An all-solid-state cell of In/Li_2S composites is charged and then discharged at 25°C. The initial reversible capacity is 800 mAh g⁻¹ at the current density of 0.064 mA cm⁻² (Nagao et al., 2012b). The cell retained 750 mAh g⁻¹ for 10 cycles. Charge–discharge reaction mechanisms were examined using high-resolution TEM observation (Nagao et al., 2015). **Figure 9** shows TEM images for the Li_2S electrodes; **Figure 9A** before charge–discharge test, **Figure 9B** after the initial charge, and **Figure 9C** after the initial discharge. **Figure 9A** shows that nanoparticles of *ca.* 5 nm in size with different crystal orientations are distributed randomly in the matrix consisting of amorphous SE and AB. Those nanoparticles are attributable to crystalline Li_2S. **Figure 9B** shows that no lattice fringes because of the crystalline Li_2S are apparent, and

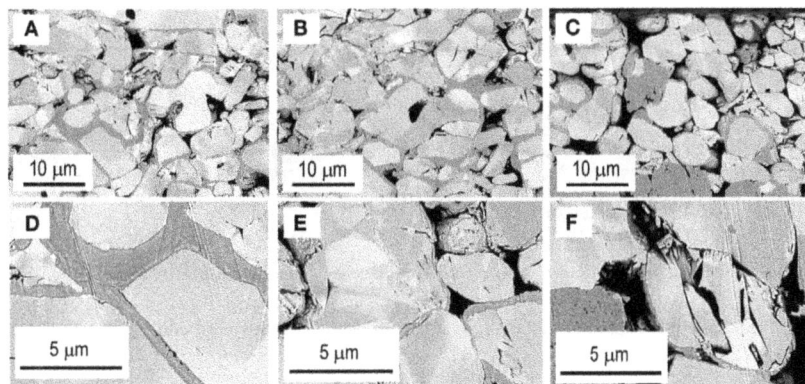

FIGURE 7 | Cross-sectional SEM images of SE-coated LiCoO₂ positive electrodes (A) as-prepared, (B) after the initial charge–discharge (cutoff voltage: 4.3 V), and (C) after the initial charge–discharge (cutoff voltage: 4.6 V). Images of (D–F), respectively, depict enlarged images of (A–C).

there exists the characteristic contrast attributable to amorphous structure in the whole region after the initial charge process. As shown in **Figure 9C**, lattice fringes with spacing of about 3.9 Å are clearly apparent after the initial discharge, suggesting that amorphous sulfur is converted into crystalline nanoparticles during discharge reaction. Reversible transformation between crystallization and amorphization of sulfur-based active nanoparticles is responsible for the high capacity and its retention.

The utilization of Li₂S is ca. 50% in the composite electrode. To increase the utilization of Li₂S, one strategy is the fundamental enhancement of ionic conductivity in Li₂S. A partial substitution of more polarizable iodide anion with larger ionic radii for sulfide anion in Li₂S is expected to increase conductivity by introducing lithium vacancies and by increasing the lattice constant. Solid solutions in the system Li₂S-LiI are therefore prepared using mechanical milling (Hakari et al., 2015a). Only the XRD peaks attributable to Li₂S are observed; LiI peaks disappear completely in the composition range of 0 < LiI (mol%) < 20. **Figure 10A** presents the composition dependence of lattice constant and conductivity at 25°C for the prepared Li₂S-LiI materials. The lattice constant increases monotonically with increased LiI content, suggesting that Li₂S-based solid solutions are prepared using a mechanical milling process. Conductivity is also enhanced by increasing the LiI content. The solid solution with 20 mol% LiI has conductivity of 2.2×10^{-6} S cm^{-1}, which is two orders of magnitude higher than that for Li₂S itself without the addition of LiI. A composite positive electrode with the 80Li₂S·20LiI (mol%) solid solution, vapor-grown carbon fiber (VGCF), and Li₃PS₄ glass electrolyte with a weight ratio of 50/10/40 is applied to all-solid-state cells. **Figure 10B** shows that an all-solid-state cell (Li-In/Li₃PS₄/80Li₂S·20LiI) is charged and discharged at a current density of 0.13 mA cm^{-2} (0.07 C) at 25°C. The cell shows a reversible capacity of 930 mAh g^{-1} for 50 cycles. The capacity corresponds to 80% utilization of Li₂S. It is noteworthy that the enhancement of conductivity of Li₂S is effective for increasing the utilization of the active material. This strategy is beneficial for the development of all-solid-state cells with a higher energy density.

A SE Li₃PS₄ is ball-milled with nanocarbon, such as AB. The prepared Li₃PS₄–AB materials as a mixed conductor are useful

as positive electrode acting not only as a SE but also as an active material. An all-solid-state cell with the Li₃PS₄–AB composite positive electrode is charged and discharged. Its operation voltage of ca. 2.6 V vs. Li⁺/Li is somewhat higher than that of Li₂S (Hakari et al., 2015b). Redox-active electrolytes, such as CuCl₂-dissolved solution in porous carbons, are also reported to apply to supercapacitors (Mai et al., 2013). The use of the SEs as an active material in electrode layers is effective at increasing the reversible capacity per gram of the total mass of positive electrodes.

Transition Metal Sulfide Positive Electrodes

Typical MS$_x$, such as TiS₂, are used as active materials in all-solid-state Li and Na batteries with sulfide SEs. Decreasing the particle size of MS$_x$ and forming a wide contact area with both SEs and conductive additives are important for increasing the MS$_x$ utilization.

Monodispersed MS$_x$ nanoparticles are prepared using a so-called "hot-soap" technique using high-boiling point solvents as a reaction medium. Particle morphology and size can be controlled by choosing the reaction conditions and combinations of coordinating or non-coordinating solvents. NiS particles of 50 nm size were prepared using thermal decomposition of nickel acetylacetonate in a mixed solution of 1-dodecanethiol as a sulfur source and 1-octadecene as a non-coordinating solvent at 280°C for 5 h (Aso et al., 2011). The NiS nanoparticles are crystallized directly on a carbon fiber (VGCF) by adding VGCF to a liquid medium. Good adhesion between NiS and carbon is achieved (Aso et al., 2012). Sulfide electrolyte coating on NiS–VGCF was produced using the PLD method. All-solid-state cells with the prepared NiS composite electrode operate as a secondary battery at 25°C, suggesting that electron and Li⁺ ion conduction paths are formed in the composite electrodes (Aso et al., 2013).

To increase the positive electrode capacity, sulfur-rich MS$_x$ are desired. For example, titanium trisulfide TiS₃ shows a higher capacity than that of TiS₂, because additional sulfurs in TiS₃ contribute to the redox reaction that occurs during charge–discharge processes (Hayashi et al., 2012a). Amorphous NbS$_x$ (x = 3, 4, 5)

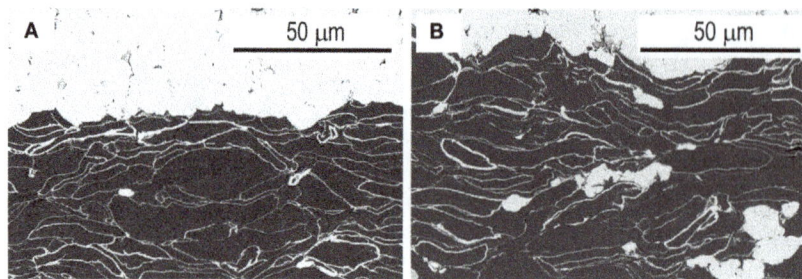

FIGURE 8 | Cross-sectional SEM images of SE-coated graphite negative electrodes (A) as-prepared and (B) after the initial charge–discharge (cutoff voltage: 4.3 V).

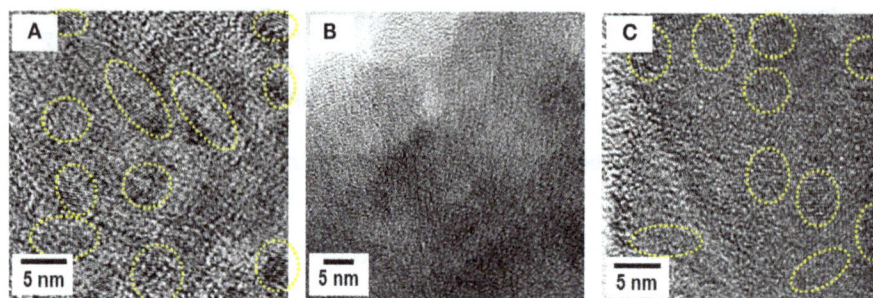

FIGURE 9 | TEM images for the Li₂S electrodes (A) before charge–discharge test, (B) after the initial charge, and (C) after the initial discharge.

are prepared mechanochemically. Electrochemical cells with an organic liquid electrolyte using the amorphous NbS_x ($x = 3, 4, 5$) show higher discharge capacities with an increase in the sulfur content of NbS_x (Sakuda et al., 2014). Amorphous TiS_3 (a-TiS_3) retains a higher capacity than crystalline TiS_3 in all-solid-state lithium cells. The crystal structure of TiS_3 is partially deteriorated at the initial cycle, leading to an irreversible capacity in the cell with crystalline TiS_3 (Matsuyama et al., 2016a,b). It is noteworthy that sulfur-rich amorphous MS_x with electrical conductivity are promising for use as positive electrodes instead of sulfur active materials.

Figure 11A shows the first and tenth charge–discharge curves of all-solid-state lithium cell Li-In/a-TiS_3 at 0.013 mA cm⁻² at 25°C. The right side ordinate axis represents the electrode potential vs. Li⁺/Li, as calculated based on the potential difference between the Li-In and Li electrode (0.62 V). The cell with a-TiS_3 positive electrode including no carbon-conductive additives and SEs shows a reversible capacity for 10 cycles of about 510 mAh g⁻¹ of a-TiS_3, which equals the weight of the total positive electrode. The capacity corresponds to storage of about 3M Li to a-TiS_3. Electronic structural analyses using S2p XPS and S K-edge XANES reveal that a reversible sulfur redox in a-TiS_3 appears mainly during charge–discharge processes and contributes to its good capacity retention (Matsuyama et al., 2016a,b).

Amorphous TiS_3 is also applicable to all-solid-state sodium batteries. An all-solid-state sodium cell using a-TiS_3 shows capacity higher than 300 mAh g⁻¹ at the first discharge process,

as presented in **Figure 11B** (Tanibata et al., 2015a). The composite positive electrode consisting of a-TiS_3 and the cubic Na_3PS_4 electrolyte with a weight ratio of 40/60 is used. $Na_{15}Sn_4$ alloy and the cubic Na_3PS_4 electrolyte are used, respectively, as a negative electrode and a separator layer. The cell capacity decreases gradually during discharge–charge cycles. SEM–EDX analysis reveals that the a-TiS_3 particles aggregate after the cycles, and that resistance in the a-TiS_3 composite electrode increases during cycles. To secure electron conduction paths to a-TiS_3, 6 wt% AB is added to the positive electrode. Good capacity retention is achieved in the cell using a-TiS_3 electrode with AB. The added AB particles might prevent the shutoff of electron conduction paths in the composite electrode and give a buffer space for the volume change of a-TiS_3 particles for preserving adhesion among the particles. The addition of AB to the $Na_{15}Sn_4$ electrode also suppresses cell resistance after charge–discharge. In addition, the replacement of Na_3PS_4 glass–ceramic by $94Na_3PS_4 \cdot 6Na_4SiS_4$ (mol%) glass–ceramic with higher conductivity decreases cell resistance and increases the rate performance of all-solid-state cells (Tanibata et al., 2015b). Reversible capacity in all-solid-state Na cells is less than that in all-solid-state Li cells, as shown in **Figures 11A,B**. The lower conductivity of SEs and loss of particle contacts because of the larger volume change are responsible for a smaller capacity for the Na cells. Approaches that are used to prepare composite positive electrodes suitable for insertion/de-insertion of Na⁺ ion with larger ionic radius must be developed to improve battery performance in all-solid-state Na batteries.

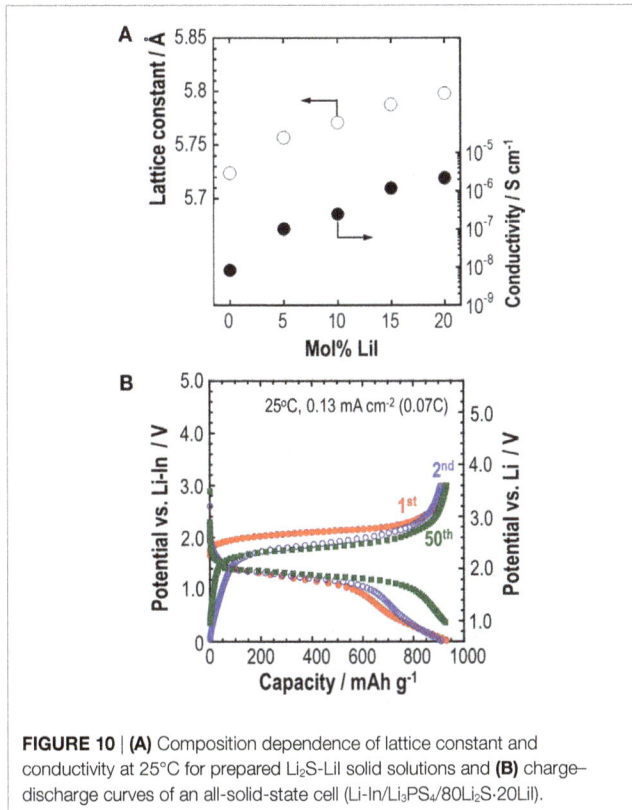

FIGURE 10 | **(A)** Composition dependence of lattice constant and conductivity at 25°C for prepared Li_2S-LiI solid solutions and **(B)** charge–discharge curves of an all-solid-state cell (Li-In/Li_3PS_4/$80Li_2S·20LiI$).

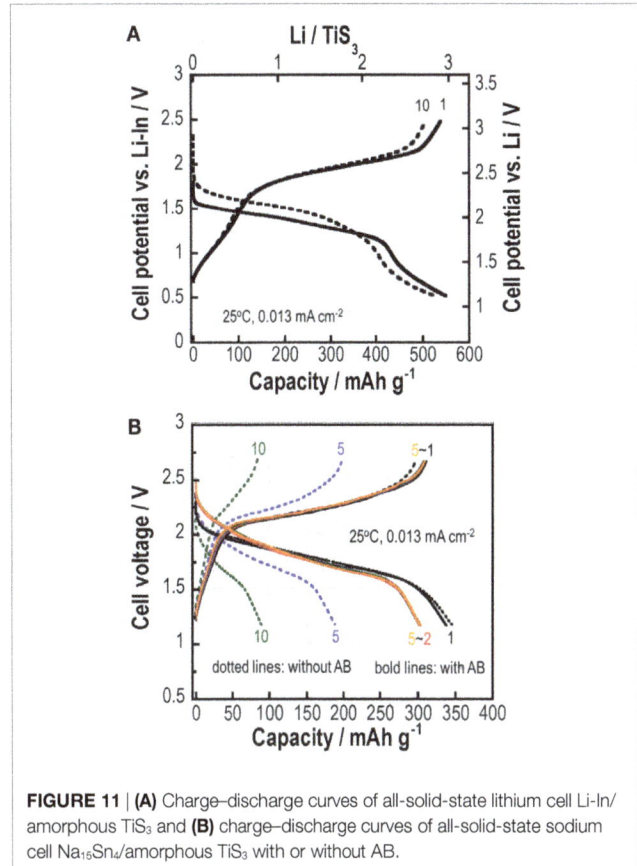

FIGURE 11 | **(A)** Charge–discharge curves of all-solid-state lithium cell Li-In/amorphous TiS_3 and **(B)** charge–discharge curves of all-solid-state sodium cell $Na_{15}Sn_4$/amorphous TiS_3 with or without AB.

Interface Modification for Li Metal Negative Electrode

To achieve high energy density of all-solid-state Li batteries, the final goal is the use of lithium metal as a negative electrode. Lithium metal is an ultimate negative electrode because of a large theoretical capacity of 3861 mAh g^{-1} and the lowest electrochemical potential of −3.04 V vs. SHE. However, a possibility of fatal problems emerged by short-circuit with dendrite formation prevent the practical use of Li metal negative electrode in lithium cells with conventional liquid and polymer electrolytes. Combination with inorganic SEs is expected to resolve the problem. In fact, thin-film solid-state batteries with Li negative electrode have excellent cycle life without capacity fading (Bates et al., 1993). The rate of utilization of Li electrodes is not high in thin-film batteries because positive electrodes such as $LiCoO_2$ have lithium sources, which are mainly used for a charge–discharge process.

Compared with thin-film batteries fabricated by gas-phase deposition, insufficient Li–SE interfaces in bulk-type all-solid-state batteries prepared by cold pressing are important issues that must be resolved. It has been revealed by *in situ* SEM observation that lithium is deposited through grain boundaries and voids in the SE (Nagao et al., 2013). Insertion of a Li-alloy thin layer at the interface between the Li electrode and SE layers brought about stable Li dissolution and deposition in the all-solid-state Li metal cells (Hiratani et al., 1988; Okita et al., 2011; Nagao et al., 2012a). These interface modifications are effective at establishing homogeneous interfaces between the Li metal and SEs.

Intensive utilization of Li is important for achieving high energy density of all-solid-state lithium metal batteries. Insertion of a Au thin film at the Li–SE interface is effective for increasing Li utilization (Kato et al., 2016a). Li and Au thin films were formed on a pelletized Li_3PS_4 glass electrolyte by vacuum evaporation. Galvanostatic cycling tests for the Li/Li_3PS_4/Li cell are presented in **Figure 12A**. At the initial cycle, the utilization of Li metal is about 40%, which is a higher rate of utilization of Li metal for thin-film batteries (about 20%). However, utilization of Li metal for the cell decreases rapidly after five cycles. The Li thin-film morphology is rough after galvanostatic cycling tests, as shown in the SEM image (**Figure 12C**), indicating that inhomogeneous Li dissolution–deposition reactions occur. The cell Li/Au/Li_3PS_4/Au/Li has about 35% Li utilization at the initial cycle and retains about 25% after the fifth cycle, as presented in **Figure 12B**. The morphology of Li metal after Li dissolution–deposition reaction became more uniform, as shown in the SEM image of panel (**Figure 12D**), compared with the cell without Au thin films. The insertion of Au film to a Li–SE interface is a first step for improving the cyclability of Li deposition–dissolution reactions with high Li utilization in all-solid-state lithium metal batteries. Intensive studies are in progress.

CONCLUDING REMARKS

We have reviewed recent developments related to sulfide SEs and interface formation processes for all-solid-state rechargeable batteries. The conductivity of sulfide Li^+ ion conductors, such as

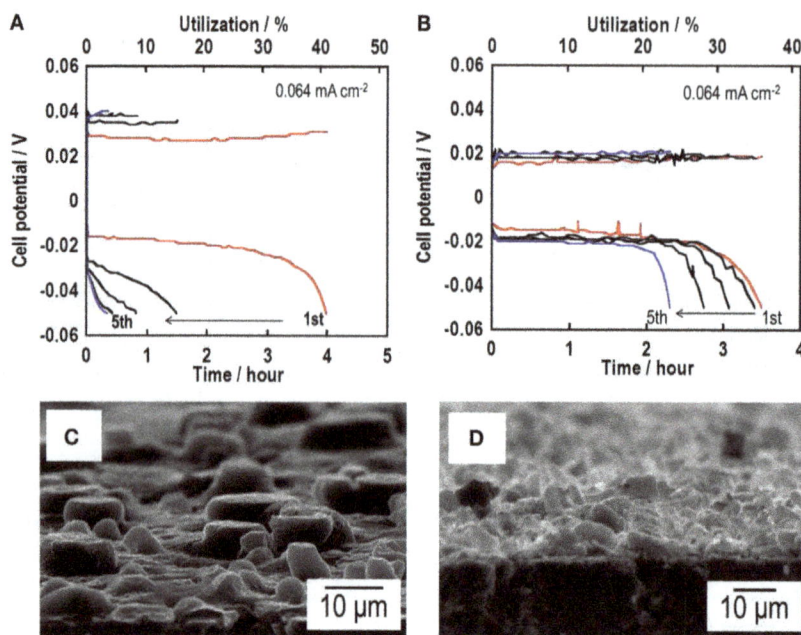

FIGURE 12 | Galvanostatic cycling tests for (A) a Li symmetric Li/Li$_3$PS$_4$/Li cell and (B) a Li/Au/Li$_3$PS$_4$/Au/Li cell. SEM images of the surface of (C) a Li/Li$_3$PS$_4$/Li cell and (D) a Li/Au/Li$_3$PS$_4$/Au/Li cell after galvanostatic cycling tests.

Li$_{10}$GeP$_2$S$_{12}$, Li$_7$P$_3$S$_{11}$, and Li$_{9.54}$Si$_{1.74}$P$_{1.44}$S$_{11.7}$Cl$_{0.3}$, has already reached 10^{-2} S cm^{-1} at room temperature. Conductivities of sulfide Na$^+$ ion conductors are lower than those of Li$^+$ ion conductors at the present stage. The highest conductivity of 10^{-3} S cm^{-1} was obtained for Na$_3$PSe$_4$. Sulfide glass electrolytes with high alkali content were prepared by high-energy ball milling process. Metastable phases, such as Li$_7$P$_3$S$_{11}$ and cubic Na$_3$PS$_4$, with high conductivity were formed by careful crystallization of the prepared glasses. Sulfide electrolytes can be synthesized *via* liquid-phase processing, which is useful for coating application. Sulfide glass electrolytes have favorable formability, and Young's modulus for forming good electrode–electrolyte interfaces achieving rapid charge transfer in bulk-type all-solid-state batteries. Chemical stability in air is a great shortcoming of using a sulfide electrolyte. The composition of Li$_3$PS$_4$ has superior chemical stability in the Li$_2$S–P$_2$S$_5$ binary system. The higher chemical stability is achieved by partial substitution of oxygen for sulfur. Selecting composition and designing structure for sulfide electrolytes is expected to improve conductivity, mechanical properties, and chemical stability further. As Na$^+$ ion conductors, higher conductivity of more than 10^{-2} S cm^{-1} is predicted for Sn-substituted cubic Na$_3$PS$_4$, but the conductivity has not been achieved experimentally. Further studies seeking new electrolytes and suitable preparation processes must be undertaken.

Coating of SE and preparation of nanocomposites are useful for forming favorable solid–solid interface in an electrode layer for bulk-type all-solid-state batteries. Coating of sulfide electrolytes on LiCoO$_2$ or graphite particles using gas-phase or liquid-phase techniques is effective for increasing solid–solid contact area using extremely small amounts of electrolytes. Preparation of nanocomposites using high-energy ball milling is useful for sulfur or Li$_2$S active materials with an insulative nature. Conductivity enhancement of Li$_2$S by combination with LiI contributes to the improvement of Li$_2$S utilization. Amorphous MS$_x$, such as amorphous TiS$_3$, are attractive as mixed conductors with large capacity in all-solid-state batteries. Interface modification for Li metal negative electrode with Au thin film improves the cycle performance of Li dissolution–deposition while maintaining a high rate of utilization. For further improvement of electrochemical performance of the batteries, facile approaches achieving favorable electrode–electrolyte interfaces with large contact area will be developed. Controlling the size, morphology, and dispersibility of both electrolyte and electrode particles and selecting suitable electrolytes for maintaining close solid–solid contacts during charge–discharge processes will be assessed in future studies.

AUTHOR CONTRIBUTIONS

AH contributed to the preparation of the manuscript. AS and MT contributed to the discussions about research results.

ACKNOWLEDGMENTS

This research is supported by a Grant-in-Aid for Scientific Research from the Ministry of Education, Culture, Sports, Science and Technology (MEXT) of Japan. In particular, the research about all-solid-state Li batteries was financially supported by the Japan Science and Technology Agency (JST), Advanced Low Carbon Technology Research and Development Program (ALCA), Specially Promoted Research for Innovative Next Generation Batteries (SPRING) Project, while the research about all-solid-state Na batteries was supported by the MEXT program "Elements Strategy Initiative for Catalysts and Batteries (ESICB)."

REFERENCES

Aotani, N., Iwamoto, K., Takada, K., and Kondo, S. (1994). Synthesis and electrochemical properties of lithium ion conductive glass, Li3PO4-Li2S-SiS2. *Solid State Ionics.* 68, 35–39. doi:10.1016/0167-2738(94)90232-1

Aso, K., Hayashi, A., and Tatsumisago, M. (2012). Synthesis of NiS-carbon fiber composites in high-boiling solvent to improve electrochemical performance in all-solid-state lithium secondary batteries. *Electrochim. Acta* 83, 448–453. doi:10.1016/j.electacta.2012.07.088

Aso, K., Kitaura, H., Hayashi, A., and Tatsumisago, M. (2011). Synthesis of nanosized nickel sulfide in high-boiling solvent for all-solid-state lithium secondary batteries. *J. Mater. Chem.* 21, 2987–2990. doi:10.1039/c0jm02639e

Aso, K., Sakuda, A., Hayashi, A., and Tatsumisago, M. (2013). All-solid-state lithium secondary batteries using NiS-carbon fiber composite electrodes coated with Li2S-P2S5 solid electrolytes by pulsed laser deposition. *ACS Appl. Mater. Interfaces* 5, 686–690. doi:10.1021/am302164e

Bates, J., Dudney, N., Gruzalski, G., Zuhr, R., Choudhury, A., and Luck, C. (1993). Fabrication and characterization of amorphous lithium electrolyte thin films and rechargeable thin-film batteries. *J. Power Sources* 4, 103–110. doi:10.1016/0378-7753(93)80106-Y

Boron, P., Johansson, S., Zick, K., Gunne, J., Dehnen, S., and Roling, B. (2013). Li10SnP2S12: an affordable lithium superionic conductor. *J. Am. Chem. Soc.* 135, 15694–15697. doi:10.1021/ja407339y

Boulineau, S., Courty, M., Tarascon, J. M., and Viallet, V. (2012). Mechanochemical synthesis of Li-argyrodite Li6PS5X (X=Cl, Br, I) as sulfur-based solid electrolytes for all solid state batteries application. *Solid State Ionics.* 221, 1–5. doi:10.1016/j.ssi.2012.06.008

Hakari, T., Hayashi, A., and Tatsumisago, M. (2015a). Highly utilized lithium sulfide active material by enhancing conductivity in all-solid-state batteries. *Chem. Lett.* 44, 1664–1666. doi:10.1246/cl.150758

Hakari, T., Nagao, M., Hayashi, A., and Tatsumisago, M. (2015b). All-solid-state lithium batteries with Li3PS4 glass as active material. *J. Power Sources* 293, 721–725. doi:10.1016/j.jpowsour.2015.05.073

Hayashi, A., Hama, S., Minami, T., and Tatsumisago, M. (2003). Formation of superionic crystals from mechanically milled Li2S-P2S5 glasses. *Electrochem. Commun.* 5, 111–114. doi:10.1016/S1388-2481(02)00555-6

Hayashi, A., Matsuyama, T., Sakuda, A., and Tatsumisago, M. (2012a). Amorphous titanium sulfide electrode for all-solid-state rechargeable lithium batteries with high capacity. *Chem. Lett.* 41, 886–888. doi:10.1246/cl.2012.886

Hayashi, A., Noi, K., Sakuda, A., and Tatsumisago, M. (2012b). Superionic glass-ceramic electrolytes for room-temperature rechargeable sodium batteries. *Nat. Commun.* 3, 1–5. doi:10.1038/ncomms1843

Hayashi, A., Muramatsu, H., Ohtomo, T., Hama, S., and Tatsumisago, M. (2013). Improvement of chemical stability of Li3PS4 glass electrolytes by adding MxOy (M=Fe, Zn, and Bi) nanoparticles. *J. Mater. Chem. A* 1, 6320–6326. doi:10.1039/c3ta10247e

Hiratani, M., Miyauchi, K., and Kudo, T. (1988). Effect of a lithium alloy layer inserted between a lithium anode and a solid electrolyte. *Solid State Ionics.* 2, 1406–1410. doi:10.1016/0167-2738(88)90394-3

Hori, S., Kato, M., Suzuki, K., Hirayama, M., Kato, Y., and Kanno, R. (2015). Phase diagram of the Li4GeS4-Li3PS4 quasi-binary system containing the superionic conductor Li10GeP2S12. *J. Am. Ceram. Soc.* 98, 3352–3360. doi:10.1111/jace.13694

Ito, S., Nakakita, M., Aihara, Y., Uehara, T., and Machida, N. (2014). A synthesis of crystalline Li7P3S11 solid electrolyte from 1, 2-dimethoxyethane solvent. *J. Power Sources* 271, 342–345. doi:10.1016/j.jpowsour.2014.08.024

Jansen, M., and Henseler, U. (1992). Synthesis, structure determination, and ionic conductivity of sodium tetrathiophosphate. *J. Solid State Chem.* 99, 110–119. doi:10.1016/0022-4596(92)90295-7

Ji, X., and Nazar, L. (2010). Advances in Li-S batteries. *J. Mater. Chem.* 20, 9821–9826. doi:10.1039/b925751a

Kamaya, N., Homma, K., Yamakawa, Y., Hirayama, M., Kanno, R., Yonemura, M., et al. (2011). A lithium superionic conductor. *Nat. Mater* 10, 682–686. doi:10.1038/nmat3066

Kato, A., Hayashi, A., and Tatsumisago, M. (2016a). Enhancing utilization of lithium metal electrodes in all-solid-state batteries by interface modification with gold thin films. *J. Power Sources* 309, 27–32. doi:10.1016/j.jpowsour.2016.01.068

Kato, Y., Hori, S., Saito, T., Suzuki, K., Hirayama, M., Mitsui, A., et al. (2016b). High-power all-solid-state batteries using sulfide superionic conductors. *Nat. Energy* 1, 16030. doi:10.1038/nenergy.2016.30

Liu, Z., Fu, W., Payzant, E., Yu, X., Wu, Z., Dudney, N., et al. (2013). Anomalous high ionic conductivity of nanoporous β-Li3PS4. *J. Am. Chem. Soc.* 135, 975–978. doi:10.1021/ja3110895

Mai, L. Q., Minhas-Khan, A., Tian, X., Hercule, K. M., Zhao, Y. L., Lin, X., et al. (2013). Synergistic interaction between redox-active electrolyte and binder-free functionalized carbon for ultrahigh supercapacitor performance. *Nat. Commun.* 4, 2923. doi:10.1038/ncomms3923

Matsuyama, T., Deguchi, M., Mitsuhara, K., Ohta, T., Mori, T., Orikasa, Y., et al. (2016a). Structure analyses using X-ray photoelectron spectroscopy and X-ray absorption near edge structure for amorphous MS3 (M: Ti, Mo) electrodes in all-solid-state lithium batteries. *J. Power Sources* 313, 104–111. doi:10.1016/j.jpowsour.2016.02.044

Matsuyama, T., Hayashi, A., Ozaki, T., Mori, S., and Tatsumisago, M. (2016b). Improved electrochemical performance of amorphous TiS3 electrodes compared to its crystal for all-solid-state rechargeable lithium batteries. *J. Ceram. Soc. Jpn.* 124, 242–246. doi:10.2109/jcersj2.15299

Minami, K., Hayashi, A., and Tatsumisago, M. (2010). Characterization of solid electrolytes prepared from Li2S-P2S5 glass and ionic liquids. *J. Electrochem. Soc.* 157, A1296–A1301. doi:10.1149/1.3489352

Minami, K., Hayashi, A., Ujiie, S., and Tatsumisago, M. (2011). Electrical and electrochemical properties of glass–ceramic electrolytes in the systems Li2S-P2S5-P2S3 and Li2S-P2S5-P2O5. *Solid State Ionics.* 192, 122–125. doi:10.1016/j.ssi.2010.06.018

Mizuno, F., Hayashi, A., Tadanaga, K., and Tatsumisago, M. (2005). New, highly ion-conductive crystals precipitated from Li2S-P2S5 glasses. *Adv. Mater.* 17, 918–921. doi:10.1002/adma.200401286

Mizuno, F., Hayashi, A., Tadanaga, K., and Tatsumisago, M. (2006). High lithium ion conducting glass-ceramics in the system Li2S-P2S5. *Solid State Ionics.* 177, 2721–2725. doi:10.1016/j.ssi.2006.04.017

Muramatsu, H., Hayashi, A., Ohtomo, T., Hama, S., and Tatsumisago, M. (2011). Structural change of Li2S-P2S5 sulfide solid electrolytes in the atmosphere. *Solid State Ionics.* 182, 116–119. doi:10.1016/j.ssi.2010.10.013

Nagao, M., Hayashi, A., and Tatsumisago, M. (2011). Sulfur-carbon composite electrode for all-solid-state Li/S battery with Li2S-P2S5 solid electrolyte. *Electrochim. Acta* 56, 6055–6059. doi:10.1016/j.electacta.2011.04.084

Nagao, M., Hayashi, A., and Tatsumisago, M. (2012a). Bulk-type lithium metal secondary battery with indium thin layer at interface between Li electrode and Li2S-P2S5 solid electrolyte. *Electrochem.* 80, 734–736. doi:10.5796/electrochemistry.80.734

Nagao, M., Hayashi, A., and Tatsumisago, M. (2012b). High-capacity Li2S-nanocarbon composite electrode for all-solid-state rechargeable lithium batteries. *J. Mater. Chem* 22, 10015–10020. doi:10.1039/c2jm16802b

Nagao, M., Hayashi, A., Tatsumisago, M., Ichinose, T., Ozaki, T., Togawa, Y., et al. (2015). Li2S nanocomposites underlying high-capacity and cycling stability in all-solid-state lithium-sulfur batteries. *J. Power Sources* 274, 471–476. doi:10.1016/j.jpowsour.2014.10.043

Nagao, M., Hayashi, A., Tatsumisago, M., Kanetsuku, T., Tsuda, T., and Kuwabata, S. (2013). In situ SEM study of a lithium deposition and dissolution mechanism in a bulk-type solid-state cell with a Li2S-P2S5 solid electrolyte. *Phys. Chem. Chem. Phys.* 15, 18600–18606. doi:10.1039/c3cp51059j

Nose, M., Kato, A., Sakuda, A., Hayashi, A., and Tatsumisago, M. (2015). Evaluation of mechanical properties of Na2S-P2S5 sulfide glass electrolytes. *J. Mater. Chem. A* 3, 22061–22065. doi:10.1039/C5TA05590C

Oh, D. Y., Nam, Y. J., Park, K. H., Jung, S. H., Cho, S. J., Kim, Y. K., et al. (2015). Excellent compatibility of solvate ionic liquids with sulfide solid electrolytes: toward favorable ionic contacts in bulk-type all-solid-state lithium-ion batteries. *Adv. Energy Mater.* 5, 1500865. doi:10.1002/aenm.201570120

Ohtomo, T., Hayashi, A., Tatsumisago, M., and Kawamoto, K. (2013a). Glass electrolytes with high ion conductivity and high chemical stability in the system LiI-Li2O-Li2S-P2S5. *Electrochemistry* 81, 428–431. doi:10.5796/electrochemistry.81.428

Ohtomo, T., Hayashi, A., Tatsumisago, M., and Kawamoto, K. (2013b). All-solid-state batteries with Li2O-Li2S-P2S5 glass electrolytes synthesized by two-step mechanical milling. *J. Solid State Chem.* 17, 2551–2557. doi:10.1007/s10008-013-2149-5

Ohtomo, T., Hayashi, A., Tatsumisago, M., and Kawamoto, K. (2013c). All-solid-state lithium secondary batteries using the 75Li2S·25P2S5 glass and the

70Li$_2$S·30P$_2$S$_5$ glass-ceramic as solid electrolytes. *J. Power Sources* 233, 231–235. doi:10.1016/j.jpowsour.2013.01.090

Okita, K., Ikeda, K., Sano, H., Iriyama, Y., and Sakaebe, H. (2011). Stabilizing lithium plating-stripping reaction between a lithium phosphorus oxynitride glass electrolyte and copper thin film by platinum insertion. *J. Power Sources* 196, 2135–2142. doi:10.1016/j.jpowsour.2010.10.014

Phuc, N., Morikawa, K., Totani, M., Muto, H., and Matsuda, A. (2016). Chemical synthesis of Li$_3$PS$_4$ precursor suspension by liquid-phase shaking. *Solid State Ionics*. 285, 2–5. doi:10.1016/j.ssi.2015.11.019

Sahu, G., Lin, Z., Li, J., Liu, Z., Dudney, N., and Liang, C. (2014). Air-stable, high-conduction solid electrolytes of arsenic-substituted Li$_4$SnS$_4$. *Energy Environ. Sci.* 7, 1053–1058. doi:10.1039/C3EE43357A

Sakuda, A., Hayashi, A., Ohtomo, T., Hama, S., and Tatsumisago, M. (2010). LiCoO$_2$ electrode particles coated with Li$_2$S-P$_2$S$_5$ solid electrolyte for all-solid-state batteries. *Electrochem. Solid-State Lett.* 13, A73–A75. doi:10.1149/1.3376620

Sakuda, A., Hayashi, A., Ohtomo, T., Hama, S., and Tatsumisago, M. (2011). All-solid-state lithium secondary batteries using LiCoO$_2$ particles with pulsed laser deposition coatings of Li$_2$S-P$_2$S$_5$ solid electrolyte. *J. Power Sources* 196, 6735–6741. doi:10.1016/j.jpowsour.2010.10.103

Sakuda, A., Hayashi, A., Takigawa, Y., Higashi, K., and Tatsumisago, M. (2013a). Evaluation of elastic modulus of Li$_2$S-P$_2$S$_5$ glassy solid electrolyte by ultrasonic sound velocity measurement and compression test. *J. Ceram. Soc. Jpn.* 121, 946–949. doi:10.2109/jcersj2.121.946

Sakuda, A., Hayashi, A., and Tatsumisago, M. (2013b). Sulfide solid electrolyte with favorable mechanical property for all-solid-state lithium battery. *Sci. Rep.* 3, 2261. doi:10.1038/srep02261

Sakuda, A., Taguchi, N., Takeuchi, T., Kobayashi, H., Sakaebe, H., Tatsumi, K., et al. (2014). Amorphous niobium sulfides as novel positive-electrode materials. *ECS Electrochem. Lett.* 3, A79–A81. doi:10.1149/2.0091407eel

Seino, Y., Ota, T., Takada, K., Hayashi, A., and Tatsumisago, M. (2014). A sulphide lithium super ion conductor is superior to liquid ion conductors for use in rechargeable batteries. *Energy Environ. Sci.* 7, 627–631. doi:10.1039/C3EE41655K

Souquet, J. L., Robinel, E., Barrau, B., and Ribes, M. (1981). Glass formation and ionic conduction in the M$_2$S-GeS$_2$ (M=Li, Na, Ag) systems. *Solid State Ionics*. 3-4, 317–321. doi:10.1016/0167-2738(81)90105-3

Takada, K. (2013). Progress and prospective of solid-state lithium batteries. *Acta Mater.* 61, 759–770. doi:10.1016/j.actamat.2012.10.034

Tanibata, N., Hayashi, A., and Tatsumisago, M. (2015a). Improvement of rate performance for all-solid-state Na$_{15}$Sn$_4$/amorphous TiS$_3$ cells using 94Na$_3$PS$_4$·6Na$_4$SiS$_4$ glass-ceramic electrolytes. *J. Electrochem. Soc.* 162, A793–A795. doi:10.1149/2.0011506jes

Tanibata, N., Matsuyama, T., Hayashi, A., and Tatsumisago, M. (2015b). Improvement of rate performance for all-solid-state Na$_{15}$Sn$_4$/amorphous TiS$_3$ cells using 94Na$_3$PS$_4$·6Na$_4$SiS$_4$ glass-ceramic electrolytes. *J. Power Sources* 275, 284–287. doi:10.1016/j.jpowsour.2014.10.193

Tanibata, N., Noi, K., Hayashi, A., Kitamura, N., Idemoto, Y., and Tatsumisago, M. (2014). X-ray crystal structure analysis of sodium-ion conductivity in 94Na$_3$PS$_4$·6Na$_4$SiS$_4$ glass-ceramic electrolytes. *Chem. Electro. Chem.* 1, 1130–1132. doi:10.1002/celc.201402016

Tatsumisago, M., and Hayashi, A. (2008). All-solid-state lithium secondary batteries using sulfide-based glass-ceramic electrolytes. *Funct. Mater. Lett.* 1, 31–36. doi:10.1142/S1793604708000071

Tatsumisago, M., and Hayashi, A. (2014). Sulfide glass-ceramic electrolytes for all-solid-state lithium and sodium batteries. *Int. J. Appl. Glass. Sci.* 5, 226–235. doi:10.1111/ijag.12084

Tatsumisago, M., Nagao, M., and Hayashi, A. (2013). Recent development of sulfide solid electrolytes and interfacial modification for all-solid-state rechargeable lithium batteries. *J. Asian. Ceram. Soc.* 1, 17–25. doi:10.1016/j.jascer.2013.03.005

Tatsumisago, M., Shinkuma, Y., and Minami, T. (1991). Stabilization of superionic α-AgI at room temperature in a glass matrix. *Nature* 354, 217–218. doi:10.1038/354217a0

Teragawa, S., Aso, K., Tadanaga, K., Hayashi, A., and Tatsumisago, M. (2014a). Liquid-phase synthesis of a Li$_3$PS$_4$ solid electrolyte using N-methylformamide for all-solid-state lithium batteries. *J. Mater. Chem. A* 2, 5095–5099. doi:10.1039/c3ta15090a

Teragawa, S., Aso, K., Tadanaga, K., Hayashi, A., and Tatsumisago, M. (2014b). Preparation of Li$_2$S-P$_2$S$_5$ solid electrolyte from N-methylformamide solution and application for all-solid-state lithium battery. *J. Power Sources* 248, 939–942. doi:10.1016/j.jpowsour.2013.09.117

Ujiie, S., Hayashi, A., and Tatsumisago, M. (2014). Preparation and electrochemical characterization of (100-x)(0.7Li$_2$S·0.3P$_2$S$_5$)·xLiBr glass-ceramic electrolytes. *Mater. Renew. Sustain. Energy* 3, 1–8. doi:10.1007/s40243-013-0018-x

Wada, H., Menetrier, M., Levasseur, A., and Hagenmuller, P. (1983). Preparation and ionic conductivity of new B$_2$S$_3$-Li$_2$S-LiI glasses. *Mater. Res. Bull.* 18, 189–193. doi:10.1016/0025-5408(83)90080-6

Yabuchi, N., Kubota, K., Dahbi, M., and Komaba, S. (2015b). Research development on sodium-ion batteries. *Chem. Rev.* 114, 11636–11682. doi:10.1021/cr500192f

Yamada, A. (2014). Iron-based materials strategies. *MRS Bull.* 39, 423–428. doi:10.1557/mrs.2014.89

Yamauchi, A., Sakuda, A., Hayashi, A., and Tatsumisago, M. (2013). Preparation and ionic conductivities of (100-x)(0.75Li$_2$S·0.25P$_2$S$_5$)·xLiBH$_4$ glass electrolytes. *J. Power Sources* 244, 707–710. doi:10.1016/j.jpowsour.2012.12.001

Yubuchi, S., Hayashi, A., and Tatsumisago, M. (2015a). Sodium-ion conducting Na$_3$PS$_4$ electrolyte synthesized *via* a liquid-phase process using N-methylformamide. *Chem. Lett.* 44, 884–886. doi:10.1246/cl.150195

Yubuchi, S., Teragawa, S., Aso, K., Tadanaga, K., Hayashi, A., and Tatsumisago, M. (2015b). Preparation of high lithium-ion conducting Li$_6$PS$_5$Cl solid electrolyte from ethanol solution for all-solid-state lithium batteries. *J. Power Sources* 293, 941–945. doi:10.1016/j.jpowsour.2015.05.093

Zhang, L., Yang, K., Mi, J., Lu, L., Zhao, L., Wang, L., et al. (2015). Solid electrolytes: Na$_3$PSe$_4$: a novel chalcogenide solid electrolyte with high ionic conductivity. *Adv. Energy Mater* 5, 1501294. doi:10.1002/aenm.201501294

Zhang, Z., and Kennedy, J. H. (1990). Synthesis and characterization of the B$_2$S$_3$-Li$_2$S, the P$_2$S$_5$-Li$_2$S and the B$_2$S$_3$-P$_2$S$_5$-Li$_2$S glass systems. *Solid State Ionics*. 38, 217–224. doi:10.1016/0167-2738(90)90424-P

Zhu, Z., Chu, I.-H., Deng, Z., and Ong, S. P. (2015). Role of Na$^+$ interstitials and dopauts in enhancing the Na$^+$ conductivity of the cubic Na$_3$PS$_4$ superionic conductor. *Chem. Mater.* 27, 8318–8325. doi:10.1021/acs.chemmater.5b03656

Conflict of Interest Statement: The authors declare that the research was conducted in the absence of any commercial or financial relationships that could be construed as a potential conflict of interest.

16

General Business Model Patterns for Local Energy Management Concepts

Emanuele Facchinetti* and Sabine Sulzer

Lucerne Competence Center for Energy Research, Lucerne University of Applied Science and Arts, Horw, Switzerland

The transition toward a more sustainable global energy system, significantly relying on renewable energies and decentralized energy systems, requires a deep reorganization of the energy sector. The way how energy services are generated, delivered, and traded is expected to be very different in the coming years. Business model innovation is recognized as a key driver for the successful implementation of the energy turnaround. This work contributes to this topic by introducing a heuristic methodology easing the identification of general business model patterns best suited for Local Energy Management concepts such as Energy Hubs. A conceptual framework characterizing the Local Energy Management business model solution space is developed. Three reference business model patterns providing orientation across the defined solution space are identified, analyzed, and compared. Through a market review, a number of successfully implemented innovative business models have been analyzed and allocated within the defined solution space. The outcomes of this work offer to potential stakeholders a starting point and guidelines for the business model innovation process, as well as insights for policy makers on challenges and opportunities related to Local Energy Management concepts.

Keywords: local energy management, energy hub, business models, business innovation, decentralized energy systems, distributed generation, renewable energy, energy market

Edited by:
Léo Benichou,
The Shift Project, France

Reviewed by:
Pierre Serkine,
KIC InnoEnergy SE, Belgium
Eric Vidalenc,
Agence de l'Environnement et de la
Maitrise de l'Energie, France

*Correspondence:
Emanuele Facchinetti
emanuele.facchinetti@hslu.ch

Specialty section:
This article was submitted to Energy
Systems and Policy,
a section of the journal
Frontiers in Energy Research

Citation:
Facchinetti E and Sulzer S (2016)
General Business Model
Patterns for Local Energy
Management Concepts.
Front. Energy Res. 4:7.

INTRODUCTION

The diffusion of decentralized energy systems is expected to provide a significant contribution toward a more sustainable global energy system and to the achievement of the international greenhouse gases reduction targets (Viral and Khatod, 2012; IEA, 2014). A fast growing interest on the transition from the conventional centralized power generation toward distributed generation can be observed in many industrialized countries (IEA, 2002; Karger and Hennings, 2009; Allan et al., 2015). Having set very ambitious sustainability targets within the European Commission's Energy Road Map 2050 (European Commission, 2011), Europe emerges as a leader of the energy transition. Appropriate measures and policies have been deployed to achieve these targets (European Commission, 2012, 2014). In particular, the recent adoption of the Energy Union strategy (European Commission, 2015a), with the aim to fully integrate the internal energy market, represents a major step ahead toward the transformation of Europe's energy system.

The major driving forces behind the growth of distributed generation are: the liberalization of electricity and gas markets, facilitated by the significant cost reduction for information and communication technologies (ICT); the trend toward energy services densification driven by the increasing urban density; the significant increase of the intrinsically distributed renewable energy; and the avoidance of new high voltage transmission lines (Viral and Khatod, 2012; Allan et al., 2015). The distributed generation development is expected to provide major impacts on energy efficiency increase and emission reduction especially on the building sector that according to the International Energy Agency (IEA, 2014) accounts for more than 30% of the world total energy consumption. This share strongly increases in industrialized countries, being for instance more than 45% in Switzerland (SFOE, 2012).

At urban scale, many scientific works have acknowledged the environomic (i.e., thermodynamic, economic, and environmental) advantage of decentralized advanced multi-energy systems with respect to conventional centralized energy systems (Capuder and Mancarella, 2014; Mancarella, 2014; Orehounig et al., 2014). In the future, advanced multi energy systems are expected to harmonize the integration of increasing share of fluctuating renewable energy generation together with the interactions of many advanced energy conversion and storage technologies, operating across different energy carriers, and capable of a more rational conversion of bio and fossil fuels (Weber and Favrat, 2010; Graves et al., 2011; Facchinetti et al., 2014). In the literature, these concepts of Local Energy Management – the management of energy supply, demand and storage within a given geographical area – are referred to in many different ways (Mancarella, 2014). A concept well established within the scientific community and comprehensively encompassing every possible declination of Local Energy Management approach is the Energy Hub (Geidl et al., 2007; Parisio et al., 2012; Orehounig et al., 2014, 2015). The Energy Hub concept was first introduced by Geidl et al. (2007) as a conceptual model of an energy system operating across multi energy carriers (i.e., electricity, thermal, and chemical energies) through the optimal management and integration of energy conversion and storage technologies.

The transition toward more sustainable energy systems, significantly relying on the deployment of Energy Hub concepts, requires a deep reorganization of the power generation sector. The power generation sector and its stakeholders stand today at the starting point of a challenging revolution: the way how energy services are generated, delivered, and traded is expected to be very different in the coming years (Frei, 2008; Schleicher-Tappeser, 2012).

Currently, utility companies substantially control the energy market, especially regarding the electricity sector. With their dominant market position and influence on policy makers (i.e., continuous lobby activities), utilities are expected to play a major role on the energy transition (Sühlsen and Hisschemöller, 2014). In order to compete in this changing environment, they need to face the challenge of creating new successful business models (Frei, 2008; Schleicher-Tappeser, 2012).

In this regard, a number of scientific works have recently focused on new business models for renewable energies and distributed generation. Richter (2012) reviewed the current state of the literature on utilities business models for renewable energy, proposed two generic business models suitable for renewable energy, and highlighted the emerging barriers and opportunities for utilities. The study concludes that despite the fact that business models related to large-scale projects currently comprise less risks and better returns, utilities should strategically increase their capability in the field of business

model innovation to exploit the business opportunities offered by small-scale distributed generation. The same author recently analyzed the applicability and potential on the German market of the two generic business models interviewing a large number of German utility managers (Richter, 2013). The outcome of this work outlines how utility managers developed business model for large-scale renewable based business while at the same time they struggled to identify adequate business models at small scales typical of distributed generation. Loock (2012) reported the results of choice experiments with investment managers for renewable energy aiming to identify their investment preferences. A clearly emerging outcome is the fact that business models proposing best services are considered more attractive than business models oriented to low price and best technologies. Curtius et al. (2012) explored the customer segmentation for smart grids on the basis of a European study and derived a number of generic business models best suited to address the different customer segments. Furthermore, this work highlighted that no single business model can guarantee the successful penetration of smart grids. Instead, various business model characterized by optimized value propositions matching the heterogeneous customer value perceptions should be developed.

In the last years, the scientific community clearly outlined the strategic role of business model innovation to guarantee the efficient implementation of the energy turnaround. This work contributes to this topic by introducing a heuristic methodology easing the identification of general business model patterns best suited for Energy Hub concepts. A conceptual framework characterizing the Energy Hub business model solution space is developed. Three reference business model patterns providing orientation across the defined solution space are identified, analyzed, and compared.[1] Through a market review, a number of successfully implemented innovative business models have been analyzed and allocated within the defined solution space. The conceptual framework has been developed to comprehensively embrace and be generally applicable to any form of Local Energy Management concepts spanning from basic single energy carrier Energy Hubs (e.g., a platform connecting photovoltaic prosumers to the energy market), to very complex multi energy carriers Energy Hubs (e.g., involving conversion across electricity, thermal, and chemical carriers). The outcomes of this work offer to potential Energy Hub stakeholders a starting point and guidelines for the required business model innovation process, as well as insights for policy makers on challenges and opportunities related to Energy Hub concepts.

The paper is organized as follows: in Section "Methods," the applied heuristic methodology is described and the Energy Hub business model solution space is defined. In Section "Results and Discussion," the selected reference business model patterns are presented and discussed. Furthermore, the outcomes of the market review are presented. Finally, in Section "Conclusion," energy policy implications are outlined.

---

[1] The evaluation of their market efficiency is out of the scope of this work.

METHODS

Business model innovation is recognized to be a challenging multidimensional and interdisciplinary process strongly reliant on experimentation (Chesbrough, 2010; Richter, 2013; Gassmann et al., 2014). Objective of the heuristic methodology presented here is to define a conceptual framework supporting the business model innovation process for Energy Hub at the early stage. The methodology enables to identify potentially interesting business model ideas and organize them in patterns aligned with the Energy Hub value chain. The obtained business model patterns are meant to be further developed, refined, and adapted through implementation and testing in order to complete the business model innovation process.

The Energy Hub and Its Value Chain

The Energy Hub is a broad concept encompassing every possible form of Local Energy Management approach. From the energy perspective, the Energy Hub is an entity operating within a given geographical area and across multiple energy carriers (i.e., electricity, thermal, and chemical) that guarantees the energy supply to meet the demand through the optimized management of internal flexibilities and energy market participation. From the business perspective the Energy Hub is an entity potentially operated by utilities, aggregators, ESCOS or public bodies, which connects consumers, prosumers, and partners (i.e., neighbor Energy Hub, industry, utilities) to each other and with the wholesale energy market (**Figure 1**).

The Energy Hub value chain is used as a key driver through the heuristic methodology presented in the next section and therefore it is characterized first. The value chain of a business is defined as the number of activities to be performed to generate the value proposition offered to the customer (Porter, 1985). From the business point of view, the Energy Hub is a provider of energy services that guarantees the energy supply to meet the demand through the optimized management of the internal flexibilities

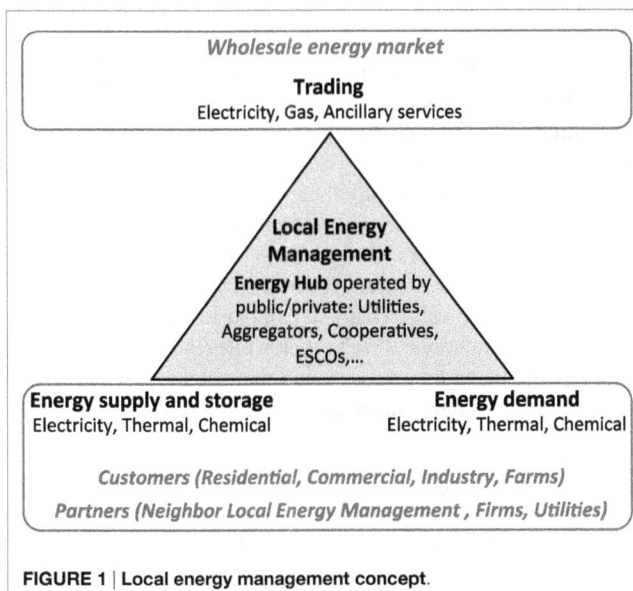

FIGURE 1 | Local energy management concept.

and through the participation in the energy market. Possibly, the Energy Hub offers value-added services going beyond the energy services supply. The Energy Hub value chain characterized by the authors can be described by five activities, which are detailed hereafter.

The *Acquisition/Loyalty* refers to the establishment of relationships with customers and partners of the business. Customers of Energy Hubs could be consumers or prosumers of energy services. Partners can be companies dealing with energy generation, conversion, storage, or transmission technologies as well as financing or trading institutes or other Energy Hubs operating in the proximity.

The *Procurement of Infrastructure* refers to the need of exploiting infrastructures for the production, storage, conversion, and delivery of energy services. ICT and ancillary systems are also required.

The *Operation and Control of Infrastructures* refers to the need of constantly operate, control, and maintain the exploited infrastructure. This activity includes also the balancing within the Energy Hub grid and with the external grids by offering ancillary services.

The *Delivery of Energy Services* comprises the secure delivery of energy to customers from the generation within or outside an Energy Hub. Delivery of complementary services going beyond energy supply, e.g., domestic services and mobility solutions, is also part of this activity. Furthermore, this activity includes the metering intended as the accounting of the energy exchanged.

The *Pricing* comprises administrative tasks such as, the establishment of prices, the communication with the customers and the partners, the contracting, the billing, and the account of trading costs and revenues.

The Heuristic Methodology

In the rather young research field of business model innovation the Business Model Navigator approach, recently introduced by Gassmann et al. (2014), emerges as an original and promising attempt to develop a comprehensive and systematic conceptual framework. Such framework has been used as starting point for the developed heuristic methodology.

Relying on an extensive experience on business innovation including collaborations with a variety of partner firms (Gassmann et al., 2010; Bucherer et al., 2012), within the Business Model Navigator Gassmann et al. (2014) analyzed the vast majority of all successfully deployed business models across all market sectors in the last 25 years. As an outcome of this analysis, they recognized that about 90% of such business models were not created from disruptive ideas, but they were rather a recombination of 55 recurring basic business model ideas. With this significant conclusion, Gassmann et al. showed that, even though creativity should not be put aside, the 55 business model ideas represent a consistent starting point for the business model innovation process across all market sectors.

Based on this list of business model ideas, the developed heuristic methodology enabled in five steps the selection and organization of the applicable business model ideas into a conceptual business model solution space for Energy Hub concepts. As suggested by Gassmann et al. (2014), the developed methodology progresses through steps requiring alternatively diverging or converging thinking. The diverging phases promote the creation of a large number of possible solutions and thus an extension of the solution space. The converging phases aim to identify within the extended solution phase the most pertinent solutions. A number of consecutive diverging–converging iterations guide through the multidimensional problem and enable to find an appropriate solution. Hereafter, the five steps of the developed methodology (**Figure 2**) are explained and the results of their application presented.

I. Filtering (Converging Thinking)

In the first step, the 55 business model ideas are filtered considering two criteria. The first criterion is the general compatibility with the Energy Hub business: only the business model ideas potentially applicable to the Energy Hub business are kept.[2]

The second criterion was applied only to the ideas selected with the first criterion. This criterion regards the affinity with the different value chain activities of the business: the ideas were assigned to the value chain activities in which they are potentially applicable.[3]

As a result of these two criteria, 32 business model ideas are filtered and assigned to the different value chain activities, as represented in **Table 1**. The largest number of ideas refers to the *Delivery of energy services* and *Pricing* activities. The *Operation and Control of infrastructures* is characterized by only one business model idea and in the following will be associated with the *Procurement of Infrastructures* activity due to the close link between the two activities. Two business model ideas, *Aikido* and *Performance based contracting*, are relevant to two activities; all the other ideas are associated with only one activity. A short description of each business model ideas, taken from Gassmann et al. (2014), is available in the Supplementary Material.

II. Compatibility (Diverging Thinking)

Within the second step, the compatibility between the filtered business model ideas is systematically evaluated: the potential application of each business model idea in combination with every other individual business model idea within the same value chain activity is verified. The different value chain activities are considered fully independent to each other. As a result, a large number of possible combination of ideas are identified. These

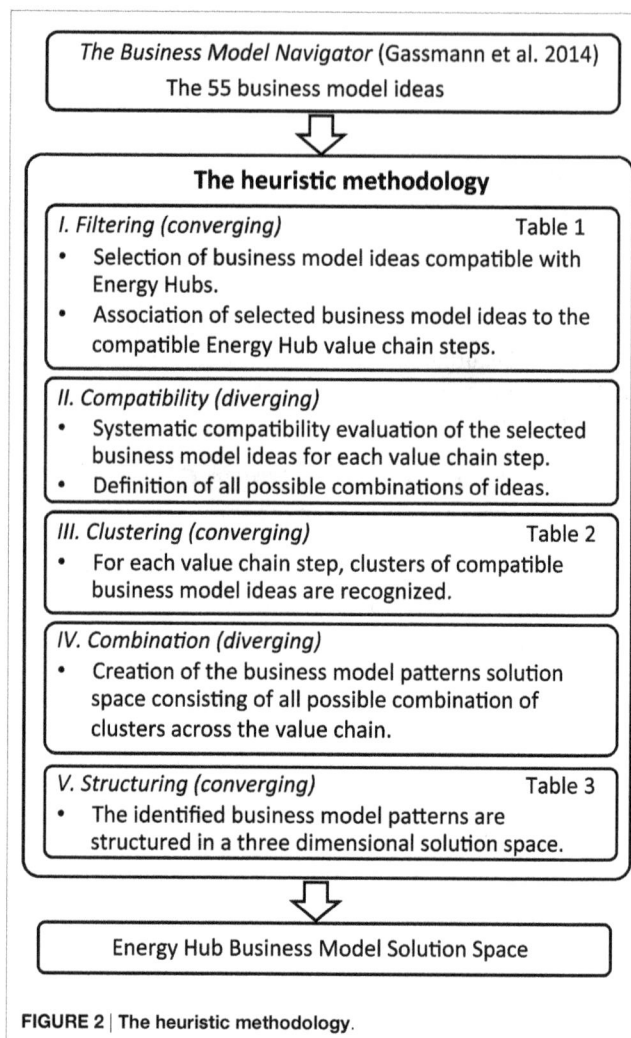

FIGURE 2 | The heuristic methodology.

combinations create the solution space for the next convergent step of the methodology.

III. Clustering (Converging Thinking)

Based on the compatibility step II, in the third step, clusters containing only ideas compatible to each other are recognized per each value chain activity. The ideas bearing the largest number of incompatibilities and thus differentiating and strongly characterizing the clusters are recognized as dominant and are used to label the clusters.[4] The results of this converging step are presented in **Table 2**. Reading **Table 2** in vertical direction (i.e., going through the value chain steps), clusters contain business model ideas compatible to each other; while reading in horizontal direction (focusing on a single value chain step) clusters include incompatible business model ideas.

---

[2] As an example, the business model idea *E-Commerce*, referring to the delivery of the value proposition through online channel only to reduce sales and distribution costs, has not been considered applicable for Energy Hubs and has not been retained. A further example on the application of the first criterion is the following one: the business model idea *Open Source*, referring to the free offer of a product with the purpose of gaining money with support and consulting; *Razor and Blades*, related to a free offer of a basic product requiring expensing consumable; and *Lock-in*, in which the customers are kept locked through high switching costs; have been considered substantially equivalent from the Energy Hub view point and only the idea *Lock-in* has been retained.

[3] For instance, the business model idea *Crowdfunding*, in which a crowd of investors finances the required investments, is assigned to the *Procurement of Infrastructures* activity. Likewise the idea *Barter*, referring to exchanging with customers goods with goods instead of money, is associate to the *Pricing* activity.

[4] Taking as example, the *Delivery of Energy Services* activity, the ideas *No Frills*, referring to delivering as basic as possible services; *Experience Selling*, referring to enrich the core value proposition with a comprehensive customer experience; and *User Designed*, in which the customers can tailor the value proposition, are recognized as not compatible to each other and are chosen to lead three different clusters.

TABLE 1 | Filtered business model ideas.

Value chain activities	Filtered business models		
Acquisition/loyalty by customers, partners, or EH	Affiliation Customer loyalty Make more of it		
Procurement of infrastructures by customers, partners, or EH	Crowdfunding Fractional ownership	Open business model Performance-based contracting	Rent instead of buying
Operation and control of infrastructures by customers, partners, or EH	Orchestrator		
Delivery of energy services by customers, partners, or EH	Aikido Cross selling Direct selling Experience selling	Guaranteed availability Ingredient branding No frills Peer to peer	Solution provider Two-sided-market Ultimate luxury User designed
Pricing managed by EH	Add-on Aikido Auction Barter Cash machine	Freemium Hidden revenue Lock-in Pay per use Pay what you want	Performance-based contracting Revenue sharing Subscription

TABLE 2 | Clusters of business model ideas.

Value chain activities	Clusters		
Acquisition/loyalty by customers, partners, or EH	Affiliation Customer loyalty Make more of it		
Procurement of infrastructure/operation and control by customers, partners, or EH	*Fractional ownership* Crowdfunding Open business model	*Orchestrator* Open business model Performance-based contracting	*Rent instead of buying* Open business model Performance-based contracting
Delivery of energy services by customers, partners, or EH	*Experience selling* Aikido Cross selling Direct selling Guaranteed availability Ingredient branding Solution provider Two-sided-market Ultimate luxury	*No frills* Aikido Direct selling Peer to peer Two-sided-market	*User designed* Aikido Cross selling Direct selling Guaranteed availability Ingredient branding Solution provider Two-sided-market
Pricing managed by EH	*Add-on* Aikido Auction Barter Cash machine Freemium Hidden revenue Lock-in Pay what you want Performance-based contracting Revenue sharing	*Pay per use* Aikido Auction Barter Freemium Hidden revenue Lock-in Pay what you want Performance-based contracting Revenue sharing	*Subscription* Aikido Cash machine Freemium Hidden revenue Lock-in Pay what you want Performance-based contracting Revenue sharing

No cluster has been identified in *Acquisition/Loyalty,* due to the small number of business model ideas and their mutual compatibility. Three clusters per activity are identified on the other activities.

IV. Combination (Diverging Thinking)

Combining a single cluster per each activity of the value chain it is possible to identify a number of potential business model patterns going across the whole Energy Hub value chain. All possible combinations of clusters represent the business model patterns solution space, which is structured within the last convergent step (V) of the methodology.

V. Structuring (Converging Thinking)

The possible clusters combinations identified in the previous step constitute the business model solution space. Within the last step of the methodology, the latter is structured in three dimensions: one represented by the *Procurement and Control of Infrastructure*, one represented by the *Delivery of Energy Services*, and one represented by the *Pricing*. In **Table 3**, the business model solution space is depicted in two of these three dimensions: the *Procurement and Control of Infrastructures* and the *Delivery of Energy Services* dimensions. On the *Procurement of Infrastructures* dimension (vertical axis), the business model options span from full Energy Hub's ownership of the infrastructure, through shared ownership, to full customer's (or partner's) ownership. On the *Delivery of Energy Service* dimension (horizontal axis), the available options span from high quality comprehensive services, through customized offers, to basic services.

Aiming to provide a reference orientation across the defined solution space, three reference business model patterns are selected (Pattern I, II, and III in **Table 3**). The third solution space dimension, the *Pricing*, is presented for each reference pattern within **Table 3**. Due to the higher flexibility characterizing the *Pricing* activity and for the sake of generality, each reference pattern features more than one *Pricing* cluster option.

In the next section, the three reference patterns are thoroughly analyzed and compared in order to put the developed conceptual framework into context through the use of examples.

RESULTS AND DISCUSSION

In the first part of Section "Results and Discussion," the reference business model patterns introduced in the previous section are thoroughly described and compared. In the second part, a number of examples of innovative business models recently implemented in the energy market are presented and their location within the defined solution space is discussed. Finally, at the end of the section, general outcomes are discussed.

The Reference Business Model Patterns

The conceptual business model solution space defined in the previous section is meant to support the Energy Hub business model innovation process at early stages. Potential stakeholders should determine the portion of the defined business model solution space more suitable for their intended Energy Hub concept through the selection of the most appropriate general business model pattern(s). This selection should be performed considering the determinants characterizing the intended Energy Hub concept, such as the customer segmentation, the available technical and financial resources, and external determinants (e.g., regulatory framework). Future works are expected to address the identification and impact of such determinants. Each general business model pattern comprises a collection of compatible business model ideas organized per value chain activity. Through combination, refinement, and adaptation of the identified business model ideas, the selected general business model pattern(s) can be tailored to the needs of the intended Energy Hub concept.

The reference business model patterns identified in the previous section aim to offer an orientation within the defined business model solution space. They have been selected to span across the whole solution space and thus to provide references easing the solution space characterization through examples.

Pattern I is based on the cluster *Orchestrator* associated with the *Procurement and control of infrastructures* value chain activities and on the cluster *No Frills* associated with the activity *Delivery of energy services*. Selected *Pricing* options are the clusters *Pay per use* and *Subscription*. Bottom line of this business model is to run the Energy Hub focusing on the operation and control and outsourcing the procurement of infrastructures. The Energy Hub's investment costs are low. The services offered to the customers focus on the essential. The related cost savings can be shared with the customers, which can benefit from low prices.

Pattern II is based on the cluster *Fractional Ownership* associated with the *Procurement and control of infrastructures* value chain activities, and on the cluster *User Designed* associated with the activity *Delivery of energy services*. All *Pricing* options are available: *Add-on*, *Pay per use*, and *Subscription*. Within this pattern, the Energy Hub shares the ownership of the infrastructures with one or multiple customers and offers the possibility of tailored energy services. The Energy Hub benefits from the infrastructure availability and from reduced investment costs and risks, which are partly or fully taken over by the customers. The customers can benefit from the possible valorization of their partly owned infrastructure while having access to complementary energy services provided by the Energy Hub.

Pattern III is based on the cluster *Rent instead of buying* associated with the *Procurement and control of infrastructures* value chain activities, and on the cluster *Experience Selling* associated with the activity *Delivery of energy services*. Selected *Pricing* options are the clusters *Pay per use* and *Subscription*. Within this business model pattern, the Energy Hub offers to the customer the possibility to lease all-inclusive turnkey solutions. The customers benefit from avoiding investments in the required infrastructure and from a complete and high quality customer

TABLE 3 | The business model patterns solution space and the reference patterns.

		Delivery of energy services				
		No frills	**User designed**		**Experience selling**	
Procurement of infrastructure/operation and control	**Rent instead of buying**				*Pattern III* Pay per use Subscription	
	Fractional ownership		*Pattern II* Pay per use Subscription	Add on		
	Orchestrator	*Pattern I* Pay per use Subscription				

experience. The Energy Hub is exposed to high investment costs that are met by the potentially high margins expected from the offered high quality services.

Hereafter, the reference patterns are analyzed following the business model terminology and conceptualization proposed by Osterwalder and Pigneur (2010) and Osterwalder et al. (2014). Their description of business models is coherent, extensively applied in the literature, and in particular in the energy sector (Okkonen and Suhonen, 2010; Richter, 2013). Osterwalder et al. (2014) defined a business model as "the rationale of how an organization creates, delivers, and captures value" and identified four main elements fully characterizing it: *Value Proposition*; *Customers*, including customer relationship, customer segments, and channels; *Infrastructures*, including key activities, key resources, and key partners; and *Financial Viability*, including cost structure and revenue stream.

The main features characterizing the three reference business model patterns analyzed and compared hereafter are summarized in **Table 4**.

Value Proposition

Pattern I main value proposition is to offer a multisided platform connecting consumers, prosumers, partners, and the energy market. The Energy Hub manages the energy exchanges and guarantees the availability of energy services to its customers.

Pattern II proposes tailored solutions fully adaptable to the customer infrastructures and needs. Depending on the customer owned infrastructure and preferences, the Energy Hub offers basic and complementary energy services with different level of participation.

Pattern III offers comprehensive turnkey solutions owned by the Energy Hub potentially going beyond traditional energy services and covering also closely related business opportunity such as domestic and mobility services. These products are offered to the customers with a complete service package to guarantee a high quality customer experience.

Customers of Pattern II and III could also benefit from a reduced exposition on external determinants (e.g., wholesale electricity markets) and improved energy security. This could be enabled by the higher level of self-sufficiency and flexibility ensured by the Energy Hub (co-)owned energy conversion/storage infrastructures.

Customers

Energy Hubs typically deal with segmented customers, potentially including residential buildings, commercial buildings, industries, and farms. Due to the required physical links with the customer, e.g. district heating grids, the geographical location of the Energy Hub represents its main attractiveness to customers.

Within this customer segmentation, Pattern I targets cost sensitive customers and prosumers owning the energy conversion infrastructures. The relationship with customers is regulated by automated exchanges based on established contracts. The multisided platform represents the main interface between customers and Energy Hub.

Pattern II addresses customers willing to participate to the investment in infrastructures. The customer relationship includes dedicated assistance in tailoring the value proposition offer and the creation of a community atmosphere fostering involvement of customers into the Energy Hub activity and development.

Pattern III aims to customers inclined to pay higher prices in order to profit of a high quality customer experience. The relationship with customers includes comprehensive support covering all aspects from the installation to the operation, maintenance, and replacement of the infrastructures in order to guarantee the promised comfort level. Within Pattern II and III customers are bound by long-term contracts aligned to the long amortization time of the investment in infrastructures afforded by the Energy Hub (Pattern III) or both the Energy Hub and the customers (Pattern II).

Customer's activities comprise consumption, production, and storage of energy, as well as internal energy trading if not contracted to the Energy Hub. Particularly for Pattern I and II customer's activities also include the maintenance of the owned infrastructure if not contracted to the Energy Hub.

Within Pattern I and II consumer's costs regard only the purchase of energy services. Prosumer's costs are associated with the amortization and maintenance of own (Pattern I) or partially own (Pattern II) infrastructures and to the purchasing costs of complementary energy services. Pattern III customer's costs simply regard the all-inclusive contract with the Energy Hub.

Infrastructure

Key resources of the Energy Hub are the energy conversion/storage and internal grid infrastructures, the required ICT systems, the personnel necessary to operate and maintain the infrastructure and responsible for the customer support, and the energy

TABLE 4 | Main features of the identified general business model patterns.

	Pattern I	Pattern II	Pattern III
	Orchestrator-No frills	*Fractional ownership – User designed*	*Rent instead of buying – Experience selling*
Value proposition	Multiside platform connecting consumers, prosumers and energy market	Tailored energy services adaptable and complementary to customer infrastructures	Comprehensive turnkey solutions going beyond energy services
Customers	Cost sensitive customers Prosumers owning the infrastructures	Customers participating to the infrastructure investments	Customers inclined to pay higher prices to get the best service quality
Infrastructures	No investment in infrastructures Strong partnerships	Infrastructure ownership shared with customers	Owned infrastructures leased to customers
Financial viability	Revenues from energy trading only	Revenues from energy trading and service on infrastructures	Revenues from energy trading, leasing, and additional services

trading. Depending on the business model pattern, the resources are (co-)owned, contracted from partners, or out of the scope of the Energy Hub.

In Pattern I, the Energy Hub does not invest in infrastructures and can focus its resources on strengthening its core activities of supplying energy services while keeping a more flexible structure compared to the other patterns.

In Pattern II, the ownership of the required infrastructures is shared between the Energy Hub and one or multiple customers. The Energy Hub develops tailored and flexible solutions enabling to incorporate and complement customer's infrastructures.

In Pattern III, the Energy Hub owns the required infrastructures and leases them to the customers. Additional infrastructures going beyond the supply of energy services and enriching the value proposition, e.g., domestic or mobility-related infrastructures could be additional key resources.

The Energy Hub key activities generally comprise the exchange of energy with and among customers (consumers and prosumers) and partners; and the continuous optimization of the operation of customer's, partners, and (co-)owned (Patterns II and III) infrastructures. The optimization aims to keep the internal grid balanced and to maximize the profit from energy trading. Specific activity of Pattern I is the continuous development of the multisided platform used as interface with the customers. For both Pattern II and III, additional key activities are the maintenance of the (co-)owned infrastructures and the strategic development of the Energy Hub infrastructures to continuously adapt to customer and partner acquisitions.

Key partners of the Energy Hub comprise: grid owners, ICT system providers, energy conversion/storage infrastructure providers, and suppliers of energy services/resources operating in the proximity (e.g., other Energy Hubs, or industries). Pattern I relies more on partnerships than Pattern II and III due to its peculiarity of not investing in infrastructures. Finally also financial institutions are key partners that offer financial instruments to support the investment in infrastructures of the Energy Hub (Patterns II and III), and of the customers (Patterns I and II).

Financial Viability

In the cost structure of the Energy Hub, fixed costs are related to: the overhead, the fixed expenses due to the contracts with partners, the maintenance of the multisided platform (Pattern I), the amortization and maintenance costs of the (partially-)owned or contracted infrastructures (Patterns II and III), and to the dedicated support services offered to the clients (Patterns II and III). Variable costs are mainly related to: the results of the energy trading, the grid charges, the exploitation of the partners' infrastructures (Pattern I), and to the energy resources used to operate the (partially-)owned or contracted infrastructures (Patterns II and III).

The Energy Hub revenue stream comprises: the revenues from the energy trading, which can be shared with prosumers and partners, the income related to the delivered energy to customers, and to the potential hidden revenues from partners interested in, e.g., accessing the demand and production profiles of the Energy Hub customers. Additionally, Pattern II benefits by the revenues from the provided maintenance of infrastructure, and Pattern III

profits by the incomes from the infrastructure leased and from the comprehensive support services offered.

From a customer's and partner's perspective, all three patterns offer the possibility of valorizing investments in owned infrastructures through the access to the internal energy trading, or through external trading carried out by the Energy Hub organization. Within Pattern I and II consumer's costs regard only the purchase of energy services. Prosumer's costs are associated with the amortization and maintenance of own (Pattern I) or partially own (Pattern II) infrastructures and to the purchasing costs of complementary energy services. Pattern III customer's costs simply regard the all-inclusive contract with the Energy Hub.

Innovative Market Implemented Business Models

The need to face the currently very challenging energy market and its expected short-term evolution pushes utilities and more generally energy related companies to adapt their organizations and business models to gain competitiveness. The trend is especially noticeable in developed countries already characterized by a partially or fully liberalized market such as US and Germany. Policy makers urge to adapt the outdated existing regulatory framework to allow the exploitation of the emerging business opportunities. At European level a new Energy Market Design proposal is expected to come in the forthcoming months (European Commission, 2015b). Aim of this section is to present examples of innovative business models recently implemented in the market and allocate them to the most closely related region of the defined business model solution space (**Table 5**). The market review, partly based on the recent work of Fratzscher (2015), is not meant to be exhaustive. Instead, it aims to provide an overview across the different fields of opportunities related to the energy market and put into context the introduced conceptual framework. The analyzed business models are organized in four categories: focusing on demand management, focusing on generation, focusing on distribution, and focusing on value-added services.

Demand Management

The demand management side certainly represents a very active area offering a large spectrum of business opportunities aiming to support customers improving their energy efficiency. Even largest IT companies such as IBM, Oracle, and Google recognized the market potential on this sector and plunged into the market profiting on their capability of collecting and analyzing large number of data from the customers. Nest (2015) bought in 2015 by Google, is a well known example of innovative company offering smart devices providing energy management services. Other examples of innovative business model approaches focusing on demand management are as follows: Opower (2015) which offers to utility the interpretation of customers data with the aim to improve energy efficiency; Viridity Energy (2015) which support customers in valorizing demand side management capability in the wholesale energy market. Two interesting examples of companies providing innovative solutions through demand response and load management arise from Switzerland: Misurio (2015) an IBM supported company, provide solutions to

TABLE 5 | Allocation of emerging business models within the business model solution space.

		Delivery of energy services		
		No frills	*User designed*	*Experience selling*
Procurement of infrastructure/ operation and control	*Rent instead of buying*		E ON (E ON, 2015) E ON (E ON, 2015)	NRG (NRG, 2015) Regio Energie Solothurn (Regio Energie Solothurn, 2015) NRG (NRG, 2015)
	Fractional ownership	Trianel Group (Trianel Group, 2015)	Clean Energy Collectives (Clean Energy Collectives, 2015) Schwäbisch-Hall Stadtwerke (Schwäbisch-Hall Stadwerke, 2015) Stroomversnelling (Stroomversnelling, 2015)	
	Orchestrator	Tiko (Tiko, 2015) National Grid (National Grid, 2015) Sun (SUN, 2015) Next Kraftwerke (Next Kraftwerke, 2015) Lichtblick (Lichtblick, 2015) E2m (e2m, 2015) Clean Energy Sourcing Statkraft (Statkraft, 2015)	Opower (Opower, 2015) Viridity Energy (Viridity Energy, 2015) Misurio (Misurio, 2015) Austin Energy (Austin Energy, 2015) Repower (Repower, 2015) WGL (WGL Energy, 2015) YelloStrom (EnBW, 2015)	Nest (Nest, 2015)

Business model categories: demand management/distribution/generation/added services.

optimize the operation of multi-energy networks of distributed energy conversion systems Tiko (2015); a spinoff of the largest Swiss telecommunication company Swisscom, offer the possibility to small to large customers to connect in a storage network and valorized their aggregated flexibility into the power reserve and balancing market.

As highlighted in **Table 5**, innovative business models focusing on demand management are commonly characterized by lean structures, associable to Pattern I, addressing the operation of customer's owned infrastructures. However on the Delivery of Energy Services side, they extensively cover the range of possibilities offering from basic to tailored comprehensive packages of services.

Distribution

On the distribution side, new business opportunities arise from the needs of low voltage distributed smart grids bidirectionally connecting prosumers and consumers to the higher voltage grid infrastructure. However, the success of this typology of business models strongly depends upon the development of a regulatory framework more suitable to competitiveness. A number of utility companies substantially reoriented their core business activity toward the distribution of energy services. In US, an emblematic example is the strategy adopted by National Grid (2015) of upgrading the grid infrastructure to foster the integration of smart technologies appealing to customers. In Germany, Trianel Group (2015) aims to bundling resources purchasing networks and strengthening through aggregation small municipality utilities. The local utility SUN (2015) supports the integration of distributed renewable generation units providing network capability and management options to small players characterized by limited resources.

Innovative distribution focused business models aim to provide to customers infrastructures access to the distribution network. They provide rather basic services and operate as a two-sided platform connecting distributed generation to grid operators (**Table 5**). They can be generally associated with Pattern I.

Generation

The transition from centralized generation toward distributed generation implies the development of innovative business model approaches focused on the generation side. A first example is the development of business model to promote the diffusion and integration of many small-scale distributed generation units. In US, utilities offer leasing solutions for solar generation and storage systems including, installation, operation, and maintenance. Alternatively they lease customer's roofs to install and operate distributed small-scale solar-based generation units (NRG, 2015). In Switzerland, Regio Energie Solothurn (2015) offers long-term leasing solution for photovoltaic installations. At the contract expiration, the equipment ownership passes to the customer without additional fee.

Another diffused option in US and Germany [e.g., Schwäbisch-Hall Stadwerke (2015); Clean Energy Collectives (2015)] is solar or wind community projects led by utilities where customers share investments and benefits of common infrastructures operated by the utilities. Furthermore in Germany, large utilities [e.g., E ON (2015)] offer to completely handle regional renewable energy projects on behalf of customers such as local utilities.

Many examples can be found also on new business model concepts focusing on supporting customer sited small-scale generation units. In US, the municipality utility company of Austin (Austin Energy, 2015) proposes remuneration mechanism going beyond the per kilowatt-hour fees and including other benefits such as avoided fuel and infrastructure costs, energy price hedging, and environmental values. In Germany, new companies [e.g., (Next Kraftwerke, 2015; e2m, 2015; Lichtblick, 2015)] offer the possibility of connecting customers in virtual power plants balancing the volatility characterizing renewable energies and reaching the critical size needed to participate to the power reserve and balancing market.

Generation-oriented business models span across the entire solution space and in particular reflect the three reference business patterns identified. However, the tendency toward business models focusing on operation of customer's owned infrastructures and to offer customized services emerges clearly (**Table 5**).

Value-Added Services

Traditional utility business models successfully operated for a long time without the need of focusing on customer's needs. Today, a large number of business opportunities focusing on added-value services tailored to the customers are emerging.

The diversification of the offer allows targeting more customer segments and improving the market positioning with respect to the competitors. Yello Strom of EnBW (2015) in Germany, Repower (2015) in Switzerland and WGL Energy (2015) in US are examples of utilities offering a large variety of green power packages allowing the customer to personalize their electricity procurements and tariffs depending on their need and conviction.

Opening up the offer portfolio to additional services and targeting niche of customers with very specific needs emerge as another promising business strategy. Providing e-mobility services and support in the installation of distributed energy conversion infrastructures is the strategy applied by NRG (2015) in US. The largest utilities in Germany provide tailored turnkey solutions to commercial and industrial customers providing planning, construction, operation and maintenance of the energy related infrastructures [e.g., E ON (2015)]. On a different scale, smaller utilities such as Statkraft (2015) and Clean Energy Sourcing (2015) provide direct marketing to facilitate small customer to access the wholesale market.

The Stroomversnelling (2015) program in The Netherlands is an interesting example of innovative business model focusing on building retrofit. The renovation of existing buildings appears as a high potential and very challenging market for the years to come. Within Stroomversnelling program, a consortium of construction companies and housing cooperatives retrofits in a very short time poorly efficient rental housing from the 1950s–70s to net-zero buildings. With the applied business model, the investment is paid back only through the energy savings realized and with no additional charges on the tenants.

Added-value services oriented business models follow the same tendency as generation oriented business models. They cover the whole solution space with a slight tendency on favoring customized services and management of third party owned infrastructures (**Table 5**).

Discussion

The presented market analysis allowed identifying the regions of the business model solution space most commonly adopted for each different category of emerging business successfully implemented at this stage of the energy transition.

Demand management and distribution oriented businesses favor operation and control focused business models and to provide basic (distribution) or tailored (demand management) services to the customers. Generation and added-value services-oriented business models cover more uniformly the whole business model solution space. However, a tendency on favoring business models focusing on operation and control of third party owned infrastructures and providing solution tailored to the customers emerge clearly in all business model categories. As a result, the intersection between Pattern I and II appears as the area of the business model solution space where the energy transition is principally addressing the market at the moment. This tendency is in contrast with traditional vertical integrated centralized generation business models, which mainly focus on owned infrastructures and delivery of standard services (Bhattacharyya, 2011). The energy transition is decentralizing the energy conversion and storage infrastructures at the customer's place and centralizing the customers at the core of the business model. This consideration is fully aligned with the findings proposed by Richter (2012, 2013) suggesting that utilities should strategically orient toward customer-side business models, and with the vision of the recently adopted European Energy Union Strategy (European Commission, 2015a). This initiative and the consequently related activities (European Commission, 2015b) emerge as a suitable framework to tackle the challenge of designing new flexible regulatory frameworks removing the existing market barriers that hinder the costumers from becoming protagonists in fostering the energy transition.

Interpreting the results from a different perspective, the empty (or less crowded) spots of the business model solution space depicted in **Table 5** represent the currently untapped opportunities. The fact that emerging businesses generally prefer not to focus on the ownership of the infrastructures and on potentially more profitable high quality services suggests that, at present, large investments from firms are discouraged. Needless to say, this consideration is compatible with the current global economic slowdown and changing regulatory frameworks (Masini and Menichetti, 2012; Wüstenhagen and Menichetti, 2012). Assuming the next step of the energy transition being characterized by a less uncertain and more suitable regulatory framework, and in the context of a global economic recovery, the potentially more profitable business opportunities focusing on infrastructure leasing solutions and high quality services are expected to become more appealing.

CONCLUSION

The present study introduces a heuristic methodology easing the identification of general business model patterns best suited for general Local Energy Management concepts such as Energy Hubs. A conceptual framework characterizing the Energy Hub business model solution space is developed. The conceptual framework aims to ease the business model innovation process at early stages. Stakeholders should identify within the defined solution space the most appropriate general pattern(s) matching the intended Energy Hub concept. Each general business model pattern consists in an organized collection of business model ideas that can be combined to develop a specific business model tailored to the intended Energy Hub concept.

The defined business model solution space is put into context through the analysis and comparison of three reference business model patterns. The three reference patterns are selected to provide an orientation across the whole business model solution space.

Furthermore, a review of successfully market implemented innovative business models in the different value chain steps of the energy services sector is proposed. The location of the presented business models within the defined conceptual business model solution space is discussed.

On the one hand, the analysis outlined as in contrast with traditional centralized generation business models, focusing on procurement of infrastructure and delivery of standardized service as core activities, emerging business models mainly focus on operation and control of third party owned infrastructures and on customized services. The energy transition is bringing the customers at the core of the business models. On the other hand, the analysis highlighted as at present emerging business models disfavor the options of offering infrastructure leasing solutions and high quality comprehensive services. These untapped business opportunities, requiring higher investments and potentially more profitable, are expected to become more appealing when a more suitable regulatory frameworks will be in place and the global economic situation will improve.

Potential Energy Hub stakeholders should benefit from this contribution by finding inspiration to move forward from the attempts of adapting business models conceived for the traditional centralized generation, to instead create from scratch innovative business model valorizing the peculiarities of decentralized energy systems.

The outcomes of this work offer insights for policy makers on challenges and opportunities related to Energy Hub concepts. Policies establish the boundary conditions for business model developments. Therefore, they have a preeminent impact on the business model innovation process, especially at early stage. The role of policy makers is twofold. On the one hand, existing energy policies conceived to regulate the traditional energy market should be reviewed in order to avoid to hinder the business model innovation process. On the other hand, future policies should embrace the heterogeneity of distributed generation by creating an appropriately flexible regulatory framework. In this perspective, the proposed conceptual business model solution space aim to provide to policy makers indications on the full spectrum of business model patterns potentially applicable in Energy Hub concepts. The development of new policies fostering the penetration of Local Energy Management should primarily focus on the preeminent role of customers. The regulatory framework coordinating the relation between the wholesale and retail energy markets should be redefined to enable/facilitate the access of customers (or aggregation of customers) to the new arising business opportunities. New policies should enable the development of business models focusing on customer tailored solutions and potentially including services going across other market segments (e.g., including mobility, home automation and security, telecommunication services). Furthermore, the dissemination of Local Energy Managements is expected to take major advantages from shared ownership based business models. For this reason, policy makers should encourage such approaches that were not considered, or considered to a minor extend, within the in force regulatory frameworks. Finally, policy makers should improve the attractiveness of Local Energy Management to large investments from private companies. In this perspective, the development of new regulatory frameworks clear, stable, and consistent across the time period necessary to achieve the long-term sustainability targets is expected to be beneficial.

AUTHOR CONTRIBUTIONS

EF: main author, main contributions to the conception and realization of the work. SS: substantial contribution to conception and realization of the work, and critical revision of the work.

ACKNOWLEDGMENTS

This work has been accomplished in the frame of the Swiss Competence Center for Energy Research on Future Energy Efficient Buildings & Districts SCCER FEEB&D, funded by The Commission for Technology and Innovation of the Swiss Confederation.

REFERENCES

Allan, G., Eromenko, I., Gilmartin, M., Kockar, I., and McGregor, P. (2015). The economics of distributed energy generation: a literature review. *Renew. Sustainable Energy Rev* 42, 543–556. doi:10.1016/j.rser.2014.07.064

Austin Energy. (2015). Available at: https://www.austinenergy.com/ [accessed July 21, 2015].

Bhattacharyya, S. C. (2011). *Energy Economics Concepts, Issues, Markets and Governance*. London: Springer.

Bucherer, E., Eisert, U., and Gassmann, O. (2012). Towards systematic business model innovation: lessons from product innovation management. *Creat. Innovat. Manage.* 21, 183–198. doi:10.1111/j.1467-8691.2012.00637.x

Capuder, T., and Mancarella, P. (2014). Techno-economic and environmental modelling and optimization of flexible distributed multi-generation options. *Energy* 71, 516–533. doi:10.1016/j.energy.2014.04.097

Chesbrough, H. (2010). Business model innovation: opportunities and barriers. *Long Range Plann.* 43, 354–363. doi:10.1016/j.lrp.2009.07.010

Clean Energy Collectives. (2015). Available at: http://www.easycleanenergy.com [accessed July 21, 2015].

Clean Energy Sourcing. (2015). Available at: http://www.clens.eu [accessed July 21, 2015].

Curtius, H. C., Künzel, K., and Loock, M. (2012). Generic customer segments and business models for smart grids. *Der Markt* 51, 63–74. doi:10.1007/s12642-012-0076-0

e2m. (2015). Available at: http://www.energy2market.de [accessed July 21, 2015].

EnBW. (2015). Available at: https://www.enbw.com [accessed July 21, 2015].

E ON. (2015). Available at: https://www.eon.com [accessed July 21, 2015].

European Commission. (2011). *COM/2011/0885 Energy Roadmap 2050*. Brussels: EU.

European Commission. (2012). *Directive 2012/27/EU of the European Parliament and of the Council of 25 October 2012 on Energy Efficiency*. Brussels: EU.

European Commission. (2014). *COM/2014/015 A Policy Framework for Climate and Energy in the Period from 2020 to 2030*. Brussels: EU.

European Commission. (2015a). *COM/2015/080 A Framework Strategy for a Resilient Energy Union with a Forward-Looking Climate Change Policy*. Brussels: EU.

European Commission. (2015b). *Commission Proposes "New Deal" for Energy Consumers, Redesign of Electricity Market and Revision of Energy Label for More Clarity. Press Release.* Available at: https://ec.europa.eu/energy/en/news/new-electricity-market-consumers [accessed July 15, 2015].

Facchinetti, E., Favrat, D., and Marechal, F. (2014). Design and optimization of an innovative solid oxide fuel cell-gas turbine hybrid cycle for small scale distributed generation. *Fuel Cells* 14, 595–606. doi:10.1002/fuce.201300196

Fratzscher, S. (2015). *The Future of Utilities: Extinction or Re-Invention? A Transatlantic Perspective.* Berlin: Heinrich Böll Stift.

Frei, C. W. (2008). What if…? Utility vision 2020. *Energy Policy* 36, 3640–3645. doi:10.1016/j.enpol.2008.07.016

Gassmann, O., Enkel, E., and Chesbrough, H. (2010). The future of open innovation. *Res. Dev. Manage.* 40, 213–221. doi:10.1111/j.1467-9310.2010.00605.x

Gassmann, O., Frankenberger, K., and Csik, M. (2014). *The Business Model Navigator.* Upper Saddle River, NJ: FT Press.

Geidl, M., Koeppel, G., Favre-Perrod, P., Klockl, B., Andersson, G., and Frohlich, K. (2007). Energy hubs for the future. *IEEE Power Energy Mag.* 5, 24–30. doi:10.1109/MPAE.2007.264850

Graves, C., Ebbesen, S. D., Mogensen, M., and Lackner, K. S. (2011). Sustainable hydrocarbon fuels by recycling CO2 and H2O with renewable or nuclear energy. *Renew. Sustain. Energy Rev.* 15, 1–23. doi:10.1016/j.rser.2010.07.014

IEA. (2002). *Distributed Generation in Liberalised Electricity Markets.* Paris: OECD/International Energy Agency (IEA).

IEA. (2014). *World Energy Outlook 2014.* Paris: OECD/International Energy Agency (IEA).

Karger, C. R., and Hennings, W. (2009). Sustainability evaluation of decentralized electricity generation. *Renew. Sustain. Energy Rev.* 13, 583–593. doi:10.1016/j.rser.2007.11.003

Lichtblick. (2015). Available at: http://www.lichtblick.de [accessed July 21, 2015].

Loock, M. (2012). Going beyond best technology and lowest price: on renewable energy investors' preference for service-driven business models. *Energy Policy* 40, 21–27. doi:10.1016/j.enpol.2010.06.059

Mancarella, P. (2014). MES (multi-energy systems): an overview of concepts and evaluation models. *Energy* 65, 1–17. doi:10.1016/j.energy.2013.10.041

Masini, A., and Menichetti, E. (2012). The impact of behavioural factors in the renewable energy investment decision making process: conceptual framework and empirical findings. *Energy Policy* 40, 28–38. doi:10.1016/j.enpol.2010.06.062

Misurio. (2015). Available at: http://www.misurio.ch [accessed July 21, 2015].

National Grid. (2015). Available at: https://www.nationalgridus.com/ [accessed July 21, 2015].

Nest. (2015). Available at: https://nest.com [accessed July 21, 2015].

Next Kraftwerke. (2015). Available at: https://www.next-kraftwerke.de [accessed July 21, 2015].

NRG. (2015). Available at: http://www.nrg.com [accessed July 21, 2015].

Okkonen, L., and Suhonen, N. (2010). Business models of heat entrepreneurship in Finland. *Energy Policy* 38, 3443–3452. doi:10.1016/j.enpol.2010.02.018

Opower. (2015). Available at: http://www.opower.com [accessed July 21, 2015].

Orehounig, K., Evins, R., and Dorer, V. (2015). Integration of decentralized energy systems in neighbourhoods using the energy hub approach. *Appl. Energy* 154, 277–289. doi:10.1016/j.apenergy.2015.04.114

Orehounig, K., Mavromatidis, G., Evins, R., Dorer, V., and Carmeliet, J. (2014). Towards an energy sustainable community: an energy system analysis for a village in Switzerland. *Energy Build.* 84, 277–286. doi:10.1016/j.enbuild.2014.08.012

Osterwalder, A., and Pigneur, Y. (2010). *Business Model Generation: A Handbook for Visionaries, Game Changers, and Challengers.* Hoboken, NJ: Wiley.

Osterwalder, A., Pigneur, Y., Bernarda, G., and Smith, A. (2014). *Value Proposition Design: How to Create Products and Services Customers Want.* Hoboken, NJ: Wiley.

Parisio, A., Del Vecchio, C., and Vaccaro, A. (2012). A robust optimization approach to energy hub management. *Int. J. Electr. Power Energy Syst.* 42, 98–104. doi:10.1016/j.ijepes.2012.03.015

Porter, M. E. (1985). *The Competitive Advantage: Creating and Sustaining Superior Performance.* New York, NY: Free Press.

Regio Energie Solothurn. (2015). Available at: http://www.regioenergie.ch [accessed July 21, 2015].

Repower. (2015). Available at: http://www.repower.com [accessed July 21, 2015].

Richter, M. (2012). Utilities' business models for renewable energy: a review. *Renew. Sustain. Energy Rev.* 16, 2483–2493. doi:10.1016/j.rser.2012.01.072

Richter, M. (2013). Business model innovation for sustainable energy: German utilities and renewable energy. *Energy Policy* 62, 1226–1237. doi:10.1016/j.enpol.2013.05.038

Schleicher-Tappeser, R. (2012). How renewables will change electricity markets in the next five years. *Energy Policy* 48, 64–75. doi:10.1016/j.enpol.2012.04.042

Schwäbisch-Hall Stadtwerke. (2015). Available at: http://www.stadtwerke-hall.de [accessed July 21, 2015].

SFOE. (2012). *Analyse des schweizerischen Energieverbrauchs 2000–2012 nach Verwendungszwecken* [Energy Consumption in Switzerland According to Purpose 2000–2012]. Bern: Swiss Federal Office of Energy.

Statkraft. (2015). Available at: http://www.statkraft.de [accessed July 21, 2015].

Stroomversnelling. (2015). Available at: http://www.stroomversnelling.net [accessed July 21, 2015].

Sühlsen, K., and Hisschemöller, M. (2014). Lobbying the "Energiewende." Assessing the effectiveness of strategies to promote the renewable energy business in Germany. *Energy Policy* 69, 316–325. doi:10.1016/j.enpol.2014.02.018

SUN. (2015). Available at: http://www.sun-stadtwerke.de [accessed July 21, 2015].

Tiko. (2015). Available at: https://tiko.ch [accessed July 21, 2015].

Trianel Group. (2015). Available at: http://www.trianel.com [accessed July 21, 2015].

Viral, R., and Khatod, D. K. (2012). Optimal planning of distributed generation systems in distribution system: a review. *Renew. Sustain. Energy Rev.* 16, 5146–5165. doi:10.1016/j.rser.2012.05.020

Viridity Energy. (2015). Available at: http://viridityenergy.com [accessed July 21, 2015].

Weber, C., and Favrat, D. (2010). Conventional and advanced CO2 based district energy systems. *Energy* 35, 5070–5081. doi:10.1016/j.energy.2010.08.008

WGL Energy. (2015). Available at: http://www.wges.com [accessed July 21, 2015].

Wüstenhagen, R., and Menichetti, E. (2012). Strategic choices for renewable energy investment: conceptual framework and opportunities for further research. *Energy Policy* 40, 1–10. doi:10.1016/j.enpol.2011.06.050

Conflict of Interest Statement: The authors declare that the research was conducted in the absence of any commercial or financial relationships that could be construed as a potential conflict of interest.

17

Energy Data Visualization Requires Additional Approaches to Continue to be Relevant in a World with Greater Low-Carbon Generation

I. A. Grant Wilson*

Environmental and Energy Engineering Group, Department of Chemical and Biological Engineering, The University of Sheffield, Sheffield, UK

The hypothesis described in this article proposes that energy visualization diagrams commonly used need additional changes to continue to be relevant in a world with greater low-carbon generation. The diagrams that display national energy data are influenced by the properties of the type of energy being displayed, which in most cases has historically meant fossil fuels, nuclear fuels, or hydro. As many energy systems throughout the world increase their use of electricity from wind- or solar-based renewables, a more granular display of energy data in the time domain is required. This article also introduces the shared axes energy diagram that provides a simple and powerful way to compare the scale and seasonality of the demands and supplies of an energy system. This aims to complement, rather than replace existing diagrams, and has an additional benefit of promoting a whole systems approach to energy systems, as differing energy vectors, such as natural gas, transport fuels, and electricity, can all be displayed together. This, in particular, is useful to both policy makers and to industry, to build a visual foundation for a whole systems narrative, which provides a basis for discussion of the synergies and opportunities across and between different energy vectors and demands. The diagram's ability to wrap a sense of scale around a whole energy system in a simple way is thought to explain its growing popularity.

Keywords: energy system visualization, energy demand comparisons, energy data visualization, seasonal energy demands, whole systems visualization

Edited by:
Fu Zhao,
Purdue University, USA

Reviewed by:
Payman Dehghanian,
Texas A&M University, USA
Hakan Caliskan,
Uşak University, Turkey

*Correspondence:
I. A. Grant Wilson
grant.wilson@sheffield.ac.uk

Specialty section:
This article was submitted
to Energy Systems and Policy,
a section of the journal
Frontiers in Energy Research

Citation:
Grant Wilson IA (2016) Energy Data
Visualization Requires Additional
Approaches to Continue to be
Relevant in a World with Greater
Low-Carbon Generation.
Front. Energy Res. 4:33.

INTRODUCTION

The need to reduce the amount of greenhouse gas emissions entering the atmosphere from human activity is well understood and exemplified by the UN Framework Convention on Climate Change to hold "the increase in the global average temperature to well below 2°C and to pursue efforts to limit the temperature increase to 1.5°C" through the Paris Agreement (UNFCCC, 2015). To achieve this, many policy makers around the world will continue to focus on limiting the greenhouse gas emissions from their energy systems, with electrical systems, in particular, being an area of initial effort. The recent increases in global deployment of wind and solar electrical generation demonstrate this, with their output increasing significantly over the time period from 2006 to 2014 (136747–910923 GWh) (IRENA, 2016). This increase in deployment has provided cost reductions through economies of scale and deployment experience (Rubin et al., 2015), and these technologies are now considered mainstream choices for electrical generation alongside conventional forms of thermal generation or hydro. Having the ability to harvest the primary energy resource of wind or solar within a national or energy system boundary has an appeal not only from a low-carbon perspective but also from an energy-import dependence perspective. The main driver, so far, has been to control carbon emissions, rather than offset energy imports, but, over time, this may change, especially if the cost of imported fuels achieves greater political importance.

Having a greater level of policy support from the end user, customer base of an electrical system is a desirable outcome for policy makers charged with effecting system wide changes. This support can be influenced by end users having access to empirical evidence provided by trusted energy systems professionals. In Great Britain, a thorough resource for energy data is provided by teams at the Department of Energy and Climate Change (merged into the Department for Business, Energy, and Industrial Strategy in July 2016), the energy team at the Scottish Government, and the statistical team at BP, to name a few. These teams provide sources of energy data and utilize diagrams to convey information, with timeframes generally of a month, a quarter, or a year. The bar charts, pie charts, and line graphs in common use today were all invented by the Scotsman William Playfair around 1800 (Friendly, 2008), just over 200 years ago, and have clearly stood the test of time. These are the graphs most utilized in national level energy data visualizations. However, as increasing levels of primary energy in Great Britain is derived from wind and solar resources, wider stakeholders are becoming increasingly interested in energy data over shorter time periods, such as the energy changes between days, precisely because the nature of weather-dependent primary energy resources results in an intermittent resource.

This article proposes a hypothesis that energy visualization diagrams commonly used, need additional changes to continue to be relevant, and then describes a novel energy diagram that has been growing in importance in Great Britain. The history, methods, and considerations when creating this shared axes energy diagram (SAED) are explained, in order to allow others to create their own.

The contribution of this article is to present the hypothesis of why energy diagrams need to change, and, then, to provide a description on the creation of a SAED. The importance of the graphical representation of statistical data has not diminished since William Playfair stated in The Statistical Breviary in 1801, that "making an appeal to the eye when proportion and magnitude are concerned, is the best and readiest method of conveying a distinct idea" (Spence, 2005). This is especially true of energy data and, arguably, will increase in importance as the energy systems themselves undergo profound changes over the timeframe to 2050.

BACKGROUND

Accommodating the increasing amounts of wind and solar electrical generation onto electrical grids that have historically been designed to accommodate a limited number of larger and controllable power stations has a number of challenges. One of the greatest is how to accommodate the variability inherent in weather-dependent primary energy. Wind and solar resources are unlikely to match the electrical demand for large parts of the year, and, therefore, generation from these sources has to be augmented from other sources, when there is too little (Wilson, 2016), or, indeed, local demand increased through storage or demand side management, if there is a local surplus. Electrical generation, including wind and solar, can also be curtailed in order to limit the amount of electrical energy being generated onto parts of an electrical network that is unable to accommodate or export the surplus for certain periods. This is to keep the voltage on the network within defined limits and protect equipment connected to the network and the network itself. When the source of the primary energy is a fuel or from a hydro resource, the primary energy is able to be stockpiled before it is transformed into electricity. This allows a significant buffer against supply chain disruptions, and usually allows significant load following capability (dependent on the type of power stations).

Electricity for all its marvelous benefits, is prohibitively expensive to store at scale, and it is, therefore, ordinarily changed into another form of energy, for ease and cost of storage, e.g., hydro-pumped storage (Deane et al., 2010; Wilson et al., 2010; Hittinger et al., 2012; Castillo and Gayme, 2014; Hittinger and Lueken, 2015) or power-to-gas (Schiebahn et al., 2015; Götz et al., 2016) and, then, changed back to electricity at a later stage, if required. From a whole systems perspective, primary electricity can be transformed into another form of energy and be utilized in the energy system without the need to be transformed back into electricity, which may offer multiple benefits.

To reiterate, an important difference between historical fuel/hydro-based electrical systems and future primary electricity-based systems is in the ability to load follow, i.e., to control the output of power plants (thermal or hydro) to match the electrical demand. This controllability of electrical output from certain power plants is itself based on accessing the underlying fuel or hydro resource (no fuel = no electricity), as the energy is stored in the fuels or water resources themselves. Therefore, if a nation wished to have a greater security of supply through greater levels of "stored" electrical energy, it could simply mandate greater levels of fuel stockpiles to be readily available to electrical generators. Natural gas and coal are internationally traded commodities, and a country, such as Great Britain, with sufficient import capacity for either should not expect to run out of these fossil fuels over the medium term. The fuels should continue to be available for importation, as and when required. However, natural gas supplies can become tight, as the level of gas in storage is difficult to forecast from year to year. The major use of natural gas in Great Britain is for domestic heating, where the demand is intrinsically linked to the weather patterns over the winter period. These weather patterns are highly complex and difficult to forecast for a season in advance (Slingo et al., 2014; Bauer et al., 2015).

Electrical generation based on the wind or the sun has a different level of control that generators and system operators can call upon. Wind or solar generation can be curtailed (their outputs can be turned down) or, perhaps, their outputs can be turned up if they were originally part loaded. All generation types share a similarity that they cannot produce electricity if their energy inputs are unavailable and is true of fuel-based generation as it is of wind- or solar-based generation. The crucial difference is that, unlike fuels, the wind or the sun cannot be stored prior to conversion to electricity or indeed purchased and imported from other countries. This lack of control over the input to wind or solar generation and, therefore, the output from wind or solar generation to match electrical demand, is the basis of a major concern for policy makers and wider publics.

HYPOTHESIS

The hypothesis proposes that energy visualization diagrams commonly used need additional changes to continue to be relevant in a world with greater low-carbon generation.

Energy analysis has several diagrams that are popular due to their ability to present information in a clear and concise visual manner. The majority of energy data visualizations are bar charts, pie charts, or line graphs (BP, 2016), which typically show the changes in energy data over a year, a quarter, or even a month. These are perfectly suitable for energy systems that are based on fossil and hydro sources of primary energy, where the data being displayed show the trends over annual timeframes. Implicit in this is an assumption that these primary energy sources continue to be available over the timeframes being shown. **Figure 1** (DECC, 2014) shows an shows an Energy Flow Chart (a Sankey diagram) for 2014 for the United Kingdom. This is a more complex diagram than a simple bar or line chart and is useful to see the scale of energy vectors and demands. However, these are aggregated over a year, which, therefore, masks the seasonality of primary energy supply and demands that change throughout the year. As the primary energy sources for the UK change to incorporate greater amounts of wind- and solar-derived electricity, this masking of supply and demand variation becomes more of a problem. Additional diagrams are, therefore, needed to provide insights into this changing energy system, with a level of resolution that is appropriate to the nature of the primary energy source itself.

THE SHARED AXES ENERGY DIAGRAM

A SAED was created to overcome some of the limitations with the typical bar and line charts with monthly or yearly resolutions.

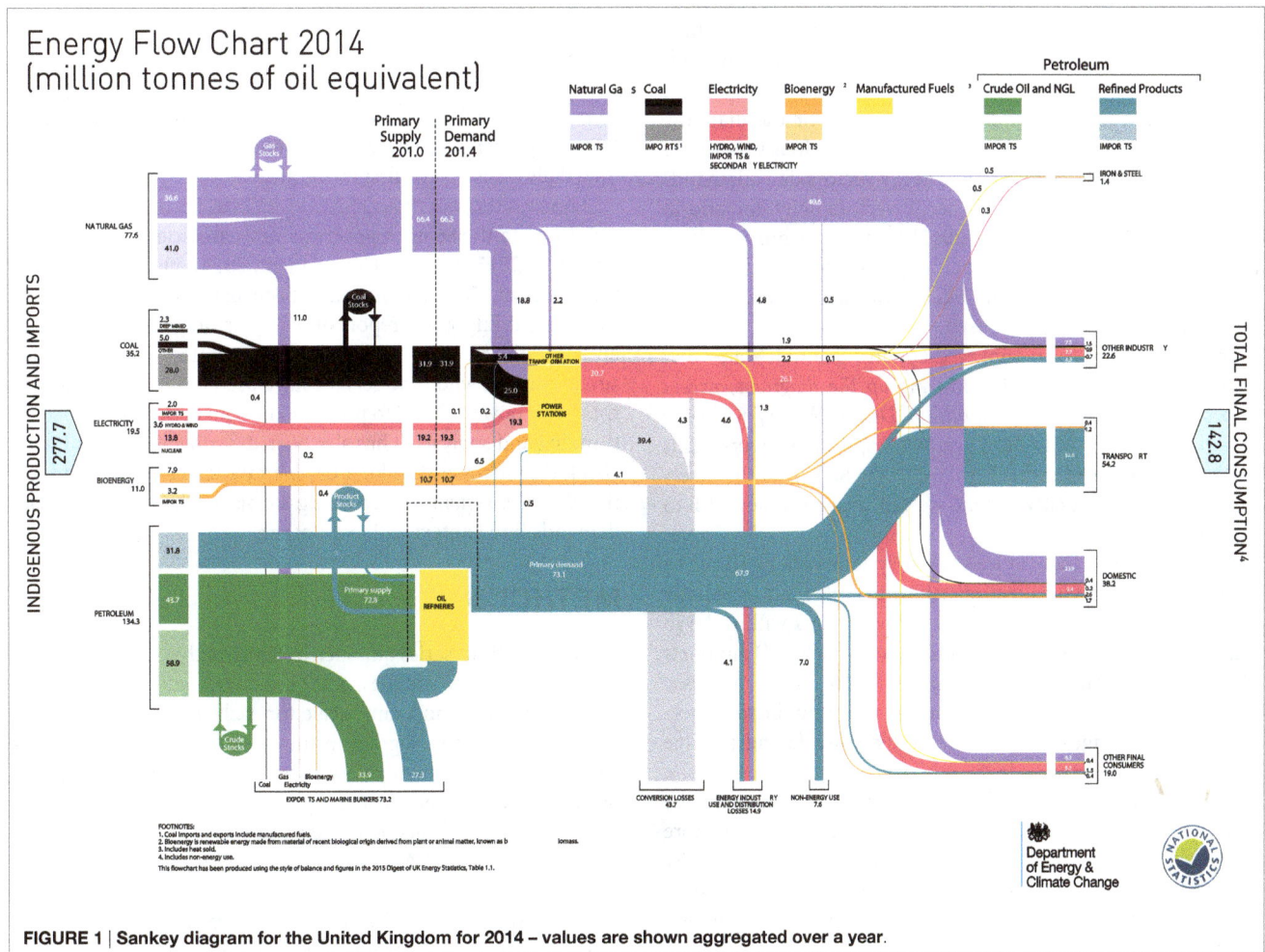

FIGURE 1 | Sankey diagram for the United Kingdom for 2014 – values are shown aggregated over a year.

It aims to provide the continued relevance required of energy data visualization, by allowing a greater insight into the scale and seasonality of the generation from weather-dependent electrical generation. The diagram also provides the ability to visualize low-carbon generation against various different supplies on the same axes, which provides a sense of the scale and timing of how one relates to another.

Another major aim of the SAED is to allow insights in a whole systems approach. It seems obvious that in a northern European country such as Great Britain that the heating demand would be significantly greater than the heating demand in the summer, but it is interesting to understand how this might compare to the demand for electricity, and to transport fuels. This whole systems approach of thinking about energy systems (rather than in silos) is crucial to the transition to greater amounts of primary energy coming from weather-dependent renewable energy. The historical separation of energy market legislation along energy vector boundaries, e.g., liquid fuels, electricity, natural gas, will eventually become open to question, as energy service boundaries become blurred in the future. This whole systems approach is driven by the increase in low-carbon generation and also by technology innovation, e.g., in the transport sector, where electric vehicles will transfer an energy demand from the liquid fuels network over to the electrical network.

The diagram has been growing in popularity in recent years and is complementary to other forms of energy data visualization, as it provides additional insights into the scale and seasonality of energy supplies and demands. Much consideration was given to a number of alternative names for the diagram, such as the titles given in publications using various versions of the diagram, e.g., "Great Britain's energy vectors daily demand" (Wilson et al., 2013a) or "UK energy vectors daily demand in TWh per day" (Wilson et al., 2013b), or "Transmission level daily GB Energy – in TWh per day" (Wilson et al., 2014). However, for the sake of simplicity, the name "shared axes energy diagram" has been chosen.

In his 2008 paper on Sankey diagrams, Schmidt states that "there have been no rules for drawing up the diagrams, except those of visual perception and intuition. Despite this, a few aspects of Sankey's diagram have been assumed implicitly by users." This article aims to provide a background to the SAED, so that users do not need to "implicitly assume" certain aspects if they wish to create their own.

The rationale is that future energy systems will continue to need energy data viusalizations that are relevant, especially given the profound changes that energy systems will undergo to limit the emissions of greenhouse gases. The SAED supports whole systems thinking (Strbac et al., 2016) and provides a sense of scale for policy makers and the wider public to understand the challenges that energy systems are facing (Bridge et al., 2013).

A SAED provides a method to contrast and compare the energy supplies and demands in an energy system over more relevant resolutions. This provides insights into the seasonality and volatility of supplies and demands, which is becoming more important with the increased deployment of weather-dependent electrical renewable energy generation.

HISTORY OF THE SAED

The first version of the SAED (**Figure 2**) was presented to the Scottish Hydrogen and Fuel Cell Association on energy storage on the 9th of March, 2012. It showed the non-daily metered (NDM) natural gas demand alongside the electrical demand for Great Britain (see Methods). The NDM gas demand is the aggregate natural gas demand from customers that are too small to need a daily meter, e.g., household properties, small industrial units, and small commercial units. The seasonality of NDM demand is felt to be a good proxy for *household* natural gas demand in Great Britain although the absolute amount would be less, due to the other non-household demands in the overall aggregate value. The values for electricity are for the entire electrical demand termed Initial Demand Outturn (INDO) and are available for each 30-min period over a day, however, the gas data are only available on a resolution of a daily basis.

The initial diagram used a daily resolution to show the natural gas and electrical demands and how these compared to each other, to get a sense of the challenge of moving the heat demand of Great Britain over to the electrical network. In this regard, the diagram provides a simple visual comparison not only of the scale but also of the seasonal variation within and between the separate demands. Shown with a daily resolution, the seasonality of the daily gas demand becomes very obvious, which are masked by values aggregated over a year.

After this initial diagram was created, the additional demand of transport fuel was added (**Figure 3**). This diagram presented two major energy vectors and an energy demand in Great Britain on a daily basis for the first ever time.

Figure 4 shows a typical SAED for Great Britain that has been published in the Scottish government's statistical publication for energy in 2015 (Scottish Government, 2015) and 2016 (Scottish Government, 2016). It has also been used in the Institute of Mechanical Engineer's report on energy storage (IMechE, 2014) and also their report on heat (IMechE, 2015). Feedback from several parties who have seen or have used the diagram suggests that it is a powerful tool to promote the concept of whole systems thinking and analysis. It has also been described as a useful sense check that focusses thinking on *energy systems* rather than just the *electrical system*, and in doing so, places the scale of various demands into context with each other.

METHODS

A SAED is a time-series plot such as **Figure 4** that shows a number of primary energy supplies, energy carriers, or energy demands on the same axes and same scale for each plotted variable. It is not an energy flow diagram in the same technical manner as a Sankey diagram, as there may well be a degree of double counting or overlap between primary energy supplies, energy carriers such as electricity, and final energy demands.

The creation of a SAED depends on access to reliable underlying energy data that have sufficient detail to provide a degree of insight into the scale and variation of the variables being considered. Great Britain is fortunate in its publically available energy

Energy Data Visualization Requires Additional Approaches to Continue to be Relevant in a World with Greater...

167

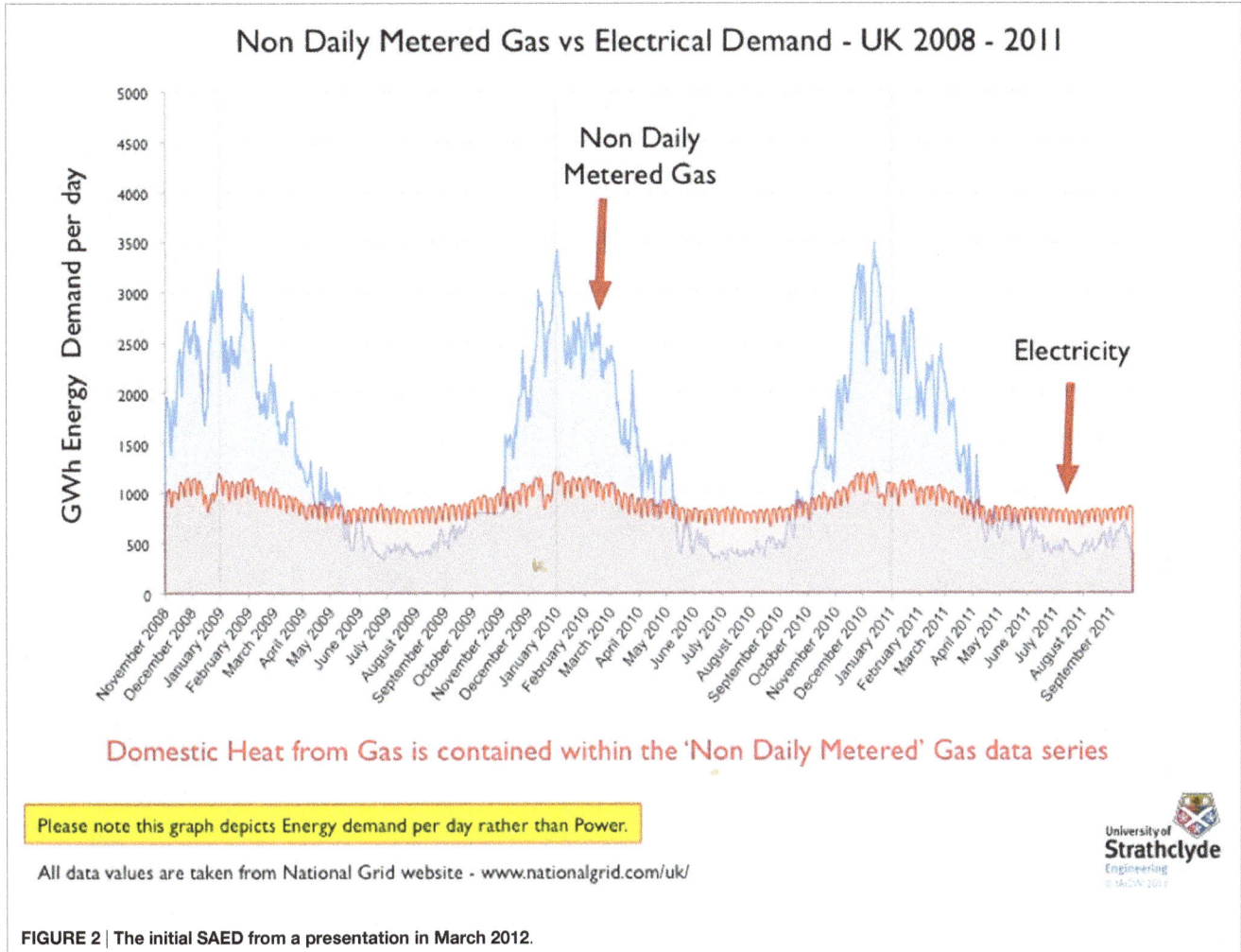

FIGURE 2 | The initial SAED from a presentation in March 2012.

system data through the Transmission System operator's website for electricity and for natural gas. The data sources for many of the diagrams in this article are:

- Natural gas data from National Grid's data explorer (National Grid, 2016a) due to its resolution down to a single day, as well as helpful supply and demand categories.
- Transport fuel data from the energy trends spreadsheet "Deliveries of petroleum products for inland consumptions (ET 3.13)" (DECC, 2016). This is available at a resolution of 1 month.
- Electricity data from the "Metered half-hourly electricity demands" data from National Grid's website (National Grid, 2016b).

The various data sources are recalculated into units of kilowatt hour/day in order to be comparable on a similar resolution. The natural gas and electricity data are already in kilowatt hour, but the transport fuel data was converted from the original units of 1000 ton of fuel into kilowatt hour by using energy content values of: motor spirit = 47.09 GJ/ton; derv = 45.64 GJ/ton; aviation turbine fuel = 46.19 GJ/ton. Once the data have been calculated on a daily time-series for each variable, they can

then be plotted using graphing packages that users are most comfortable with.

Access to the underlying data is key to creating a SAED, and if this is not readily available within a particular energy system or country, it is recommended that the energy system stakeholders, such as the electrical system network operator, should be engaged to allow access to data for historical daily comparisons.

MAJOR CONSIDERATIONS THAT ARE TYPICALLY ENCOUNTERED WHEN CREATING A SAED

Common Elements

At its basic level, the SAED is a time-series plot of energy data and, therefore, has:

1. A horizontal axis in units of time
2. A vertical axis in units of energy per unit of time

The choice of units for the time-series for a *daily* diagram is a day. However, as the horizontal axis is a unit of time, this could

UK Energy vectors daily demand - Gas vs Electricity vs Transport Fuel GWh/day

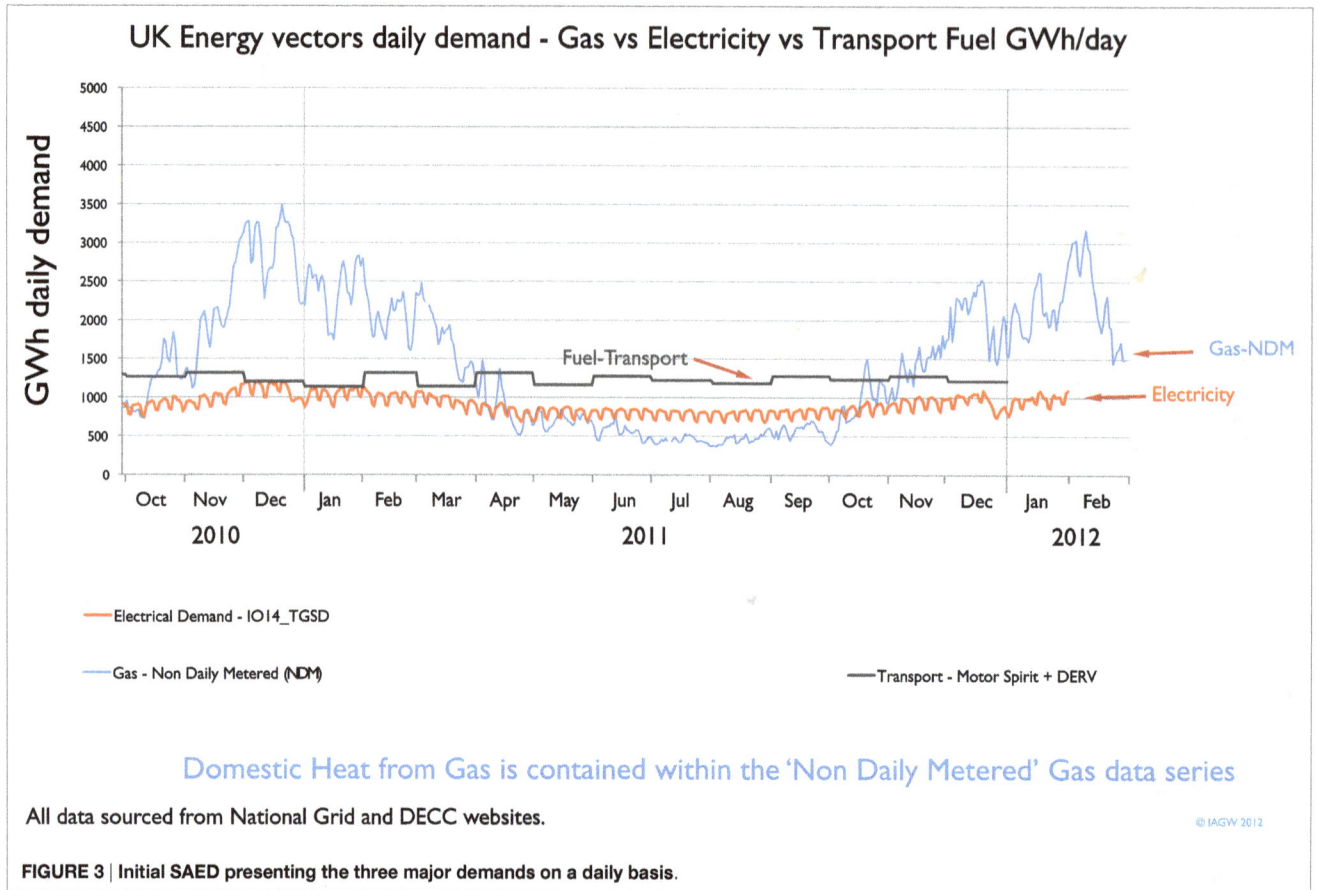

Electrical Demand - IO14_TGSD

Gas - Non Daily Metered (NDM)

Transport - Motor Spirit + DERV

Domestic Heat from Gas is contained within the 'Non Daily Metered' Gas data series

All data sourced from National Grid and DECC websites.

© IAGW 2012

FIGURE 3 | Initial SAED presenting the three major demands on a daily basis.

be changed to present non-daily variations too, e.g., sub-daily. To show an inter-seasonal variation over a couple of years, a daily time resolution is highly suitable, but this itself masks an intra-daily variation.

The choice of units for the vertical axis energy is open, but as electricity is likely to be one of the variables to be plotted, kilowatt hours, megawatt hours, gigawatt hours, or terawatt hours have typically been the energy units of choice. These are combined with the unit of time measurement from the horizontal axis, to give a vertical axis unit such as gigawatt hours/day. This energy per unit time is, therefore, actually a unit of power.

It does not particularly matter which units of time or energy are used, the important aspect of a SAED is that *all plotted variables share the same units*. This provides the simple visual comparison that gives the diagram its clarity.

It is useful for a SAED to display the major energy vectors; in the case of Great Britain these are electricity (both primary and from fuels), natural gas, and transport fuels, but could also include biomass and geothermal heat or other energy inputs to the energy system being represented, even including embodied energy in imported products (Barrett et al., 2013; Sakai and Barrett, 2016).

As the diagram presents units of energy per unit of time plotted against units of time, the integral of a plotted variable is also the amount of energy demand or supply by that variable over a period of time. This can also be helpful in terms of a

simple visualization to gain insight into the scale of energy demands.

In addition to these common elements, there are choices to be made regarding the number of variables to be shown, e.g., the disaggregated sources of electricity or natural gas demand.

How Detailed Should the Time Axis Be?

This is influenced by the level of detail in the underlying data, but the time-series nature of the data means it can be aggregated up over greater time windows, e.g., weeks or months, or indeed averaged out over smaller time windows such as each hour or half hour. The choice is also influenced by the time horizon that the particular diagram will present. Showing inter-seasonal variation requires that a minimum of a full year of data are shown, and potentially over 2 or 3 years such as **Figure 4**. Given a time window of several years, choosing a data resolution of an hour will most likely not show up when the SAED is presented on a page or a screen. The choice of how detailed the time axis should be is, therefore, a matter of presentation as well as the underlying data, and as such, in common with the Sankey diagram, there are no firm rules, only esthetic judgment calls to be made.

This can be seen in **Figure 4** where the underlying data for electricity were available at a 30 min resolution, the natural gas data were available at a daily resolution, and the transport fuels data were available at a monthly resolution. A choice

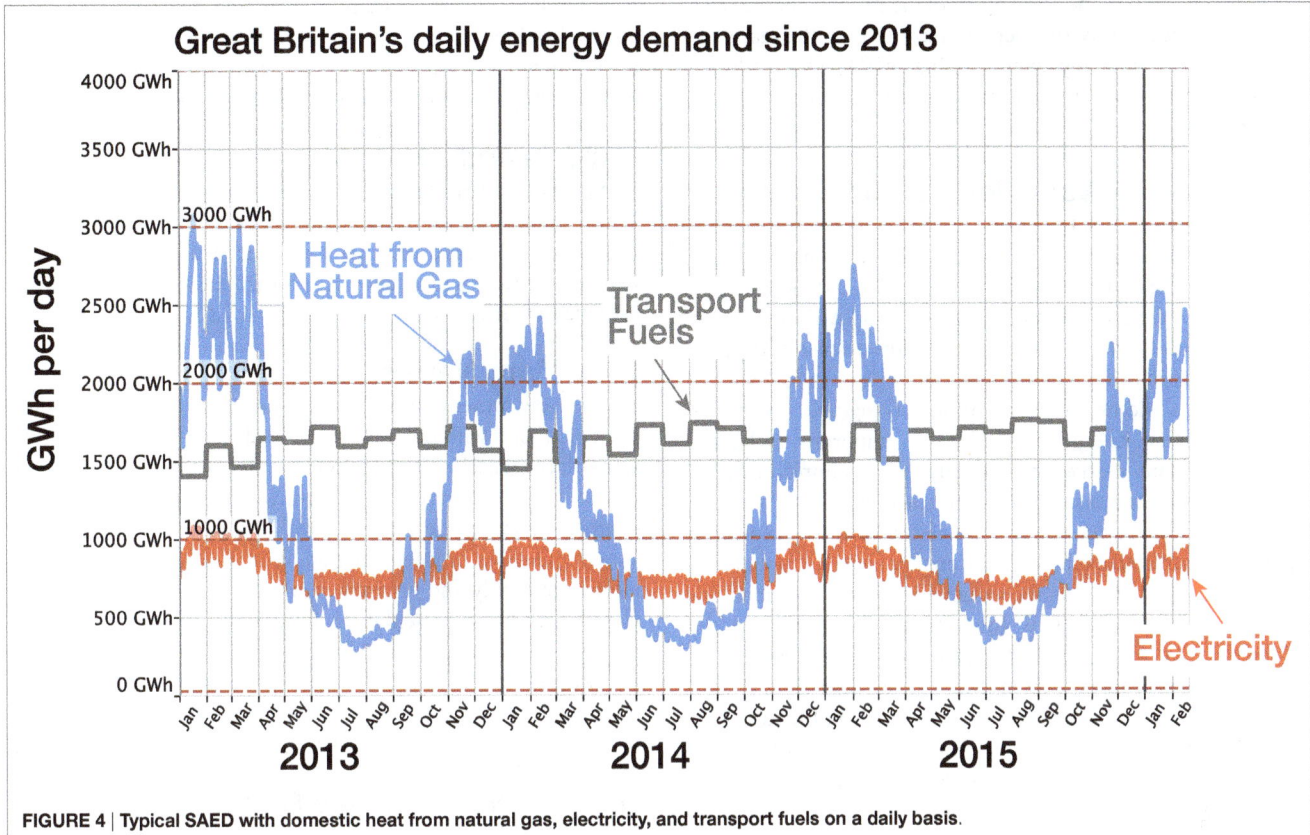

FIGURE 4 | Typical SAED with domestic heat from natural gas, electricity, and transport fuels on a daily basis.

was made to normalize all values to a daily time window, so the 30-min electrical data were summed over each day, and the monthly transport fuels data were divided by the days in each month. Choosing a resolution of a week or a month would lose a level of detail from the resulting diagram due to the summing of the electrical and natural gas data over the week or month, so, e.g., weekends would not be visible in the electrical data. However, choosing a time window with a resolution less than a day would only show the changes in the electrical data, as the natural gas and transport fuels would have to be averaged over a day. This may be interesting to look at in certain circumstances, but is lost in the scale of showing a few years of data at a time.

The compromise of a daily time window to display the different variables is not, however, without its risks; the actual demand for natural gas varies significantly within a day (Newborough and Augood, 1999; Cockroft and Kelly, 2006; Hawkes et al., 2009) and the actual demand for transport fuels will also vary significantly throughout a month (Anable et al., 2012) and within a day. This is demonstrably not a problem to the natural gas or transport fuel networks, which are designed and built to cope with this within day and within month variation, but it would most definitely be a problem when this within-day and within-month variability is transferred over to the electrical network through the greater installation of resistive heating, heat pumps, and electric vehicles. The daily natural gas data mask this issue, so caution is required in the interpretation of a SAED with a

resolution that may mask an underlying variation. This is a basic tenet of information theory.

Comparison to a Sankey Diagram

Sankey diagrams (Schmidt, 2008) are a versatile and popular visualization of the flows of something around a system, e.g., energy, materials, or value. When used with energy data, Sankey diagrams are a true reflection of the energy flows through a system when a conservation of energy approach is applied. They commonly show the energy flows from a primary energy source, through its energy transformation, to other energy carriers, eventually through to the final energy demand. A United Kingdom Sankey diagram for 2014 is shown in **Figure 1** (DECC, 2014), where the units of energy are a million tonnes of oil equivalent. The strength of the Sankey diagram is in the simplicity of its visual presentation, such as the *width* of the arrows being proportional to the *value* of the energy flows. The wider the arrow the larger the value of the energy flow. Efficiencies between each transformative step are, therefore, easy to see, as is the overall scale of the primary energy sources and final demands.

Sankey diagrams are a powerful way to visualize data flow in a simple and effective manner, which explains their ongoing popularity in many areas. However, a weakness of Sankey diagrams is their ability to indicate the flows of energy of a system throughout time, as they show the total values summed over a particular timeframe. It is common for national energy Sankey diagrams to present data summed over an annual basis.

The peaks and troughs between hours of the day, days of the week, and, even, over seasons are, therefore, lost. This is the main area of weakness of the Sankey diagram that a SAED seeks to address, by providing additional visual information to complement the Sankey diagram.

What Types of Variables Can Be Shown?

An additional benefit of the SAED is in its flexibility to show various forms of energy on the same axes. In many other visualizations, the primary energy sources and the final end use demands are likely to be clearly separated, however, on a SAED, the benefit of plotting these together is precisely to show a sense of the scale and variability in order to aid comparison.

As supplies and demands can be shown together, the definition of "primary" energy is arguably less important as long as the variables are marked accordingly. For example, whether nuclear heat that is transformed into nuclear electricity is used as the "primary" energy or whether the electrical output of nuclear generators is treated as the primary energy source is not critical to the diagram, as long as the variable is clearly described and marked. The United Nations' International Recommendations for Energy Statistics (United Nations, 2016) provides a sound background to the definitions of the types of energy.

Having a mixture of supplies and demands on the same diagram also means that the sum of the variables would *not* sum to a total value for Primary energy, as the energy flows such as natural gas that produces electricity could be counted twice (as natural gas), and also as the electricity produced from natural gas too. The magenta line called primary energy demand in **Figure 5** is a sum of the transport fuels, coal, natural gas, and the electricity

supplied by low-carbon sources, including nuclear, wind, solar, and hydro. Biomass has not been added to the total, as the values are unknown on a daily basis with the publicly available data sets. Primary energy, therefore, can be presented on a SAED too.

Imports and Exports

If exports of a primary or secondary energy source are shown on the diagram, they could be shown below the horizontal axis to indicate the exporting nature of the demand. However, as demands are also shown above the horizontal axis, this is another esthetic matter of choice in the creation of a SAED.

How Many Variables?

A slideshow presentation allows the flexibility to present a slide-pack of SAEDs with different variables. If these are located in the same place on each slide this simulates the adding (or removal) or different variables between consecutive slides. In this way, a more complex story can be told than with a static diagram. **Figure 5** shows a number of interesting additional variables (low-carbon electricity supply and primary energy) that are simple and clear to show during a presentation, however, a static diagram risks becoming too complex, and losing the impact that it may have had with fewer variables. The choice is sometimes difficult, as it not only depends on the reason why a diagram has been created, but the space available to display it and the intended audience too.

How Precise Does the Diagram Need To Be?

It is not the intention of a SAED to be a power flow diagram, as the underlying data may themselves not be precise. However,

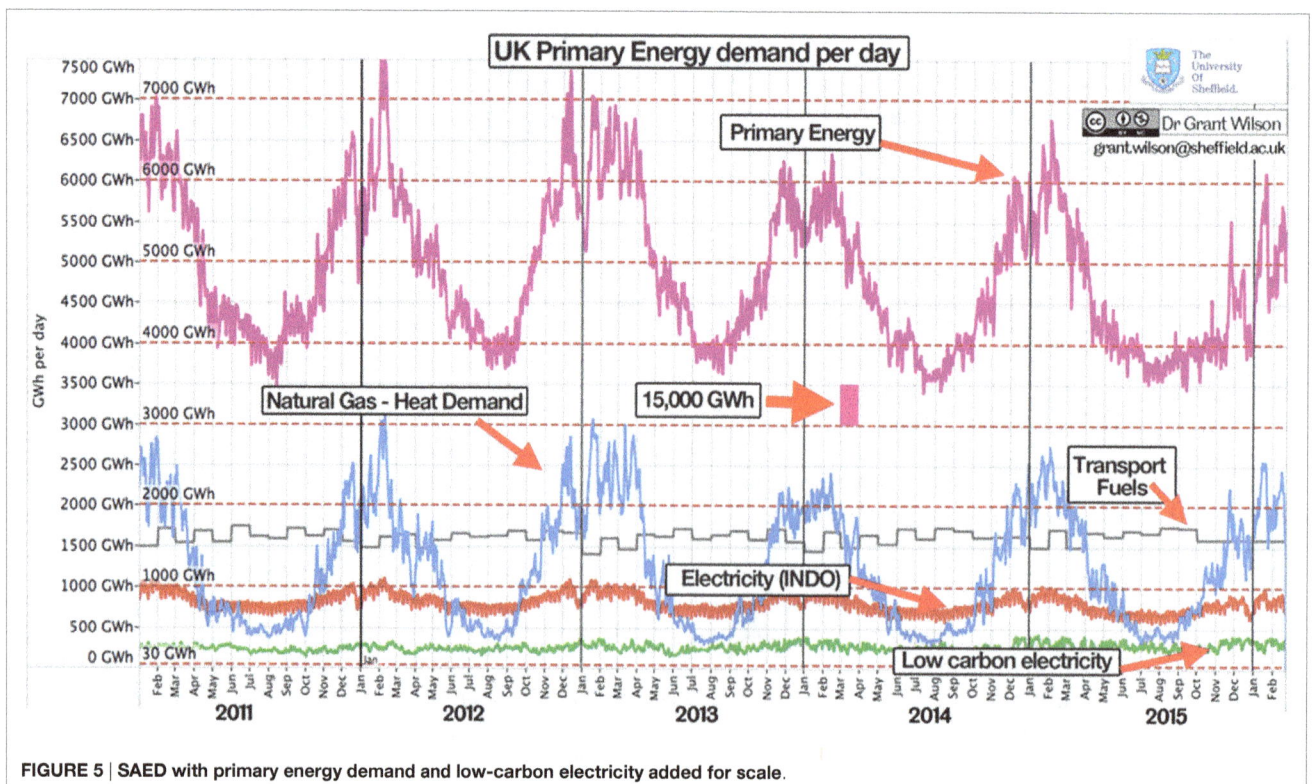

FIGURE 5 | SAED with primary energy demand and low-carbon electricity added for scale.

in order to compare the scale and variation of several supplies and demands against each other, the diagram needs to be correct in terms of matching the time-series axis. Given the lack of detailed data on certain demands, inferences are likely to be made from data that are available. This is felt to be entirely suitable, as it flags up areas that would require further detail to be more robust, but at the same time allows some insights to be gained with existing data.

Gaps are an inevitable part of energy systems data sets, and should be dealt with in a common sense manner, e.g., choosing to use the previous values in a data set, or choosing to use a linear fit between two data points.

CONCLUSION

The hypothesis that energy visualization diagrams commonly used need additional changes to continue to be relevant is brought about by the changing nature of the energy itself that the diagrams seek to show. As energy systems incorporate greater amounts of variable renewable generation from wind and solar sources, the resolution of the time-series data requires to be improved.

The SAED is an important addition to the toolbox of energy visualization for a number of reasons. It is a simple and readily understood visualization to compare various supplies and demands of an energy system, and it helps to promote whole systems thinking to consider demands other than electrical demands. A SAED can help to put a scale on the amount of flexibility required in the energy system over an inter-seasonal basis, which is something that Sankey diagrams are not designed to do. This scale is important to help frame the different infrastructure, balancing (Elliott, 2016), storage (Strbac et al., 2016), and flexibility (Pfenninger and Keirstead, 2015) options for energy systems.

The creation of a SAED clearly depends on access to the underlying data, and, if this is not available in sufficient detail to allow even a daily visualization to be created, this would be indicative of an area that should require attention and discussion with data providers.

Future energy systems are going to become more and more reliant on primary electricity harvested through variable renewable generators powered by the wind and the sun, i.e., weather-dependent renewable generation. As now, future electrical systems will still require the demand and supply to be balanced on a sub-second level. So if an energy system is currently unable to understand its whole system demands on at least a daily level, then this would indicate an area that it should certainly look to resolve.

The creation of a SAED with a daily level of detail, therefore, provides a very simple test to see whether an energy system indeed has this level of detail, or requires a change to its data-collection strategies to allow it to better prepare for future energy system challenges with a greater evidence base.

If SAEDs with a daily level of detail were able to be created for a range of differing local, national, and regional energy systems, then, this would help to embed whole systems thinking in energy system decision-making. The more that people know how to create and understand this diagram the better, from a whole systems analysis point of view. It would also provide a wider understanding of some of the network challenges that appear when primary energy sources move from fossil fuels (with their intrinsic storage of energy) to renewable electricity sources that require additional flexibility options to match supply and demand, especially over seasonal timeframes.

Part of the process of understanding energy systems in a whole systems manner is aided by the presentation of energy data on a SAED. This novel diagram has been utilized more and more in Great Britain to help make this whole systems viewpoint – to allow a better understanding of the scale and seasonality of the different energy vectors and demands in Great Britain. Understanding the recent historical energy demands and supplies also helps to frame a wider appreciation of the energy challenges of a particular system, and the SAED is a useful addition to the typical energy flow (Sankey) diagrams used by policy makers to understand and present energy systems.

If one considers that future energy systems will have to provide enough energy to end users for their final energy demand at the right time, then one can start to understand the benefit of considering an energy system as an energy system, not just as the electrical system or the transport fuels system. Historically the planning of electrical, natural gas, and transport fuel systems has been largely separate as these have themselves been based on separate fuels. In future, the move to provide greater levels of primary energy from variable renewable energy sources means that more primary energy will be in the form of primary electricity. The problem with electricity, however, is that it is an expensive form of energy to store, it is, therefore, usually turned into something else that is cheaper to store (which has an accompanying energy penalty). The sheer scale of inter-seasonal storage becomes clearer with a daily SAED and suggests an ongoing role for fuels of some sort in Great Britain to provide the Terawatt hours required for inter-seasonal storage.

Whole systems thinking is a prerequisite to whole systems planning, which is an important step in the transition to low-carbon energy systems. The SAED is a useful diagram to developing thinking in a whole systems manner and it is hoped will, therefore, become a more common method to display energy data alongside existing energy data visualizations. The insights that these range of diagrams provide will continue to be useful for policy makers, industry, and customers, as energy systems undergo profound changes in order to limit greenhouse gas emissions. The diagrams help policy makers and industry to find a common language to inform legislation.

AUTHOR CONTRIBUTIONS

The author confirms being the sole contributor of this work and approved it for publication.

FUNDING

This work was funded by grant EP/K007947/1 – CO$_2$Chem Grand Challenge Network by EPSRC in the United Kingdom.

REFERENCES

Anable, J., Brand, C., Tran, M., and Eyre, N. (2012). Modelling transport energy demand: a socio-technical approach. *Energy Policy* 41, 125–138. doi:10.1016/j.enpol.2010.08.020

Barrett, J., Peters, G., Wiedmann, T., Scott, K., Lenzen, M., Roelich, K., et al. (2013). Consumption-based GHG emission accounting: a UK case study. *Clim. Policy* 13, 451–470. doi:10.1080/14693062.2013.788858

Bauer, P., Thorpe, A., and Brunet, G. (2015). The quiet revolution of numerical weather prediction. *Nature* 525, 47–55. doi:10.1038/nature14956

BP. (2016). Available at: https://www.bp.com/content/dam/bp/pdf/energy-economics/statistical-review-2016/bp-statistical-review-of-world-energy-2016-full-report.pdf

Bridge, G., Bouzarovski, S., Bradshaw, M., and Eyre, N. (2013). Geographies of energy transition: space, place and the low-carbon economy. *Energy Policy* 53, 331–340. doi:10.1016/j.enpol.2012.10.066

Castillo, A., and Gayme, D. F. (2014). Grid-scale energy storage applications in renewable energy integration: a survey. *Energy Convers. Manag.* 87, 885–894. doi:10.1016/j.enconman.2014.07.063

Cockroft, J., and Kelly, N. (2006). A comparative assessment of future heat and power sources for the UK domestic sector. *Energy Convers. Manag.* 47, 2349–2360. doi:10.1016/j.enconman.2005.11.021

Deane, J. P., Gallachóir, B. P. Ó, and McKeogh, E. J. (2010). Techno-economic review of existing and new pumped hydro energy storage plant. *Renewable Sustainable Energy Rev.* 14, 1293–1302. doi:10.1016/j.rser.2009.11.015

DECC. (2016). *Deliveries of Petroleum Products for Inland Consumptions (ET 3.13)*. Available at: https://www.gov.uk/government/statistics/oil-and-oil-products-section-3-energy-trends

DECC. (2014). *Energy Flow Diagram*. Available at: https://www.gov.uk/government/statistics/energy-flow-chart-2014

Elliott, D. (2016). A balancing act for renewables. *Nat. Energy* 1, 15003. doi:10.1038/nenergy.2015.3

Friendly, M. (2008). The golden age of statistical graphics on JSTOR. *Stat. Sci.* 23, 502–535. doi:10.2307/20697655

Götz, M., Lefebvreb, J., Mörsa, F., McDaniel Kocha, A., Grafa, F., Bajohrb, S., et al. (2016). Renewable power-to-gas: a technological and economic review. *Renew. Energy* 85, 1371–1390. doi:10.1016/j.renene.2015.07.066

Hawkes, A., Staffell, I., Brett, D., and Brandon, N. (2009). Fuel cells for micro-combined heat and power generation. *Energy Environ. Sci.* 2, 729–744. doi:10.1039/b902222h

Hittinger, E., and Lueken, R. (2015). Is inexpensive natural gas hindering the grid energy storage industry? *Energy Policy* 87, 140–152. doi:10.1016/j.enpol.2015.08.036

Hittinger, E., Whitacre, J. F., and Apt, J. (2012). What properties of grid energy storage are most valuable? *J. Power Sources* 206, 436–449. doi:10.1016/j.jpowsour.2011.12.003

IMechE. (2014). Available at: https://www.imeche.org/docs/default-source/reports/imeche-energy-storage-report.pdf

IMechE. (2015). Available at: https://www.imeche.org/docs/default-source/reports/imeche-heat-report.pdf

IRENA. (2016). *Renewable Energy Statistics 2016*. Abu Dhabi: The International Renewable Energy Agency.

National Grid. (2016a). *National Grid Data Item Explorer*. Available at: http://www2.nationalgrid.com/data-item-explorer/

National Grid. (2016b). *Metered Half Hourly Electricity Demands Data*. Available at: http://www2.nationalgrid.com/UK/Industry-information/Electricity-transmission-operational-data/Data-explorer/

Newborough, M., and Augood, P. (1999). Demand-side management opportunities for the UK domestic sector. *IEE Proc. Generat. Transm. Distrib.* 146, 283–293. doi:10.1049/ip-gtd:19990318

Pfenninger, S., and Keirstead, J. (2015). Renewables, nuclear, or fossil fuels? Scenarios for Great Britain's power system considering costs, emissions and energy security. *Appl. Energy* 152, 83–93. doi:10.1016/j.apenergy.2015.04.102

Rubin, E. S., Azevedo, I. M. L., Jaramillo, P., and Yeh, S. (2015). A review of learning rates for electricity supply technologies. *Energy Policy* 86, 198–218. doi:10.1016/j.enpol.2015.06.011

Sakai, M., and Barrett, J. (2016). Border carbon adjustments: addressing emissions embodied in trade. *Energy Policy* 92, 102–110. doi:10.1016/j.enpol.2016.01.038

Schiebahn, S., Grube, T., Robinius, M., Tietze, V., Kumar, B., and Stolten, D. (2015). Power to gas: technological overview, systems analysis and economic assessment for a case study in Germany. *Int. J. Hydrogen Energy* 40, 4285–4294. doi:10.1016/j.ijhydene.2015.01.123

Schmidt, M. (2008). The Sankey diagram in energy and material flow management. *J. Ind. Ecol.* 12, 82–94. doi:10.1111/j.1530-9290.2008.00015.x

Scottish Government. (2015). *Energy in Scotland 2015*. Available at: http://www.gov.scot/Resource/0046/00469235.pdf

Scottish Government. (2016). *Energy in Scotland 2016*. Available at: http://www.gov.scot/Topics/Statistics/Browse/Business/Energy/EIS2016

Slingo, J., Belcher, S., Scaife, A., McCarthy, M., Saulter, A., McBeath, K., et al. (2014). *The Recent Storms and Floods in the UK*. Available at: http://www.metoffice.gov.uk/media/pdf/1/2/Recent_Storms_Briefing_Final_SLR_20140211.pdf

Spence, I. (2005). No humble pie: the origins and usage of a statistical chart. *J. Educ. Behav. Stat.* 30, 353–368. doi:10.3102/10769986030004353

Strbac, G., Konstantelos, I., Pollitt, M., and Green, R. (2016). *Delivering a Future-Proof Energy Infrastructure*. Imperial College London and University of Cambridge Energy Policy Research Group. Available at: https://www.gov.uk/government/uploads/system/uploads/attachment_data/file/507256/Future-proof_energy_infrastructure_Imp_Cam_Feb_2016.pdf

UNFCCC. (2015). *Paris Agreement English*. 1–32. Available at: http://unfccc.int/resource/docs/2015/cop21/eng/l09r01.pdf

United Nations. (2016). *International Recommendations for Energy Statistics, Series M No. 93*. Available at: http://unstats.un.org/unsd/energy/ires/

Wilson, I. (2016). *Energy-charts dot org – Charting Great Britain's Energy Transition*. Available at: http://www.energy-charts.org

Wilson, I. A. G., McGregor, P. G., and Hall, P. J. (2010). Energy storage in the UK electrical network: estimation of the scale and review of technology options. *Energy Policy* 38, 4099–4106. doi:10.1016/j.enpol.2010.03.036

Wilson, I. A. G., Rennie, A. J. R., and Hall, P. J. (2014). Great Britain's energy vectors and transmission level energy storage. *Energy Proc.* 62, 619–628. doi:10.1016/j.egypro.2014.12.425

Wilson, I. A. G., Renniea, A. J. R., Dingb, Y., Eamesc, P. C., Halla, P. J., and Kelly, N. J. (2013a). Historical daily gas and electrical energy flows through Great Britain's transmission networks and the decarbonisation of domestic heat. *Energy Policy* 61, 301–305. doi:10.1016/j.enpol.2013.05.110

Wilson, I. A. G., Rennie, A. J. R., Hall, P. J., and Kelly, N. (2013b). A daily representation of Great Britain's energy vectors: natural gas, electricity and transport fuels. *ISEREE 2013*. Available at: http://www.iner.gob.ec/wp-content/uploads/downloads/2015/06/ISEREE_A-daily-representation-of-Great-Britains-energy-vectors.pdf

Conflict of Interest Statement: The author declares that the research was conducted in the absence of any commercial or financial relationships that could be construed as a potential conflict of interest.

18

Numerical Simulation of Simultaneous Electrostatic Precipitation and Trace Gas Adsorption: Electrohydrodynamic Effects

Herek L. Clack*

Department of Civil and Environmental Engineering, University of Michigan, Ann Arbor, MI, USA

Edited by:
Kalpit V. Shah,
RMIT University, Australia

Reviewed by:
Milinkumar Shah,
Curtin University, Australia
Laltu Chandra,
Indian Institute of Technology
Jodhpur, India

*Correspondence:
Herek L. Clack
hclack@umich.edu

Specialty section:
This article was submitted to
Advanced Fossil Fuel Technologies,
a section of the journal
Frontiers in Energy Research

Citation:
Clack HL (2017) Numerical
Simulation of Simultaneous
Electrostatic Precipitation and Trace
Gas Adsorption:
Electrohydrodynamic Effects.
Front. Energy Res. 5:3.

Electrostatic precipitators (ESPs) are now being tasked with simultaneously removing particulate matter (PM) and trace gas-phase pollutants such as mercury released during coal combustion. This represents a significant expansion of their original operational mission, one which is not captured by decades old quasi-1-D analytical expressions developed from first principles for predicting PM removal alone. At the same time, technological advances in ESP power supplies have led to steady increases over the years in the applied voltage achievable in new or refurbished ESPs. In light of these industry trends, the present study extends our previous study to examine the multiphase flow phenomena that may occur during such ESP operations, specifically the effects of electrohydrodynamic (EHD) fluid flow phenomena that can emerge when electrical current densities are high and/or fluid velocities are low. The results show good agreement at low current densities between the present numerical simulation results and ESP performance predictions obtained from classical analytical expressions, with increasing divergence in predicted performance at higher current densities. Under the influence of EHD phenomena, the acceleration of the fluid by electric body forces effectively increases average fluid velocities through the ESP channel with a commiserate reduction in PM removal efficiency. The impact on trace gas-phase pollutant removal is mixed, with EHD phenomena found to variously promote or inhibit gas-phase pollutant removal.

Keywords: electrohydrodynamics, electrostatic precipitator, ESP, particulate matter, mercury, activated carbon

INTRODUCTION

Over decades of use in controlling particulate emissions, electrostatic precipitators (ESPs) have been proven to be robust devices with electronic control systems that often operate virtually independently with minimal human intervention. Especially for well-defined processes for which particulate emissions vary little over time, such as stationary power generation, well-designed ESPs have operated reliably for decades after initial installation, typically undergoing periodic repair and maintenance or receiving power supply upgrades. When significant changes in particulate control performance have been needed, often due to changes in particle electrical properties, flue gas conditioning additives such as sulfur trioxide, and less commonly ammonia, have been injected upstream of ESPs (Shanthakumar et al., 2008), altering particle electrical properties and improving the particulate matter (PM) removal performance of the device. Operating temperature has long been known to affect ESP performance, and the perceived advantages of high-temperature electrostatic largely drove the installation of a number of so-called hot-side ESPs (Calvert and Englund, 1984), and researchers continue to study how elevated gas temperatures may be used to mitigate ESP performance fluctuations due to variable coal quality (Noda and Makino, 2010). The development of combustion systems capable of postcombustion carbon capture has also driven studies of ESP operation during oxy-fueled combustion (Han et al., 2010; Kim et al., 2014). Increasingly, ESPs have been the focus not for PM emissions but rather for simultaneously controlling the emissions of a second pollutant. Upstream injection of lime, limestone, or trona to neutralize acid gases and powdered mercury sorbents such as powdered activated carbon (PAC) to adsorb toxic trace metals such as mercury has grown in the United States as new and more stringent regulations take effect.

Wet and dry ESPs operate on the same basic particle collection principles, but differ in how they handle the bulk collected dust. Dry ESPs periodically mechanically or acoustically dislodge the bulk collected dust (dustcake) from the collection electrodes, causing it to fall by gravity into collection hoppers below. Wet ESPs direct a flow of water, either continuously or intermittently, over the surfaces of the collection electrodes, thereby capturing particles in the falling liquid sheet that are extracted from the liquid in a separate process. Because of the added complexity of the separate liquid handling system and the liquid waste stream produced by wet ESPs, their main use has been in controlling liquid mists or sticky particulates, explosive aerosol suspensions, and as mist recovery processes in sulfuric acid production (Cooper and Alley, 2011; Seetharama et al., 2013). The use of ESPs for two-way (or even three-way) air pollutant emissions control is advancing. Compared to dry ESPs, the lower operating temperatures and elimination of particle resuspension into the flue gas during mechanical rapping offered by wet ESPs increase the removal efficiencies of fine particles (PM$_{2.5}$) (Cooper and Alley, 2011). Because of their high surface area per unit mass, these smallest particles are also most likely to harbor compounds condensed from the flue gas such as heavy metals (Seames and Wendt, 2000) including mercury (Reynolds, 2004; Seetharama et al., 2013), providing modest, incidental removal of such semivolatile pollutants. In addition to higher PM$_{2.5}$ removal efficiency, wet ESPs are also superior to dry ESPs at removing liquid particles, droplets, and mists (Cooper and Alley, 2011). Although flue gas from coal combustion nominally contains no condensate, the moist, low-temperature operating environment within wet ESPs promotes condensation of acid gases such as sulfuric acid as temperatures approach the acid dew point of the flue gas. In this way, wet ESPs act to reduce acid gas emissions along with PM emissions.

Dry ESPs used for two- or three-way air pollutant emissions control have made advances largely through full-scale field tests and extrapolating from decades of ESP operating experience. Given the dominance of dry ESPs in use at electric utilities

in the United States and elsewhere, the remaining discussion focuses on this primary population of devices; references to ESPs beyond this point imply dry ESPs. The reliance on ESP PM removal principles to interpret trace gaseous pollutant emissions reductions observed from ESP field tests, however, can lead to erroneous inferences of causal relationships. For example, ESP PM removal efficiency is understood to increase with increasing collection electrode area; however, mass transfer limitations prevent the accumulated dustcake on collection electrodes from contributing significantly to the overall reduction in trace pollutant concentrations across an ESP, thereby greatly weakening the influence of collection electrode area on trace pollutant removal. Suspended adsorbent particles in the flue gas contribute to a greater degree to trace pollutant removal in ESPs of conventional design. Even the implicit assumption that multiple mechanisms of trace pollutant adsorption can be dealt with in an additive fashion is questionable. Results from a recent numerical simulation of trace mercury adsorption by suspended aerosols and the accumulated dustcake on the collection electrodes of an ESP channel (Clack, 2015) indicate that the mercury-lean boundary layers that develop over the adsorbent dustcake suppress adsorption by suspended aerosols as they drift toward the collection electrodes. Because all aerosols must traverse these boundary layers before they are collected on the plate electrodes, this suppression has an outsized impact on in-flight adsorption. In the extreme case of collection of electrodes covered by a highly adsorbent dustcake layer, the deeply mercury-lean concentration boundary layers that are produced actually result in *lower* mercury removal by the combined in-flight and wall-bounded mechanisms that would be predicted by summing the separate contributions of the two adsorption mechanisms.

Nevertheless, anecdotal evidence from full-scale field tests suggests that larger ESPs as measured by space velocity broadly tend to exhibit higher levels of mercury removal. While increased space velocity can reflect greater collection electrode, and thus dustcake, surface area, it can also reflect reduced flue gas velocities. Reduced gas velocity increases the potential for electrohydrodynamic (EHD) phenomena to occur within ESP channels by reducing fluid inertial forces compared to EHD forces, the ratio of which is represented by EHD number or Masuda number and which can be used to predict the onset of prominent EHD phenomena and flow features (IEEE-DEIS-EHD Technical Committee, 2003). Study of EHD phenomena in principle began in the 1980s (Leonard et al., 1983; Yamamoto and Sparks, 1986; Kallio and Stock, 1992) with detailed simulations (Zhao and Adamiak, 2008) and experimental measurements (Zouzou et al., 2011) arising more recently that focus on conventional ESPs and PM control. This study seeks to shed light on how EHD phenomena might affect, either incidentally or intentionally, the performance of ESPs operated as multipollutant collectors that simultaneously remove PM and trace toxic metals from combustion flue gas. Considerations of EHD effects have particular relevance given operating and design conventions that call for increases in ESP-specific collection area (SCA) and/or power supply upgrades to improve PM collection efficiency; both actions have the effect of increasing the relative influence of electric body forces and the relative extent in the flow field of EHD effects.

MATERIALS AND METHODS

The methodology of the numerical simulation has been presented in detail previously (Clack, 2015) and thus will be summarized here. Solved are the steady conservation equations for the electric and fluid fields (including EHD forces); charged-induced particle motion; and associated convective gas-particle mass transfer rates. There are two 2-D computational domains representing partial segments of a complete ESP channel: a 3-wire segment measuring 2 m × 0.3 m (L × W) or a 9-wire segment measuring 5 m × 0.3 m, each containing 1 mm diameter wire discharge electrodes spaced 0.5 m apart. **Figure 1** presents a schematic view of the 3-wire segment; the 9-wire segment is longer but of similar design to the 3-wire segment, both having the same spacing between the wire discharge electrodes and the same distance separating the inlet from the first electrode and the last electrode from the outlet.

Electric Field

The electric field, assumed to depend only on the continuous phase fluid properties, derives from Eq. 1 for the electric potential:

$$-\nabla\Phi = \vec{E} \tag{1}$$

where Φ is voltage [V] and $\vec{E}$ is the electric field vector [V/m]. The solution to Eq. 1 must also satisfy Poisson's equation (Eq. 2) and current continuity (Eq. 3) assuming Fickian ion diffusion is neglected compared to fluid and electric field-driven ion advection:

$$\nabla^2\Phi = -\frac{q_i}{\varepsilon} \tag{2}$$

$$\nabla \bullet \left(\left(b_i\nabla\Phi + \vec{u}\right)q_i + \alpha\nabla q_i\right) = 0 \tag{3}$$

where q_i is the local space charge density [C/m³], ε is the electrical permittivity of the fluid, b_i is the ionic mobility in air (1.6e−4 [m²/V − s]), α is the ion diffusivity coefficient, and $\vec{u}$ is the fluid velocity vector. The space charge density distribution $q_i(x,y)$ that

constrains $\Phi(x,y)$ such that $\nabla q_i(x,y)$ and $\nabla\Phi(x,y)$ satisfy Eqs 2 and 3 is determined through repeated, iterative solutions of the electric field, as described previously (Clack, 2015). The discharge corona constitutes a negligible fraction of the computational domain, and thus, it is not numerically simulated.

Fluid/Continuous Phase

The incompressible Reynolds-averaged Navier–Stokes equations (Eqs 4 and 5) modified to include a term representing the electric body force ($q_i\nabla\Phi$) are as follows:

$$\rho\left(\nabla \bullet \vec{U}\right) = 0 \tag{4}$$

$$\rho\left(\vec{U} \bullet \nabla\vec{U}\right) = -\nabla P + \mu\nabla^2\vec{U} + q_i\nabla\Phi \tag{5}$$

In Eqs 4 and 5, $\vec{U}$ is local gas velocity vector [m/s], P is local pressure [Pa], ρ is gas density [kg/m³], μ is the gas dynamic viscosity [Pa − s], q_i is local space charge density [C/m³], and Φ is local electric potential [V]. The material property database that is native in COMSOL includes the temperature dependency of fluid properties and the native $k - \varepsilon$ turbulence model; its default parameters are used without modification.

Particle/Dispersed Phase

Particle volume fraction ϕ and number density ND_p are treated as scalar quantities, spatially distributed in the fluid flow, i.e., $\phi(x,y)$ and $ND_p(x,y)$. Both because particle charging and particle dynamic motion are highly dependent on particle size, both quantities must be tracked by particle size, i.e., $\phi(x,y,d_p)$ and $ND_p(x,y,d_p)$. Considered here are two particle size distributions entering each ESP channel segment: a log-normal distribution and an algebraic representation of a measured size distribution reported previously (Clack, 2015). **Figure 2** compares the two size distributions and **Table 1**. The two particle size distributions are discretized into 11 size bins ($1.5 < d_p < 125$ μm; see **Table 1**) for the log-normal distribution and 10 size bins ($1.5 < d_p < 85$ μm) for the measured size distribution. Particles are assumed to be PAC, the measured size distribution being that of Norit FGD (Prabhu et al., 2012).

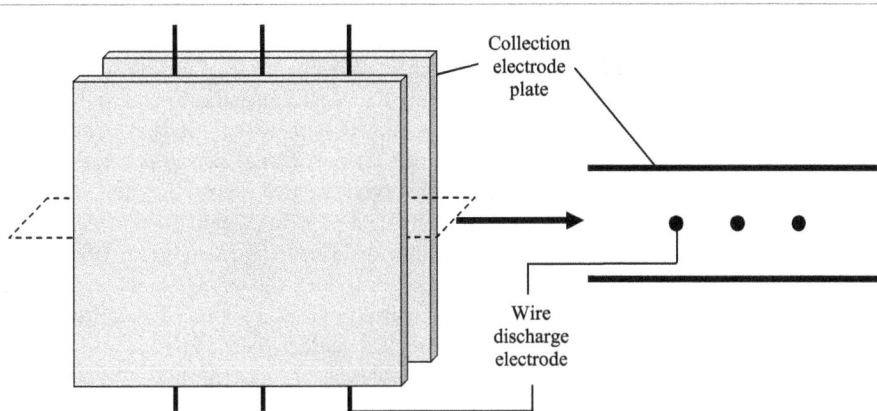

FIGURE 1 | Left: Schematic of a portion of wire-plate electrostatic precipitators in typical electrode arrangement, showing 3-wire channel segment. Right: Corresponding schematic of 2-D computational domain. Schematic not to drawn to scale.

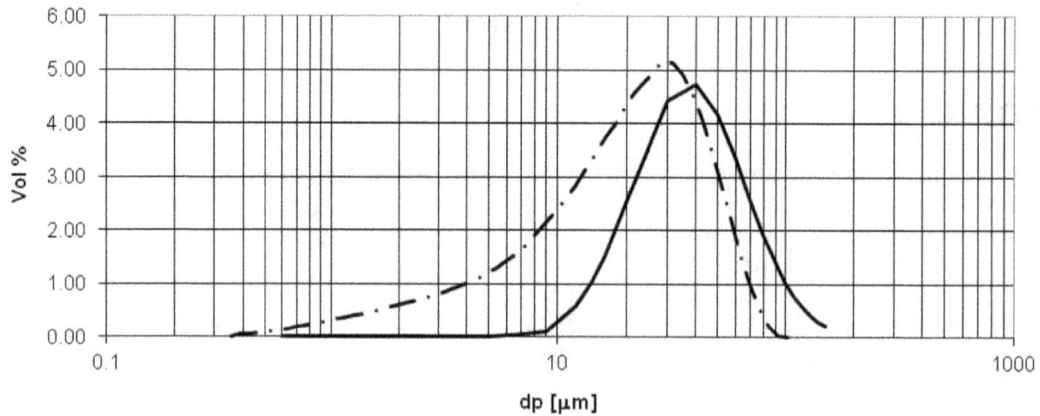

FIGURE 2 | Particle size distributions assumed in present analysis. Solid line: Log-normal particle size distribution, $\overline{d_p}$ = 20 μm, σ_g = 1.75. Dashed line: Measured size distribution (Prabhu et al., 2012) of FGD powdered activated carbon product, d_{50} = 17.32 μm.

TABLE 1 | Boundary conditions, by physical phenomenon.

Physics (COMSOL model)	Inlet BC	Outlet BC	Collection Electrode BC	Discharge Electrode BC	Properties[a]
Electric field (.es)	$D = \varepsilon\nabla\Phi = 0$	$D = \varepsilon\nabla\Phi = 0$	$\Phi = 0$	$\Phi = -50$ kV $q_i = -A(R/r)^n$	3DE: $A \sim 1.8 \times 10^{-3}$ C/m³ $n \sim 1.723$ 9DE: $A \sim 2.2 \times 10^{-2}$ C/m³ $n \sim 1.8$ Current density[a]: 0.89 mA/m²
Fluid/continuous phase (.spf2)	Dry air $T = 180°C$ $P = 101.325$ kPa $U_0 = 0.2$ m/s	Open boundary, $P = 95$ kPa	Solid boundary, no slip	Solid boundary, no slip	
Particle/dispersed phase (.mm)	Particle mass loading of 0.1 g/m³, a result of specified particle volume fraction flux ($\phi'' = U_0\phi$) and number density flux $\left(ND_p'' = U_0ND_p\right)$	Specified particle volume fraction flux ($\phi'' = u\phi$), number density flux $\left(ND_p'' = uND_p\right)$	Specified particle volume fraction flux ($\phi'' = u\phi$), number density flux $\left(ND_p'' = uND_p\right)$	No particle flux, particle bounce	$\rho_p = 0.51$ g/cc $\varepsilon = 4$ Log-normal distribution ($\overline{d_p}$ = 20 μm σ_g = 1.75) or skewed distribution, d_{50} = 17.32 μm
Transport of dilute species (.chds)	Specified species flux $\left(N_i' = U_0C_0\right)$	Specified species flux $\left(N_i' = uC\right)$	Specified species concentration ($C = 0$)	No species flux $\left(N_i' = 0\right)$	$C_0 = 4 \times 10^{-7}$ mol/m³ MW = 201 g/gmol $D_{ab} = 3.4 \times 10^{-5}$ m²/s

[a]Properties corresponding to current density of 0.11 mA/m² as specified in the study by Clack (2015).

Although in practice PAC along with mineral fly ash is suspended, the flue gas entering an ESP, under most circumstances, the contribution of fly ash to adsorbed mercury is negligible, as discussed previously (Clack, 2015). Particle–particle interactions and two-way particle-fluid coupling are beyond the scope of this analysis. The boundary condition for particles intercepting the wire discharge electrodes is an elastic bounce, while particles intercepting the planar collection electrodes are removed from the computational domain permanently. For particles of the size considered here (<150 μm), displacements due to gravitational acceleration are negligible compared to that of fluid drag and Coulombic (charge-driven) forces.

Solutions for the size-dependent spatial distributions of particle number density $ND_p(x,y,d_p)$ and particle volume fraction $\phi(x,y,d_p)$ for all size bins are computed simultaneously to accurately render

their collective rates of gas-particle mass transfer and influence on the local concentration of the gas-phase pollutant $C(x,y)$, assumed here to be mercury. The gas-particle mass transfer rate for particles of size d_p is a function of the relative velocity between the two phases, i.e., the particle slip velocity $\vec{U}_{slip}$. For charged particles within an electric field, EHD phenomena can introduce strong velocity gradients in the fluid and Coulombic forces induce slip velocities between the fluid and the particle typically at least an order of magnitude greater than gravitational settling and spatially varying. Thus, inclusion of such effects in calculating gas-particle mass transfer represents an important advance beyond previously reported Deutsch-Anderson (D-A)-based analyses (Clack, 2006a,b, 2009).

At the inlet boundary, the two-phase gas-particle flow enters the computational domain with a uniform specified gas (continuous phase) velocity, zero particle slip velocity, and a particle mass loading concentration of 0.1 g/m³, the equivalent of 6 lbs/MMacf (pounds of PAC per million actual cubic feet of flue gas). Trajectories of particles of size d_p are governed by the x- and y-components of $\vec{U}_{\text{slip}}$, u_{slip}, and v_{slip}, representing the balance between viscous and Coulombic forces on the particles:

$$u_{\text{slip}} = \frac{N_e(-e)E_x C_c}{3\pi\mu d_p} \tag{6}$$

$$v_{\text{slip}} = \frac{N_e(-e)E_y C_c}{3\pi\mu d_p} \tag{7}$$

where N_e is the number of elementary charges on each particle, e is the elementary charge of an electron (1.6E−19 C), E_x and E_y are the x- and y-components of the electric field vector $\vec{E}$, respectively, and μ and d_p are as previously defined. C_c is the Cunningham slip correction factor:

$$C_c = 1 + \text{Kn}\left[1.257 + 0.4\left(\exp\left(\frac{-1.1}{\text{Kn}}\right)\right)\right] \tag{8}$$

where Kn, Knudsen number, is defined as λ/d_p and where λ is the gas mean free path evaluated identically as our previous analyses (Clack, 2006a,b). As described previously (Clack, 2015), particle charging occurs very rapidly under typical ESP conditions. Given the computational expense that would be required in using a Lagrangian approach to explicitly calculate particle charge as a function particle position for ~10⁹ particles, assigning a constant saturation charge for each particle size class that represents N_e induced by the average $\vec{E}$ and $q_i(x,y)$ in the domain was previously deemed (Clack, 2015) to be an acceptable compromise between accuracy and computational expense. Because only super-micron size bins are considered in the present analysis, only field charging is considered in the determination of N_e [Eq. 9, from the study by Friedlander (2000)]:

$$N_e = \left[1 + 2\frac{\varepsilon-1}{\varepsilon+2}\right]\frac{Ed_p^2}{4e} \tag{9}$$

where $E = \left|\vec{E}(x,y)\right|$, ε is the particle relative permittivity (dielectric constant) [-], approximated as that of graphite for PAC, and d_p and e are as defined previously.

Eq. 10 (Frössling equation) is a correlation between Reynolds number based on slip velocity and the Sherwood number, a non-dimensional mass transfer parameter the definition of which can be used to solve for the mean convective mass transfer coefficient between a flowing fluid and a spherical particle $(\overline{h_m})$:

$$\overline{Sh_d} \equiv \frac{\overline{h_m}d_p}{D_{ab}} = 2 + 0.552\left(\frac{\rho\left|\vec{U}_{\text{slip}}\right|d_p}{\mu}\right)^{1/2}\left(\frac{\mu}{\rho D_{ab}}\right)^{1/3} \tag{10}$$

where D_{ab} is the binary mass diffusivity of a species in a dilute mixture, here assumed to be elemental mercury diffusing in air [3.4E−5 m²/s, taken from the study by Clack (2006a,b)], and μ, ρ, and d_p are as defined previously. The spatial distribution of the collective rate of gas-particle mass transfer for all particles is subsequently used to calculate the distribution of mercury concentration.

Adsorption of Trace Mercury

The spatial distribution of mercury concentration $C(x,y)$ [mol/m³] in the domain reflects the inlet mercury concentration, reduced by the collective gas-particle mass transfer of the suspended particles; adsorption occurring due to the collected dustcake of particles on the surface of the collection electrodes (Clack, 2015) is not considered. The mass transfer rate of mercury between the gas and the surface of each particle ($\dot{M}_{Hg}$) is given by:

$$\dot{M}_{Hg} = 4\pi\left(\frac{d_p}{2}\right)^2\overline{h_m}\text{MW}(C - C_s) \tag{11}$$

where C and C_s are the mercury concentrations locally in the gas and at the surface, respectively; all other quantities are as defined previously. Concentration of mercury at the particle surface is assumed to be zero, based on the high reactivity and high adsorption capacity of current-generation PAC products (Clack, 2015). Mercury removal efficiency in all cases is the percentage difference between inlet species flux and outlet species flux for each channel segment.

COMSOL Multiphysics™ has been previously demonstrated to be suitable for simulating gas-particle phenomena within ESPs (Back and Cramsky, 2012; Clack, 2013), and here version 4.4 was used. An example of the computational mesh is presented in **Figure 3**, for the 3-wire channel segment. Grid independence was established from the 9-wire channel segment results in which a nearly twofold increase in the number of computational elements from 24,000 to 47,000 yielded a 0.42% change in the PM₂.₅ particle volume flux at the channel segment outlet (1.089e−16 to 1.0936e−16 m³/s). Computational times on a 64-bit HP Xeon workstation (2 2.93 GHz processors, 6 Gb RAM) were typically ~60 min for the 3-wire channel segment. COMSOL automatically meshes the computational domain based on the physical scale of the computational domain, the boundary conditions applied, and the order of the PDEs being solved.

Table 1 summarizes the physical mechanisms, their governing equations and boundary and initial conditions, and parameter and variable definitions treated by the separate computational modules. The methodology, constraints, and assumptions are similar to those described in the study by Clack (2015) with two exceptions. First, the earlier work considered three trace pollutant removal mechanisms: so-called in-flight (gas-particle) adsorption, wall-bounded adsorption by an adsorbent dustcake layer covering the collection electrodes, and coupling of both in-flight and wall-bounded adsorption. The present work only considers in-flight and wall-bounded trace pollutant adsorption separately. Second, the previous work only considered fluid and electrical conditions (fluid inlet velocity; current density and electric potential) unlikely to produce EHD phenomena. In the present work, fluid and electrical conditions are specifically chosen to promote the onset of EHD phenomena. EHD phenomena become significant for values of EHD number, N_{EHD} (Eq. 12)

$$N_{EHD} = \left(\frac{I_0 L^3}{\rho v^2 b_i A}\right)^{1/2} \tag{12}$$

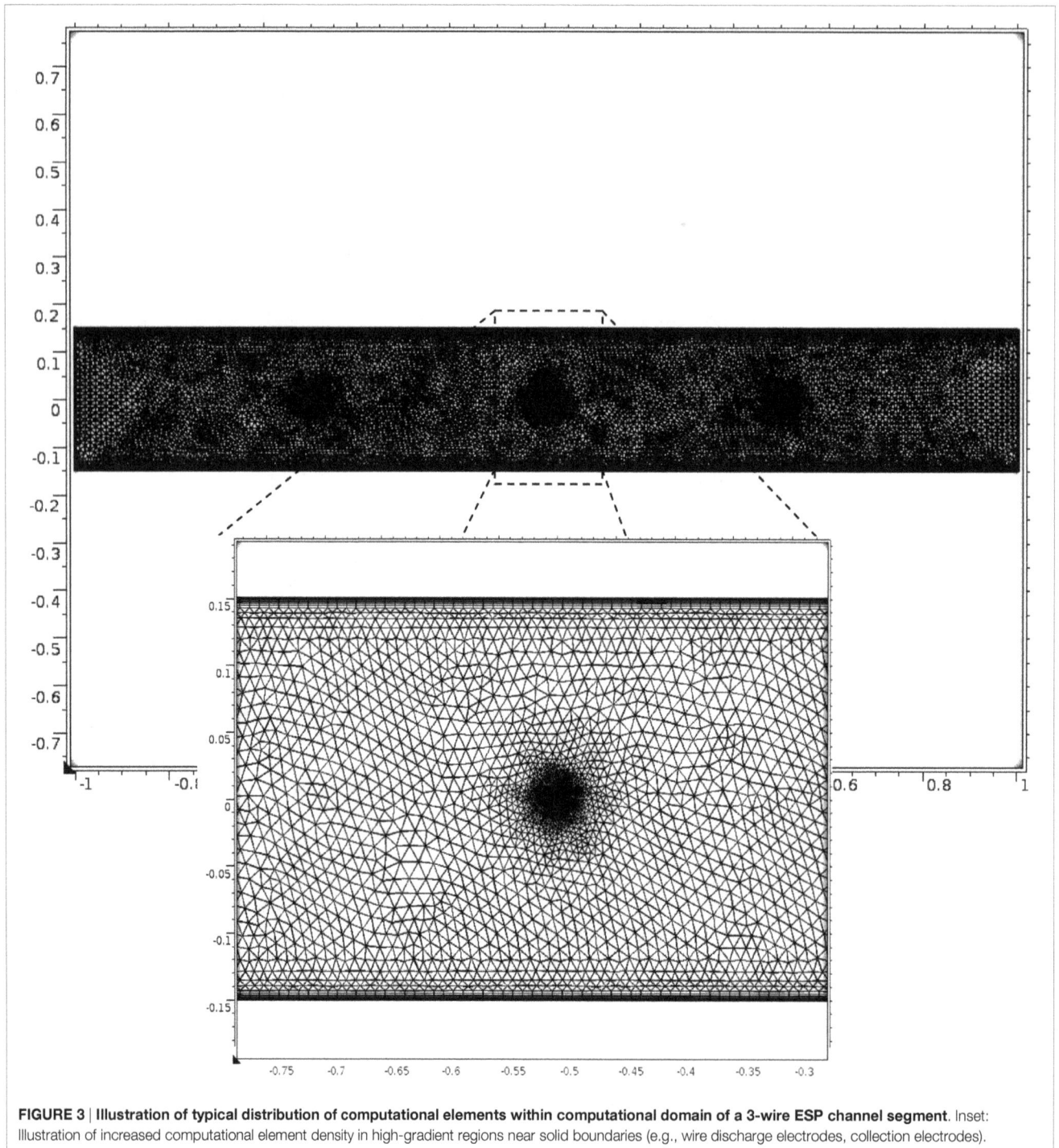

FIGURE 3 | Illustration of typical distribution of computational elements within computational domain of a 3-wire ESP channel segment. Inset: Illustration of increased computational element density in high-gradient regions near solid boundaries (e.g., wire discharge electrodes, collection electrodes).

approximately equal to or exceeding ($\approx>$) the square of the Reynolds number, $Re_L{}^2$ (IEEE-DEIS-EHD Technical Committee, 2003), in which L, A, and I_0 are, respectively, the characteristic dimension, surface area, and electrical current of the relevant component, in this case the wire discharge electrodes. For air, b_i is the ionic mobility (assumed value of 1.6e−4 m^2 V^{-1} s^{-1}) and ν is the kinematic viscosity. For the square of the Reynolds

number, $Re_L{}^2 = U_0{}^2 L^2/\nu^2$, where U_0 is the gas velocity entering the channel segment, $Re_L{}^2 = 154$ for the present analysis. At the higher of the two current densities (0.89 mA/m^2), values of N_{EHD} for the 3-wire ($N_{EHD} = 57.5$) and 9-wire ($N_{EHD} = 125$) ESP channel segments are somewhat lower than the value of $Re_L{}^2$, but by a much smaller margin than in our previous work (Clack, 2015), in which $N_{EHD} = 25.7 < 9,260 = Re_L{}^2$ for current density of 0.11 mA/m^2 and velocity $U_0 = 1.55$ m/s. Computed fluid streamline patterns at 0.89 and 0.11 mA/m^2 (see **Figure 4**, see Results and Discussion) are consistent with the principle that EHD phenomena become increasingly significant as N_{EHD} approaches and exceeds the value of Re$_L^2$.

Total computational times on a 64-bit HP Xeon workstation (2 2.93 GHz processors, 6 Gb RAM) were approximately 60 min for the 3-wire channel segment (13,340 computational elements).

RESULTS AND DISCUSSION

Under high EHD conditions, the major EHD-induced flow feature within both the 3- and 9-wire channel segments for $U_0 = 0.2$ m/s is a repeating diverging-converging fluid flow pattern interspersed between the wire electrodes. **Figure 4** compares the velocity distributions within two 9-wire channel segments under high and low EHD conditions as reflected by their different imposed space charge density distributions and the different current densities (0.89 vs. 0.11 mA/m^2) that result. The higher current density condition produces a more pronounced diverging-converging flow pattern at these velocities; at higher velocities (and thus higher Re$_L^2$), the momentum of the bulk flow dominates, preventing the establishment by the electric body forces of any discernible secondary flows. This repeating diverging-converging flow pattern strongly influences the suspended particle motion as well, given that charge-driven particle drift velocities are several orders of magnitude smaller than gas velocities. The diverging regions of this flow pattern would be expected to enhance PM collection through particle advection toward the collection electrodes, with the greatest benefit apparent in the fine fraction whose charge-driven drift velocities are slowest. Accordingly, diminished PM

collection would be expected in converging regions of the flow that carry particles away from the walls.

PM Removal: Influences of EHD Phenomena

Figure 5 presents six data points for PM$_{2.5}$ collection efficiency obtained from numerical simulations of the log-normal particle suspension: four results representing the conditions of the current analysis and two results from previous analyses (Clack, 2015) at the same electrical condition (0.11 mA/m^2) but a much higher inlet velocity ($U_0 = 1.55$ m/s) that are presented to assist in interpreting the four EHD-influenced results. The four results from the current analysis represent the two ESP channel segment lengths (3 and 9 wires) and the two electrical conditions (0.11 and 0.89 mA/m^2), all for $U_0 = 0.2$ m/s. The higher electrical condition led to higher PM$_{2.5}$ collection efficiency, with greater performance enhancement occurring for the 3-wire geometry (increased from 13 to 90%) than the 9-wire geometry (increased from 84 to 99.6%). Although detailed results for the skewed PSD are not presented, overall trends were similar to those for the log-normal PSD. Generally, PM$_{2.5}$ collection efficiency for the skewed PSD was lower (10.7% under high EHD conditions, increasing to 83.3% in the 9-wire channel segment) than results for the log-normal PSD under the same conditions. Lower PM$_{2.5}$ collection efficiency for the skewed PSD likely reflects its larger fine fraction.

The differences in computed PM$_{2.5}$ collection efficiency attributable to EHD phenomena in **Figure 5** can be only properly interpreted if they are distinguished from effects resulting from changes in other conventional ESP operating parameters. In comparing two electrical conditions, the higher electrical condition would be expected to produce higher PM collection efficiencies in a conventional ESP through more highly charged particles and stronger electric fields. In comparing the 3-wire and 9-wire channel segments having the same inlet gas velocity, the longer of the two channels would be expected to have a higher collection efficiency due to its higher SCA. The 3- and 9-wire channel segments operating at the same current density also have different distributions of space charge density and electric field: The average $|E|$ in the 9-wire channel segment (1.37×10^5 V/m) is about twice that in the 3-wire segment (6.57×10^4 V/m) when both are operating at 0.11 mA/m^2. Even for both 3-wire and 9-wire channel segments energized to 0.11 mA/m^2, the higher E-field of the longer channel segment increases particle saturation charge, particle drift velocities, and rates of PM collection independent of the increased collection surface area and SCA. In light of these considerations, along with the six PM$_{2.5}$ collection efficiency results determined from present and previous simulations, **Figure 5** also presents four continuous collection efficiency curves predicted from D-A analyses of $d_p = 2.5$ μm particles. The four D-A-predicted collection efficiency curves represent the two ESP channel segment lengths (3 and 9 wire) and two values of current density (0.11 and 0.89 mA/m^2). Each combination presents a different particle charging environment because different space charge density values are needed to achieve the same current density in ESP channel segments of different lengths. Each of the four D-A-predicted collection efficiency curves for $d_p = 2.5$ μm

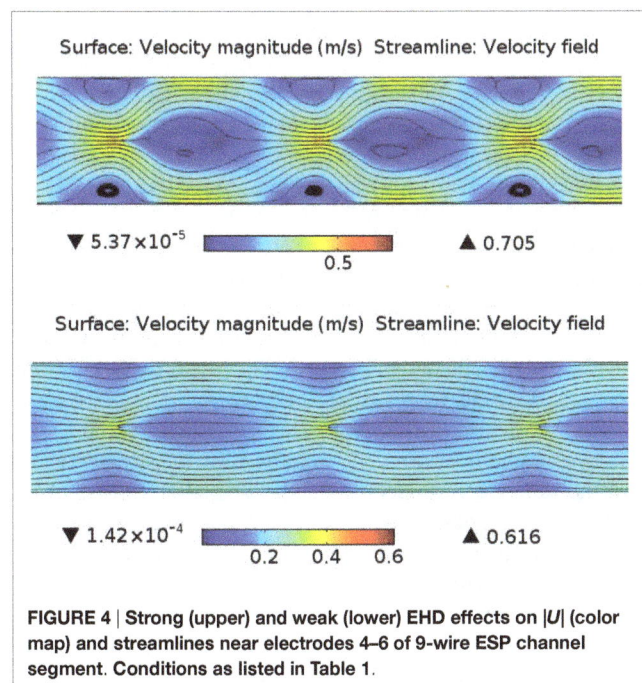

Surface: Velocity magnitude (m/s) Streamline: Velocity field

▼ 5.37×10^{-5} ▲ 0.705
0.5

Surface: Velocity magnitude (m/s) Streamline: Velocity field

▼ 1.42×10^{-4} ▲ 0.616
0.2   0.4   0.6

FIGURE 4 | Strong (upper) and weak (lower) EHD effects on |U| (color map) and streamlines near electrodes 4–6 of 9-wire ESP channel segment. Conditions as listed in Table 1.

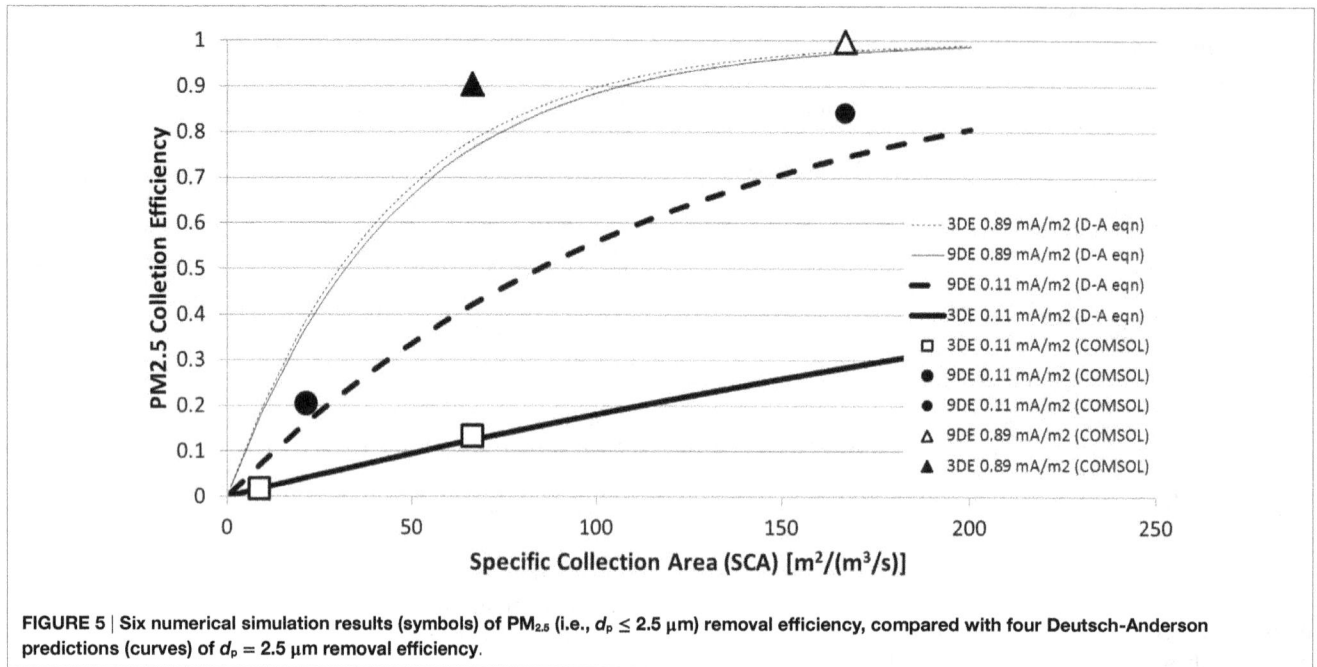

FIGURE 5 | Six numerical simulation results (symbols) of $PM_{2.5}$ (i.e., $d_p \leq 2.5$ µm) removal efficiency, compared with four Deutsch-Anderson predictions (curves) of $d_p = 2.5$ µm removal efficiency.

is based on the spatially averaged transverse drift velocity of all 2.5 µm particles as determined from the corresponding numerical simulation. Because a D-A analysis assumes that both the electric field and particle transverse drift velocity (for a specified d_p) are uniform, the effects of fluid–electric coupling cannot be represented, making it a useful point of comparison both for PM collection efficiency and for interpreting the effects of EHD phenomena.

The two highest collection efficiency curves for $d_p = 2.5$ µm predicted by D-A analyses correspond to the higher current density (0.89 mA/m²) and the lowest collection efficiency curve corresponds to the lower current density (0.11 mA/m²), irrespective of the channel segment length (**Figure 5**). These D-A-predicted results confirm the known relationships between higher current densities and higher particle collection through higher particle saturation charge irrespective of EHD considerations. This is evident when considering the example of a 7- µm particle: assuming the conditions in **Table 1**, increasing current density from 0.11 to 0.89 mA/m² increases its saturation charge on average by more than an order of magnitude (from 112e⁻ to 3810e⁻), with corresponding increases in transverse drift velocity, rates of collection, and overall collection efficiency. Saturation charges on particles of $d_p = 2.5$ µm and their resulting transverse drift velocities are different for the 3-wire and 9-wire geometries energized to 0.11 mA/m², a consequence of the different space charge density and electric field distributions needed to impose the same current density within the two domains having different numbers of discharge electrodes (by a factor of 3). Under these conditions, the combined differences in particle saturation charge and the distribution of the gradient of the electric potential yield average $d_p = 2.5$ µm transverse drift velocities computed from the COMSOL results that differ by a factor of 4

between the 3-wire (0.002 m/s) and 9-wire (0.008 m/s) channel segments at 0.11 mA/m2. These same combined effects yielded more similar transverse drift velocities between the two channel segments at 0.89 mA/m² (0.022 and 0.023 m/s). The impact of these differences on the D-A-predicted removal efficiencies is a more pronounced performance difference between the two channel segments at 0.11 mA/m² than at 0.89 mA/m² for all values of SCA.

Comparing the six $PM_{2.5}$ (i.e., $d_p < 2.5$ µm) numerical simulation results to the D-A-predicted results, there is very good agreement, less than 1% point difference, for the 3-wire channel segment at the lower current density value (0.11 mA/m²) for both values of SCA (8.6 and 66.7 s/m). Under these fluid and electrical conditions, EHD effects are essentially non-existent, and the electric field is not sufficiently strong that spatial variations in particle drift velocities translate into deviations from D-A-predicted collection efficiencies. For the longer 9-wire channel segment at 0.11 mA/m² current density, the increased SCA compared to the 3-wire channel segment leads to correspondingly higher predictions of particle removal efficiency. For equivalent SCA, the numerical simulation results for the 9-wire channel segment are marginally to significantly higher than the corresponding D-A-predicted results: several percentage points higher at SCA = 21.5 s/m and 10% points higher at SCA = 166.7 s/m. The two methods employ the same values for particle charge and the same average electric field (a spatially uniform value for the D-A analysis, derived from the spatially resolved E-field calculated in the numerical simulation). Consequently, the differences between the two methods in predicting particle removal efficiency are not likely attributable to SCA, particle charge, current density, or electric field. Because the D-A analysis assumes $d_p = 2.5$ µm, whereas the numerical simulation considers $PM_{2.5}$

(i.e., $d_p \leq 2.5$ µm), it would be expected that the D-A analysis would predict higher PM removal efficiencies than the numerical simulation and its contingent of 1.5 µm particles and their slower drift velocities. The opposite trend is evident in **Figure 5**, with higher PM removal efficiencies predicted by the numerical simulation. The answer to this apparent contradiction may be the result of the uniform E-field assumption inherent in the D-A analysis.

Figure 6 presents D-A-predicted and numerically simulated particle fluxes for $d_p = 2.5$ µm particles to the walls of a 3-wire channel segment energized to 0.11 mA/m². The D-A analysis and its assumption of a spatially uniform electric field yield a gradually and monotonically decreasing particle flux, while the particle flux derived from the numerical simulation is much more dynamic, starting near 0 at the channel inlet and rapidly rising to exceed the D-A-predicted particle flux in regions surrounding each of the three discharge electrodes. At 0.11 mA/m² in the 3-wire channel segment (**Figure 6**, upper), the predicted $d_p = 2.5$ µm collection efficiency determined from D-A analysis (1.7%) is nearly the same as that determined by the numerical simulation (1.8%), confirming that D-A analyses and their assumption of spatially uniform E-field introduce negligible error in predicted PM collection efficiencies at low current densities. However, considering the 9-wire channel segment still at 0.11 mA/m² current density, predicted $d_p = 2.5$ µm

removal efficiencies for the two methods begin to diverge with the numerical simulation predicting higher removal efficiency (20.5%) than the D-A analysis (16.2%). Even at values of current density unlikely to produce EHD effects, **Figure 6** shows that non-uniformity of the electric field leads to local maxima in particle wall fluxes, maxima whose influence on overall PM collection efficiency grows with E-field strength even when holding SCA, current density, and particle charge constant and absent the influence of EHD phenomena. Thus, PM removal efficiencies determined at high current density conditions likely represent the influences of both EHD phenomena and the non-uniform particle fluxes driven by non-uniform E-fields, none of which is captured by D-A analyses and their assumptions of spatially uniform E-fields.

At the higher value of current density (0.89 mA/m²), EHD-driven fluid flow patterns become significant, potentially contributing to any differences between the numerical simulation results and the D-A-predicted results for PM removal efficiency, particularly given that the D-A-predicted results categorically cannot capture EHD effects. **Figure 6**, lower, shows particle fluxes for the 3-wire channel segment energized to 0.89 mA/m². Particle fluxes at 0.89 mA/m² are of much greater magnitude [$O(10^8)$] than at 0.11 mA/m² [$O(10^3)$]. The numerical simulation results more clearly exhibit a streamwise decay trend in particle flux comparable to the D-A-predicted flux, most evident in the dramatic decay in peak particle flux in the vicinity of the three discharge electrodes. Particularly noticeable are the sharp inflexions in particle flux immediately upstream of the three electrodes. These correspond to the recirculation zones (see **Figure 7**, lower) and the regions of reverse flow, wall impingement, and flow separation from the wall that they engender. The effects of these recirculation zones on particle flux are superimposed on the periodic behavior corresponding to the peaks in electric field around the discharge electrodes. The discrepancies between D-A-predicted $d_p = 2.5$ µm removal efficiency and the numerical simulation results likely reflect, in part, the spatial resolution in particle flux that is not captured in the D-A analysis.

Figure 5 shows that compared to the predicted results at 0.11 mA/m², both numerical simulation and D-A analysis predict markedly higher collection efficiencies at 0.89 mA/m² for all values of SCA. Although D-A analyses predicted markedly higher collection efficiencies for the 9-wire channel segment compared to the 3-wire segment when both are energized to 0.11 mA/m² due to the former's generally higher electric field and higher particle saturation charge, at 0.89 mA/m², D-A-predicted results are quite similar between both channel segment lengths. As was the case at 0.11 mA/m², at 0.89 mA/m², numerical simulation results are higher than D-A-predicted results: higher by 12% points at SCA = 66.7 s/m and by 2% points at SCA = 166.7 s/m. The latter is still a significant increase in the context of removal efficiencies exceeding 97% in an asymptotic region where incremental performance increases require disproportionate increases in operating conditions. **Figure 7** shows numerical simulation results of the distributions of PM$_{2.5}$ concentration in a 9-wire channel segment at 0.11 and 0.89 mA/m². **Figure 7** considers only PM$_{2.5}$ concentrations because particles of such size present the greatest

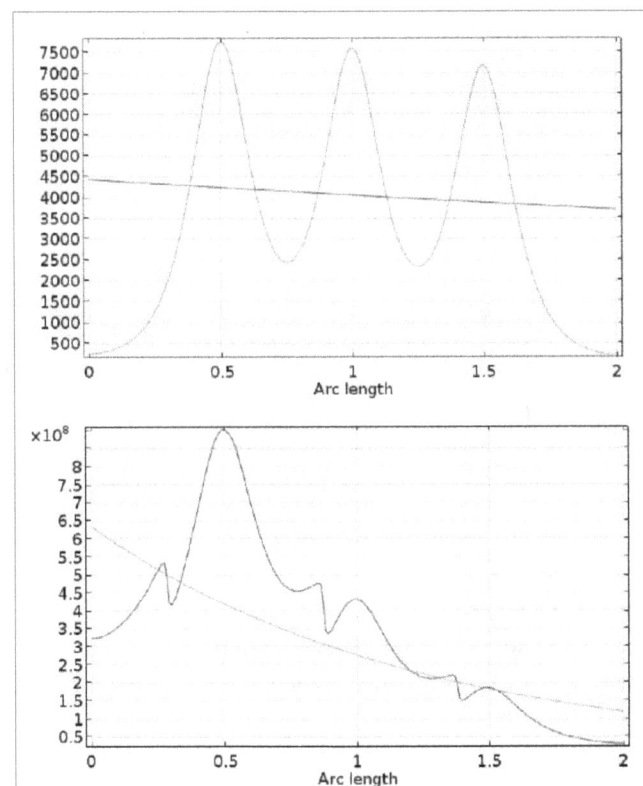

FIGURE 6 | Numerical simulation and Deutsch-Anderson analysis of the flux of $d_p = 2.5$ µm particles to collection electrodes of a 3-wire ESP channel segment at 0.11 mA/m² (upper) and 0.89 mA/m² (lower) current density. Conditions as listed in Table 1.

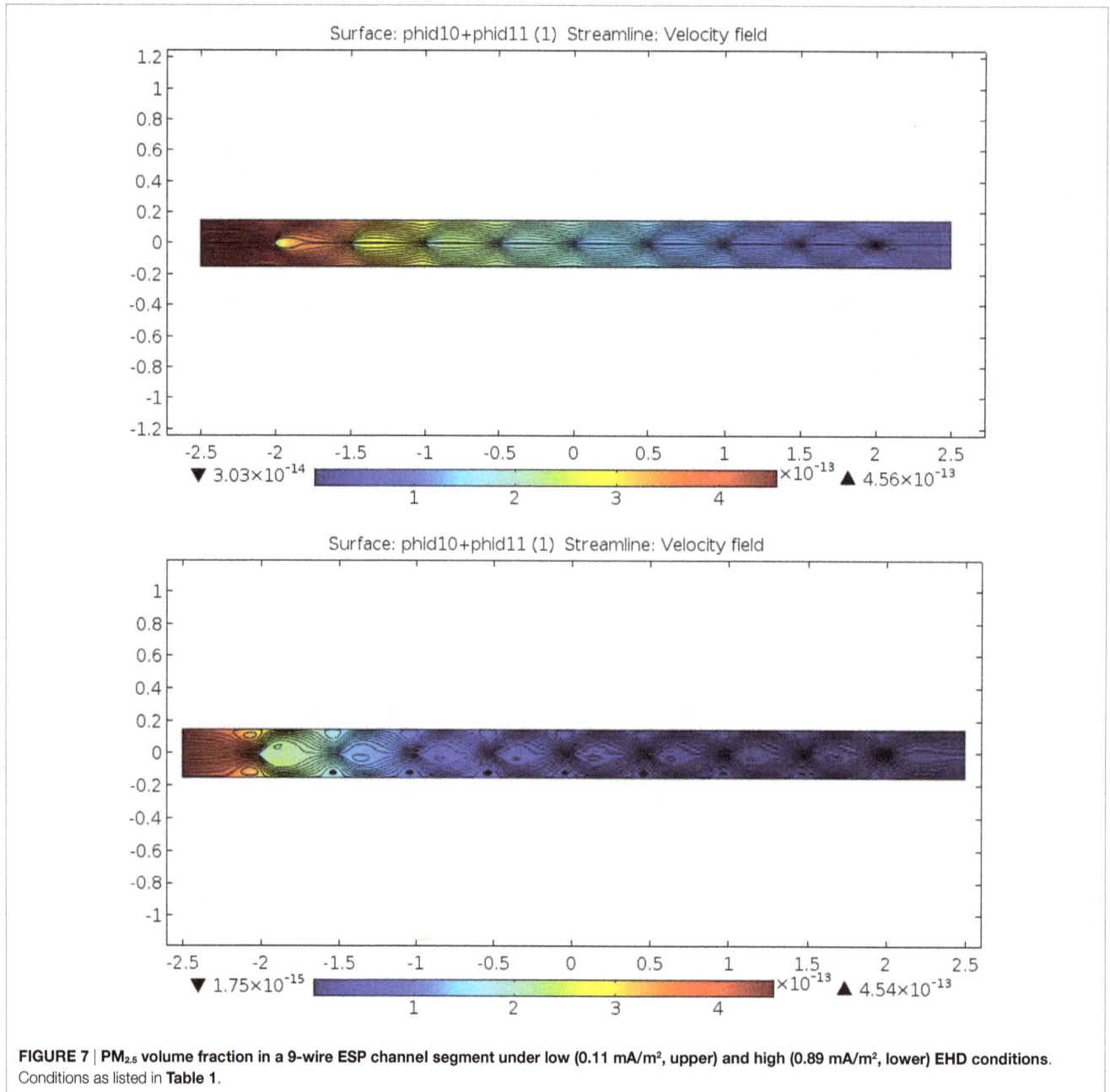

FIGURE 7 | PM₂.₅ volume fraction in a 9-wire ESP channel segment under low (0.11 mA/m², upper) and high (0.89 mA/m², lower) EHD conditions. Conditions as listed in **Table 1**.

challenge to ESP performance, are most relevant to current air quality regulations, and to distinguish their behavior from that of the coarser PM fraction that is collected more rapidly and for which total PM collection efficiency exerts an outsized influence on overall PM collection efficiency. Compared to 0.11 mA/m², at 0.89 mA/m², PM₂.₅ concentration reduces more rapidly in the along the length of the channel segment (**Figure 4**). Also shown in **Figure 4** are streamline patterns that reveal the prominence of EHD phenomena in the two fluid flows; the characteristic periodic converging-diverging patterns that are more evident under the 0.89 mA/m² conditions could promote more rapid collection of PM₂.₅.

Trace Gaseous Pollutant Adsorption: Influences of EHD Phenomena

Figure 5 shows mercury (Hg) concentration distributions resulting from numerical simulations of a 9-wire ESP channel segment energized to 0.11 (upper) and 0.89 (lower) mA/m². Both results illustrate the rapid reduction in Hg concentration at the channel segment entrance where particle concentrations, and therefore gas-particle mass transfer, are highest and downstream of which the majority of particles have been removed and streamwise Hg concentration gradients become negligible. The key feature distinguishing the results at the higher current density condition is the appearance of Hg-lean regions in a periodic pattern along the channel segment walls. The periodic regions of low Hg concentration are generally more prominently featured than similar regions in the $PM_{2.5}$ concentration results and coincide with the pairs of EHD-induced recirculation zones along the channel segment walls at the higher current density condition. Distributed in between the paired, EHD-induced recirculation zones are regions of jetted fluid in which fluid velocities substantially exceed U_0 and are responsible for the transport of inlet concentrations of Hg and PM well into the channel segment. For both the 0.11 and 0.89 mA/m^2 conditions, the centerline region is generally characterized by low PM concentrations and high Hg concentration relative to areas at the same axial position but nearer to the channel walls. Overall, Hg removal efficiency in the 9-wire channel segment for the log-normal PSD is higher (17.2%) at the lower current density of 0.11 mA/m^2 than at 0.89 mA/m^2 under the influence of EHD phenomena (9.4%). Generally, as rates of PM removal increase, particles remain suspended in the gas for shorter periods of time, reducing the opportunity for in-flight adsorption of trace pollutants. Consequently, intercomparisons between the trends in $PM_{2.5}$ removal efficiency and Hg removal efficiency are illustrative. As shown in **Figure 5** and discussed previously for the 3-wire and 9-wire channel segments, D-A analysis confirms that even when two ESP channel segments are operated at comparable values of current density and SCA, the longer channel segment imposes a stronger electric field with a more intense particle charging environment leading to higher particle removal efficiencies than the shorter segment. In our recent analysis of particle and trace pollutant removal within ESPs under conditions that are unfavorable to EHD (Clack, 2015), numerical simulation results for conditions well below the threshold for onset of EHD phenomena demonstrated that, all other conditions being comparable, more rapid particle collection diminishes trace pollutant adsorption by the in-flight mechanism. For the present results, the same trends are evident when comparing results with and without EHD phenomena. Comparing results obtained for the 3-wire channel segment, increasing current density from 0.11 to 0.89 mA/m^2 causes $PM_{2.5}$ removal efficiency for the log-normal PSD to increase from 13.4 to 90.4% and Hg removal efficiency to decrease from 27.7 to 5.9%. The results for the finer skewed PSD reflect its greater concentration of fine particles: increasing current density from 0.11 to 0.89 mA/m^2 produces the same trends in $PM_{2.5}$ removal efficiency (10.7% increasing to 83.3%) and Hg removal efficiency (80% decreasing to 44.7%) involving values of each that are, respectively, lower and higher than corresponding values for the coarser log-normal PSD. In addition to the effects of more rapid particle collection as discussed previously (Clack, 2015), these results reflect the influences of EHD phenomena. It is worth noting that the decrease in Hg removal efficiency at the higher value of current density suggests that the higher gas-particle mass transfer that results from higher particle slip velocities is offset by the reduction in gas-particle mass transfer caused by the loss of particles and available particle surface area.

Mercury removal through the in-flight mechanism is strongly influenced by the collective particle surface area available in the fluid flow and the time available for gas-particle mass transfer

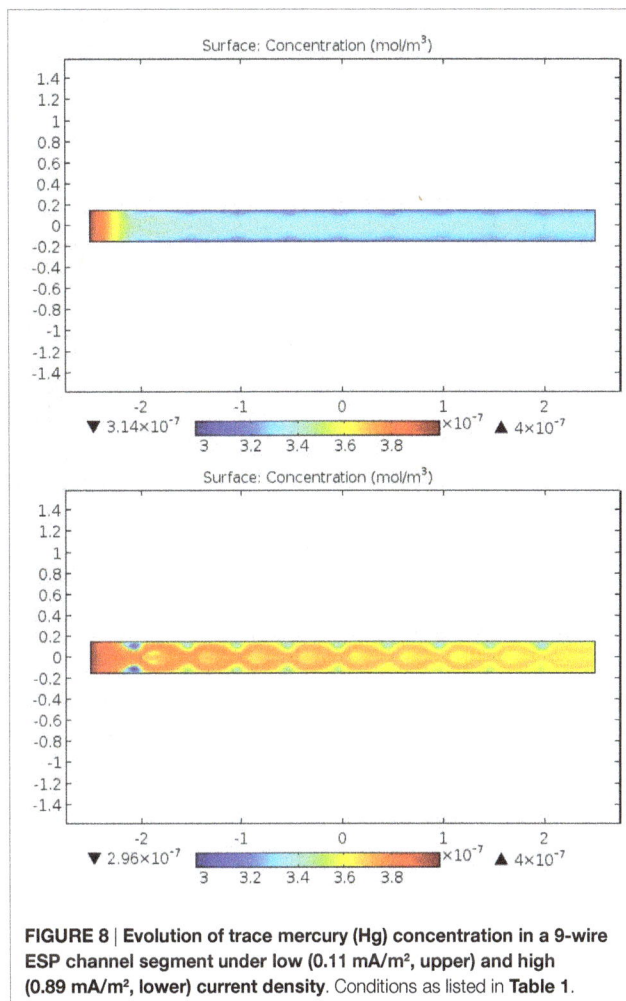

FIGURE 8 | Evolution of trace mercury (Hg) concentration in a 9-wire ESP channel segment under low (0.11 mA/m^2, upper) and high (0.89 mA/m^2, lower) current density. Conditions as listed in **Table 1**.

to that collective surface. The pronounced EHD-driven flow patterns evident at 0.89 mA/m^2 in **Figures 7** and **8** include regions of fluid convergence where fluid velocity magnitude can reach 2.5 U_0. EHD-induced fluid acceleration means shorter fluid residence or transit times through the channel and less time available for gas-particle mass transfer, ultimately producing a lower Hg removal efficiency.

The two indentified mechanisms through which high current densities and EHD phenomena impact Hg removal efficiency—more highly charged particles removed more rapidly from the flow and EHD-driven acceleration of fluid and particle transport—can be considered separately by considering numerical simulations at high current density in which the electric body force term of the Navier–Stokes equations has been disabled. For a 3-wire channel segment operating at a current density of 0.89 mA/m^2, $U_0 = 0.2$ m/s, and the electric body force term enabled, predicted $PM_{2.5}$ collection efficiencies are 90.4% for the log-normal PSD and 83.3% for the skewed PSD. The results of tracking 10 1-μm, neutrally charged and neutrally buoyant tracer particles through this flow field revealed a range of particle transit times, from 8.25 s to as short as 6.3 s, times that are 17 to 37% shorter than the 10 s average residence time that would be determined from a plug flow analysis given the inlet velocity ($U_0 = 0.2$ m/s) and the channel segment length (2 m). These results provide strong evidence of the non-uniform fluid acceleration induced by EHD phenomena. Disabling the electric body force to eliminate EHD phenomena increased $PM_{2.5}$ collection efficiencies: 97.4% for the log-normal PSD and 87.3% for the skewed PSD. With the electric body force disabled, systematic increases in U_0 decreased $PM_{2.5}$ collection efficiency in numerical simulations: When U_0 was increased from 0.2 m/s to 0.24 (+ 17%) and 0.32 (+ 37%) m/s, predicted $PM_{2.5}$ removal efficiencies decreased to 94.3 and 86.3%, respectively, for the log-normal PSD and to 86.2 and 76.3%, respectively, for the skewed PSD. The corresponding Hg removal efficiency results suggest that the presence of EHD phenomena, despite the negative impacts on $PM_{2.5}$ removal efficiency, actually enhances Hg removal. For the log-normal PSD, numerical simulation results with electric body force terms enabled and $U_0 = 0.2$ m/s yielded the highest Hg removal efficiency (5.9%), higher than the three results in which EHD phenomena were eliminated: $U_0 = 0.2$ m/s (5.7% Hg removal efficiency), 0.24 m/s (5.45%), and 0.32 m/s (5%). For these results, it appears that EHD-induced fluid acceleration and jetting through the channel segment were offset by higher Hg removal in other portions of the channel segment, most likely in the Hg-lean paired recirculation zones (**Figure 8**). In the absence of EHD phenomena, these regions are eliminated, and plug flow simulations through the channel achieve approximately the same or lower Hg removal depending on what value of U_0 is used (or not used) to achieve fluid transit times through the channel segment that are comparable to those induced in the presence of EHD phenomena. For the skewed PSD, numerical simulation results with electric body force terms enabled and $U_0 = 0.2$ m/s achieved an Hg removal efficiency (44.7%) that was greater than two of the three results in which EHD phenomena were eliminated; the highest Hg removal efficiency (46.3%) was predicted in the simulation without EHD phenomena and in which there was no increase in U_0 to achieve comparable fluid transit times through the channel segment (i.e., $U_0 = 0.2$ m/s). Taking the log-normal and skewed PSD results together, the results suggest that EHD phenomena both promote and diminish Hg removal, with particle size distribution helping to determine the net overall effect. Finally, it bares noting that spatial non-uniformities in the E-field, velocity field, and concentration fields (whether particle or gas) cannot be accommodated within the assumptions underlying classical analytical or quasi-1-D approaches to modeling ESP performance.

CONCLUSION

This study advances fundamental understanding of EHD phenomena and their influences on particulate removal and trace gaseous pollutant removal within industrial ESPs. 2-D numerical simulations reveal differences in the collection patterns of PM and gaseous pollutants as a function of imposed current density, length of the wire-plate ESP channel, and the particle size distribution of the suspended particles. Comparisons to the classical D-A ESP performance prediction equation show that the high current densities required to induce EHD phenomena also lead to higher electric fields and greater saturation charge on particles, both of which promote more rapid particle collection independent of the onset of EHD phenomena. However, even when accounting for such enhancement, numerical simulations of fine PM collection generally agree with classical equations for predicting ESP performance at low current densities but predict higher collection efficiency than the classical equations at higher current densities. Trace pollutant removal by the in-flight mechanism of gas-particle adsorption is shown to be both promoted and diminished by the presence of EHD phenomena. Accelerations of the fluid caused by the electric body force reduce the time available for gas-particle mass transfer. However, this is offset to varying degrees by the greater degree of trace pollutant adsorption that occurs in key regions of the flow in which fluid recirculation occurs. The relative magnitude of the promotion and diminution mechanisms appears to be dependent on the initial distribution of particle sizes and likely will also depend on geometric factors characterizing the ESP channel.

AUTHOR CONTRIBUTIONS

HC is wholly responsible for the entire content of this manuscript.

FUNDING

No sponsored research funds were used in the completion of this work.

REFERENCES

Back, A., and Cramsky, J. (2012). Comparison of numerical and experimental results for the duct-type electrostatic precipitator. *Int. J. Plasma Environ. Sci. Technol.* 6, 33–42.

Calvert, S., and Englund, H. M. (1984). *Handbook of Air Pollution Technology.* New York, NY: John Wiley and Sons.

Clack, H. L. (2006a). Mass transfer within ESPs: in-flight adsorption of mercury by charged suspended particulates. *Environ. Sci. Technol.* 40, 3617–3622. doi:10.1021/es050246+

Clack, H. L. (2006b). Particle size distribution effects on gas-particle mass transfer within electrostatic precipitators. *Environ. Sci. Technol.* 40, 3929–3933. doi:10.1021/es051649c

Clack, H. L. (2009). Mercury capture within coal-fired power plant electrostatic precipitators: model evaluation. *Environ. Sci. Technol.* 43, 1460–1466. doi:10.1021/es8015183

Clack, H. L. (2013). "Computational modeling of electrohydrodynamically-influenced mercury adsorption within ESPs," in *11th International Conference on Electrostatic Precipitation* (Bangalore, India).

Clack, H. L. (2015). Simultaneous removal of particulate matter and gas-phase pollutants within electrostatic precipitators: coupled in-flight and wall-bounded adsorption. *Aerosol Air Qual. Res.* 15, 2445–2455. doi:10.4209/aaqr.2015.06.0280

Cooper, C. D., and Alley, F. C. (2011). *Air Pollution Control: A Design Approach.* Long Grove, IL: Waveland Press.

Friedlander, S. K. (2000). *Smoke, Dust and Haze, Fundamentals of Aerosol Dynamics.* Oxford: Oxford University Press.

Han, B., Kim, H. J., and Kim, Y. J. (2010). Fine particle collection of an electrostatic precipitator in CO_2-rich gas conditions for oxy-fuel combustion. *Sci. Total Environ.* 408, 5158–5164. doi:10.1016/j.scitotenv.2010.07.028

IEEE-DEIS-EHD Technical Committee. (2003). Recommended international standard for dimensionless parameters used in electrohydrodynamics.

IEEE Trans. Dielectr. Electr. Insul. 19, 3–6. doi:10.1109/TDEI.2003.
1176545

Kallio, G. A., and Stock, D. E. (1992). Interaction of electrostatic and fluid dynamic fields in wire-plate electrostatic precipitators. *J. Fluid Mech.* 240, 133–166. doi:10.1017/S0022112092000053

Kim, H., Han, B., Woo, C., Kima, Y., Ono, R., and Oda, T. (2014). Performance evaluation of dry and wet electrostatic precipitators used in an oxygen-pulverized coal combustion and a CO_2 capture and storage pilot plant. *J. Aerosol Sci.* 77, 116–126. doi:10.1016/j.jaerosci.2014.07.003

Leonard, G. L., Mitchner, M., and Self, S. A. (1983). An experimental study of the electro-hydrodynamic flow in electrostatic precipitators. *J. Fluid Mech.* 127, 123–140. doi:10.1017/S0022112083002657

Noda, N., and Makino, H. (2010). Influence of operating temperature on performance of electrostatic precipitator for pulverized coal combustion boiler. *Adv. Powder Technol.* 21, 495–499. doi:10.1016/j.apt.2010.04.012

Prabhu, V., Kim, T., Khakpour, Y., Serre, S., and Clack, H. L. (2012). On the electrostatic precipitation of fly ash-powdered mercury sorbent mixtures. *Fuel Process. Technol.* 93, 8–12. doi:10.1016/j.fuproc.2011.09.006

Reynolds, J. (2004). "Multi-pollutant control using membrane-based up-flow west electrostatic precipitation," in *National Energy Technology Laboratory Report on Wet ESP Performance at First Energy's Bruce Mansfield Plant*. U.S. Department of Energy.

Seames, W. S., and Wendt, J. (2000). Partitioning of arsenic, selenium, and cadmium during the combustion of Pittsburgh and Illinois #6 coals in a self-sustained combustor. *Fuel Process. Technol.* 63, 179–196. doi:10.1016/S0378-3820(99)00096-X

Seetharama, S., Benedict, A., Reynolds, J. (2013). "Comparison of wet and dry electrostatic precipitator (ESP) technologies," in *13th International Conference on Electrostatic Precipitation* (Bangalore, India).

Shanthakumar, S., Singh, D. N., and Phadke, R. C. (2008). Flue gas conditioning for reducing suspended particulate matter from thermal power stations. *Prog. Energy Combust. Sci.* 34, 685–695. doi:10.1016/j.pecs.2008.04.001

Yamamoto, T., and Sparks, L. E. (1986). Numerical simulation of three-dimensional tuft corona and electrohydrodynamics. *IEEE Trans. Ind. Appl.* 22, 880–885. doi:10.1109/TIA.1986.4504808

Zhao, L., and Adamiak, K. (2008). Numerical simulation of the electrohydrodynamic flow in a single wire-plate electrostatic precipitator. *IEEE Trans. Ind. Appl.* 44, 683–691. doi:10.1109/TIA.2008.921453

Zouzou, N., Dramane, B., Moreau, E., and Touchard, G. (2011). EHD flow and collection efficiency of a DBD ESP in wire-to-plane and plane-to-plane configurations. *IEEE Trans. Ind. Appl.* 47, 336–343. doi:10.1109/TIA.2010.2091473

Conflict of Interest Statement: The author declares that the research was conducted in the absence of any commercial or financial relationships that could be construed as a potential conflict of interest.

19

Indacenodithienothiophene-Based Ternary Organic Solar Cells

Nicola Gasparini[1]*, Amaranda García-Rodríguez[2], Mario Prosa[3], Şebnem Bayseç[2], Alex Palma-Cando[2], Athanasios Katsouras[4], Apostolos Avgeropoulos[4], Georgia Pagona[5,6], Vasilis G. Gregoriou[5,6], Christos L. Chochos[4,5], Sybille Allard[2], Ulrich Scherf[2], Christoph J. Brabec[1,7] and Tayebeh Ameri[1]*

[1] Institute of Materials for Electronics and Energy Technology (I-MEET), Friedrich-Alexander-University Erlangen-Nuremberg, Erlangen, Germany, [2] Macromolecular Chemistry Group (buwmakro), Institute for Polymer Technology, BergischeUniversität Wuppertal, Wuppertal, Germany, [3] Istituto per lo Studio dei Materiali Nanostrutturati (ISMN), Consiglio Nazionale delle Ricerche (CNR), Bologna, Italy, [4] Department of Materials Science Engineering, University of Ioannina, Ioannina, Greece, [5] Advent Technologies SA, Patras Science Park, Patra, Greece, [6] National Hellenic Research Foundation (NHRF), Athens, Greece, [7] Bavarian Center for Applied Energy Research (ZAE Bayern), Erlangen, Germany

Edited by:
Amlan J. Pal,
Indian Association for the
Cultivation of Science, India

Reviewed by:
Lau Sing Liong,
Universiti Tunku Abdul Rahman,
Malaysia
Praveen C. Ramamurthy,
Indian Institute of Science, India

*Correspondence:
Nicola Gasparini
nicola.gasparini@fau.de;
Tayebeh Ameri
tayebeh.ameri@fau.de

Specialty section:
This article was submitted
to Solar Energy,
a section of the journal
Frontiers in Energy Research

Citation:
Gasparini N, García-Rodríguez A,
Prosa M, Bayseç Ş, Palma-Cando A,
Katsouras A, Avgeropoulos A,
Pagona G, Gregoriou VG,
Chochos CL, Allard S, Scherf U,
Brabec CJ and Ameri T (2017)
Indacenodithienothiophene-Based
Ternary Organic Solar Cells.
Front. Energy Res. 4:40.

One of the key aspects to achieve high efficiency in ternary bulk-heterojunction solar cells is the physical and chemical compatibility between the donor materials. Here, we report the synthesis of a novel conjugated polymer (P1) containing alternating pyridyl[2,1,3] thiadiazole between two different donor fragments, dithienosilole and indacenodithienothiophene (IDTT), used as a sensitizer in a host system of indacenodithieno[3,2-b] thiophene,2,3-bis(3-(octyloxy)phenyl)quinoxaline (PIDTTQ) and [6,6]-phenyl C_{70} butyric acid methyl ester ($PC_{71}BM$). We found that the use of the same IDTT unit in the host and guest materials does not lead to significant changes in the morphology of the ternary blend compared to the host binary. With the complementary use of optoelectronic characterizations, we found that the ternary cells suffer from a lower mobility-lifetime ($\mu\tau$) product, adversely impacting the fill factor. However, the significant light harvesting in the near infrared region improvement, compensating the transport losses, results in an overall power conversion efficiency enhancement of ~7% for ternary blends as compared to the PIDTTQ:$PC_{71}BM$ devices.

Keywords: organic solar cells, ternary devices, OPV, IDTT, organic electronics

INTRODUCTION

During the last decades, the power conversion efficiency (PCE) of organic bulk-heterojunction (BHJ) solar cells based on donor/acceptor blends surpassed the 10% threshold, mainly due to the discovery of novel materials as well as device structure engineering (Liu et al., 2014; He et al., 2015; Holliday et al., 2016; Huang et al., 2016; Spyropoulos et al., 2016; Zhao et al., 2016). Polymers and/or small molecules, used as donor materials, in combination with fullerene derivatives, used as acceptor, are the common active components in BHJ devices (Zhang et al., 2014; Lu et al., 2015c; Min et al., 2015; Squeo et al., 2015). Due to the narrow absorption of the donor materials, one of the main challenges in order to further boost the PCE of organic solar cells is to achieve better absorption match to the solar irradiance spectrum. In this regard, two main concepts have been developed: tandem and ternary organic solar cells (Ameri et al., 2009, 2013a,b; Li et al., 2013; You et al., 2013; Spyropoulos et al., 2014; Lu et al., 2015b; Yang et al., 2015; Cheng et al., 2016; Goh et al., 2016;

Keawsongsaeng et al., 2016; Lee et al., 2016; Nian et al., 2016). The former is based on a complex multi-layer stack with the main challenge of designing a robust solution-processed intermediate layer. The latter, made of two donors and one acceptor, mixed together in a unique solution, overcomes the complexities of the tandem device architecture, maintaining the easy processability of a single-junction organic BHJ solar cell. To date, polymers (Lu et al., 2014, 2015a; Gasparini et al., 2015b; Yang et al., 2015), small molecules (Zhang et al., 2015), dyes (Ke et al., 2016), quantum dots (Itskos et al., 2011), and fullerene derivatives (Cheng et al., 2014) have been adopted as "guest" in the polymer-fullerene "host" system. In addition to the need for donor materials with the complementary absorption, one of the key points to surpass the performance of binary cells in ternary devices is to find donor materials with compatible physical and chemical nature (Yang et al., 2015). This can prevent the formation of recombination centers or morphological traps, which deteriorate the photovoltaic properties.

Here, we report a ternary organic solar cell system processed in air that shows a pronounced sensitization effect, resulting in a PCE of more than 4.6%. As sensitizer, we incorporate the near infrared (NIR) polymer P1 containing alternating pyridyl[2,1,3]thiadiazole between two different donor fragments, dithienosilole and indacenodithienothiophene (IDTT), into a host system of indacenodithieno[3,2-b]thiophene,2,3-bis(3-(octyloxy)phenyl)quinoxaline (PIDTTQ) (Gasparini et al., 2015a) blended with [6,6]-phenyl C_{70} butyric acid methyl ester ($PC_{71}BM$). Indeed, in order to have components with a similar chemical nature in the ternary blend system, we used two polymers with the same backbone IDTT unite for the host as well as the guest donors.

The polymer P1 was synthesized by Stille-type aromatic cross-coupling reaction of a stoichiometric balance ratio of the distannyl derivative of para-hexyl-phenyl substituted IDTT (M2) and 4,4'-(4,4-bis(2-ethylhexyl)-4H-silolo[3,2-b:4,5-b'] dithiophene-2,6-diyl)bis(7-bromo-[1,2,5]thiadiazolo[3,4-c] pyridine) (M1), in the presence of tris(dibenzylideneacetone) dipalladium(0) (Pd_2dba_3) and tri(o-tolyl)phosphine (P(o-tol)$_3$) as the catalytic system (Scheme 1). After soxhlet extraction the polymer was obtained from the o-dichlorobenzene fraction with a number average molecular weight Mn of 36,800 g mol^{-1} and a polydispersity index of 3.3.

Figure S1 in Supplementary Material shows the absorption spectrum of P1 in DCB solution and as solid. The copolymer for both cases shows a single band in the high energy region, which is assigned to a localized $\pi-\pi^*$ transition and another absorption band in the low energy region (up to 1,000 nm), which is assigned to an intramolecular charge-transfer transition. The maximum of the NIR absorption band of P1 in the solid state is bathochromic shifted (738 nm) in comparison to the corresponding UV–VIS solution (695 nm). The optical band gap energy estimated from the absorption edge of film spectrum was estimated to be 1.87 eV. Based on the onsets of the oxidation and reduction peaks in cyclic voltammetry (CV) measurements, the electrochemical highest occupied molecular orbital (HOMO) and lowest unoccupied molecular orbital (LUMO) energies were estimated to be −5.34 and −3.71 eV, respectively, corresponding to an electrochemical band gap energy of 1.63 eV (Gedefaw et al., 2016).

Next, we analyzed the device performances of the ternary devices. The device architecture used in this work is based on ITO/ZnO/active layer/MoOx/Ag. PIDTTQ [its lower molecular weight version of PIDTTQ-LMW (Gasparini et al., 2015a)] has been previously presented in the literature. All the solution-processed layers are doctor bladed in air. **Figure 1A** depicts the energy levels, measured with CV of the polymers and the fullerene derivate. **Figure 1C** shows the current density–voltage characteristics of the binary PIDTTQ:$PC_{71}BM$ (1:2 wt/wt) as well as ternary PIDTTQ:P1:$PC_{71}BM$ (different composition) under 1 sun illumination (100 mW cm^{-2}). In agreement with previous reports, binary cells delivered a PCE of 4.3% with an open circuit voltage (V_{oc}) of 0.84 V, a short circuit current (J_{sc}) of 8.62 mA cm^{-2}, and a fill factor (FF) of 60%. Adding 15 wt% of the NIR sensitizer delivers the highest short-circuit current, reaching 10.60 mA cm^{-2}, V_{oc} of 0.84 V, and FF of 52%, increases the overall efficiency of the ternary system 4.6% under 1 sun conditions. As shown in **Table 1**, J_{sc} increased monotonically by increasing the amount of P1, due to the better harvesting of the ternary system in the NIR region, in the best case, an improvement in J_{sc} of ~20% is achieved in the ternary system PIDTTQ:P1:$PC_{71}BM$ (0.85:0.15:2). Notably, the V_{oc} obtained in the ternary cells is identical to the binary PIDTTQ:$PC_{71}BM$, reflecting an energy cascade between the HOMO and LUMO energy levels of the three components (**Figure 1A**). Unfortunately, we observed a continuously decreased in FF by introducing higher amount of P1, which indeed inhibits the higher improvements of the ternary device performance compared to its reference.

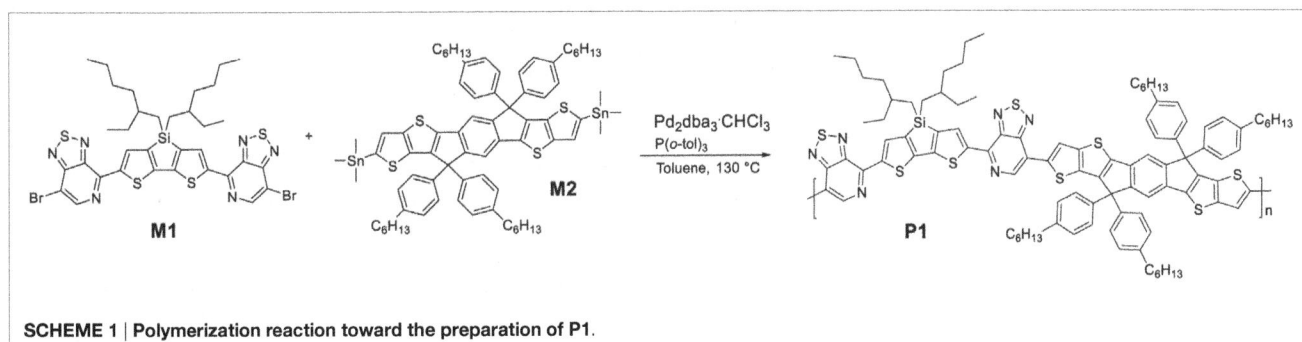

SCHEME 1 | Polymerization reaction toward the preparation of P1.

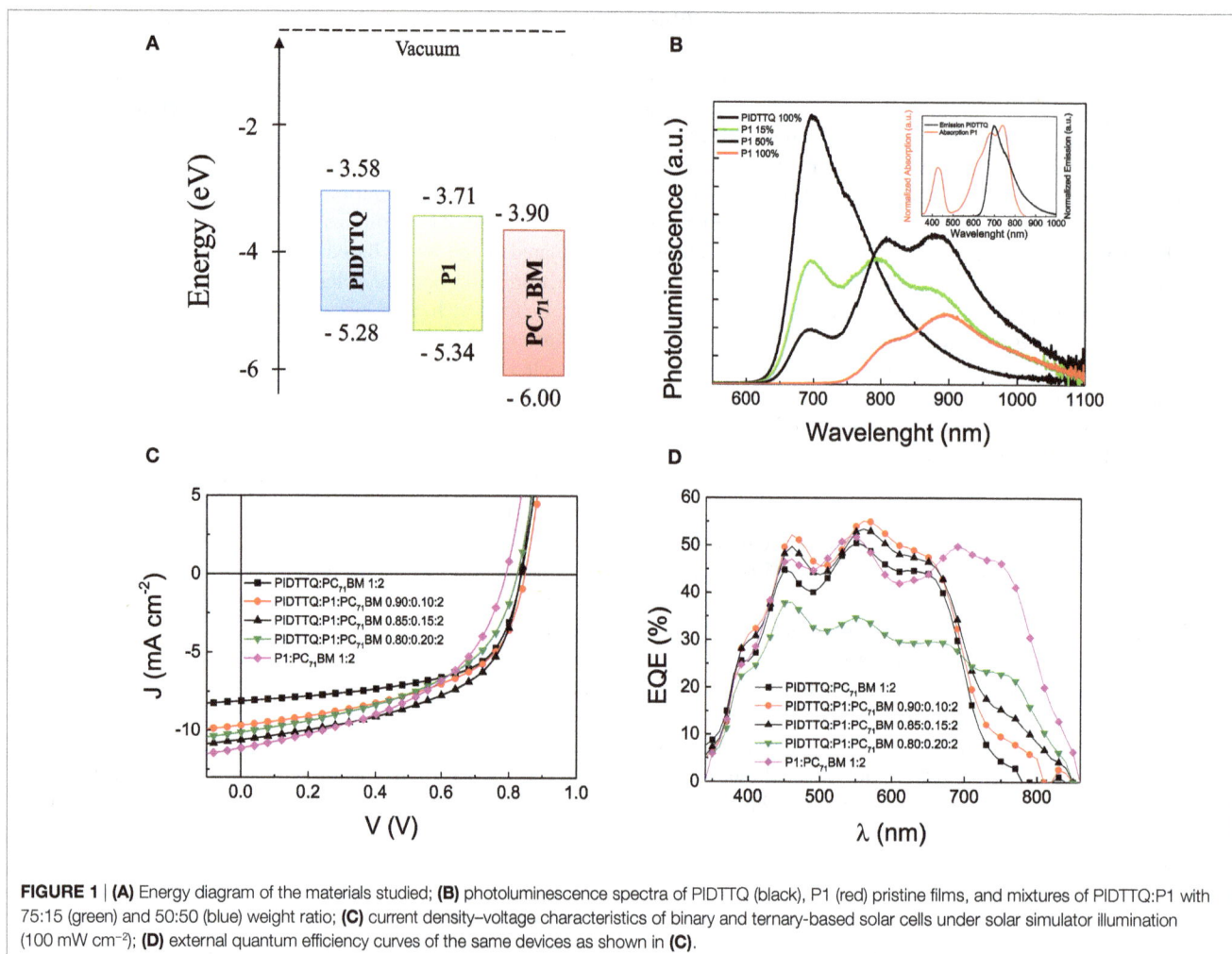

FIGURE 1 | (A) Energy diagram of the materials studied; (B) photoluminescence spectra of PIDTTQ (black), P1 (red) pristine films, and mixtures of PIDTTQ:P1 with 75:15 (green) and 50:50 (blue) weight ratio; (C) current density–voltage characteristics of binary and ternary-based solar cells under solar simulator illumination (100 mW cm^{-2}); (D) external quantum efficiency curves of the same devices as shown in (C).

TABLE 1 | Photovoltaic device parameters of low, medium, and high molecular weight PIDTTQ-based inverted solar cells under 1 sun illumination (100 mW cm^{-2}).

PIDTTQ:P1:PC$_{71}$BM	V_{oc} (V)	J_{sc} (mA cm^{-2})	Fill factor (%)	Power conversion efficiency (%)
1:0:2	0.84 (0.84 ± 0.01)	8.62 (8.49 ± 0.23)	60.33 (59.72 ± 0.65)	4.32 (4.24 ± 0.10)
0.90:0.10:2	0.84 (0.84 ± 0.01)	9.69 (9.50 ± 0.18)	53.14 (52.48 ± 0.55)	4.29 (4.20 ± 0.10)
0.85:0.15:2	0.84 (0.84 ± 0.01)	10.60 (10.43 ± 0.22)	51.87 (50.64 ± 1.17)	4.63 (4.45 ± 0.19)
0.80:0.20:2	0.84 (0.84 ± 0.01)	10.14 (9.64 ± 0.44)	48.86 (48.64 ± 0.20)	4.04 (3.86 ± 0.17)
0:1:2	0.81 (0.81 ± 0.01)	10.87 (10.37 ± 0.41)	46.61 (45.57 ± 0.69)	3.95 (3.79 ± 0.10)

We further measured external quantum efficiency (EQE) spectra of OPV devices made from PIDTTQ:P1:PC$_{71}$BM and PIDTTQ:PC$_{71}$BM (**Figure 1D**). Photoaction spectra of active layers with increasing P1 content show improved photoresponse particularly around 800 nm, i.e., the enhancement in J_{sc} originates dominantly from the NIR polymer absorption regime. We note that the integrated EQE for these devices matches the measured short circuit current within a margin of 5%.

In order to shine light into the mechanism, we performed photoluminescence (PL) measurements. PL is widely used in ternary BHJ solar cells to discriminate between energy and charge transfer between host and guest materials (Lu et al., 2014,

2015a; Gasparini et al., 2015b). In principle, if the charges are transfer from the wide to the low band gap material, the PL of the host should decrease while the PL of guest material should not increase. On the other hands, if the energy transfer is the main mechanism, a quenching of the host PL is associated with an increase of guest PL. Moreover, in order to have an energy transfer, the absorption of the guest polymer should overlap with the emission of the guest. The inset of **Figure 1B** confirms the aforementioned requirement. Thus, we mixed together PIDTTQ and P1 in different weight ratio. As depicted in **Figure 1B**, the PL of PIDTTQ is quenched of 53 and 79% upon introduction of P1 (85–15 and 50–50, respectively). In addition, we observed a clear enhancement of P1 PL compared to the pristine one, indicating an efficient energy transfer.

Before getting insight into the optical and electrical behavior of the binary and ternary devices, we performed intermittent contact mode atomic force microscopy (AFM, **Figure 2**). In agreement with our previous study (Gasparini et al., 2015a), the topography of PIDTTQ:PC$_{71}$BM layer shows spherical features with domains of ~100 nm. Interestingly, we observed similar morphology for ternary active layers. We calculated a root mean square roughness of 0.46, 0.51, and 0.60 nm for PIDTTQ:PC$_{71}$BM, PIDTTQ:P1:PC$_{71}$BM (0.85:0.15:2) and P1:PC$_{71}$BM blends, respectively. Thus, the low FF calculated for PIDTTQ:P1:PC$_{71}$BM cannot be ascertained to changes in the microstructure upon addition of the guest sensitizer.

In order to understand the lower FF obtained in the ternary BHJ solar cells, we first studied the charge transport properties by employing the technique of photoinduced charge carrier

extraction by linearly increasing voltage (photo-CELIV) (Mozer et al., 2005; Clarke et al., 2015; Min et al., 2015). From the measured photocurrent transients, the charge carrier mobility (μ) is calculated using the following equation:

$$\mu = \frac{2d^2}{3At_{max}^2\left[1+0.36\frac{\Delta j}{j(0)}\right]} \quad \text{if } \Delta j \leq j(0), \quad (1)$$

where d is the active layer thickness, A is the voltage rise speed $A = dU/dt$, U is the applied voltage, t_{max} is the time corresponding to the maximum of Δj of the extraction peak, and $j(0)$ is the displacement current.

Figure 3A shows the transient recorded by applying a 2 V/60 μs linearly increasing reverse bias and a delay time (td) of 10 μs. Analysis of the photo-CELIV traces extracts charge carrier

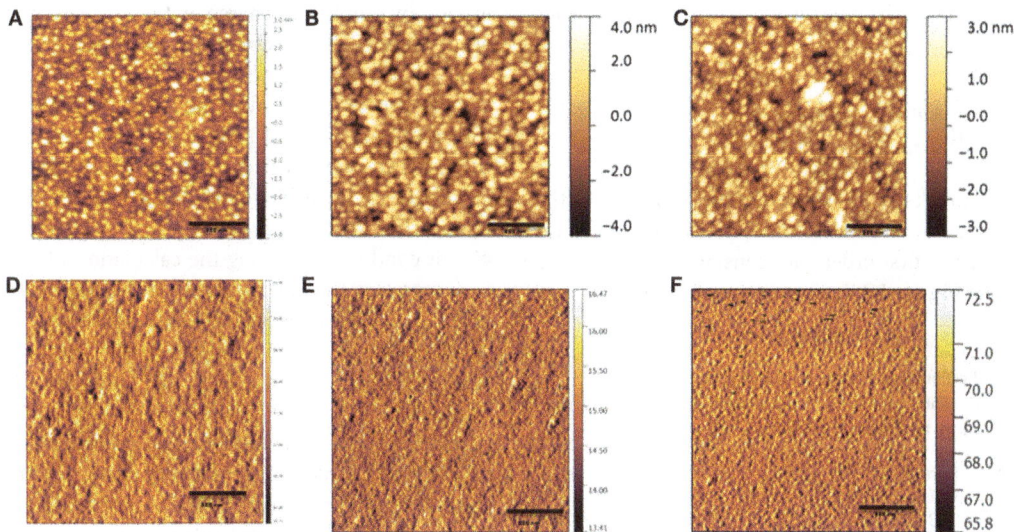

FIGURE 2 | Topography and phase images of films of PIDTTQ:PC$_{71}$BM (1:2) (A–D), PIDTTQ:P1:PC$_{71}$BM (0.85:0.15:2) (B–E), and P1:PC$_{71}$BM (1:2) (C–F) on top of a layer of ZnO, as measured by intermittent contact mode atomic force microscopy.

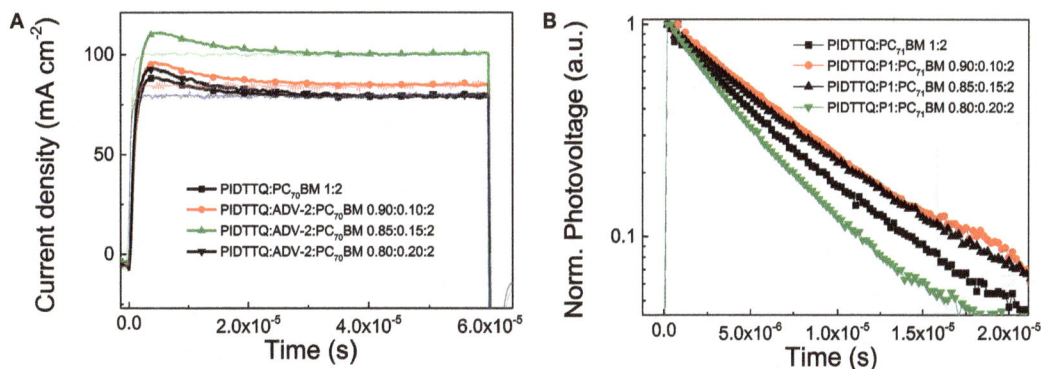

FIGURE 3 | Time-dependent photo-CELIV traces under light (solid lines) and dark (semitransparent traces) conditions (A) and transient photovoltage decays (B) of PIDTTQ:PC$_{71}$BM binary and PIDTTQ:P1:PC$_{71}$BM ternary devices.

mobility values 1.13×10^{-4} cm^2 V^{-1} s^{-1}, 8.54×10^{-5} cm^2 V^{-1} s^{-1}, 7.53×10^{-5} cm^2 V^{-1} s^{-1}, and 7.42×10^{-5} cm^2 V^{-1} s^{-1} for PIDTTQ:PC$_{71}$BM (1:2)-, PIDTTQ:P1:PC$_{71}$BM (0.90:0.10:2)-, PIDTTQ:P1:PC$_{71}$BM (0.85:0.15:2)-, PIDTTQ:P1:PC$_{71}$BM (0.80:0.20:2)-based devices, respectively. We also calculated the charge mobility of P1:PC71BM (Figure S5 in Supplementary Material), and we found a value of 5.02×10^{-5} cm^2 V^{-1} s^{-1}. The lower FFs obtained in the ternary cells are in agreement with the lower charge carrier mobility calculated with photo-CELIV technique. We then analyze the lifetime of charge carriers by employing transient photovoltage technique (TPV) (Shuttle et al., 2008). The samples were connected to the terminal of an oscilloscope with the input impedance of 1 MΩ and illuminated with a continuous background laser to control the V_{oc}. A small optical perturbation was applied using a blue laser ($\lambda = 405$ nm). The pulse intensity was controlled to keep the height of the photovoltage transient smaller than 10 mV resulting in a voltage transient with amplitude $\Delta V \ll V_{oc}$. The measured transient decays show the form of single exponentials, as expected for the pseudo-first-order kinetic (Hamilton et al., 2010; Heumueller et al., 2015).

$$\frac{d\Delta V}{dt} \propto \frac{d\Delta n}{dt} = -k_{eff} = -\frac{\Delta n}{\tau_{\Delta n}}, \qquad (2)$$

where V is the photovoltage, t is the time, Δn is change in the density of photogenerated carriers due to the perturbation pulse, k_{eff} is the pseudo-first order rate constant, and $\tau_{\Delta n}$ is the carrier lifetime. The **Figure 3B** depicts normalized photovoltage

decays as a function of time for the binary and ternary devices at 1 sun condition by exciting with a blue laser ($\lambda = 405$ nm). As reported in **Table 2**, the lifetime of charge carriers is similar for PIDTTQ:PC$_{71}$BM (1:2), PIDTTQ:P1:PC$_{71}$BM (0.90:0.10:2), PIDTTQ:P1:PC$_{71}$BM (0.85:0.15:2), 6.72, 7.35, and 7.23 µs, respectively, suggesting that these ternary blends are not limited by the short lifetime of charge carriers. Otherwise, a reduce τ is observed for the PIDTTQ:P1:PC$_{71}$BM (0.80:0.20:2) based ternary system (4.72 µs). Thus, with the combination of photo-CELIV and TPV techniques, we were able to calculate the mobility-lifetime product ($\mu\tau$). As collected in **Table 2**, $\mu\tau$ decreases from 7.59×10^{-10} cm^2 V^{-1} to 6.72×10^{-10} cm^2 V^{-1}, 5.44×10^{-10} cm^2 V^{-1}, and 3.50×10^{-10} cm^2 V^{-1} for PIDTTQ:PC$_{71}$BM (1:2), PIDTTQ:P1:PC$_{71}$BM (0.90:0.10:2), PIDTTQ:P1:PC$_{71}$BM (0.85:0.15:2), PIDTTQ:P1:PC$_{71}$BM (0.80:0.20:2)-based devices, reason of the poorer transport properties in the ternary blends compared to the binary BHJ devices.

Understood the limitation in the transport properties, we then focus on the better ability of the ternary system in the photogeneration by employing charge extraction (CE) and photoinduced absorption (PIA) spectroscopy (Salvador et al., 2012; Löslein et al., 2013; Gasparini et al., 2015b). In CE measurements, the samples were connected to the terminal of an oscilloscope with the input impedance of 1 MΩ and illuminated with a continuous background laser to keep it in the V_{oc} condition (Heumueller et al., 2015). In order to study the transient decay, we used a nanosecond switch that shifts the cell from open circuit to short circuit condition, allowing the calculation of charge density (n) by integrating the curves, respectively (**Figure 4A**) (Heumueller et al., 2015; Gasparini et al., 2016). In agreement with the J_{sc} values obtained, we calculate n as 2.97×10^{16} cm^{-3}, 3.02×10^{16} cm^{-3}, 3.66×10^{16} cm^{-3}, and 1.15×10^{16} cm^{-3} for PIDTTQ:PC$_{71}$BM (1:2), PIDTTQ:P1:PC$_{71}$BM (0.90:0.10:2), PIDTTQ:P1:PC$_{71}$BM (0.85:0.15:2), PIDTTQ:P1:PC$_{71}$BM (0.80:0.20:2) based devices, respectively (**Table 2**). The ability of charge generation in the BHJ solar cells is also studied with PIA spectroscopy. We employed steady-state and frequency-dependent PIA spectroscopy at a pump energy of 2.33 eV to gain further insight into the underlying photophysical steps of the sensitization process. **Figure 4B**

TABLE 2 | Summary of calculated charge carrier mobility (µ), bimolecular lifetime (τ), mobility-lifetime product (µτ), and charge carrier concentration (n) of binary and ternary devices.

PIDTTQ:P1:PC^{71}BM	µ [cm^2 V^{-1}s^{-1}]	τ [s]	µτ [cm^2 V^{-1}]	n [cm^{-3}]
1:0:2	1.13×10^{-4}	6.72×10^{-6}	7.59×10^{-10}	2.97×10^{16}
0.90:0.10:2	8.54×10^{-5}	7.35×10^{-6}	6.72×10^{-10}	3.02×10^{16}
0.85:0.15:2	7.53×10^{-5}	7.23×10^{-6}	5.44×10^{-10}	3.66×10^{16}
0.80:0.20:2	7.42×10^{-5}	4.72×10^{-6}	3.50×10^{-10}	1.15×10^{16}

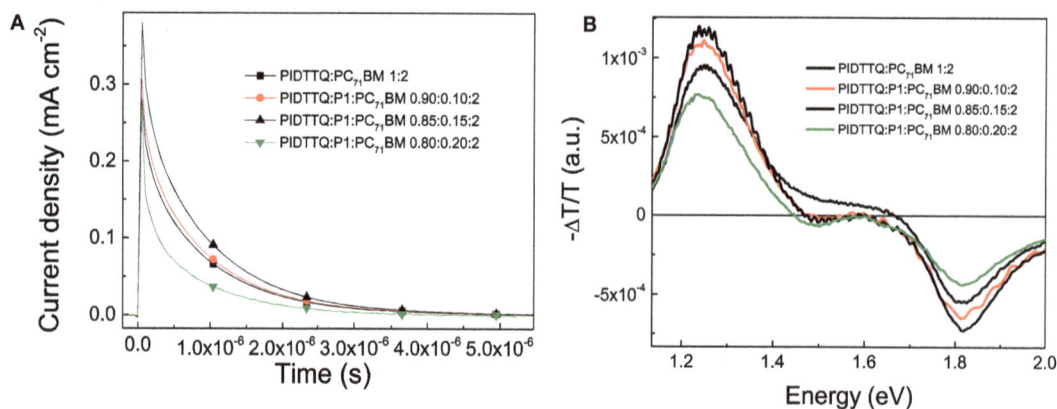

FIGURE 4 | Charge extraction curves (A) photo-induced absorption spectra (B) of PIDTTQ:PC$_{71}$BM binary and PIDTTQ:P1:PC$_{71}$BM ternary devices.

depicts the PIA spectra of binary and ternary devices measured at 60 mW cm^{-2} pump intensity at 10 K. All spectra show a pronounced transmission minimum (bleach) around 1.81 eV and a PIA feature around 1.24 eV. Contrary, the ternary blends show a novel bleaching feature around 1.58 eV. We associate the two transmission maxima at 1.81 and 1.58 eV to the photobleaching of the electronic ground states of PIDTTQ and P1, respectively. As shown in **Figure 4B**, higher polaron signal is observed for the ternary PIDTTQ:P1:PC$_{71}$BM (0.90:0.10:2), PIDTTQ:P1:PC$_{71}$BM (0.85:0.15:2)-based devices compared with the binary cells, confirming the higher photogeneration ability by adding 10–15% of P1 into the host PIDTTQ:PC$_{71}$BM system.

CONCLUSION

In conclusion, we reported a novel ternary system with a clear contribution in the incident photon-to-current efficiency in the NIR region. A J_{sc} improvement of around 20% was obtained for PIDTTQ:P1:PC$_{71}$BM (0.85:0.15:2) ternary devices compared to PIDTTQ:PC$_{71}$BM binary cells. However, the low FF limited the performances of the ternary BHJ solar cells. We studied the transport mechanism of the organic solar cells by employing photo-CELIV and TPV techniques. We found that by adding P1 into the host system of PIDTTQ:PC$_{71}$BM the $\mu\tau$ product is reduced, explaining the lower FFs. Despite the poorer transport properties, the complementary results of CE and PIA spectroscopy showed an improved charge generation in PIDTTQ:P1:PC$_{71}$BM (0.85:0.15:2) ternary solar cells, leading to a PCE of more than 4.6%.

EXPERIMENTAL SECTION

Materials and instruments: all reagents and starting materials were received from commercial suppliers and used without further purification. Anhydrous toluene was purchased from Sigma-Aldrich. Monomers M1 (Gasparini et al., 2015a) and M2 (Welch et al., 2013) were prepared according to literature procedures. The reactions were carried out under argon with standard and Schlenk techniques. Gel permeation chromatography (GPC) measurement was carried out at 135°C on a Waters Alliance 2000 GPC System equipped with PL-Guard and PL-mixed-B columns using trichlorobenzene as solvent. The UV spectrum was recorded on a Jasco V-670 spectrometer at room temperature. The HOMO energy level was determined by atmospheric pressure photoelectron spectroscopy using a photoelectron spectrometer model AC-2. The optical band gap was calculated using the formulae Eg = h(c/λ) + 0.3 eV. Cyclic voltammetry was executed in chloroform with 0.1 M (n-Bu)4NClO$_4$ against a standard calomel electrode. The electrochemical HOMO and LUMO energy levels were calculated using the formulae HOMO = −(Eox + 4.7) eV and LUMO = −(Ered + 4.7) eV, respectively.

Fabrication of Photovoltaic Devices

All devices were fabricated using doctor blading under ambient conditions with the structure of **Figure 1A**. Pre-structured ITO substrates were cleaned with acetone and isopropyl alcohol in an ultrasonic bath for 10 min each. After drying, the substrates

were successively coated with 40 nm of zinc oxide (ZnO), 10 nm of Ba(OH)$_2$, and finally a 80- to 90-nm-thick active layer based on PIDTTQ:PC$_{71}$BM and PIDTTQ:P1:PC$_{71}$BM (20 g L^{-1}). To complete the fabrication of the devices 10 nm of MoOx and 100 nm of Ag were thermally evaporated through a mask (with a 10.4 mm^2 active area opening) under a vacuum of ~2 × 10^{-6} mbar.

Nuclear Magnetic Resonance (NMR)

^{1}H-NMR and ^{13}C-NMR measurements were carried out in CDCl$_3$solutions on a BruckerAVANCE III 600 spectrometer using a resonance frequency of ^{1}H-250 MHz. The NMR system was controlled by the TopSpin 2.1 software by Bruker (Figure S2 in Supplementary Material).

Thermogravimetric Analysis

Thermogravimetric analysis measurements were performed on a Perkin–Elmer Pyris Diamond TG/DTA. Samples of approximately 5 mg were heated in air from 25 to 9°C, at a rate of 5°C/min.

Cyclic Voltammetry (CV)

Cyclic voltammetry studies were performed using a standard three-electrode cell. A platinum disk electrode was used as working electrode, a platinum wire as the counter electrode, and silver as the quasi-reference electrode. The oxidation and reduction potentials were calibrated against a ferrocene/ferrocenium (Fc/Fc$^+$) redox couple, then they were referenced against saturated calomel electrode. Tetrabutylammonium perchlorate (TBAP; 99%) was used as supporting electrolyte. Measurements were recorded using a PAR potensiostat/galvanostat Model VersaSTAT4, which was connected to a personal computer running the VersaStudio software version 2.44. In a typical experiment, around 2 mg of the material was diluted in chloroform in the presence of 0.1 M TBAP. The cyclic voltammetry graphs were recorded at a potential scan rate of 100 mV s^{-1} under argon atmosphere at 25°C.

J–V Measurements

The J–V characteristics were measured using a source measurement unit from BoTest. Illumination was provided by a solar simulator (Oriel Sol 1A, from Newport) with AM1.5G spectrum at 100 mW/cm^2. UV–VIS absorption was performed on a Lambda 950, from Perkin Elmer. EQEs were measured using an integrated system from Enlitech, Taiwan. In order to study the light intensity dependence of current density, we used a series of neutral color density filters. The intensity of light transmitted through the filter was independently measured *via* a power meter. All the devices were tested in ambient air.

Photo-CELIV

In photo-CELIV measurements, the devices were illuminated with a 405-nm laser diode. Current transients were recorded across an internal 50 Ω resistor of an oscilloscope (Agilent Technologies DSO-X 2024A). We used a fast electrical switch to isolate the cell and prevent CE or sweep out during the laser pulse and the td. After a variable td, a linear extraction ramp is applied *via* a function generator. The ramp, which was 20-μs long and 2 V in amplitude, was set to start with an offset matching the V_{oc} of the cell for each td.

TPV and CE Measurements

A 405-nm laser diode was settled for keeping the solar cells in approximately V_{oc} condition. Driving the laser intensity with a waveform generator Agilent 33500B and measuring the light intensity with a highly linear photodiode allowed to reproducibly adjust the light intensity with an error below 0.5% over a range of 0.2–4 suns. A small perturbation was induced with a second 405 nm laser diode driven by a function generator from Agilent. The intensity of the short (50 ns) laser pulse was adjusted to keep the voltage perturbation below 10 mV, typically at 5 mV. After the pulse, the voltage decays back to its steady state value in a single exponential decay. The characteristic decay time was determined from a linear fit to a logarithmic plot of the voltage transient and returned the small perturbation charge carrier lifetime. In CE measurements, a 405-nm laser diode illuminated the solar cell for 200 µs, which was sufficient to reach a constant open-circuit voltage with steady state conditions. At the end of the illumination period, an analog switch was triggered that switched the solar cell from open-circuit to short-circuit (50 Ω) conditions within less than 50 ns.

Photoinduced Absorption

Photoinduced absorption studies were performed by exciting the sample with a 405-nm laser while simultaneously probing the sample with a white lamp. The PIA spectra of the sample were dispersed by a 1,200 lines/mm grating monochromator (iHR320) and detected by a silicon detector through lock-in technique.

Atomic Force Microscopy

Atomic force microscopy measurements were performed on a solver nano from NT-MDT using 300 kHz single crystal silicon cantilevers (Nt-MDT, NSG30).

AUTHOR CONTRIBUTIONS

NG, TA, and CB conceived and developed the ideas. NG designed the experiments and performed device fabrication, electrical characterization, and data analysis. NG performed photo-CELIV, TPV, CE, and PIA measurements. AG-R synthesized the polymer under the supervision of AK, AA, GP, VG, CC, SA, US, SB, and AP-C performed CV measurements. NG and MP performed PL measurements and contributed to revision the manuscript. The projects were supervised by TA and CB.

ACKNOWLEDGMENTS

This project has received funding from the European Community's Seventh Framework Programme (FP7/2007-2013) under the Grant Agreement no 607585 project OSNIRO. In addition, this project has received funding from the European Community's Seventh Framework Programme (FP7/2007-2013) under the Grant Agreement no. 331389. CC acknowledges the financial support of a Marie Curie Intra European Fellowship (FP7-PEOPLE-2012-IEF) project ECOCHEM. GP would like to thank the Ministry of Education and Religious Affairs in Greece for the financial support of this work provided under the co-operational program "AdvePol: E850." The authors gratefully acknowledge the support of the Cluster of Excellence "Engineering of Advanced Materials" at the University of Erlangen-Nuremberg, which is funded by the German Research Foundation (DFG) within the framework of its "Excellence Initiative," Synthetic Carbon Allotropes (SFB953) and Solar Technologies go Hybrid (SolTech).

REFERENCES

Ameri, T., Dennler, G., Lungenschmied, C., and Brabec, C. J. (2009). Organic tandem solar cells: a review. *Energy Environ. Sci.* 2, 347. doi:10.1039/b817952b

Ameri, T., Li, N., and Brabec, C. J. (2013a). Highly efficient organic tandem solar cells: a follow up review. *Energy Environ. Sci.* 6, 2390–2413. doi:10.1039/c3ee40388b

Ameri, T., Khoram, P., Min, J., and Brabec, C. J. (2013b). Organic ternary solar cells: a review. *Adv. Mater.* 25, 4245–4266. doi:10.1002/adma.201300623

Cheng, P., Li, Y., and Zhan, X. (2014). Efficient ternary blend polymer solar cells with indene-C60 bisadduct as an electron-cascade acceptor. *Energy Environ. Sci.* 7, 2005. doi:10.1039/c3ee44202k

Cheng, P., Yan, C., Wu, Y., Wang, J., Qin, M., An, Q., et al. (2016). Alloy acceptor: superior alternative to PCBM toward efficient and stable organic solar cells. *Adv. Mater.* 28, 8021–8028. doi:10.1002/adma.201602067

Clarke, T. M., Lungenschmied, C., Peet, J., Drolet, N., and Mozer, A. J. (2015). A comparison of five experimental techniques to measure charge carrier lifetime in polymer/fullerene solar cells. *Adv. Energy Mater.* 5, 1401345. doi:10.1002/aenm.201401345

Gasparini, N., Jiao, X., Heumueller, T., Baran, D., Matt, G. J., Fladischer, S., et al. (2016). Designing ternary blend bulk heterojunction solar cells with reduced carrier recombination and a fill factor of 77%. *Nat. Energy* 1, 16118. doi:10.1038/nenergy.2016.118

Gasparini, N., Katsouras, A., Prodromidis, M. I., Avgeropoulos, A., Baran, D., Salvador, M., et al. (2015a). Photophysics of molecular-weight-induced losses in indacenodithienothiophene-based solar cells. *Adv. Funct. Mater.* 25, 4898–4907. doi:10.1002/adfm.201501062

Gasparini, N., Salvador, M., Fladischer, S., Katsouras, A., Avgeropoulos, A., Spiecker, et al. (2015b). An alternative strategy to adjust the recombination mechanism of organic photovoltaics by implementing ternary compounds. *Adv. Energy Mater.* 5, 1501527. doi:10.1002/aenm.201570132

Gedefaw, D., Tessarolo, M., Prosa, M., Bolognesi, M., Henriksson, P., Zhuang, W., et al. (2016). Induced photodegradation of quinoxaline based copolymers for photovoltaic applications. *Sol. Energy Mater. Sol. Cells* 144, 150–158. doi:10.1016/j.solmat.2015.08.015

Goh, T., Huang, J.-S., Yager, K. G., Sfeir, M. Y., Nam, C.-Y., Tong, X., et al. (2016). Quaternary organic solar cells enhanced by cocrystalline squaraines with power conversion efficiencies >10%. *Adv. Energy Mater* 6, 1600660. doi:10.1002/aenm.201600660

Hamilton, R., Shuttle, C. G., O'Regan, B., Hammant, T. C., Nelson, J., Durrant, J. R., et al. (2010). Recombination in annealed and nonannealed polythiophene/fullerene solar cells: transient photovoltage studies versus numerical modeling. *J. Phys. Chem. Lett.* 1, 1432–1436. doi:10.1021/jz1001506

He, Z., Xiao, B., Liu, F., Wu, H., Yang, Y., Xiao, S., et al. (2015). Single-junction polymer solar cells with high efficiency and photovoltage. *Nat. Photonics* 9, 174–179. doi:10.1038/nphoton.2015.6

Heumueller, T., Burke, T. M., Mateker, W. R., Sachs-Quintana, I. T., Vandewal, K., Brabec, C. J., et al. (2015). Disorder-induced open-circuit voltage losses in organic solar cells during photoinduced burn-in. *Adv. Energy Mater.* 5, 1500111. doi:10.1002/aenm.201500111

Holliday, S., Ashraf, R. S., Wadsworth, A., Baran, D., Yousaf, S. A., Nielsen, C. B., et al. (2016). High-efficiency and air-stable P3HT-based polymer solar cells with a new non-fullerene acceptor. *Nat. Commun.* 7, 11585. doi:10.1038/ncomms11585

Huang, J., Carpenter, J. H., Li, C.-Z., Yu, J.-S., Ade, H., and Jen, A. K. (2016). Highly efficient organic solar cells with improved vertical donor-acceptor compositional gradient via an inverted off-center spinning method. *Adv. Mater.* 28, 967–974. doi:10.1002/adma.201504014

Itskos, G., Othonos, A., Rauch, T., Tedde, S. F., Hayden, O., Kovalenko, M. V., et al. (2011). Optical properties of organic semiconductor blends with near-infrared quantum-dot sensitizers for light harvesting applications. *Adv. Energy Mater.* 1, 802–812. doi:10.1002/aenm.201100182

Ke, L., Min, J., Adam, M., Gasparini, N., Hou, Y., Perea, J. D., et al. (2016). A series of pyrene-substituted silicon phthalocyanines as near-IR sensitizers in organic ternary solar cells. *Adv. Energy Mater.* 6, 1502355. doi:10.1002/aenm.201502355

Keawsongsaeng, W., Gasiorowski, J., Denk, P., Oppelt, K., Apaydin, D. H., Rojanathanes, R., et al. (2016). Systematic investigation of porphyrin-thiophene conjugates for ternary bulk heterojunction solar cells. *Adv. Energy Mater.* 6, 1600957. doi:10.1002/aenm.201600957

Lee, T. H., Uddin, M. A., Zhong, C., Ko, S.-J., Walker, B., Kim, T., et al. (2016). Investigation of charge carrier behavior in high performance ternary blend polymer solar cells. *Adv. Energy Mater.* 6, 1600637. doi:10.1002/aenm.201600637

Li, N., Baran, D., Forberich, K., Machui, F., Ameri, T., Turbiez, M., et al. (2013). Towards 15% energy conversion efficiency: a systematic study of the solution-processed organic tandem solar cells based on commercially available materials. *Energy Environ. Sci.* 6, 3407–3413. doi:10.1039/c3ee42307g

Liu, Y., Zhao, J., Li, Z., Mu, C., Ma, W., Hu, H., et al. (2014). Aggregation and morphology control enables multiple cases of high-efficiency polymer solar cells. *Nat. Commun.* 5, 5293. doi:10.1038/ncomms6293

Löslein, H., Ameri, T., Matt, G. J., Koppe, M., Egelhaaf, H. J., Troeger, A., et al. (2013). Transient absorption spectroscopy studies on polythiophene-fullerene bulk heterojunction organic blend films sensitized with a low-bandgap polymer. *Macromol. Rapid Commun.* 34, 1090–1097. doi:10.1002/marc.201300354

Lu, L., Chen, W., Xu, T., and Yu, L. (2015a). High-performance ternary blend polymer solar cells involving both energy transfer and hole relay processes. *Nat. Commun.* 6, 7327. doi:10.1038/ncomms8327

Lu, L., Kelly, M. A., You, W., and Yu, L. (2015b). Status and prospects for ternary organic photovoltaics. *Nat. Photonics* 9, 491–500. doi:10.1038/nphoton.2015.128

Lu, L., Zheng, T., Wu, Q., Schneider, A. M., Zhao, D., Yu, L., et al. (2015c). Recent advances in bulk heterojunction polymer solar cells. *Chem. Rev.* 115, 12666–12731. doi:10.1021/acs.chemrev.5b00098

Lu, L., Xu, T., Chen, W., Landry, E. S., and Yu, L. (2014). Ternary blend polymer solar cells with enhanced power conversion efficiency. *Nat. Photonics* 8, 716–722. doi:10.1038/nphoton.2014.172

Min, J., Luponosov, Y. N., Gasparini, N., Xue, L., Drozdov, F. V., Peregudova, S. M., et al. (2015). Integrated molecular, morphological and interfacial engineering towards highly efficient and stable solution-processed small molecule solar cells. *J. Mater. Chem. A* 3, 22695–22707. doi:10.1039/C5TA06706E

Mozer, A. J., Sariciftci, N. S., Lutsen, L., Vanderzande, D., Österbacka, R., Westerling, M., et al. (2005). Charge transport and recombination in bulk heterojunction solar cells studied by the photoinduced charge extraction in linearly increasing voltage technique. *Appl. Phys. Lett.* 86, 112104. doi:10.1063/1.1882753

Nian, L., Gao, K., Liu, F., Kan, Y., Jiang, X., Liu, L., et al. (2016). 11% efficient ternary organic solar cells with high composition tolerance via integrated near-IR sensitization and interface engineering. *Adv. Mater.* 28, 8184–8190. doi:10.1002/adma.201602834

Salvador, M., MacLeod, B. A., Hess, A., Kulkarni, A. P., Munechika, K., Chen, J. I. L., et al. (2012). Electron accumulation on metal nanoparticles in plasmon-enhanced organic solar cells. *ACS Nano* 6, 10024–10032. doi:10.1021/nn303725v

Shuttle, C. G., O'Regan, B., Ballantyne, A. M., Nelson, J., Bradley, D. D. C., De Mello, J., et al. (2008). Experimental determination of the rate law for charge carrier decay in a polythiophene: fullerene solar cell. *Appl. Phys. Lett.* 92, 90–93. doi:10.1063/1.2891871

Spyropoulos, G. D., Kubis, P., Li, N., Baran, D., Lucera, L., Salvador, M., et al. (2014). Flexible organic tandem solar modules with 6% efficiency: combining roll-to-roll compatible processing with high geometric fill factors. *Energy Environ. Sci.* 7, 3284–3290. doi:10.1039/C4EE02003K

Spyropoulos, G. D., Ramirez Quiroz, C. O., Salvador, M., Hou, Y., Gasparini, N., Schweizer, P., et al. (2016). Organic and perovskite solar modules innovated by adhesive top electrode and depth-resolved laser patterning. *Energy Environ. Sci.* 9, 2302–2313. doi:10.1039/C6EE01555G

Squeo, B. M., Gasparini, N., Ameri, T., Palma-Cando, A., Allard, S., Gregoriou, V., et al. (2015). Ultra low band gap α,β-unsubstituted BODIPY-based copolymer synthesized by palladium catalyzed cross-coupling polymerization for near infrared organic photovoltaics. *J. Mater. Chem. A* 3, 16279–16286. doi:10.1039/C5TA04229A

Welch, G. C., Bakus, R. C., Teat, S. J., and Bazan, G. C. (2013). Impact of regiochemistry and isoelectronic bridgehead substitution on the molecular shape and bulk organization of narrow bandgap chromophores. *J. Am. Chem. Soc.* 135, 2298–2305. doi:10.1021/ja310694t

Yang, Y. M., Chen, W., Dou, L., Chang, W., Duan, H., Bob, B., et al. (2015). High-performance multiple-donor bulk heterojunction solar cells. *Nat. Photonics* 9, 190–198. doi:10.1038/nphoton.2015.9

You, J., Dou, L., Yoshimura, K., Kato, T., Ohya, K., Moriarty, T., et al. (2013). A polymer tandem solar cell with 10.6% power conversion efficiency. *Nat. Commun.* 4, 1446. doi:10.1038/ncomms2411

Zhang, J., Zhang, Y., Fang, J., Lu, K., Wang, Z., Ma, W., et al. (2015). Conjugated polymer-small molecule alloy leads to high efficient ternary organic solar cells. *J. Am. Chem. Soc.* 137, 8176–8183. doi:10.1021/jacs.5b03449

Zhang, Q., Kan, B., Liu, F., Long, G., Wan, X., Chen, X., et al. (2014). Small-molecule solar cells with efficiency over 9%. *Nat. Photonics* 9, 35–41. doi:10.1038/nphoton.2014.269

Zhao, W., Qian, D., Zhang, S., Li, S., Inganäs, O., Gao, F., et al. (2016). Fullerene-free polymer solar cells with over 11% efficiency and excellent thermal stability. *Adv. Mater.* 28, 4734–4739. doi:10.1002/adma.201600281

Conflict of Interest Statement: The authors declare that the research was conducted in the absence of any commercial or financial relationships that could be construed as a potential conflict of interest.

20

Economic Analysis of Improved Alkaline Water Electrolysis

Wilhelm Kuckshinrichs*, Thomas Ketelaer and Jan Christian Koj

Forschungszentrum Juelich, Institute for Energy and Climate Research – Systems Analysis and Technology Evaluation (IEK-STE), Juelich, Germany

Edited by:
Michel Feidt,
University of Lorraine, France

Reviewed by:
Rui Filipe Martins,
Centre de Recherche Public Henri Tudor, Luxembourg
Stoian Petrescu,
University Politehnica Bucharest, Romania

*Correspondence:
Wilhelm Kuckshinrichs
w.kuckshinrichs@fz-juelich.de

Specialty section:
This article was submitted to Energy Systems and Policy, a section of the journal Frontiers in Energy Research

Citation:
Kuckshinrichs W, Ketelaer T and Koj JC (2017) Economic Analysis of Improved Alkaline Water Electrolysis. Front. Energy Res. 5:1.

Alkaline water electrolysis (AWE) is a mature hydrogen production technology and there exists a range of economic assessments for available technologies. For advanced AWEs, which may be based on novel polymer-based membrane concepts, it is of prime importance that development comes along with new configurations and technical and economic key process parameters for AWE that might be of interest for further economic assessments. This paper presents an advanced AWE technology referring to three different sites in Europe (Germany, Austria, and Spain). The focus is on financial metrics, the projection of key performance parameters of advanced AWEs, and further financial and tax parameters. For financial analysis from an investor's (business) perspective, a comprehensive assessment of a technology not only comprises cost analysis but also further financial analysis quantifying attractiveness and supply/market flexibility. Therefore, based on cash flow (CF) analysis, a comprehensible set of metrics may comprise levelised cost of energy or, respectively, levelized cost of hydrogen (LCH) for cost assessment, net present value (NPV) for attractiveness analysis, and variable cost (VC) for analysis of market flexibility. The German AWE site turns out to perform best in all three financial metrics (LCH, NPV, and VC). Though there are slight differences in investment cost and operation and maintenance cost projections for the three sites, the major cost impact is due to the electricity cost. Although investment cost is slightly lower and labor cost is significantly lower in Spain, the difference can not outweigh the higher electricity cost compared to Germany. Given the assumption that the electrolysis operators are customers directly and actively participating in power markets, and based on the regulatory framework in the three countries, in this special case electricity cost in Germany is lowest. However, as electricity cost is profoundly influenced by political decisions as well as the implementation of economic instruments for transforming electricity systems toward sustainability, it is hardly possible to further improve electricity price forecasts.

Keywords: alkaline water electrolysis, levelized cost of hydrogen, net present value, variable cost, weighted average cost of capital

INTRODUCTION

Alkaline water electrolysis (AWE) is a mature hydrogen production technology (Ursua et al., 2012) and there exists a range of economic assessments for available technologies. In most cases, these assessments focus on typical cost components such as investment, operation and maintenance

(O&M), and decommissioning, which are commonly called technology cost (Bertuccioli et al., 2014; Noack et al., 2015). For new AWEs, which may be based on novel polymer-based membrane concepts (Koj et al., 2015), it is of prime importance that development comes along with new configurations and technical and economic key process parameters for AWE that might be of interest for further economic assessments.

In many cases, economic assessments for technologies focus on cost analysis. But, from an investor's (business) perspective, a comprehensive economic assessment of a technology also comprises further financial analysis quantifying attractiveness and supply/market flexibility. Therefore, based on cash flow (CF) analysis, a comprehensible set of metrics may comprise levelized cost of energy or, respectively, levelized cost of hydrogen (LCH) for cost assessment, net present value (NPV) for attractiveness analysis, and variable cost (VC) for analysis of market flexibility.

Aiming to overcome these drawbacks and to contribute to economic assessment of AWE technology, this paper presents cost estimations for an advanced AWE technology and a financial analysis. Starting with a brief overview on the literature to AWE cost results (see AWE Cost Literature Overview), Section "Methodological Approach and Estimation of Key Parameters" defines the system boundaries for the technology assessment. Additionally, in Section "Methodological Approach and Estimation of Key Parameters," the methodological frame for the analysis is presented focusing on the financial metrics, the projection of key performance parameters of advanced AWEs, and further financial and tax parameters. The specification of an electricity scenario is of major importance. The analysis comprises projections for the technology and electricity supply at three different sites (Germany, Austria, and Spain), all representing Western European countries, though the integration into international trade and material flows is different. The results for the cost, attractiveness, and market flexibility metrics are presented in Section "Financial Analysis for 6 MW AWE Hydrogen Generation." Additionally, the relevance of key performance parameters for AWE assessment is exemplarily demonstrated by elasticities for LCH and VC. Finally, Section "Discussion" presents a discussion of main findings against the background of main assumptions and key data, and Section "Summary" identifies further methodological challenges.

AWE COST LITERATURE OVERVIEW

Alkaline water electrolysis investment cost estimations depend mainly on plant size and site-specific characteristics (Bertuccioli et al., 2014; Noack et al., 2015, Pellinger and Schmid, 2016; Shaner et al., 2016). Identifying a present range of 1,000–1,200 €2015/kW capacity, Bertuccioli estimated 1,100 €2015/kW as central investment cost and projected large potential for cost decrease up to 580 €/kW in 2030 (Bertuccioli et al., 2014).

With respect to economic analysis, many studies use the concept of LCH (Bertuccioli et al., 2014; Shaner et al., 2016). Depending on plant and site characteristics and on the terms for electricity supply, LCH in 2012 comprises 3.2 €/kg for Germany and 5.2 €/kg for UK. In the future, there is potential for cost decreases (development of technology and learning rates)

lowering the investment cost. With respect to the main operating cost component electricity, projections need to identify technological trajectories. In addition, the projections must also deal with energy and climate policy aspects, as they are making future electricity cost highly uncertain.

Even if LCH analysis prevails, most present studies neglect financial and fiscal aspects. They also represent economic analysis as pure LCH analysis, ignoring other economic aspects relevant for decision-making, either for investment or for operating decisions.

METHODOLOGICAL APPROACH AND ESTIMATION OF KEY PARAMETERS

From a methodological viewpoint, the economic assessment of a technology requires integration of the three dimensions time, space, and well-specified system boundaries equally guaranteeing relevance of the analysis and integrity of the results (**Figure 1**). The temporal system boundary reflects the time duration from "cradle to grave" which in economic terms covers investment in a technology, operation, and decommission. Investment is based on capital availability, either by equity or bank loans. Operation affords purchase of upstream products such as water or electricity. The spatial system boundary reflects the plant locations and the regional origin of upstream products. Additionally, according to the economic lifetime of the AWE technology, an electricity scenario is eminent in order to quantify cost (and possibly emission balances) for the base year and all subsequent years. The inputs have market prices and incentive schemes such as investment subsidies principally may prevail. Whereas hydrogen is the physical output, additionally, in business perspective there is also the monetary output "avoided tax on earnings," which is in any case relevant (**Figure 1**).

From private perspective, two points attract attention:

- Financial and fiscal aspects
  Financial aspects such as share of private equity or debt for investment in energy technologies deserve closer attention from a business perspective the higher the share of capital expenditure to total expenditures. On the one hand, equity and debt may require different returns, and on the other hand, only interest payments for debts are tax deductible. Furthermore, fiscal incentives such as investment subsidies or the crediting of tax redemptions need further attention (e.g., Chandrasekar and Kandpal, 2005; Simshauser, 2014).

- Discounting
  For discounting in business, the weighted average cost of capital (WACC) concept is chosen. It reflects shares and returns of equity and debts and is used in many studies on energy technology cost (Burgess, 2011; Kost et al., 2013, Peter et al., 2013; Burwen, 2015, Taylor et al., 2015). Depending on the technology and the kind of investors, WACCs may vary considerably (Stubelj et al., 2014). Small-size investments in, e.g., PV installations typically have high shares of equity. The share of equity for large-size installations such as offshore wind parks, often requiring hundreds of m€2015, is typically much

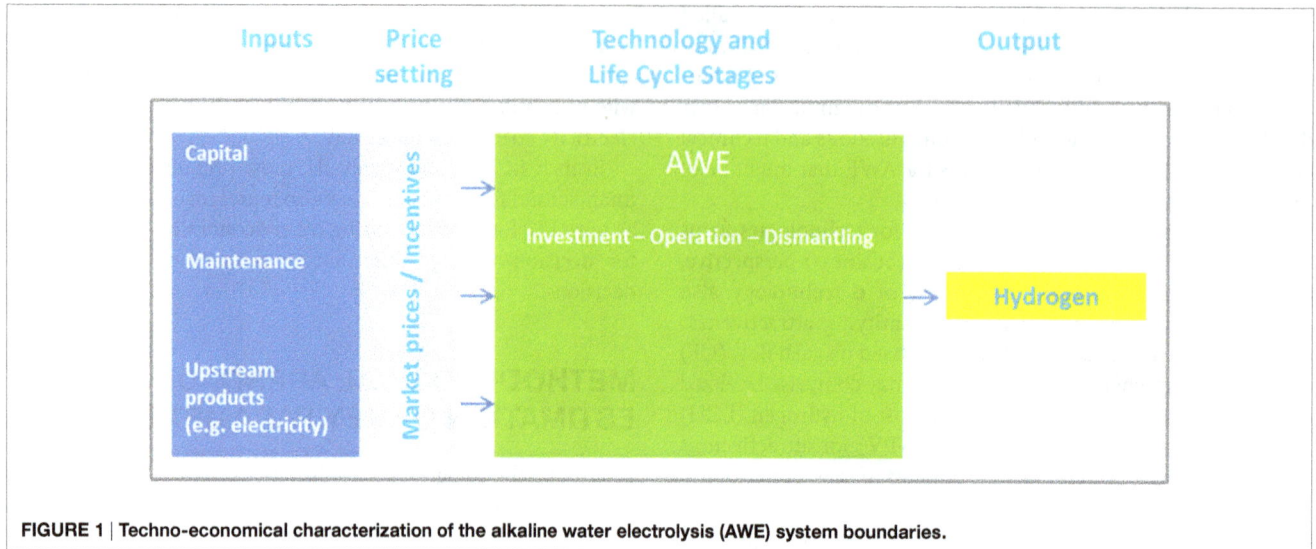

FIGURE 1 | Techno-economical characterization of the alkaline water electrolysis (AWE) system boundaries.

lower (30–40%). Accordingly, the share of debt for financing is much higher. The return on debt equals usual market conditions, whereas in the energy market the return on equity typically is higher, reflecting opportunity cost such as the expected returns on alternative investments (Peter et al., 2013). However, private household's investments in, e.g., PV installations usually do not reflect "opportunity cost" comparisons. For discounting, the post-tax Vanilla WACC is chosen. The real WACC is a function of nominal returns on equity and debt interest rates and inflation. With the post-tax Vanilla WACC, tax on earnings should be directly incorporated into the cost formula (Allen Consulting Group, 2007).

The following subsections focus on the main methodical pillars for the analysis of life-cycle cost of an AWE. Therefore, the focus is on the metrics' approach and the approaches to specify realistic projections of sensitive technical and economic parameters such as investment cost or future electricity cost. Sections "Financial Metrics for Technology Assessment" and "Key Parameters" discuss details on main metrics, key data, and their sources.

Financial Metrics for Technology Assessment

Generally, from private perspective, economic metrics assess business efforts to invest in a technology and run this technology. Accordingly, cost components such as project cost, O&M cost, financing cost, tax redemptions, and capital sources such as private equity and bank loans take center stage. In many cases, technology assessments focus on cost analysis. From an investor's (business) perspective, a comprehensive economic assessment of a technology comprises analysis of cost, attractiveness, and supply/market flexibility (**Figure 2**) (Sousa de Oliveira et al., 2011; Sousa de Oliveira and Fernandes, 2012, Lee, 2016).

Based on a CF analysis, cost metrics typically comprise indicators such as LCH (Darling et al., 2011; Harvego et al., 2012) (**Table 1**, Eq. 1). LCH represents a fundamental approach enabling to compare competing hydrogen technologies. This indicator

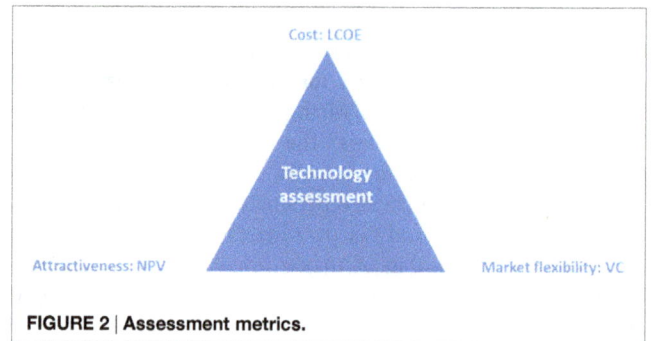

FIGURE 2 | Assessment metrics.

may be interpreted as a break-even value indicating a price that is needed as revenue over the lifetime of the technology in order to justify an investment in a particular energy generation facility and covering all expenses and the payment of an acceptable return to investors. From private perspective, the expenditures for investment and the capital's source are relevant. As private equity usually expects higher returns than is necessary for bank loans, this is important for financing cost. Also, fiscal aspects such as tax redemptions may be significant in assessing a technology investment. This is attended by the WACC approach (**Table 1**, Eq. 5), a discounting concept representing different capital sources (equity and bank loan).

In order to assess attractiveness from investor's perspective, the NPV is defined as the present value benefits less the present value cost (**Table 1**, Eq. 2) (White, 2011; Thakre, 2014, Cucchiella et al., 2015).

For operation, the flexibility of the technology in relation to market conditions is of special relevance. The VC denotes those cost components directly attributable to the operation of a plant (**Table 1**, Eq. 3) (Borenstein, 2000). VC characterizes the minimum-level market price for operating an AWE site. Lower prices result in operational losses. In contrast, fix cost covers those components attributable after investment irrespective of the operational time.

TABLE 1 | Metrics formulas.

Cost: levelized cost of hydrogen	$$\dfrac{\sum_{t=0}^{t=n}\dfrac{PCI_t}{(1+r)^t}+\sum_{t=1}^{t=n}\dfrac{\left(A_t+E_t+EC_t+LP_t\right)}{(1+r)^t}-TR\times\sum_{t=1}^{t=n}\dfrac{\left(A_t+E_t+EC_t+Dep_t+Int_t\right)}{(1+r)^t}}{\sum_{t=1}^{t=n}\dfrac{M_t}{(1+r)^t}}$$	(1)
Attractiveness: net present value	$$\sum_{t=0}^{t=n}\dfrac{PCI_t}{(1+r)^t}+\sum_{t=1}^{t=n}\dfrac{\left(A_t+E_t+EC_t+LP_t\right)}{(1+r)^t}-TR\times\sum_{t=1}^{t=n}\dfrac{\left(A_t+E_t+EC_t+Dep_t+Int_t\right)}{(1+r)^t}$$	(2)
Market flexibility: variable cost	E_t+EC_t	(3)
WACC	$$WACC^{nom}=\dfrac{e}{e+d}\times e_r+\dfrac{d}{e+d}\times e_d$$	(4)
	$$WACC^{real}=\left(\dfrac{1+WACC^{nom}}{1+infl}\right)-1$$	(5)

PCI_t: investment by equity minus invest tax credit of grant (t = 0: initial investment, t = T: replacement investment, t = N: decommissioning), A_t: fix O&M, E_t: variable O&M, EC_t: electricity cost, LP_t: loan payment (for investment by debt), TR: tax rate, Dep_t: depreciation, Int_t: interest on debt, M_t: yearly hydrogen production, r: weighted average cost of capital (WACC), e: equity, e_r: equity rate of return, d: debt, e_d: debt interest rate, infl: inflation rate.

Key Parameters

Technology and Plant Characteristics

Alkaline water electrolysis technology for hydrogen production has a long tradition and became industrialized and a mature technology within the last decades. The first pressurized AWE was already built by Zdansky/Lonza in 1948 (Kreuter and Hofmann, 1998). It is estimated to be part of a future so-called "hydrogen economy" (Bockris, 2002; McDowall and Eames, 2006, Zeng and Zhang, 2010). Centerpieces of AWE are the electrolysis cells. Inside, the cells electrolysis is enabled by an electric current between the electrodes and a circulating aqueous potassium hydroxide solution. Migration of ions and the separation of both electrolysis products, hydrogen and oxygen, is enabled by a membrane. According to Bhandari et al. (2014) following chemical reactions take place within the cells:

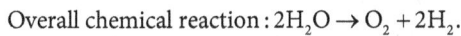

$$Anode : 4OH^- \rightarrow O_2 + 2H_2O + 4e^-$$
$$Cathode : 4H^+ + 4e^- \rightarrow 2H_2$$
$$Overall\ chemical\ reaction : 2H_2O \rightarrow O_2 + 2H_2.$$

Figure 3 shows the technical system boundaries for the AWE plant. For operation, essential upstream products beside electricity comprise deionized water, KOH solution, process steam, and nitrogen.

Koj et al. (2015) analyzed and defined key technical plant parameters for an advanced commercial scale 6 MW AWE plant in comparison to a 3.5 MW state-of-the-art pressurized electrolysis. In contrast, the new developed electrolysis plant uses advanced polymer-based membranes. The advance plant, scaled-up to 6 MW, achieves higher hydrogen outputs with the same number of cells with equal diameters. Cell stacks are essential system components framing the electrolysis cells, were the electrolysis takes place. Moreover, there are further components necessary for the system assembly and operation comprising gas separator (for separating O_2, H_2, and small amounts of KOH), lye tank for potassium hydroxide (KOH), KOH filter, heat exchangers (for cooling O_2, H_2, and KOH), pumps (pumping of KOH

and water), and power electronics. H_2 compression, storage, and use are not considered. Additionally, selling the by-product O_2 is not considered, as large-scale deployment of electrolysis capacities may go along with market saturation for O_2 and strongly decreasing prices (Hermann et al., 2014). **Table 2** shows the plant characteristics, and the functional unit is the physical output of 1 kg H_2 at 33 bar and 40°C (Koj et al., 2015).

With respect to key performance parameters, it is necessary to distinguish country-specific parameters from non-country-specific ones. Technological characteristics are by definition cross-national. Main parameters comprise expected plant lifetime (20 year), stack lifetime (83,000 h operation), operational time (8,300 h/year), and, respectively, hydrogen output (118.25 kg H_2/h).

Correspondent specific demands are summarized in **Table 3**. Additionally, specific electricity demand is of particular importance not only from a technological point of view but also from the economic perspective. The technical process and plant characteristics hold irrespective the location in Germany, Spain, or Austria (**Tables 2** and **3**).

Cost Characteristics for Financial Analysis

In order to estimate AWE cost, the capacity increases from 3.5 to 6 MW, which is the basis for the technology study in Koj et al. must be considered.

While manufacturers do not share in-house data on cost breakdown, **Figure 4** shows results gathered by literature review (Bertuccioli et al., 2014). The cost breakdown for the system cost shows that the stack contributes to half of the overall cost. In order to identify the cost impact of new polymer-based membranes, the stack level breakdown is necessary as well. Membranes contribute to 7% of the stack cost. Regarding the entire system cost, membranes only contribute 3.5%. Therefore, we argue the cost impacts introducing new membrane concepts is not of major relevance for the overall cost projection.

For plant upscaling from 3.5 to 6 MW, an engineering top-down approach for cost scaling is used (Eq. 6)

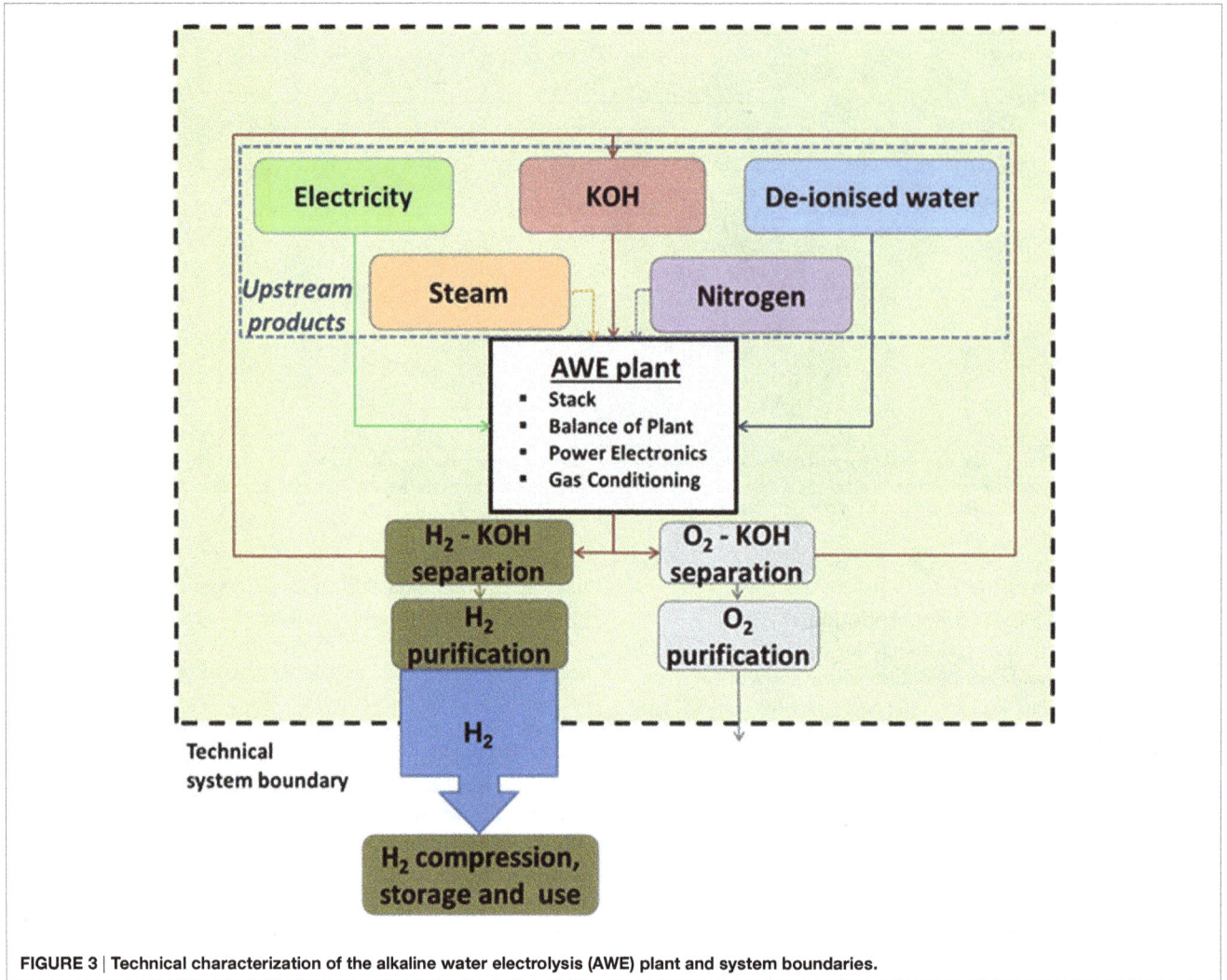

FIGURE 3 | Technical characterization of the alkaline water electrolysis (AWE) plant and system boundaries.

TABLE 2 | Alkaline water electrolysis (AWE) technology and plant characteristic (cross-national).

	Unit	AWE plant	Comments and data sources
Plant characteristics			
Capacity	MW	6.0	Advanced AWE plant, commercial scale though not yet commercially operated, plant characteristics from EU R&D project ELYGRID (Koj et al., 2015)
Plant lifetime	years	20	
Stack lifetime	H	83,000	
Operation	h/year	8,300	
Hydrogen output	kg H$_2$/h	118.25	

TABLE 3 | Alkaline water electrolysis (AWE) technology and inventory (cross-national).

	Unit	AWE plant	Comments and data sources
Upstream products			
Water, deionized	kg/kg H$_2$	10.11	Advanced AWE plant, commercial scale though not yet commercially operated, inventory results from EU R&D project ELYGRID (Koj et al., 2015)
KOH solution	10^{-4} kg/kg H$_2$	2.75	
Process steam	kg/kg H$_2$	0.038	
Nitrogen	10^{-4} kg/kg H$_2$	0.7115	
Electricity	kWh$_{el}$/kg H$_2$	53.9	

(Eerev and Patel, 2012). It enables to estimate the cost of a given investment with different capacities, in our case the increase from 3.5 to 6 MW. α denotes the scaling exponent, which is in principle unknown. Often, a value of 0.6–0.7 is used as default (also referred to as six-tenths or seven-tenths rule) (Eerev and Patel, 2012). However, as AWE technology already has a high level of maturity, we carefully set a higher scaling exponent (0.85).

$$I_x = I_{base} \times \left(\frac{Cap_x}{Cap_{base}} \right)^{\alpha} \qquad (6)$$

where x: new capacity, base: old capacity, and α: scaling exponent.

Cross-national parameter estimation only partly holds for cost specifications for the three sites (**Table 4**). We distinguish direct depreciable capital cost (investment components) from indirect

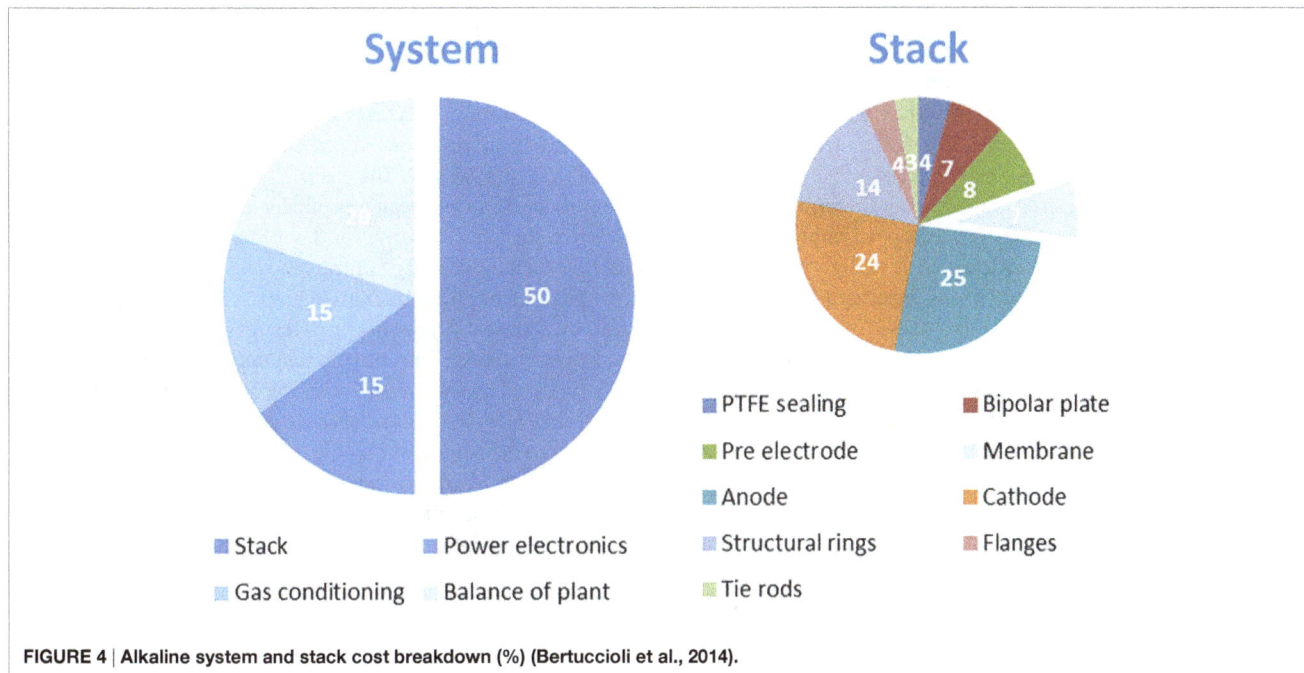

FIGURE 4 | Alkaline system and stack cost breakdown (%) (Bertuccioli et al., 2014).

TABLE 4 | Alkaline water electrolysis (AWE) technology cost and plant characteristics (except for electricity).

	Unit	Germany	Austria	Spain	Comments and data sources
Investment					
Direct depreciable capital cost (ddcc)	m€$_{2015}$/MW	0.85			• engineering scaling approach is used (Eerev and Patel, 2012)
Stack replacement	%/ddcc	50			• general investment cost data (Bertuccioli et al., 2014)
Plant decommissioning	%/ddcc	6			• distinction of ddcc and idcc (Harvego et al., 2012)
Indirect depreciable capital cost (idcc)	m€$_{2015}$/MW	0.17	0.16	0.14	• total initial capital cost for German site is consonant to Noack et al. (2015), based on bottom-up modeling of a 5 MW AWE (1.07 m€$_{2015}$/MW)
Fix O&M					
Material	%/ddcc	2.5			• material and labor cost share (Bertuccioli et al., 2014)
Labor	%/ddcc	2.5			• wage rate index 2014 (Statistisches Bundesamt, 2015)
Wage rate index		1.00	0.93	0.62	
Variable O&M, except electricity					
Water, deionized	€$_{2015}$/kg	0.0100			• best guess, representative commercial upstream product price data not available
KOH	€$_{2015}$/kg	2.5106			• price of deionized water comparatively relevant for overall variable O&M cost
Steam	€$_{2015}$/kg	0.0100			
Nitrogen	€$_{2015}$/kg	0.2783			

All cost and prices expressed in real €$_{2015}$ (nominal increase over the lifetime of technology).

ones (such as site preparation, engineering-design, and upfront permissions). The former is expected to have a world-market price whereas the latter has a high local content so that country-specific wage rates are relevant. In case of Spain, this is important as its wage rate index is approximately 0.62 (relative to Germany). Although defined by Koj et al. as commercial scale 6 MW plant, market prices for investment components are not available. Here, available information on investment components, cost structures, and average total investment cost (Bertuccioli et al., 2014) is used.

Specific investment cost ranges from 1.02 to 0.99 m€$_{2015}$/MW (Germany, Spain); the slight difference is due to the local

content for indirect depreciable capital cost. The result for top-down assessed total investment cost is consonant to results from Noack et al. (2015), who derived investment cost data based on bottom-up modeling a 5 MW AWE site in Germany for hydrogen generation (and underground storage). Stack replacement, decommissioning, and fix operation and maintenance (O&M) cost are defined as fixed percentage share of direct depreciable cost (Bertuccioli et al., 2014). VC accrues for upstream products and comprises deionized water, KOH, steam, and nitrogen. For electricity, country-specific cost is defined in the subsequent electricity scenario part.

Upstream Product: Electricity

For the upstream product electricity, a scenario approach is necessary, comprising the three dimensions time horizon, regions, and primary energy mix for electricity. The *time horizon* is from 2015 (base year) to 2035 (final year). It is chosen according to the projected economic lifetime of an AWE plant.

The *regions* that are covered are Germany, Spain, and Austria. These countries all represent Western European industries; however, the integration into international trade and material flows is different. The regional representation is important for electricity cost.

The focus on *electricity generation mix* is eminent. AWE demands considerable amounts of electricity, which in this study is supplied by the grid and which therefore shows country-specific electricity generation and emissions profiles. Focusing on the electricity mix reflects the question how to implement hydrogen generation in a present system still significantly relying—at least for some countries—on fossil electricity generation, although these systems presumably will decarbonize its electricity generation in the future. Germany and Spain both still rely extensively on fossil resources for electricity generation. On the other hand, average industrial electricity prices are higher in Spain than in Germany. Austria shows significant differences with respect to the generation mix that already mainly depends on renewable energies, i.e., hydro power. This is of special importance: there are country-specific conditions to get access to electricity and to what price.

For the *deployment of non-carbon electricity generation technologies* between 2015 and 2035, and implicitly, the electricity mix basic information is from the European "Trends to 2050" scenario (Capros et al., 2013). This study assesses a reference scenario for EU28. Based on main assumptions with respect to macroeconomic development, population, world fossil fuel prices, and further ones, a harmonized macroeconomic modeling approach is used to project the energy generation mix on EU member state level. Therefore, the results for the electricity generation mix on country level are consistently developed and harmonized against a set of macroeconomic assumptions for the European Union. Based on results for 2015 and 2035, we assume linear technology diffusion paths that signal the change from present generation to future electricity generation mix. As the countries start from different settings, the electricity generation mix in 2035 will still be different in its composition of decarbonized generation technologies.

With *respect to electricity for hydrogen production* and its base year price, the study avoids reflection on average specific industry electricity prices. These are by trend high or very high in most European countries, but also very heterogeneous across industries. Electricity prices differ considerably, which is mainly due to different generation technologies, heterogeneous regulation, and tax schemes in European countries [for an overview on different tax and pricing policies and regulation schemes in EU28 and further European countries, see Eurostat (2015a)]. For the sake of the study, instead, hydrogen plant operators are regarded as customers actively participating in power markets.

In most European countries, the electricity price can be split into three parts. A competitive part includes the cost for generation and distribution. A second part includes the cost for grid use, and a third part includes taxes and further charges.

For the three investigated countries, the competitive part is formed in a similar way. The electricity price is either a result of bilateral contracts between a producer and a consumer—in this case an industrial company—or it is a market price negotiated in different bilateral trading contract as over the counter trades or it originates from a product at an electricity exchange. It can be based on different products as futures/forward, day ahead trades and intraday trades. The composition depends on the strategy of the company. Depending on the selling organization, distribution cost differs as well. Because of these different possibilities, there is no way to determine one single price for 1 year that is valid for a whole sector. Therefore, prices for generation and distribution can only be average prices based on surveys. Eurostat is publishing such survey data every 6 month (Eurostat, 2015b).

To get a better understanding about the differences in electricity prices in Austria, Germany, and Spain, a more detailed view on their national regulation schemes is useful focusing on the regulation for grid use as well as on taxes and further charges. In Austria and Germany, these price parts depend on the electricity demand, the electricity intensity, and the grid level. In Spain, those factors do not have an effect on the calculation of the electricity prices. For the calculation of the electricity price in Austria and Germany, the specification of the electrolysis plant is necessary as to know how much electricity will be demanded and, therefore, which price category might be reasonable.

Based on data from the Austrian, regulation agency (E-Control) (E-Control, 2015), the German association for energy and water (BDEW) (BDEW, 2015), and Eurostat (Eurostat, 2015b), it can be shown that an electrolysis plant, actively participating in the electricity markets of the three respective countries differ. In Austria, the AWE can be connected to the grid levels 4–6. Therefore, the electricity price varies between 7.64 $€_{2015}$ct/kWh (grid level 4) and 10.13 $€_{2015}$ct/kWh (grid level 6). For the German site, the electricity price depends on the classification of the company that will operate the AWE. Either the company is classified as an energy-intensive company or as a non-energy-intensive company. The difference in the electricity price amounts to 6.02 $€_{2015}$ct/kWh (6.26 vs. 12.27 $€_{2015}$ct/kWh), mainly depending on the fact if the AWE operator has to pay the EEG levy. In Spain, it is not clear if the tax of the National Energy Commission is already included in the competitive part of the electricity price or not. Thus, both prices are given. The difference is less than 0.02 $€_{2015}$ct/kWh (**Table 5**).

The LCH uses the price of grid level 4 for Austria (7.64 $€_{2015}$ct/kWh), which is a common grid level for industrial companies. For Germany, the price for an energy-intensive company is taken (6.26 $€_{2015}$ct/kWh) as this will be the case if we assume that the operator only operates the AWE. The price with the taxes of the National Energy Commission not included in the competitive part is used for the calculations for Spain (8.30 $€_{2015}$ct/kWh).

The electrolysis plant is assumed to run for 20 years. Therefore, the electricity price that is needed for the LCC must be projected

TABLE 5 | Base year (2015) electricity price components for Austria, Germany, and Spain (53 GWh/a).

€_{2015}c/kWh	Austria			Germany		Spain	
	Grid level 4	Grid level 5	Grid level 6	Energy intensive	Non-Energy intensive	Tax of National Energy Commission included in competitive part	Tax of National Energy Commission not included in competitive part
Competitive part	4.53	4.53	4.53	5.95	5.95	7.9	7.9
Grid use fees	2.36	3.47	4.93	0.03	0.05	0	0
Non-recoverable taxes and further fees	0.75	0.56	0.68	0.28	6.27	0.40	0.42
Price	7.64	8.54	10.13	6.26	12.28	8.30	8.32

for this time horizon as well. The following projections can be found in the study "Trends to 2050" (Capros et al., 2013). For the time from 2010 to 2030, the trend study proposes an average electricity price increase of 1.34%/year. The increase is mainly due to capital cost, and governmentally influenced cost components such as taxes on fuels, ETS payments, and RES supporting schemes.

For our calculation, we use our own figures of 2015 prices for Austria, Germany, and Spain, and the EU figures for price increase per year. Therefore, starting from different electricity prices in 2015 the Austrian, German, and Spanish AWE sites will come up with different electricity prices in 2035.

Financial and Tax Parameters

Financial and tax parameters are in principle country specific. However, in-depth country-specific tax considerations are beyond this paper. Therefore, for financial and tax considerations, we use cross-national parameters (**Table 6**). The chosen equity-to-debt-ratio 25:75 for Western European countries is consistent to literature (Kost et al., 2013; Ondraczek et al., 2015). This also holds for return on equity and interest rates on debt (Kost et al., 2013). Inflation rate is one of those macroeconomic parameters the hardest predictable over the long term, and macroeconomic projections for inflation rates usually are short-term focused. Methodologically, this poses a challenge for financial assessment of technologies with a lifetime of 20 years. Though in macroeconomic theory and in practical policy, a targeted inflation rate of approximately 2% is regarded acceptable for price stability for our purposes we consider a lower projection of 1%, which is at the bottom line of present projections (EC Directorate-General for Economic and Financial Affairs, 2016). In case inflation projections are regarded too low, it needs to be kept in mind that higher inflation rates should result in higher nominal return and interest expectations.

Based on these parameters, WACC^{real} equals 4%, which is in good accord with considerations in literature on a wide range of energy technologies (Kost et al., 2013; Alberici et al., 2014) and also with IPCC lower range of 5% (Schlömer et al., 2014). Due to very different country-specific characteristics, IPCC additionally calculates an upper range of 10%. Ondraczek et al. (2015) use WACC parameters of 3.8–4.3% for Western European countries (Ireland, Switzerland, UK, and Netherlands).

TABLE 6 | Financial and tax on earnings parameters (cross-national).

	Unit	Germany–Austria–Spain	Comments and data sources
Share of equity	%	25	Resulting weighted average cost of capital (WACC) approximately 4% (in the range of Alberici et al., 2014)
Share of debt	%	75	
Return on equity	%	7.0	
Interest rate on debt	%	4.5	
Inflation rate	%	1.0	
Tax (on earnings) rate	%	30	Simplified approach, as different country-specific tax systems are too complex to model in detail for this purpose

FINANCIAL ANALYSIS FOR 6 MW AWE HYDROGEN GENERATION

Financial Metrics: CF, LCH, NPV, and VC

The results for CFs for an AWE site are depicted in **Figure 5** (it has to be kept in mind that the CF covers expenses and no sales). Here, the results for the AWE site in Germany are shown. The CFs for the Austrian and Spanish sites are slightly different, but similar in its structure. The investment cost is accounted for by the source of its funding, which is why investment is represented by PCI cost (investment by equity) for initial investment, stack replacement and decommissioning, and corresponding loan payment cost and interest. The electricity expenses are main cost drivers, increasing over time according to increasing (real) electricity prices. Keeping in mind that hydrogen sales are not accounted for, and the NPV is calculated for the stream of expenses, the NPV is negative (app. −66 m€_{2015}). For Austria and Spain, the corresponding (negative) NPVs are lower mainly due to higher electricity cost.

Tax deductibility of cost elements is also important, as can be seen by the negative CF elements. Negative CFs represent a form of return flows; here return flows in the form of avoided tax on earnings. Again keeping in mind that hydrogen sales are not accounted for, the NPV of the stream of expenses is negative (app. −50 m€_{2015}). Tax deductibility of cost elements has significant influence as is shown by the higher NPV.

The structure of LCH is similar for the AWE sites (**Figure 6**). In all cases, the cost for the upstream product electricity is the

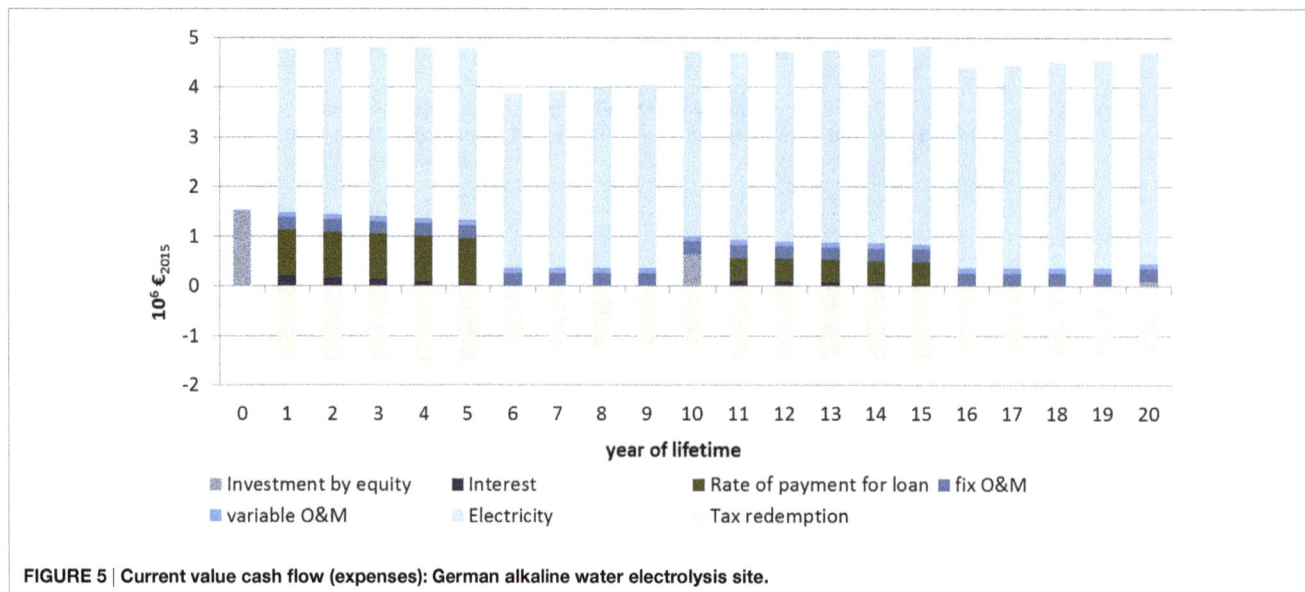

FIGURE 5 | Current value cash flow (expenses): German alkaline water electrolysis site.

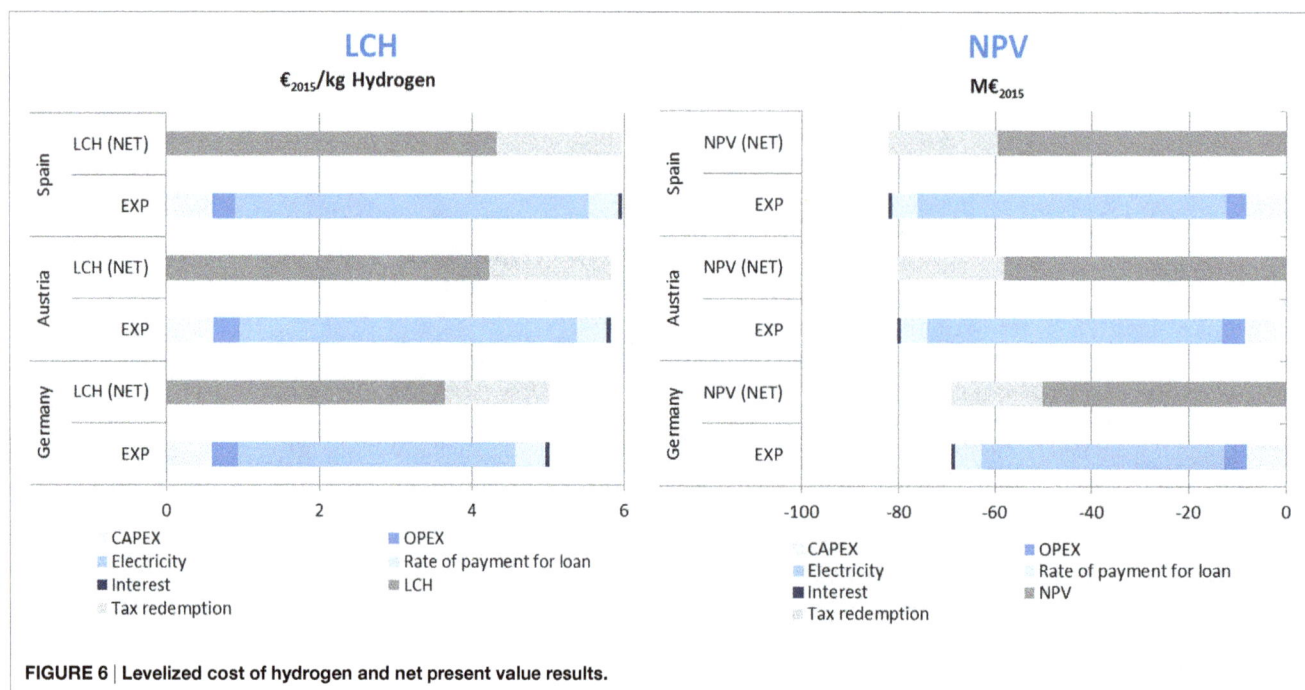

FIGURE 6 | Levelized cost of hydrogen and net present value results.

main share of LCH. In absolute terms, the results for LCH of AWE hydrogen differ significantly. Keeping in mind impacts of tax redemptions, the German site performs best with LCH of 3.64 €$_{2015}$/kg. LCHs for the Austrian and Spanish sites are approximately 15–18% higher, mainly due to higher electricity cost. In case of Spain, slightly lower investment cost do not compensate for higher electricity cost.

As market prices for hydrogen are not the focal point, NPV, representing attractiveness, is negative in all three cases. In absolute terms, Germany performs an NPV of −50 m€$_{2015}$. Again, the results for Austria and Spain are less attractive, falling below

the German value by 15–19%. Here, the result is exclusively due to cost components, mainly for electricity, as according to the technical specifications all three sites produce equal amounts of hydrogen. For both metrics, LCH (NET) and NPV (NET), tax redemptions are significant, resulting to a decrease of LCH and NPV of approximately 38%.

In case of the VC, the relative advantage of the German site is between 21 and 27% compared to the Austrian and Spanish ones. In a competitive market for hydrogen, this clearly assigns advantages to the German site considering development of sales prices for hydrogen, as the VC describes the flexibility of the site in relation to market conditions and the minimum-level price to contribute to fixed-charge coverage. Again, the dominating factor is the difference in the price of electricity in the three countries.

The range of current hydrogen market prices for industrial use is 1.4–5 €$_{2015}$/kg (Waidhas, 2015) (**Figure 7**). In case of the upper price range, for the German site there is a positive, though decreasing producer surplus over the AWE lifetime. In contrast, Austrian and Spanish sites start with a positive surplus but run into operating deficits after 15 and 12 years, respectively. With regard to the lower level hydrogen price for industrial use, neither site can prove competitiveness.

Parameter Sensitivities: LCH and VC

The generation of hydrogen based on AWE technology is cost sensitive with respect to several parameters comprising components and input product's cost, process parameters as well as financial and fiscal parameters. The chosen (cost) elasticity ε_{y,x_i} reflects the relative change of the dependent variable y (LCH and VC) induced by a relative change of a parameter value x_i (Eq. 7). We demonstrate these sensitivities by calculating elasticities for the LCH (for Germany) and VC (for all sites). For LCH, Austria and Spain perform very similar.

$$\varepsilon_{y,x_i} = \frac{\Delta y}{\Delta x_i} \times \frac{x_i}{y} \qquad (7)$$

where y: variable; x_i: parameters of respective formula.

For LCH, we tested cost parameters (initial investment cost, stack cost, and electricity price), process-engineering parameters (plant lifetime and operational hours), and financial-fiscal parameters (interest on debt and tax rate on earnings) as significant parameters. Ranging from −0.6 to 0.8 the resulting elasticities are under proportional (**Figure 8**) (Under proportionality condition: $0 < |\varepsilon| < 1$). Though significant on an absolute scale, LCH is not very sensitive to initial investment and stack replacement cost. This is also the case with respect to interest rates. Highest elasticities prevail for electricity price (0.6–0.8), plant lifetime (−0.69), and tax rate (−0.4). From a technology's point of view, tax rates may be regarded as given and immutable. Negative elasticities

for tax rates may be surprising on the first glance. However, with higher tax rate on earnings and induced higher avoided tax payments (on earnings), finally, LCH can drop. The electricity price may be partly negotiated, and plant lifetime prolongation may be influenced by technical and managerial efforts. LCH is not very sensitive toward the price of deionized water (**Table 4**). Though the price is comparatively relevant for overall non-electricity O&M cost, even a doubling does not considerably alter the LCH due to the low share of non-electricity O&M cost.

In case of VC, we tested the cost parameter electricity price (EC) according its dominating impact on VC. For all three sites, the resulting elasticities are increasing in time and nearly proportional ($|\varepsilon| = 1$) (**Figure 9**). They are increasing in time due to the increase of the relative weight of EC by increasing EC in real terms. The overwhelming relevance of electricity cost for market flexibility is evident.

DISCUSSION

The analysis identifies the German site performing best and the Austrian site second-best in all financial metrics assessing cost, attractiveness, and market flexibility (**Figure 10**). As the technical plant characteristics and financial and tax parameters are site-independent (except those for electricity, which are part of the electricity scenario calculations), the favorable performance of the German site rests upon cost characteristics comprising investment cost and O&M cost. For all metrics, the cost of electricity plays the major role, as is shown exemplarily for LCH. Although investment cost is slightly lower in Spain, the difference can not outweigh the higher electricity cost compared to Germany.

As shown, identifying a future electricity price is subject to a range of assumptions. The generation mix itself is affected by political influence. Besides the projection of generation mixes and the development of cost of the market driven components, it is the political influence on cost components such as grid transport, taxes, and contributions to finance renewable energy sources, which hardly makes it possible to improve electricity price forecasts. If, contrary to the assumptions here, electricity prices are going to adjust over the lifetime of the AWE site, the relative advantage of the German site may disappear.

This could clearly be the case, if in the course of the transformation of the German energy system governance is possibly revised. Currently, the benefits for energy-intensive industries comprise on the one hand the opportunity to purchase electricity on the spot market at low prices. On the other hand, the Renewable Energy Act (EEG) Special Equalization Scheme provides specific cost reductions for energy-intensive industries such as exemptions from renewable energies contribution. For both aspects, the current beneficial frame is a consequence of the governance of the German electricity market (Fischer et al., 2016).

However, development of the demand side of the hydrogen market is also relevant for future prospects. Besides future industrial uses of hydrogen, further uses for stationary and residential applications (fuel cell heating devices), mobility (fuel cell electric vehicles), and for hydrogen as energy storage (power-to-X) may increase (Linssen and Hake, 2016). Although there are technical options to use hydrogen, the development of demand for

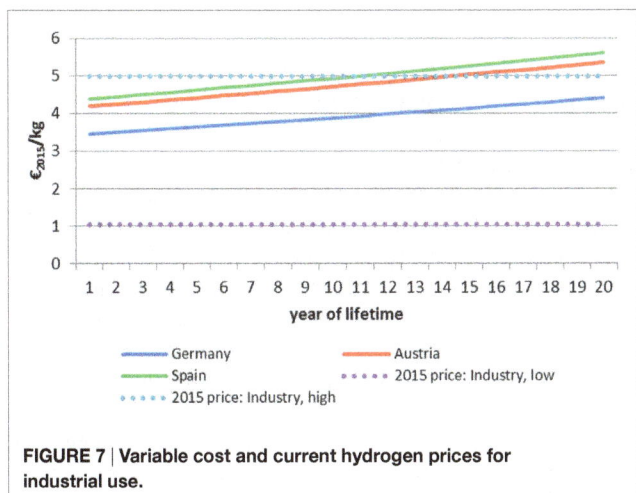

FIGURE 7 | Variable cost and current hydrogen prices for industrial use.

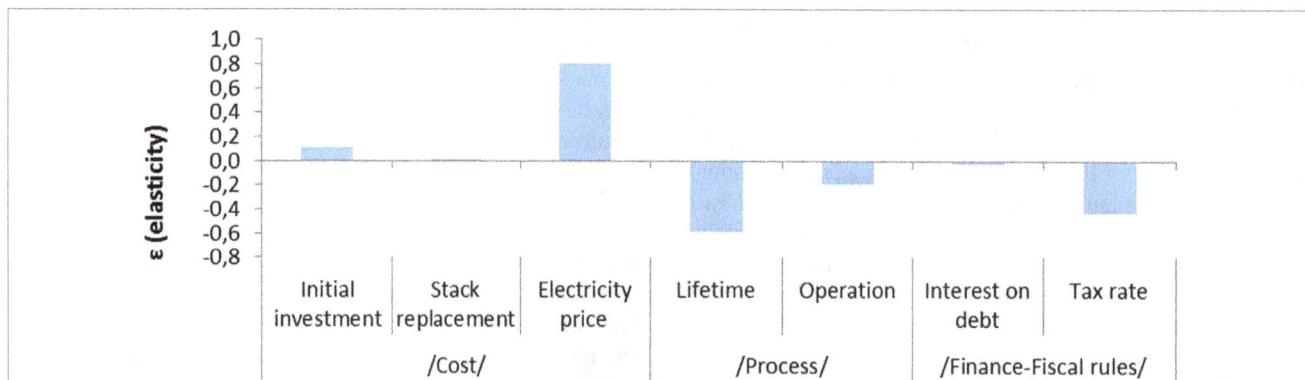

FIGURE 8 | Levelized cost of hydrogen elasticities for alkaline water electrolysis site in Germany.

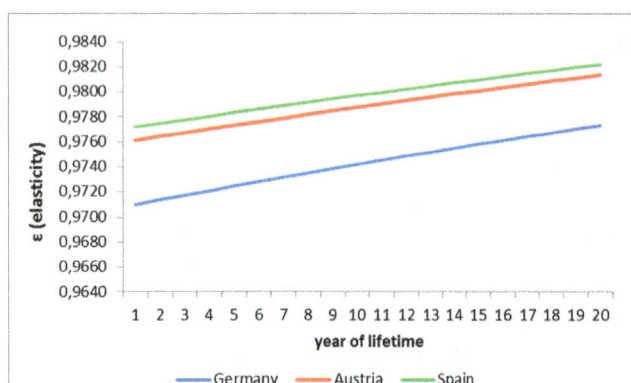

FIGURE 9 | Variable cost elasticity for alkaline water electrolysis sites.

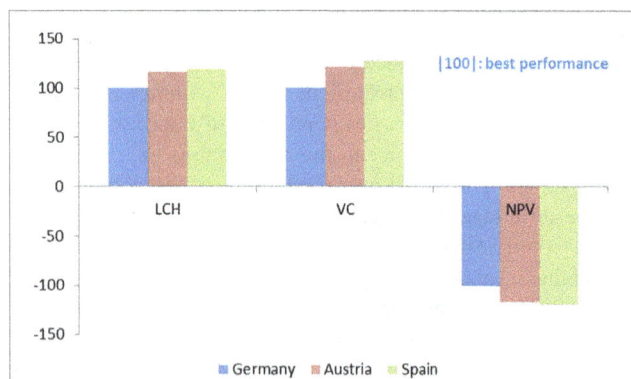

FIGURE 10 | Metrics overview.

hydrogen depends to a high degree on strategies for transforming energy systems.

SUMMARY

Aiming to contribute to economic assessment of improved AWE technology, this paper presents an economic assessment of advanced AWE technology comprising cost estimations and a financial analysis. The financial analysis not only focuses on specific cost such as LCH but also on further financial metrics quantifying attractiveness and supply/market flexibility of the technology. Therefore, besides LCH, NPV and VC are taken for assessment. The AWE technology taken in this study is a 6 MW plant scaled up from a 3.5 MW state-of-the-art pressurized electrolysis, achieving higher hydrogen output with the same number of cells with equal diameters. For comparison, three different sites in Germany, Austria, and Spain are specified. Technology and plant characteristics and upstream demand for electricity and other products (deionized water, KOH solution, and process steam) are in accordance with an EU study on advanced AWE technology. For cost characteristics, site-specific aspects for investment and operation are taken into account. This holds for investment and operational cost (O&M), labor cost, and in particular for electricity cost. The latter shows considerable differences for the three sites, mainly due to national regulation for electricity generation and the transformation toward sustainable electricity systems with high penetration of renewable and carbon-free energy sources. Therefore, electricity cost projection over the lifetime of the AWE plant is based on a European energy scenario approach, specifying characteristic framework data for the three sites.

The German AWE site turns out to perform best in all three financial metrics (LCH, NPV, and VC). Though there are slight differences in investment and O&M projections for the three sites, the major cost impact is due to the electricity cost. Although investment cost is slightly and labor cost is significantly lower in Spain, the difference can not outweigh the higher electricity cost compared to Germany. Given the assumption that the electrolysis operators are customers directly and actively participating in power markets, and based on the regulatory framework in the three countries, in this special case electricity cost in Germany is lowest. However, as electricity cost is profoundly influenced by political decisions as well as the implementation of economic instruments for transforming electricity systems toward sustainability, it is hardly possible to further improve electricity price forecasts.

Further analysis may focus on the significance to differentiate private and public economics of hydrogen technology. On the one hand, the focus is, e.g., on unpriced pollution externalities.

Keeping in mind the basis of the present analysis, an expansion toward full LCA-based inventory of relevant inputs and outputs including, e.g., emissions is obvious. On the other hand, private and public perspectives on discounting are also different. It may be the more different the more environmental externalities such as extraction of non-renewable natural resources, carbon dioxide emissions (particularly of the electricity supply), and further ones such as ozone depletion and acidification are in the spotlight. The question is whether site performances from a life-cycle perspective might change due to inclusion of public perspectives.

REFERENCES

Alberici, S., et al. (2014). *Subsidies and Cost of EU Energy*. Brussels: Ecofys. Available at: https://ec.europa.eu/energy/sites/ener/files/documents/ECOFYS%20 2014%20Subsidies%20and%20costs%20of%20EU%20energy_11_Nov.pdf

Allen Consulting Group. (2007). *Review of the Weighted Average Cost of Capital for the Purposes of Determining the Maximum Reserve Capacity Price*. Melbourne, Australia: The Allen Consulting Group. Available at: www.allenconsult.com.au

BDEW. (2015). *Industriestrompreise Ausnahmeregelungen bei Energiepreisbestandteilen (Aktualisierte Fassung) [Industrial Electricity Prices Exemptions from Energy Price Components]*. Berlin: BDEW Bundesverband der Energie- und Wasserwirtschaft e.V.

Bertuccioli, L., Chan, A., Hart, D., Lehner, F., Madden, B., and Standen, E. (2014). *Development of Water Electrolysis in the European Union. Element Energy*. Cambridge, UK; Lausanne, CH: E4tech Sarl.

Bhandari, R., Trudewind, C. A., and Zapp, P. (2014). Life cycle assessment of hydrogen production via electrolysis – a review. *J. Cleaner Prod.* 85, 151–163. doi:10.1016/j.jclepro.2013.07.048

Bockris, J. M. (2002). The origin of ideas on a hydrogen economy and its solution to the decay of the environment. *Int. J. Hydrogen Energy* 27, 731–740. doi:10.1016/S0360-3199(01)00154-9

Borenstein, S. (2000). Understanding competitive pricing and market power in wholesale electricity markets. *Electr. J.* 13, 49–57. doi:10.1016/S1040-6190(00)00124-X

Burgess, J. (2011). *New Power Cost Comparisons. Levelised Cost of Electricity for a Range of New Power Generation Technologies*. Available at: www.atse.org.au

Burwen, J. (2015). *The Impact of Three Tax-Reform Proposals on the Financial Performance of Energy Plants*. Available at: http://americanenergyinnovation. org/wp-content/uploads/2015/09/BPC_Energy-AEIC-The-Impact-of-Three-Tax-Reform-Proposals-September-2015.pdf

Capros, P., et al. (2013). *EU Energy, Transport and GHG Emissions. Trends to 2050*. Luxembourg. Available at: http://ec.europa.eu/transport/media/publications/doc/trends-to-2050-update-2013.pdf

Chandrasekar, B., and Kandpal, T. C. (2005). Effect of financial and fiscal incentives on the effective capital cost of solar energy technologies to the user. *Sol. Energy* 78, 147–156. doi:10.1016/j.solener.2004.05.003

Cucchiella, F., D'Adamo, I., and Gastaldi, M. (2015). Financial analysis for investment and policy decisions in the renewable energy sector. *Clean Technol. Environ. Policy* 17, 887–904. doi:10.1007/s10098-014-0839-z

Darling, S. B., You, F., Veselka, T., and Velosa, A. (2011). Assumptions and the levelized cost of energy for photovoltaics. *Energy Environ. Sci.* 4, 3133–3139. doi:10.1039/c0ee00698j

EC Directorate-General for Economic and Financial Affairs. (2016). *European Economic Forecast, Winter 2016*. Brussels. Available at: http://ec.europa.eu/economy_finance/publications/eeip/pdf/ip020_en.pdf

E-Control. (2015). *Auswertung der Industriepreiserhebung Strom Juli 2015 [Evaluation of Industrial Electricity Price Investigation]*. Wien: E-Control. Available at: https://www.e-control.at/marktteilnehmer/strom/strommarkt/preise/industriepreise

Eerev, S. Y., and Patel, M. K. (2012). Standardized cost estimation for new technologies (SCENT) – methodology and tool. *J. Bus.s Chem.* 9, 31–48.

Eurostat. (2015a). *Electricity Prices – Price Systems 2014*. Unit E-5: Energy, Brussels: European Commission.

Eurostat. (2015b). *Preise Elektrizität für Industrieabnehmer, ab 2007 – halbjährige Daten [Electricity Prices for Industrial Customers, 2007 – Half-Year Data]*. Brussels: EUROSTAT.

Fischer, W., Hake, J.-F., Kuckshinrichs, W., Schröder, T., and Venghaus, S. (2016). German energy policy and the way to sustainability: five controversial issues in the debate on the "Energiewende". *Energy* 115, 1580–1591. doi:10.1016/j.energy.2016.05.069

Harvego, E., O'Brien, J. E., and Mckellar, M. G. (2012). *System Evaluations and Life-Cycle Cost Analyses for High-Temperature Electrolysis Hydrogen Production Facilities*. Idaho Falls, USA. Available at: https://inldigitallibrary.inl.gov/sti/5436986.pdf

Hermann, H., Emele, L., and Loreck, C. (2014). *Prüfung der klimapolitischen Konsistenz und der Kosten von Methanisierungsstrategien [Assessment of Climate-Political Consistency and Cost of Strategies for Methanization]*. Berlin. Available at: http://www.oeko.de/oekodoc/2005/2014-021-de.pdf

Koj, J. C., Schreiber, A., Zapp, P., and Marcuello, P. (2015). Life cycle assessment of improved high pressure alkaline electrolysis. *Energy Procedia* 75, 2871–2877. doi:10.1016/j.egypro.2015.07.576

Kost, C., Mayer, J. N., Thomsen, J., Hartmann, N., Senkpiel, C., Phillips, S., et al. (2013). *Stromgestehungskosten Erneuerbare Energien [Levelized Cost of Electricity for Renewable Energies]*. Freiburg. Available at: https://www.ise.fraunhofer.de/de/veroeffentlichungen/veroeffentlichungen-pdf-dateien/studien-und-konzeptpapiere/studie-stromgestehungskosten-erneuerbare-energien.pdf

Kreuter, W., and Hofmann, H. (1998). Electrolysis: the important energy transformer in a world of sustainable energy. *Int. J. Hydrogen Energy* 23, 661–666. doi:10.1016/S0360-3199(97)00109-2

Lee, D.-H. (2016). Cost-benefit analysis, LCOE and evaluation of financial feasibility of full commercialisation of biohydrogen. *Int. J. Hydrogen Energy* 41, 4347–4357. doi:10.1016/j.ijhydene.2015.09.071

Linssen, J., and Hake, J.-F. (2016). "Hydrogen research, development, demonstration, and market deployment activities," in *Hydrogen Science and Engineering*, eds D. Stolten and B. Emonts (Weinheim, Germany: Wiley-VCH), 59–83.

McDowall, W., and Eames, M. (2006). Forecasts, scenarios, visions, backcasts and roadmaps to the hydrogen economy: a review of the hydrogen futures literature. *Energy Policy* 34, 1236–1250. doi:10.1016/j.enpol.2005.12.006

Noack, C., et al. (2015). *Studie über die Planung einer Demonstrationsanlage zur Wasserstoff-Kraftstoffgewinnung durch Elektrolyse mit Zwischenspeicherung in Salzkavernen unter Druck [Study on Planning a Demonstration Plant for Hydrogen-Fuel Generation by Electrolysis and Intermediate Storage in Salt Cavernes Under Pressure]*. Stuttgart. Available at: http://elib.dlr.de/94979/

Ondraczek, J., Komendantova, N., and Patt, A. (2015). WACC the dog: the effect of financing costs on the levelized cost of solar PV power. *Renew. Energy* 75, 888–898. doi:10.1016/j.renene.2014.10.053

Pellinger, C., and Schmid, T. (2016). *Verbundforschungsvorhaben Merit Order der Energiespeicherung im Jahr 2030. Teil 2: Technoökonomische Analyse funktionaler Energiespeicher [Research Project on Merit Order of Energy Storage in the Year 2030. Part 2: Techno-Economic Analysis of Functional Energy Storage]*. Munich: FFE. Available at: https://www.ffe.de/publikationen/pressemeldungen/615-abschlussberichtmeritorderenergiespeicherung

Peter, F., Krampe, L., and Ziegenhagen, I. (2013). *Entwicklung von Stromproduktionskosten [Development of Electricity Generation Cost]*. www.prognos.com

AUTHOR CONTRIBUTIONS

WK developed the idea of the paper and is responsible for the financial metrics calculation and for the main parts of interpretation. TK developed the scenario idea for electricity supply. JCK contributed technical and cost data for the AWE sites.

FUNDING

The work is part of the Helmholtz Programme Technology, Innovation, Society.

Schlömer, S., et al. (eds) (2014). "Annex III: technology-specific cost and performance parameters," in *Climate Change 2014: Mitigation of Climate Change. Contribution of Working Group III to the Fifth Assessment Report of the Intergovernmental Panel on Climate Change*, eds O. Edenhofer, et al. (Cambridge, UK; New York, USA: Cambridge University Press). Available at: https://www.ipcc.ch/pdf/assessment-report/ar5/wg3/ipcc_wg3_ar5_annex-iii.pdf

Shaner, M. R., Atwater, H. A., Lewis, N. S., and Mcfarland, E. W. (2016). A comparative technoeconomic analysis of renewable hydrogen production using solar energy. *Energy Environ. Sci.* 9, 2354–2371. doi:10.1030/c5ee02573g

Simshauser, P. (2014). The cost of capital for power generation in atypical capital market conditions. *Econ. Anal. Policy* 44, 184–201. doi:10.1016/j.eap.2014.05.002

Sousa de Oliveira, W., and Fernandes, A. J. (2012). Economic feasibility analysis of a wind farm in Caldas da Rainha, Portugal. *Int. J. Energy Environ.* 3, 333–346.

Sousa de Oliveira, W., Fernandes, A. J., and Gouveia, J. J. B. (2011). Economic metrics for wind energy projects. *Int. J. Energy Environ.* 2, 1013–1038.

Statistisches Bundesamt. (2015). *EU-Vergleich der Arbeitskosten 2014: Deutschland auf Rang acht [European Union Comparison of Labour Cost 2014: Germany Position Eight]*. Wiesbaden: Statistisches Bundesamt. Available at: https://www.destatis.de/DE/PresseService/Presse/Pressemitteilungen/2015/05/PD15_160_624.html

Stubelj, I., Dolenc, P., and Jerman, M. (2014). Estimating WACC for regulated industries on developing financial markets and in times of market uncertainty. *Manag. Glob. Transit.* 12, 55–77.

Taylor, M., Daniel, K., Illas, A., and So, E. (2015). *Renewable Power Generation Costs in 2014*. Available at: www.irena.org/publications

Thakre, R. (2014). Sensitivity analysis and feasibility analysis of renewable energy project. *Int. J. Eng. Innov. Technol.* 4, 230–234.

Ursua, A., Gandia, L. M., and Sanchis, P. (2012). Hydrogen production from water electrolysis: current status and future trends. *Proc. IEEE* 100, 410–426. doi:10.1109/JPROC.2011.2156750

Waidhas, M. (2015). *Power to Gas – An Economic Approach?* Erlangen, Germany: Siemens. Available at: http://www.fze.uni-saarland.de/AKE_Archiv/DPG2015-AKE_Berlin/Vortraege/DPG2015_AKE9.1_Waidhas_P2G-Economics.pdf

White, D. C. (2011). Evaluating the economics of energy-saving projects. *Chem. Eng. Prog.* 34–38.

Zeng, K., and Zhang, D. (2010). Recent progress in alkaline water electrolysis for hydrogen production and applications. *Prog. Energy Combust. Sci.* 36, 307–326. doi:10.1016/j.pecs.2009.11.002

Conflict of Interest Statement: The authors declare that the research was conducted in the absence of any commercial or financial relationships that could be construed as a potential conflict of interest.

21

An Aqueous Metal-Ion Capacitor with Oxidized Carbon Nanotubes and Metallic Zinc Electrodes

Yuheng Tian, Rose Amal and Da-Wei Wang*

School of Chemical Engineering, The University of New South Wales (UNSW), Sydney, NSW, Australia

An aqueous metal ion capacitor comprising of a zinc anode, oxidized carbon nanotubes (oCNTs) cathode, and a zinc sulfate electrolyte is reported. Since the shuttling cation is Zn^{2+}, this typical metal ion capacitor is named as zinc-ion capacitor (ZIC). The ZIC integrates the divalent zinc stripping/plating chemistry with the surface-enabled pseudocapacitive cation adsorption/desorption on oCNTs. The surface chemistry and crystallographic structure of oCNTs were extensively characterized by combining X-ray photoelectron spectroscopy, Fourier-transformed infrared spectroscopy, Raman spectroscopy, and X-ray powder diffraction. The function of the surface oxygen groups in surface cation storage was elucidated by a series of electrochemical measurement and the surface-enabled ZIC showed better performance than the ZIC with an un-oxidized CNT cathode. The reaction mechanism at the oCNT cathode involves the additional reversible Faradaic process, while the CNTs merely show electric double layer capacitive behavior involving a non-Faradaic process. The aqueous hybrid ZIC comprising the oCNT cathode exhibited a specific capacitance of 20 mF cm^{-2} (corresponding to 53 F g^{-1}) in the range of 0–1.8 V at 10 mV s^{-1} and a stable cycling performance up to 5000 cycles.

Keywords: supercapacitor, carbon nanotubes, zinc ion capacitor, functionalization, gel electrolyte

Edited by:
Guoxiu Wang,
University of Technology
Sydney, Australia

Reviewed by:
Manickam Minakshi,
Murdoch University, Australia
Shichun Mu,
Wuhan University of
Technology, China

*Correspondence:
Da-Wei Wang
da-wei.wang@unsw.edu.au

Specialty section:
This article was submitted
to Energy Storage,
a section of the journal
Frontiers in Energy Research

Citation:
Tian Y, Amal R and Wang D-W
(2016) An Aqueous Metal-Ion
Capacitor with Oxidized
Carbon Nanotubes and Metallic
Zinc Electrodes.
Front. Energy Res. 4:34.

INTRODUCTION

Supercapacitors, also called electrochemical capacitors (ECs), are such high-performance energy storage devices with excellent power capability, short charge-discharge time, long cyclic life, and outstanding reversibility (Yan et al., 2014). Charge storage in supercapacitors is principally based on either the pure electrostatic charge accumulation at the electrode–electrolyte interface, i.e., electric double layer capacitance, or the fast and reversible Faradaic processes on the electrode surface, i.e., pseudocapacitance. However, ECs store electrical charge only at the electrode surface rather than within the entire electrode so that they deliver lower energy densities compared with batteries (Dubal et al., 2015).

In conventional symmetric ECs, carbonaceous materials, such as activated carbon, carbon nanotubes (CNTs), graphene materials, and carbon aerogels, are normally used as the electrodes because of the large specific surface area, good conductivity, high chemical stability, and low cost (Frackowiak and Béguin, 2002; Vix-Guterl et al., 2005; Pandolfo and Hollenkamp, 2006; Frackowiak, 2007; Zhang et al., 2009). But they exhibit a lower capacitance, as only electric double layer behavior occurs in the system. In asymmetric ECs, one or two electrodes can be pseudocapacitive that can contribute to higher capacitance. Electroactive polymers can provide large pseudocapacitance (Snook et al., 2011;

Kurra et al., 2015). But their swelling behavior can degrade the cycling stability. Transition metal oxide electrodes have higher pseudocapacitance than carbon electrodes, whereas the stability and conductivity are inferior to those of carbons (Lokhande et al., 2011).

Driven by the need to maintain the power capability as well as enhance the capacitance and energy density performance, hybrid electrochemical capacitors (HECs) with a different asymmetric configuration have been developed in recent years (Simon and Gogotsi, 2008). HECs are generally composed of a battery-type Faradaic anode as energy source and a capacitor-type cathode as power source, which can enhance the energy and power capabilities. A particular sample is lithium-ion capacitors (LICs), which combine Li-alloying anodes with supercapacitor cathodes (Cao et al., 2014, 2015). But LICs suffer from high cost of lithium because of the scarcity in earth crust. Thereafter, sodium-ion capacitors (SICs) emerge as an attractive alternative of Li-ion capacitors for the sake of the relative abundance of sodium (Chen et al., 2012). However, safety issue is a critical concern for either LICs or SICs because the metal anodes are extremely reactive and the organic solvents are flammable. Consequently, there is an urgent need to search for environmental friendly alternative aqueous metal-ion capacitors.

Zinc as a safe, cost-effective, and eco-friendly metal has been widely studied in the batteries, e.g., zinc–air, zinc–polymer, and zinc–alkaline batteries (Rahmanifar et al., 2002; Ghaemi et al., 2003; Lee et al., 2011). Generally, the alkaline electrolytes are used in the zinc–air and zinc–alkaline batteries, where the zinc anode is oxidized to the zincate ions, $Zn(OH)_4^{2-}$, during discharge (Ghaemi et al., 2003; Lee et al., 2011). However, the zincate ions are easily decomposed to ZnO that acts as an insulator in the system, while the zincate ions exceed a saturation point in the alkaline medium. The formation of ZnO seriously degrades the cycling performance of the batteries. Regarding the zinc–polymer batteries, they typically use the zinc as the anode and the polyaniline as the cathode in the $ZnCl_2$ and NH_4Cl mixing electrolytes (pH = 4–5), where the cathodic reaction that involves the oxidation/reduction processes is accompanied by the insertion and elimination of the chloride ions (Ghanbari et al., 2007). Meanwhile, the Zn anode dissolves in terms of Zn^{2+} ions during discharging processes and deposits during charging. However, the polyaniline is electroactive merely in the acidic media and it is found to lose its electrochemical activity in the solution of pH >4 (Rahmanifar et al., 2002). Among these batteries, the shape change of the zinc anode resulting from the non-uniform deposition of zinc active material during charging and the dendrite growth due to high local current densities remain the challenges for the zinc rechargeable batteries (Vatsalarani et al., 2006).

A new type of zinc-ion batteries that are based on the shuttling of Zn^{2+} ions in the zinc sulfate electrolyte between the zinc anode and nanoparticle cathode with abundant tunnels, e.g., MnO_2 and CuHCF nanocubes, was reported recently (Xu et al., 2012; Jia et al., 2015; Trócoli and La Mantia, 2015). This novel battery operates the zinc ion insertion and extraction at near-neutral pH values, whereas the active material on the cathode is capable of accommodating Zn^{2+} in the crystalline structure. More importantly, a patent of a metal ion pseudocapacitor based on the zinc metal anode and the cathode materials, including transition metals, carbon, electro-active polymers, and combinations, has appeared (Lengsfeld and Shoureshi, 2014). It showed the excellent potential of the energy storage system through the zinc metal electro-dissolution and deposition anodic processes and Zn^{2+} ion adsorption/desorption, Faradaic or both cathodic processes. Herein, we develop a zinc ion capacitor (ZIC) using a CNT-based pseudocapacitive cathode and a zinc anode. Both liquid electrolyte and gel electrolyte containing $ZnSO_4$ were developed for this ZIC. Because of their high electrical conductivity, good mechanical strength and large surface area, CNTs have attracted significant attention as electrode materials (Choi et al., 2001; Zhai et al., 2011). We found that the oxidized carbon nanotubes (oCNTs) can increase the pseudocapacitance due to the interaction between the oxygen-containing functional groups and the Zn^{2+}, thereby enhancing the capacitance of the ZIC.

EXPERIMENTAL SECTION

Materials

Carbon nanotubes were commercially purchased. All other chemicals, including 32% hydrochloric acid, 98% sulfuric acid, potassium permanganate, 30% hydrogen peroxide solution, 5 wt.% Nafion solution, zinc sulfate, and polyvinyl alcohol (PVA), were purchased from Sigma-Aldrich.

Synthesis of oCNTs

The CNTs were treated with 20% HCl solution to dissolve metal impurities. The purified CNTs were washed with sufficient deionized (DI) water and dried in an oven at 60°C. Subsequently, the oCNTs were prepared by using an improved Hummer's oxidation as follows (Kosynkin et al., 2009). First, 1.0 g of purified CNTs were added with 120 mL of 98% concentrated H_2SO_4 and then added slowly with 1.0 g of $KMnO_4$ at room temperature upon magnetic stirring. The solution was maintained at room temperature for 15 min and then transferred to a 60°C bath for a continuous stirring of 4 h. After cooling to room temperature, 300 mL water solution containing 10 mL 30% H_2O_2 solution was slowly poured into it upon stirring of 1 h. Second, the reacted solution was centrifuged (8000 rpm for 30 min), and the supernatant was decanted away. The recovered sample was redispersed in the DI water using bath sonication and then collected by centrifugation at 10,000 rpm for 30 min. This process was repeated until the pH turned to be neutral. Finally, the resultant sample was dried in the oven at 60°C, which was noted as oCNTs.

Characterizations

Transmission electron microscopy (TEM) images of CNTs and oCNTs were collected from the Tecnai F20 field emission transmission electron microscope. X-ray photoelectron spectroscopy (XPS) was performed using a Thermo Scientific K-Alpha spectrometer with Al Kα radiation. Raman spectra were obtained on a Renishaw inVia Raman Microscope from 800 to 3600 cm^{-1} using a green excitation laser (Ar, 514 nm). X-ray diffraction (XRD) was carried out on a PANalytical Xpert Multipurpose diffraction System with a Cu Kα source. Fourier transform infrared (FTIR) spectroscopy was performed on the Varian 640 FTIR Spectrometer with a sensitive liquid nitrogen-cooled MCT detector. Scanning electron microscopy (SEM) was performed on the NanoSEM 230 field emission scanning electron microscope. The BET specific surface area measurements were conducted by Micromeritics Tristar 3030.

Electrochemical Measurements

Fifty milligrams of CNTs and oCNTs were separately dispersed in 10 mL of 0.5 wt.% Nafion solution using a probe sonication. Then, 200 μL of each sample was drop-cast onto a titanium foil as a working electrode. The electrochemical properties of CNTs and oCNTs were evaluated by using a three-electrode cell: a Pt counter electrode, a Hg/Hg_2SO_4 reference electrode in saturated K_2SO_4 solution, and a CNT or oCNT working electrode stabilized in 1M $ZnSO_4$ aqueous electrolyte (pH = 4.16) or in pH = 4.16 H_2SO_4 electrolyte at the scan rates of 10, 50, 100, 200, 500, and 800 mV s^{-1} within −1.4 to 0.2 V. Subsequently, a ZIC was constructed by using a zinc foil as the anode and the oCNTs or CNTs as the cathode. The galvanostatic charge/discharge studies of the ZICs based on the oCNT and CNT cathodes were carried out at 2, 5, and 10 mA cm^{-2} within 0–1.8 V in the 1M $ZnSO_4$ liquid electrolyte. The cyclic voltammetry of the ZICs using the oCNT cathodes was measured in 1M $ZnSO_4$ liquid electrolyte and in $ZnSO_4$–PVA gel electrolyte at the scan rates of 5, 10, 20, 50, 100, 150, 200, and 500 mV s^{-1} within 0–1 and 0–1.8 V. The gel electrolyte was prepared by soaking a filter paper into the 0.05 g mL^{-1} PVA solution overnight and then soaking the gel-treated filter paper into the 1M $ZnSO_4$ solution overnight. In order to highlight the superiority of the ZIC, a symmetric capacitor with oCNTs as both electrodes was constructed using 1M $ZnSO_4$ aqueous electrolyte and measured at 50 mV s^{-1} within 0–1.8 V. Moreover, the cycling stability of the liquid- and gel-electrolyte ZICs was tested individually within 0–1.8 V by cycling 5 times at 5 mV s^{-1} and then 1000 times at 500 mV s^{-1}. This unit process of cycling test was continuously repeated for five times. The additional cycle life test of the gel-electrolyte ZIC based on the galvanostatic charge/discharge was conducted at 2 mA cm^{-2} within 0–1.8 V for 1500 cycles. The electrochemical impedance spectroscopy (EIS) (10 mHz–10 kHz, 5 mV) was used to study the ZICs at the open circuit voltage (OCV).

RESULTS AND DISCUSSION

Surface-Enabled Zn Ion Storage

Since cyclic voltammetry is a suitable technique to characterize the capacitive behavior of electrode materials, the electrochemical performances of oCNTs and CNTs were studied in a three-electrode cell by using cyclic voltammetry. **Figure 1** exhibits the cyclic voltammograms (CV) of the oCNTs (green line) and the CNTs (black line) in 1M $ZnSO_4$ aqueous solution between −1.4 and 0.2 V (vs. Hg/Hg_2SO_4 in concentrated K_2SO_4 electrolyte) at the scan rates of 10 and 200 mV s^{-1}. The −1.4 V (vs. Hg/Hg_2SO_4) was the equilibrium potential of Zn^0/Zn^{2+} (Figure S1 in Supplementary Material). The CV curves of CNTs show a typical rectangular shape, indicating pure electric double layer capacitive behavior. However, the CV curves of oCNTs show a quasi-rectangular

shape with a broad anodic hump at 0.2 V and a cathodic hump at −0.25 V, implying the presence of pseudocapacitance. The capacitive performance of oCNTs is far superior to that of CNTs, which shows typical electric double layer capacitance. Moreover, the oCNTs exhibited distinct pseudocapacitive response at higher scan rates (Figure S2 in Supplementary Material). Additionally, as the pH of the 1M $ZnSO_4$ solution is 4.16, the pseudocapacitance could partially include the protonation capacitance. The oCNT sample was then tested in pH = 4.16 H_2SO_4 solution in order to determine the contribution of protonation to the capacitance. At the same concentration of H$^+$, the oCNTs in a pH = 4.16 H_2SO_4 solution (**Figure 1**, red line) present an inferior capacitive behavior than that in $ZnSO_4$ solution, suggesting that the presence of Zn^{2+} ions can provide higher capacitance.

It is worth noting that the atomic percentage of the Zn^{2+} ions on the oCNTs during discharge (Zn^{2+} adsorption) is higher than that during charge (Zn^{2+} desorption), as shown in the Table S1 in Supplementary Material. In the process of discharge, certain Faradaic reaction or electrochemical adsorption of the Zn^{2+} ions on oCNTs could take place, increasing the amount of Zn^{2+} ions on the oCNTs. The charge process could involve reversible electrochemical desorption of Zn^{2+} ions that decrease the atomic percentage of Zn^{2+} ions on the oCNTs.

Charge Storage Mechanism of oCNTs

In order to understand the relationship between Zn^{2+} ion storage and physicochemical properties of the oCNTs, various instrumental analyses were conducted to reveal the structural differences of oCNTs from CNTs. The inset in **Figure 2A** displays the XPS survey spectra of CNTs and oCNTs. Merely C1s and O1s peaks, without any signals of elemental impurities, are observed. The ratios of the peak intensity indicate the change in the oxygen content of the samples. Obviously, the O-to-C ratio of the oCNTs is enhanced, which suggests an increase in the oxygen amount. The high resolution C1s core-level spectra of both samples are presented in **Figure 2A**. The pristine CNT sample exhibits an asymmetric peak centered at 284.5 eV with a long tail extended to the higher energy region, representing the graphitic structure. Deconvolution of the C1s peak of the oCNTs also shows a main peak at 284.5 eV, attributed to the non-oxygenated sp^2 carbon (Stankovich et al., 2007; Yang et al., 2009). Moreover, a peak at 285.1 eV is caused by sp^3-hybridized carbon. The peaks at 286.7, 288.1, and 289.3 eV correspond to carbon atoms attached to different oxygen-containing moieties, which are assigned to C–O, C=O, and O–C=O bonds, respectively. Additionally, deconvolution of the XPS O1s of both samples is shown in Figure S3A in Supplementary Material. The O1s spectra are deconvoluted into two peaks, and these two peaks at 531.6 and 533.4 eV are assigned to C=O and C–O, respectively, which confirm the presence of some carboxylic and hydroxyl functionalities onto the nanotube surface (Stankovich et al., 2007).

Additionally, **Figure 2B** shows the FTIR spectra of CNTs and oCNTs, which further confirm the introduction of oxygen functionalities onto the CNTs. The characteristic peaks at 2934 and 2850 cm^{-1} are normally recognized as the asymmetric stretching and the symmetric stretching of CH$_2$, respectively. The vibration of CH$_2$ is still visible in the oCNT spectrum,

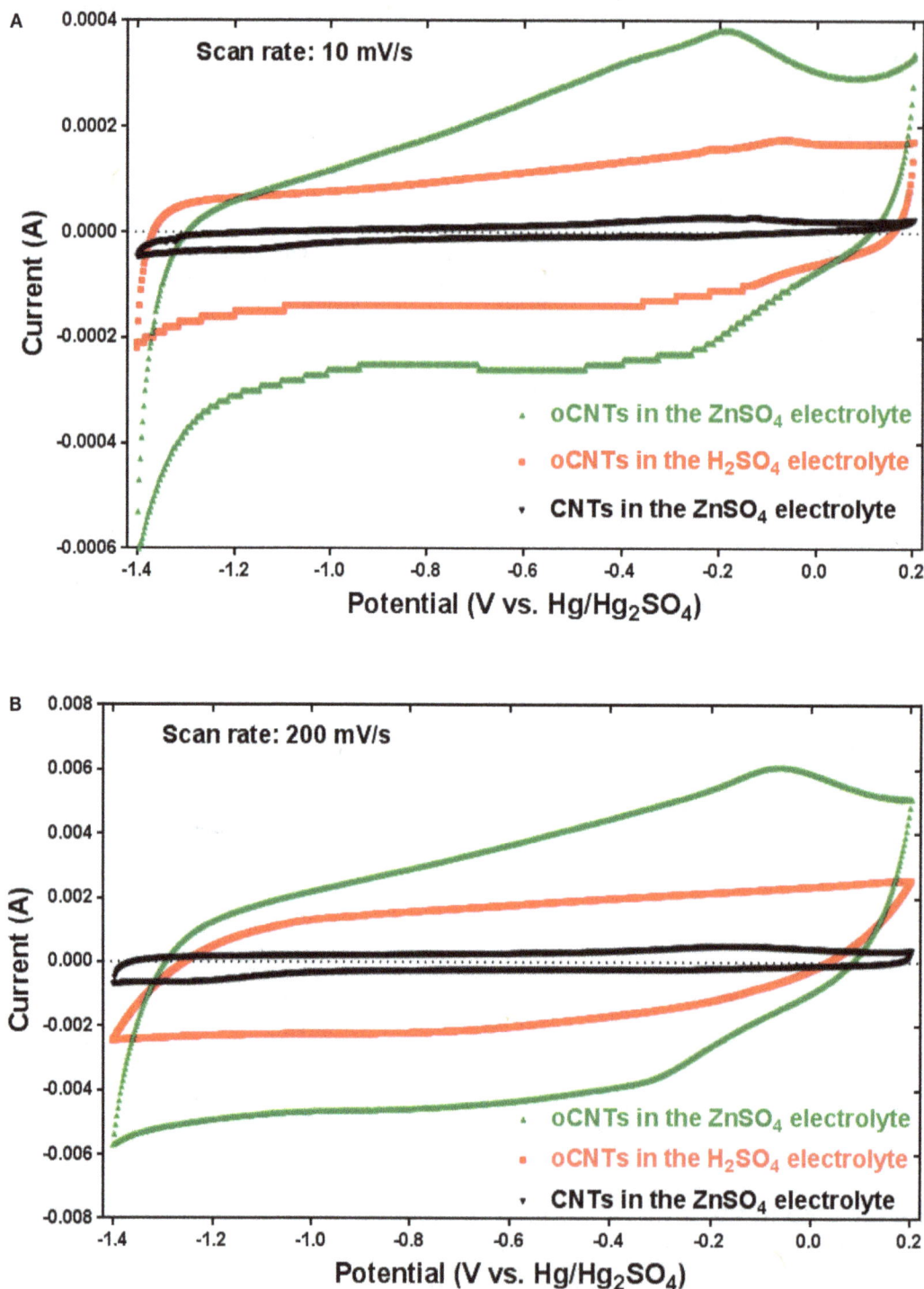

FIGURE 1 | The CV profiles of oCNTs and purified CNTs recorded in 1M $ZnSO_4$ and pH = 4.16 H_2SO_4 electrolytes at (A) 10 mV s^{-1} and (B) 200 mV s^{-1}.

suggesting that ordered graphitic domains remain in the structure (Pham et al., 2011). The major contribution to the FTIR spectra is the oxygen-containing functional groups. Near 3430 cm^{-1} in the high frequency area, there is a broad peak for the oCNTs at 3000–3700 cm^{-1} as compared with that of the CNTs,

which is caused by the stretching vibration of hydroxyl groups (Wang et al., 2012). Correspondingly, the absorption peak based on the bending vibration of –OH at 1584 cm^{-1} becomes stronger. Besides, the absorption peak occurring near 1650 cm^{-1} in the medium frequency area is caused by the stretching vibration of C=O of carboxylic acid and carbonyl groups. Finally, the absorption peaks at 1384 and 1118 cm^{-1} arise from the vibration of C–O of carboxylic acid and the stretching vibration of C–OH, respectively.

The TEM images provide the microstructural information of CNTs and oCNTs, as presented in **Figures 2C,D**. The morphology of the oCNTs did not change significantly after low-level oxidation (Figures S3B,C in Supplementary Material). But it

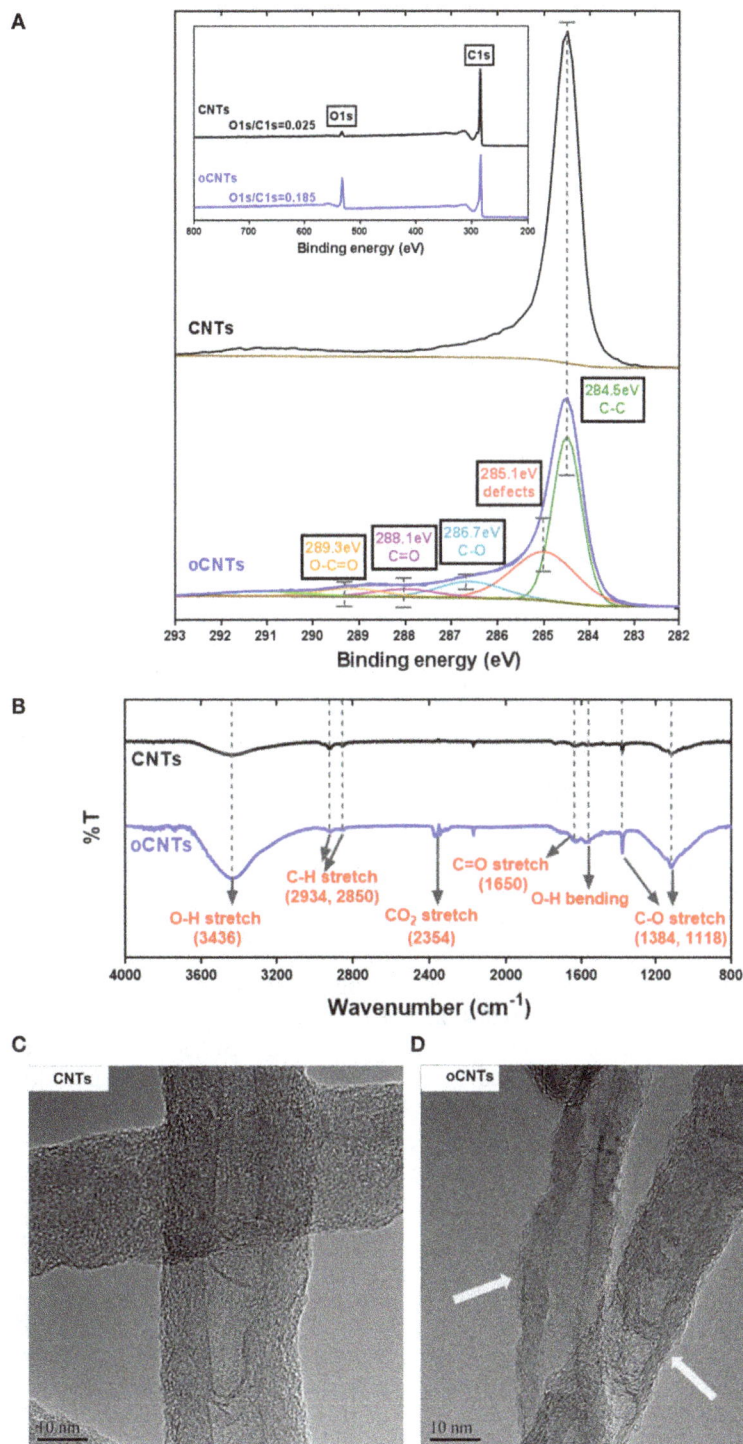

FIGURE 2 | Continued

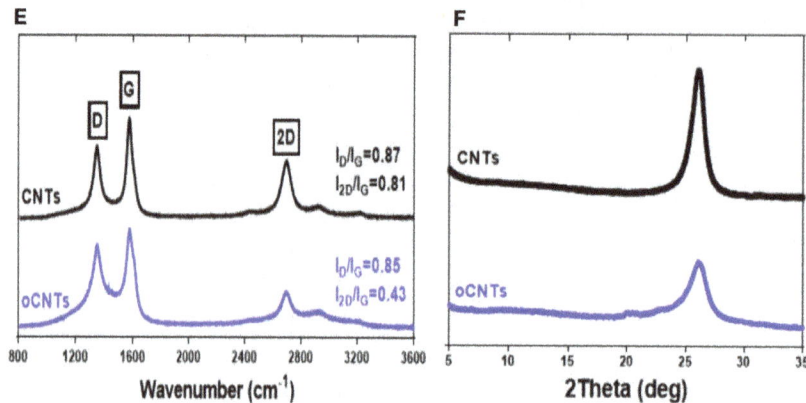

FIGURE 2 | (A) XPS C1s spectra with survey spectra inset; (B) FTIR spectra; (C,D) TEM images (the arrows point at the amorphous surface of the oCNTs), (E) Raman spectra, and (F) XRD patterns of CNTs and oCNTs.

is noticeable that, as seen in the areas pointed by the arrows in **Figure 2D**, the oCNTs were not as graphitic as the CNTs, evidenced by the amorphous fragments attached on the surfaces. Since the degree of oxidation effectively determines the extent of the CNT unzipping (Kosynkin et al., 2009), at the low oxidation level, the amorphous carbon was formed on the surface of the oCNTs with intact graphitic inner tube remaining. The TEM images are further supported by Raman and XRD results.

Figure 2E displays the Raman spectra for CNTs and oCNTs. Basically, the main features in the Raman spectra of graphitic carbon materials are the D (1354 cm^{-1}), G (1584 cm^{-1}), and 2D (2707 cm^{-1}) bands (Wu et al., 2015). Generally, the I_D/I_G value is used to represent the graphitic degree of the carbon materials. As shown in **Figure 2E**, the I_D/I_G values of the oCNTs and the CNTs are not obviously changed, indicating that after oxidation the ordering of the graphitic structure almost remained. The relative intensity of 2D band to G band is inversely associated with the doping level of oxygen functional groups on the graphene basal planes (Ferrari, 2007; Wu et al., 2015). **Figure 2E** shows the I_{2D}/I_G value of the oCNTs reduces significantly as compared to that of the CNTs, which suggests a reduced sp^2 domain size due to the occupation of introduced oxygen functional groups. Thus, the Raman spectra demonstrate that the majority of the graphitic structure in the oCNTs remained intact but with the oxygen doped to the outermost layers.

Figure 2F compares XRD patterns of both samples. It can be found that the CNTs have a 2θ value of ~26.8°, and the position of the peak of the oCNTs does not shift after oxidation treatment, which demonstrates that they have almost the same interlayered spacing. However, the peak intensity of the oCNTs is weaker than that of the CNTs, probably indicating that at such oxidation level, the outermost layers of the nanotubes have been damaged but still with the intact inner structure remaining.

Based on the structural analyses, the better capacitive performance of the oCNTs can be explained as follows. On the one hand, despite the superior electrical conductivity, CNTs do not contain abundant oxygen functional groups, implying that

its capacitive behavior merely relies on the electric double layer capacitance. On the other hand, the poor wettability of CNTs in aqueous electrolyte, resulting from the hydrophobic surfaces, can decelerate the ion transport toward the electrode–electrolyte interfaces, thereby negatively affecting the electrochemical performance. However, the oCNTs are characterized by the abundant oxygen-containing groups at the surface and the intact graphitic inner tubes that maintain the electrical conductivity. The oxygen functional groups occurring in the oCNTs are advantageous, as they can provide a large additional pseudocapacitance as well as improved wettability, giving the oCNTs a higher capacitance than that of the CNTs. According to the previous report that the oxygen functional groups of graphene oxide sheets can chemically interact with divalent metal ions (Park et al., 2008), the pseudocapacitance of oCNTs, in this case, probably comes from the electrochemical reactions, e.g., >2C–O–Zn ↔ 2C=O + Zn^{2+} + 2e$^-$ and >C–OH ↔ C=O + H$^+$ + e$^-$, and partially from the electrostatic adsorption/desorption of ions, as illustrated in **Figure 3**. As the presence of Zn^{2+} ions provides higher capacitance, the electrochemical reaction/adsorption based on the Zn^{2+} ions and oxygen-containing groups of the oCNTs should play a dominant role in the ZIC system.

Assembly of ZIC

Depending on the capacitive performance and material characterizations of the oCNTs, a safe and eco-friendly energy storage device, which consists of an oCNT cathode, a zinc anode, and a ZnSO$_4$ liquid electrolyte or a ZnSO$_4$–PVA gel electrolyte, is presented. The ZnSO$_4$ solution is a non-toxic, non-corrosive, and cost-effective mild aqueous electrolyte, and therefore, the ZIC is more environmental friendly compared to other rechargeable power sources that use either alkaline (such as Zn/MnO$_2$, Ni–Cd, or Ni–MH batteries), or acidic (lead–acid) electrolytes (Xu et al., 2012). In order to further compare the behavior of oCNTs with the CNTs, the galvanostatic charge/discharge curves of the ZICs based on the oCNT and CNT cathodes at different current densities were measured in the 1M ZnSO$_4$ liquid electrolyte within

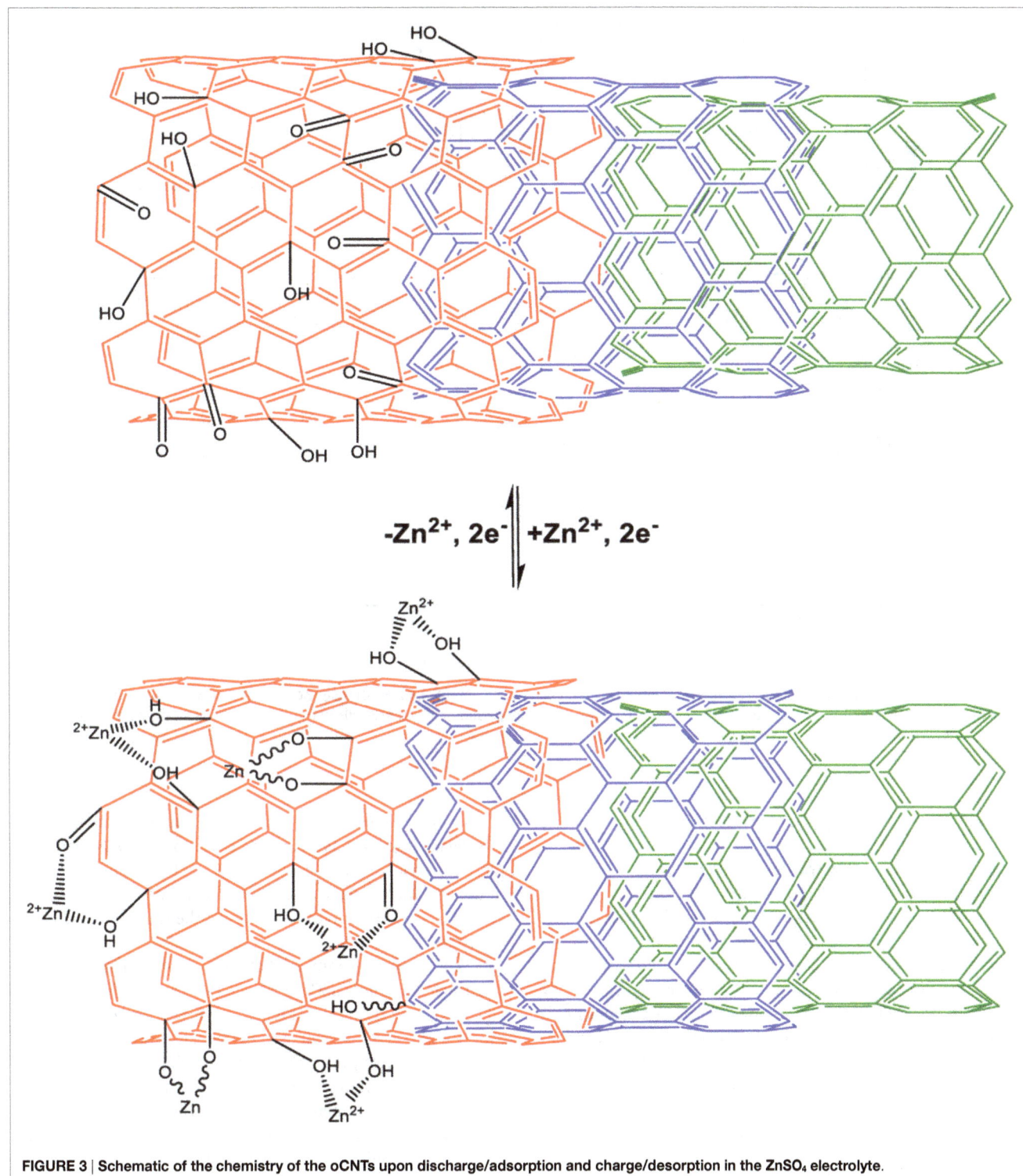

FIGURE 3 | Schematic of the chemistry of the oCNTs upon discharge/adsorption and charge/desorption in the ZnSO₄ electrolyte.

0–1.8 V, as shown in **Figures 4A–C**. The curves of the CNTs show the typical triangular shape attributed to the electric double layer capacitance, while the oCNTs exhibit the quasi-triangle shape, i.e., the V–t curve profile is not as linear as it should be, which is due to the presence of Faradaic currents. The larger area under the V–t curves of the oCNTs is the evidence of the higher capacitance than the CNTs. Also, the galvanostatic charge/discharge curves of the oCNT cathodes (**Figure 4D**) maintain the same shape in these densities, indicating the oCNTs can experience a wide range of current.

The higher capacitance of the ZIC comprising the oCNT cathode can be explained in two aspects. On the one hand, the

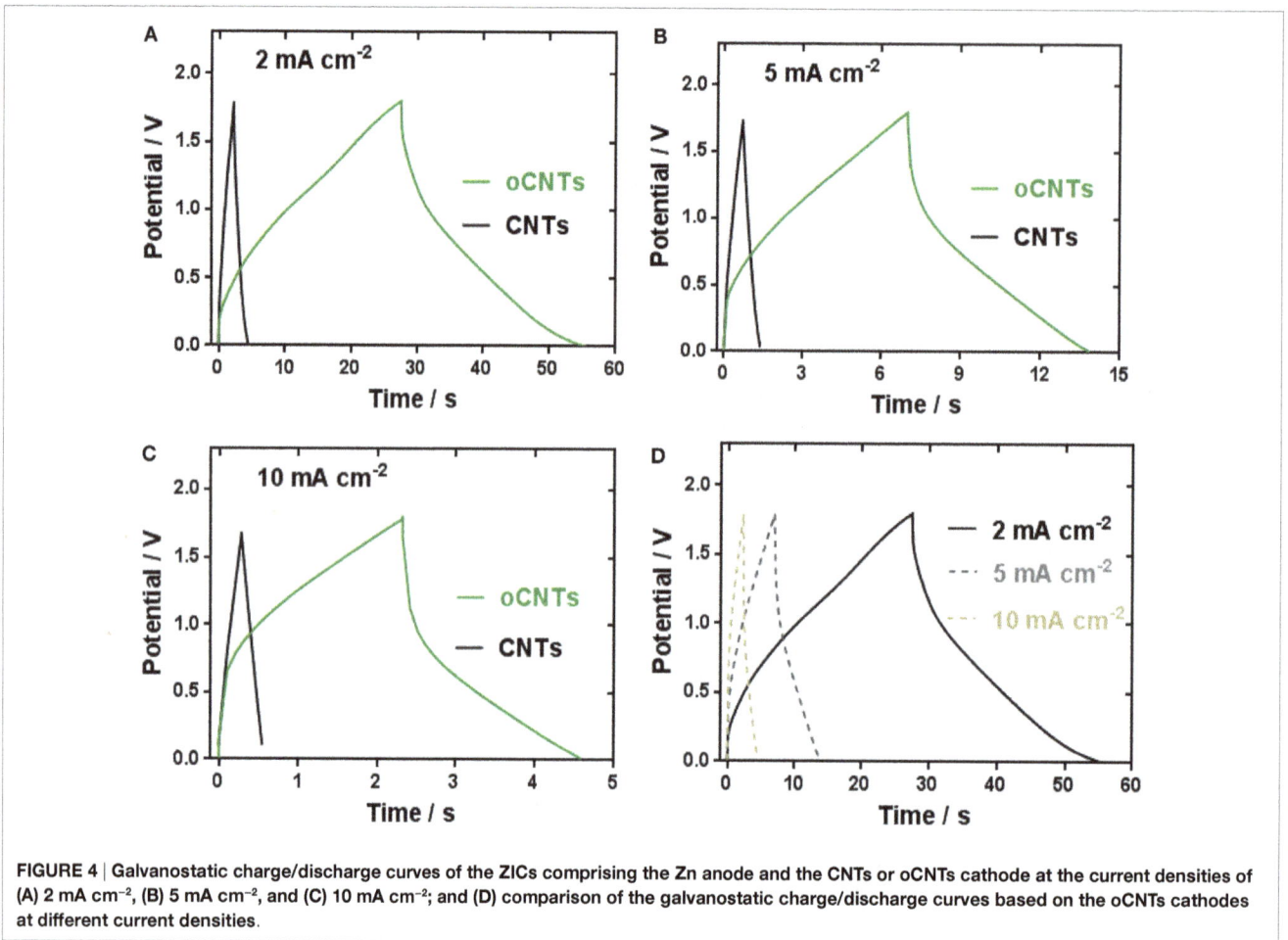

FIGURE 4 | Galvanostatic charge/discharge curves of the ZICs comprising the Zn anode and the CNTs or oCNTs cathode at the current densities of (A) 2 mA cm⁻², (B) 5 mA cm⁻², and (C) 10 mA cm⁻²; and (D) comparison of the galvanostatic charge/discharge curves based on the oCNTs cathodes at different current densities.

oCNTs by virtue of the oxygen groups decorations exhibit a larger surface area (211 m^2 g^{-1}) than the CNTs (120 m^2 g^{-1}), which can contribute more to the electric double layer capacitance. On the other hand, the oxygen groups on the oCNTs surface provide additional pseudocapacitance involving a Faradaic process. The pseudocapacitive working principle of ZIC based on the oCNT cathode is proposed here. Zinc is stripped in the form of Zn^{2+} ions during discharge and deposited during charge (Eq. 1). The Zn^{2+} ions diffuse to the oCNTs cathode where the divalent cations are electrochemically adsorbed by the oxygen functional groups upon discharge and desorbed upon charge, as depicted in Eq. 2. Since the 1.0M $ZnSO_4$ electrolyte is mildly acidic (pH = 4.16), protonation/deprotonation of oCNTs also participate in the charge/discharge of ZIC, as shown in Eq. 3.

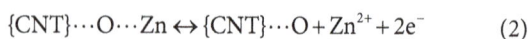

$$Zn \leftrightarrow Zn^{2+} + 2e^- \tag{1}$$

$$\{CNT\} \cdots O \cdots Zn \leftrightarrow \{CNT\} \cdots O + Zn^{2+} + 2e^- \tag{2}$$

$$\{CNT\} \cdots OH \leftrightarrow \{CNT\} \cdots O + H^+ + e^- \tag{3}$$

In terms of the full device operation, the Zn^{2+} ions shuttle between the oCNT cathode and the Zn anode to transport the charges. As the charge storage mechanism relates to the migration of Zn^{2+} ions between anode and cathode, mimicking that of LICs, this device is termed as a zinc-ion capacitor (ZIC). Nonetheless, the ZIC device is believed to be electrochemically irreversible in the alkaline electrolyte, as the formation of ZnO or $Zn(OH)_2$ cannot be converted to the Zn metal at the room temperature in an electrochemical system.

Cyclic voltammogram (CV) of the ZICs with the $ZnSO_4$ liquid electrolyte and the $ZnSO_4$–PVA gel electrolyte at a scan rate of 50 mV s^{-1} are exhibited in **Figures 5A,B** with the voltage ranges of 0–1 and 0–1.8 V. Obviously, the gel-electrolyte ZIC presents a better capacitive performance than the liquid-electrolyte one, which is also demonstrated at other scan rates as shown in Figure S4 in Supplementary Material. The better performance of the gel-electrolyte ZIC can probably be explained in this way. As the ionic conductivity of an electrolyte affects the supercapacitor performance (Yu et al., 2011), the PVA-based gel electrolyte principally provides a liquid-like medium for ion transport and therefore have the capability to enhance the ionic conductivity (Agrawal and Awadhia, 2004). Actually, folding and unfolding the polymeric chain of the PVA cause ion dissociation that enhances the charge carrier concentration; thus, a rise in ionic conductivity of the gel electrolyte can be obtained. This postulation is later verified by rate capability determination and electrochemical impedance. Moreover, the appearance of the redox peaks in **Figure 5B** at 1.0–1.6 V indicates that the existence of Zn^{2+} ions in the electrolyte causes a Faradaic process at the electrolyte–cathode interface. Although the mechanism of the ZIC is still uncertain, its pseudocapacitance is probably attributed to the electrochemical reaction/adsorption and desorption of Zn^{2+} ions on the oCNT cathode.

The influence of the scan rate on the capacitive behavior of the gel-electrolyte ZIC swept at 5–200 mV s^{-1} with the voltage range of 0–1.8 V is shown in **Figure 5C**. It is apparent that each *CV* curve has the same quasi-rectangular pattern with a broadened peak at around 1.0–1.6 V, suggesting that the oxygen-containing groups of the oCNT cathode are able to facilitate the pseudocapacitance. Also, the current response is dependent on the scan rate. With

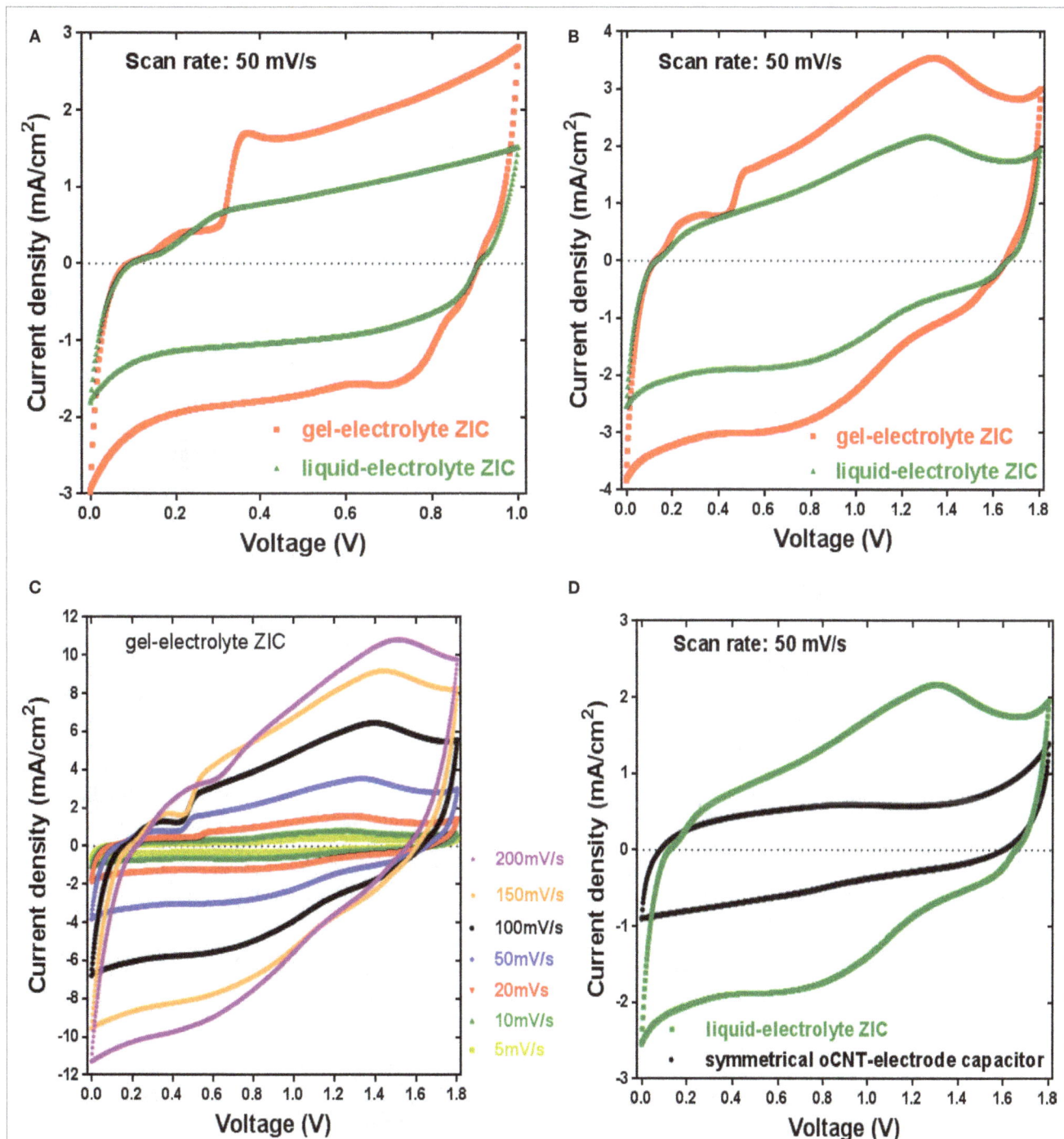

FIGURE 5 | Continued

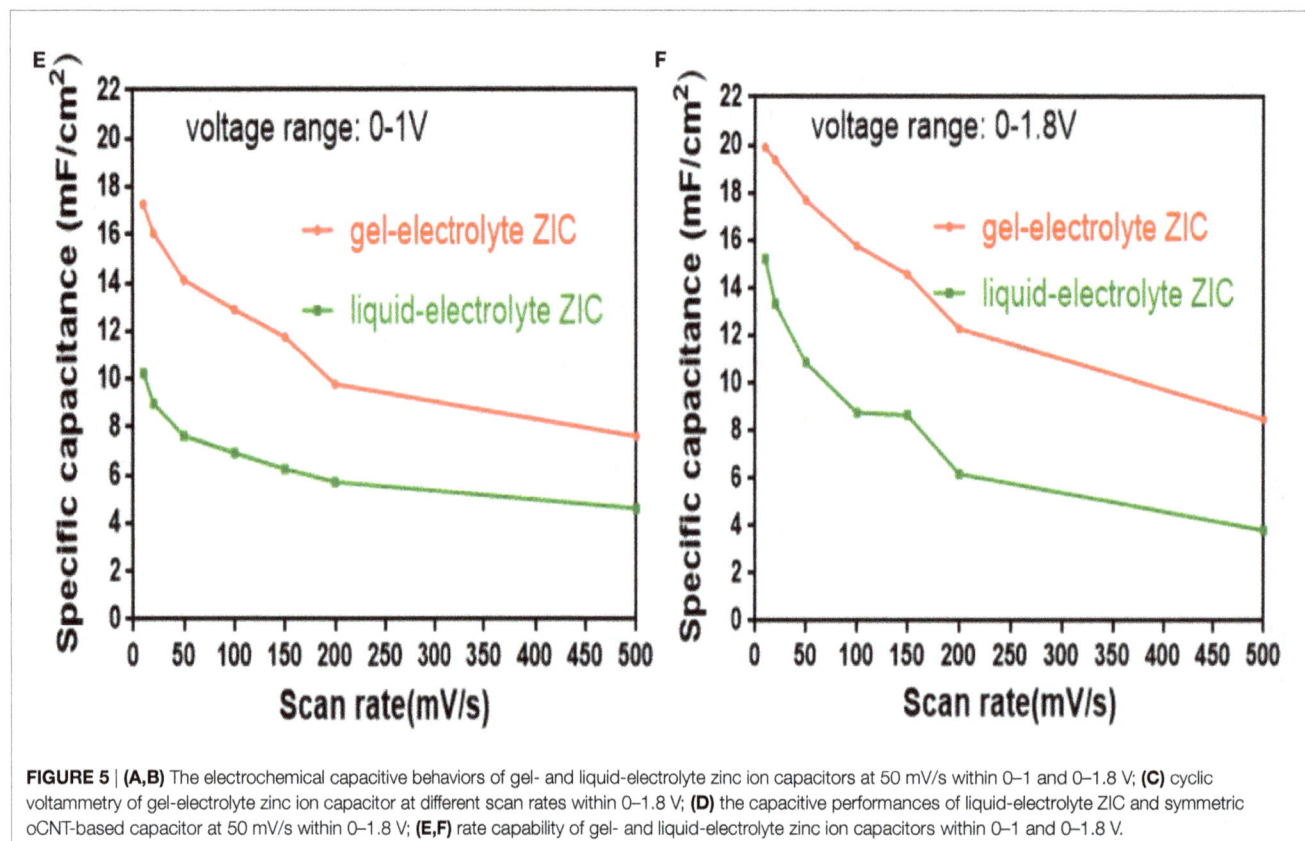

FIGURE 5 | (A,B) The electrochemical capacitive behaviors of gel- and liquid-electrolyte zinc ion capacitors at 50 mV/s within 0–1 and 0–1.8 V; (C) cyclic voltammetry of gel-electrolyte zinc ion capacitor at different scan rates within 0–1.8 V; (D) the capacitive performances of liquid-electrolyte ZIC and symmetric oCNT-based capacitor at 50 mV/s within 0–1.8 V; (E,F) rate capability of gel- and liquid-electrolyte zinc ion capacitors within 0–1 and 0–1.8 V.

an increase of the scan rate, the current is proportional to the scan rate without serious distortion in the shape of the CV curves even at the high scan rates, which demonstrates a good charge propagation behavior and ion response of the sample (Zhang et al., 2010). Furthermore, a symmetric oCNT-based capacitor using the 1M $ZnSO_4$ aqueous electrolyte was constructed. The capacitive behaviors of the symmetric capacitor as well as the liquid-electrolyte ZIC at a scan rate of 50 mV s^{-1} within 0–1.8 V are presented in **Figure 5D**. It is observed that the capacitive current of the ZIC is nearly two times higher than that of the symmetric capacitor. The symmetric capacitor exhibits a CV curve approximated to the ideal situation of electrical double layer capacitor, that is, there is no visible peak from a redox current over the voltage region (Vix-Guterl et al., 2005). This result indicates that the battery–supercapacitor combination configuration of the ZIC can accumulate charge through Faradaic electrochemical process so as to enhance the capacitance of the capacitor.

Additionally, rate capability is an important feature of supercapacitors. In **Figures 5E,F**, the specific capacitance variation with the scan rate in the range of 10–500 mV s^{-1} is illustrated (the calculation of the specific capacitance is described in Supplementary Material). It is obvious that the specific capacitance of the gel-electrolyte ZIC outperforms that with the liquid electrolyte at all scan rates, e.g., the gel-electrolyte ZIC exhibited a higher capacitance value of around 20 mF cm^{-2} than the liquid-electrolyte one (15 mF cm^{-2}) at a scan rate of 10 mV s^{-1} (**Figure 5F**). Also, when the scan rate was increased by 50 times,

the gel-electrolyte ZIC presented a capacitance retention ratio of 42.6%, which is higher than that of liquid-electrolyte one (the retention ratio of 25.0%). The area capacitance corresponds to a gravimetric capacitance of 53 F g^{-1}. The ZIC showed comparable electrochemical performance with several different types of metal ion capacitors. For example, a sodium ion capacitor based on transition metal sodium phosphate and activated carbon electrodes exhibited 45 F g^{-1} at 0.5 A g^{-1} (Minakshi et al., 2013), and the aqueous capacitors according to different fabrications of $CoMoO_4$ cathodes showed a wide capacitance range of 24–135 F g^{-1} in the NaOH electrolyte (Ramkumar and Minakshi, 2015; Ramkumar and Minakshi Sundaram, 2016).

Electrochemical impedance spectroscopy is an important technique to characterize the electrochemical frequency behavior of a device, and the obtained data are generally plotted in a Nyquist diagram that represents the imaginary part of the impedance vs. the real part (Taberna et al., 2006). **Figure 6** exhibits the Nyquist plot for two ZIC cells assembled with liquid- and gel-electrolyte. It is worth noting that at the highest frequency, the value at the real axis, called the equivalent series resistance (ESR), represents the sum of the resistance of the electrolyte and the intrinsic resistance of the electrode material. As both ZIC cells have the same electrode material, the ESR, in this case, reflects the difference of the electrolyte resistance. By comparison, the gel electrolyte exhibited a lower ESR value of 3.93 Ω than the liquid one (8.51 Ω), which should be one reason for the better capacitive behavior of the gel-electrolyte

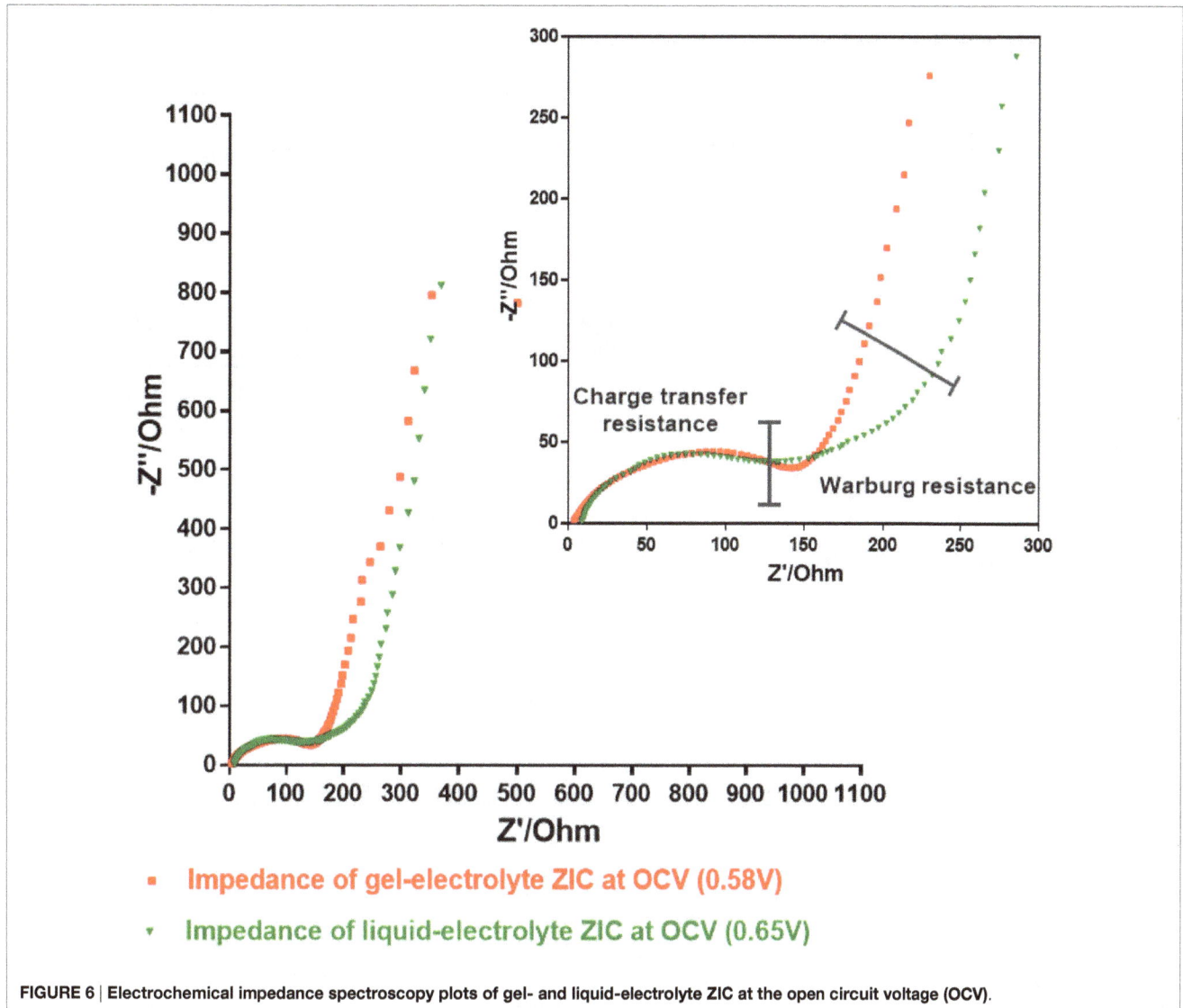

FIGURE 6 | Electrochemical impedance spectroscopy plots of gel- and liquid-electrolyte ZIC at the open circuit voltage (OCV).

ZIC. Furthermore, it is observed that for both ZICs, the imaginary parts of impedance at the low frequency region are nearly perpendicular to the real parts, which indicate that both electrolytes have satisfactory electrochemical capacitive behavior (Vix-Guterl et al., 2005). The charge transfer resistance of the gel-electrolyte ZIC is similar to that of the liquid-electrolyte one, which is deduced from the span of the single semi-circle along the real axis from high to medium frequency as shown in the close-up view of **Figure 6**. However, an apparent difference of both ZIC cells is found at medium frequency in the close-up, where the 45° sloped region of the Nyquist plots, the so-called Warburg resistance, can be seen. The Warburg resistant results from the frequency dependence of ion transport in the electrolyte (Wu et al., 2015). Since the low-frequency capacitive behavior of the liquid-electrolyte ZIC is largely shifted along the x-axis toward more resistive value (Xu et al., 2012), the smaller Warburg region of the gel-electrolyte ZIC indicates a lower ion diffusion resistance and less obstruction of the ion

movement (Wu et al., 2015). Thus, a better charge propagation and ion response of the gel-based electrolyte are demonstrated, which can explain the better electrochemical performance of the gel-electrolyte ZIC observed in the previous results.

The cycling stability is also a crucial concern for supercapacitor. To investigate the electrochemical stability of the ZIC, the charge–discharge cycling was performed at the alternant scan rates of 5 and 500 mV s^{-1} (**Figure 7**). After 5000 cycles at 500 mV s^{-1}, both liquid- and gel-electrolyte ZICs show the benign cycling stabilities, i.e., their current responses decrease slowly with the cycling numbers. This demonstrates that the pseudocapacitance effect introduced by the electrochemical reaction/adsorption and desorption between the Zn^{2+} ions and the oxygen functional groups of the oCNT cathode is stable with cycling. However, by comparing the current responses of both ZICs, the gel-electrolyte ZIC presents higher current values than the liquid-electrolyte one, which further suggests the better electrochemical capacitance behavior of the gel-electrolyte

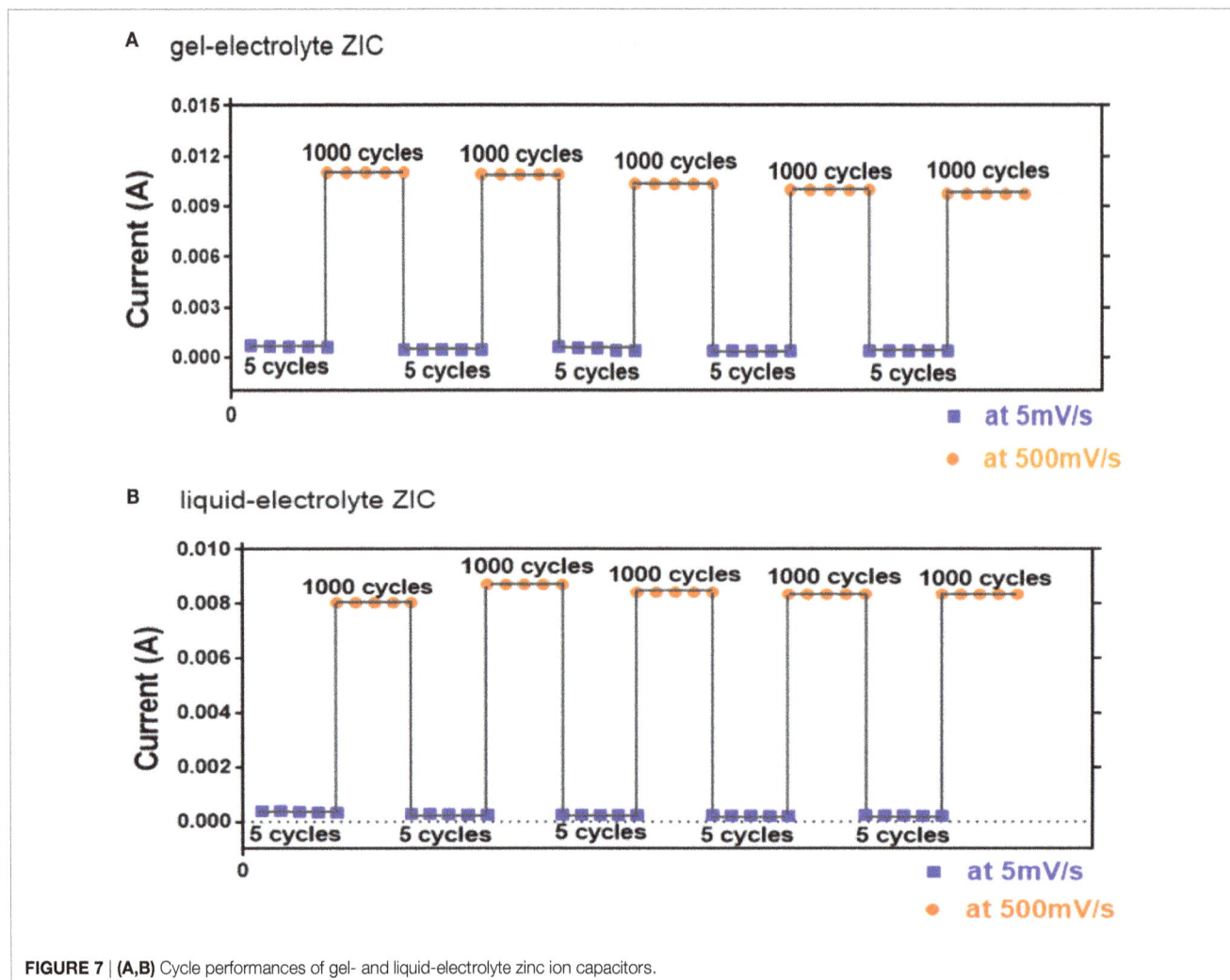

FIGURE 7 | (A,B) Cycle performances of gel- and liquid-electrolyte zinc ion capacitors.

ZIC. The additional cycle life test of the gel-electrolyte ZIC based on the galvanostatic charge/discharge at 2 mA cm⁻² is displayed in Figure S5 in Supplementary Material. The inherent change of the gel-electrolyte ZIC before and after cycling was further characterized using the EIS technique with a frequency range of 0.01 Hz–10 kHz (Figure S6 in Supplementary Material). The much smaller span of the single semi-circle at the high frequency for the ZIC after cycling indicates a smaller charge transfer resistance. But at the low-frequency region, the Nyquist plot curve of the ZIC after cycling inclines to the real axis, implying a higher ion diffusion resistance. The reduction of the ionic conductivity upon cycling might be responsible for the deterioration of the gel-electrolyte ZIC, which further explains the decrease of the current responses with the cycling numbers. Moreover, the morphology and structure of the zinc anode at different conditions were observed by SEM as shown in Figure S7 in Supplementary Material. Before cycling, the original zinc foil was flat, but its surface became rougher and more compact after cycling. The compact and uniform ripple-like zinc morphology without the formation of dendrites

contributes to the favorable cycling stability of the ZIC, because the formation of dendritic zinc is accompanied with hydrogen evolution that is negative for cyclic stability (Lengsfeld and Shoureshi, 2014).

CONCLUSION

A ZIC was assembled by using a zinc anode, an oCNT cathode, and a ZnSO₄-based electrolyte. The low-cost zinc metal and the non-toxic, non-corrosive ZnSO₄ solution enable the ZIC to be a cost-effective and environmental-friendly device. Due to the electrochemical adsorption/desorption of Zn²⁺ ions on the oxygen-containing groups of the oCNT cathode, the pseudocapacitive behavior observed in this ZIC delivered a specific capacitance of 20 mF cm⁻² at a scan rate of 10 mV s⁻¹. Additionally, this ZIC was stable during a long cycling test (up to 5000 cycles). This work demonstrated a proof-of-concept ZIC and showed that the oxygen functionalities of the oCNTs contributed to the pseudocapacitance. Further study is required to improve the performance of the ZIC.

AUTHOR CONTRIBUTIONS

D-WW supervised the project. YT conducted the experiment. D-WW and YT analyzed data. D-WW, YT, and RA wrote the manuscript.

ACKNOWLEDGMENTS

We acknowledge the support from Faculty of Engineering, The University of New South Wales, and the Australian Research Council (DP 160103244). The authors acknowledge the facilities and the scientific and technical assistance from Mark Wainwright Analytical Centre, The University of New South Wales.

REFERENCES

Agrawal, S. L., and Awadhia, A. (2004). DSC and conductivity studies on PVA based proton conducting gel electrolytes. *Bull. Mater. Sci.* 27, 523–527. doi:10.1007/BF02707280

Cao, W. J., Greenleaf, M., Li, Y. X., Adams, D., Hagen, M., Doung, T., et al. (2015). The effect of lithium loadings on anode to the voltage drop during charge and discharge of Li-ion capacitors. *J. Power Sources* 280, 600–605. doi:10.1016/j.jpowsour.2015.01.102

Cao, W. J., Shih, J., Zheng, J. P., and Doung, T. (2014). Development and characterization of Li-ion capacitor pouch cells. *J. Power Sources* 257, 388–393. doi:10.1016/j.jpowsour.2014.01.087

Chen, Z., Augustyn, V., Jia, X., Xiao, Q., Dunn, B., and Lu, Y. (2012). High-performance sodium-ion pseudocapacitors based on hierarchically porous nanowire composites. *ACS Nano* 6, 4319–4327. doi:10.1021/nn300920e

Choi, Y. C., Lee, S. M., and Chung, D. C. (2001). Supercapacitors using single-walled carbon. *Adv. Mater.* 13, 497–500. doi:10.1002/1521-4095(200104)13:7<497::AID-ADMA497>3.3.CO;2-8

Dubal, D. P., Ayyad, O., Ruiz, V., and Gómez-Romero, P. (2015). Hybrid energy storage: the merging of battery and supercapacitor chemistries. *Chem. Soc. Rev.* 44, 1777–1790. doi:10.1039/C4CS00266K

Ferrari, A. C. (2007). Raman spectroscopy of graphene and graphite: disorder, electron–phonon coupling, doping and nonadiabatic effects. *Solid State Commun.* 143, 47–57. doi:10.1016/j.ssc.2007.03.052

Frackowiak, E. (2007). Carbon materials for supercapacitor application. *Phys. Chem. Chem. Phys.* 9, 1774–1785. doi:10.1039/b618139m

Frackowiak, E., and Béguin, F. (2002). Electrochemical storage of energy in carbon nanotubes and nanostructured carbons. *Carbon N. Y.* 40, 1775–1787. doi:10.1016/S0008-6223(02)00045-3

Ghaemi, M., Amrollahi, R., Ataherian, F., and Kassaee, M. Z. (2003). New advances on bipolar rechargeable alkaline manganese dioxide-zinc batteries. *J. Power Sources* 117, 233–241. doi:10.1016/S0378-7753(03)00161-7

Ghanbari, K., Mousavi, M. F., Shamsipur, M., and Karami, H. (2007). Synthesis of polyaniline/graphite composite as a cathode of Zn-polyaniline rechargeable battery. *J. Power Sources* 170, 513–519. doi:10.1016/j.jpowsour.2007.02.090

Jia, Z., Wang, B., and Wang, Y. (2015). Copper hexacyanoferrate with a well-defined open framework as a positive electrode for aqueous zinc ion batteries. *Mater. Chem. Phys.* 149, 601–606. doi:10.1016/j.matchemphys.2014.11.014

Kosynkin, D. V., Higginbotham, A. L., Sinitskii, A., Lomeda, J. R., Dimiev, A., Price, B. K., et al. (2009). Longitudinal unzipping of carbon nanotubes to form graphene nanoribbons. *Nature* 458, 872–876. doi:10.1038/nature07872

Kurra, N., Wang, R., and Alshareef, H. N. (2015). All conducting polymer electrodes for asymmetric solid-state supercapacitors. *J. Mater. Chem. A* 3, 7368–7374. doi:10.1039/c5ta00829h

Lee, J.-S., Tai Kim, S., Cao, R., Choi, N., Liu, M., Lee, K. T., et al. (2011). Metal-air batteries with high energy density: Li-air versus Zn-air. *Adv. Energy Mater.* 1, 34–50. doi:10.1002/aenm.201000010

Lengsfeld, C. S., and Shoureshi, R. A. (2014). *Electrochemical Cell, Related Material, Process for Production, and Use Thereof.* US patent 0211370 A1.

Lokhande, C. D., Dubal, D. P., and Joo, O. S. (2011). Metal oxide thin film based supercapacitors. *Curr. Appl. Phys.* 11, 255–270. doi:10.1016/j.cap.2010.12.001

Minakshi, M., Meyrick, D., and Appadoo, D. (2013). Maricite (NaMn 1/3 Ni 1/3 Co 1/3 PO 4)/activated carbon: hybrid capacitor. *Energy Fuels* 27, 3516–3522. doi:10.1021/ef400333s

Pandolfo, A. G., and Hollenkamp, A. F. (2006). Carbon properties and their role in supercapacitors. *J. Power Sources* 157, 11–27. doi:10.1016/j.jpowsour.2006.02.065

Park, S., Lee, K.-S., Bozoklu, G., Cai, W., Nguyen, S. T., and Ruoff, R. S. (2008). Graphene oxide papers modified by divalent ions-enhancing mechanical properties via chemical cross-linking. *ACS Nano* 2, 572–578. doi:10.1021/nn700349a

Pham, V. H., Cuong, T. V., Hur, S. H., Oh, E., Kim, E. J., Shin, E. W., et al. (2011). Chemical functionalization of graphene sheets by solvothermal reduction of a graphene oxide suspension in N-methyl-2-pyrrolidone. *J. Mater. Chem.* 21, 3371. doi:10.1039/c0jm02790a

Rahmanifar, M. S., Mousavi, M. F., and Shamsipur, M. (2002). Effect of self-doped polyaniline on performance of secondary Zn-polyaniline battery. *J. Power Sources* 110, 229–232. doi:10.1016/S0378-7753(02)00260-4

Ramkumar, R., and Minakshi, M. (2015). Fabrication of ultrathin CoMoO4 nanosheets modified with chitosan and their improved performance in energy storage device. *Dalt. Trans.* 44, 6158–6168. doi:10.1039/C5DT00622H

Ramkumar, R., and Minakshi Sundaram, M. (2016). A biopolymer gel-decorated cobalt molybdate nanowafer: effective graft polymer cross-linked with an organic acid for better energy storage. *New J. Chem.* 40, 2863–2877. doi:10.1039/C5NJ02799C

Simon, P., and Gogotsi, Y. (2008). Materials for electrochemical capacitors. *Nat. Mater.* 7, 845–854. doi:10.1038/nmat2297

Snook, G. A., Kao, P., and Best, A. S. (2011). Conducting-polymer-based supercapacitor devices and electrodes. *J. Power Sources* 196, 1–12. doi:10.1016/j.jpowsour.2010.06.084

Stankovich, S., Dikin, D. A., Piner, R. D., Kohlhaas, K. A., Kleinhammes, A., Jia, Y., et al. (2007). Synthesis of graphene-based nanosheets via chemical reduction of exfoliated graphite oxide. *Carbon N. Y.* 45, 1558–1565. doi:10.1016/j.carbon.2007.02.034

Taberna, P. L., Portet, C., and Simon, P. (2006). Electrode surface treatment and electrochemical impedance spectroscopy study on carbon/carbon supercapacitors. *Appl. Phys. A* 82, 639–646. doi:10.1007/s00339-005-3404-0

Trócoli, R., and La Mantia, F. (2015). An aqueous zinc-ion battery based on copper hexacyanoferrate. *ChemSusChem* 8, 481–485. doi:10.1002/cssc.201403143

Vatsalarani, J., Geetha, S., Trivedi, D. C., and Warrier, P. C. (2006). Stabilization of zinc electrodes with a conducting polymer. *J. Power Sources* 158, 1484–1489. doi:10.1016/j.jpowsour.2005.10.094

Vix-Guterl, C., Frackowiak, E., Jurewicz, K., Friebe, M., Parmentier, J., and Béguin, F. (2005). Electrochemical energy storage in ordered porous carbon materials. *Carbon N. Y.* 43, 1293–1302. doi:10.1016/j.carbon.2004.12.028

Wang, D. W., Wu, K. H., Gentle, I. R., and Lu, G. Q. (2012). Anodic chlorine/nitrogen co-doping of reduced graphene oxide films at room temperature. *Carbon N. Y.* 50, 3333–3341. doi:10.1016/j.carbon.2011.12.054

Wu, K.-H., Wang, D.-W., and Gentle, I. R. (2015). Revisiting oxygen reduction reaction on oxidized and unzipped carbon nanotubes. *Carbon N. Y.* 81, 295–304. doi:10.1016/j.carbon.2014.09.060

Xu, C., Li, B., Du, H., and Kang, F. (2012). Energetic zinc ion chemistry: the rechargeable zinc ion battery. *Angew. Chem. Int. Ed. Engl.* 51, 933–935. doi:10.1002/anie.201106307

Yan, J., Wang, Q., Wei, T., and Fan, Z. (2014). Recent advances in design and fabrication of electrochemical supercapacitors with high energy densities. *Adv. Energy Mater.* 4, 1–43. doi:10.1002/aenm.201300816

Yang, D., Velamakanni, A., Bozoklu, G., Park, S., Stoller, M., Piner, R. D., et al. (2009). Chemical analysis of graphene oxide films after heat and chemical treatments by X-ray photoelectron and Micro-Raman spectroscopy. *Carbon N. Y.* 47, 145–152. doi:10.1016/j.carbon.2008.09.045

Yu, H., Wu, J., Fan, L., Xu, K., Zhong, X., Lin, Y., et al. (2011). Improvement of the performance for quasi-solid-state supercapacitor by using PVA-KOH-KI polymer gel electrolyte. *Electrochim. Acta* 56, 6881–6886. doi:10.1016/j.electacta.2011.06.039

Zhai, Y., Dou, Y., Zhao, D., Fulvio, P. F., Mayes, R. T., and Dai, S. (2011). Carbon materials for chemical capacitive energy storage. *Adv. Mater.* 23, 4828–4850. doi:10.1002/adma.201100984

Zhang, L. L., Zhao, S., Tian, X. N., and Zhao, X. S. (2010). Layered graphene oxide nanostructures with sandwiched conducting polymers as supercapacitor electrodes. *Langmuir* 26, 17624–17628. doi:10.1021/la103413s

Zhang, L. L., Zhou, R., and Zhao, X. S. (2009). Carbon-based materials as supercapacitor electrodes. *J. Mater. Chem.* 38, 2520–2531. doi:10.1039/c000417k

Conflict of Interest Statement: The authors declare that the research was conducted in the absence of any commercial or financial relationships that could be construed as a potential conflict of interest.

PERMISSIONS

LIST OF CONTRIBUTORS

Frank C. Krysiak and Hannes Weigt
Department of Business and Economics, University of Basel, Basel, Switzerland

Ryoji Inada, Satoshi Yasuda, Masaru Tojo, Keiji Tsuritani, Tomohiro Tojo and Yoji Sakurai
Department of Electrical and Electronic Engineering, Toyohashi University of Technology, Toyohashi, Japan

Ilya Lisenker and Conrad R. Stoldt
Department of Mechanical Engineering, University of Colorado Boulder, Boulder, CO, USA

Yuichi Aihara, Seitaro Ito, Ryo Omoda, Takanobu Yamada, Satoshi Fujiki and Taku Watanabe
Samsung R&D Institute Japan, Minoo-shi, Japan,

Youngsin Park and Seokgwang Doo
Samsung Advanced Institute of Technology, Samsung Electronics Co., Ltd, Suwon-si, South Korea

Cheng Ma and Miaofang Chi
Oak Ridge National Laboratory, Center for Nanophase Materials Sciences, Oak Ridge, TN, USA

Corinne Moser
Institute of Sustainable Development, School of Engineering, Zurich University of Applied Sciences, Winterthur, Switzerland
Natural and Social Science Interface, Institute for Environmental Decisions, Department of Environmental Systems Science, ETH Zürich, Zürich, Switzerland

Andreas Rösch
Natural and Social Science Interface, Institute for Environmental Decisions, Department of Environmental Systems Science, ETH Zürich, Zürich, Switzerland

Michael Stauffacher
Natural and Social Science Interface, Institute for Environmental Decisions, Department of Environmental Systems Science, ETH Zürich, Zürich, Switzerland
Transdisciplinarity Laboratory, Department of Environmental Systems Science, ETH Zürich, Zürich, Switzerland

Rakesh R. Narala, Sourabh Garg, Kalpesh K. Sharma, Skye R. Thomas-Hall, Miklos Deme, Yan Li and Peer M. Schenk
Algae Biotechnology Laboratory, School of Agriculture and Food Sciences, The University of Queensland, Brisbane, QLD, Australia

Atsushi Sakuda, Tomonari Takeuchi, Masahiro Shikano, Hikari Sakaebe and Hironori Kobayashi
Department of Energy and Environment, Research Institute for Electrochemical Energy, National Institute of Advanced Industrial Science and Technology (AIST), Ikeda, Japan

Shingo Ohta, Yuki Kihira and Takahiko Asaoka
Battery & Cell Division, Toyota Central R&D Labs. Inc., Nagakute, Japan

Balachandran Radhakrishnan and Shyue Ping Ong
Department of NanoEngineering, University of California San Diego, La Jolla, CA, USA

Makoto Kaneko, Hajime Iwata, Hiroyuki Shiotsu, Shota Masaki, Yuji Kawamoto, Shinya Yamasaki, Yuki Nakamatsu, Junpei Imoto, Genki Furuki, Asumi Ochiai and Satoshi Utsunomiya
Department of Chemistry, Kyushu University, Fukuoka, Japan

Kenji Nanba
Department of Environmental Management, Faculty of Symbiotic System Science, Fukushima University, Fukushima, Japan

Toshihiko Ohnuki
Advanced Science Research Center Japan Atomic Energy Agency, Tokai, Japan

Rodney C. Ewing
Department of Geological Sciences, Center for International Security and Cooperation, Stanford University, Stanford, CA, USA

Emanuele Facchinetti and Sabine Sulzer
Lucerne Competence Center for Energy Research, Lucerne University of Applied Science and Arts, Horw, Switzerland

Cherrelle Eid
Faculty of Technology, Policy and Management, Delft University of Technology, Delft, Netherlands,

Andrew Bollinger
Urban Energy Systems Laboratory, EMPA, Dübendorf, Switzerland

Yuzheng Lu, Jun Wang and Yaoming Zhang
Jiangsu Provincial Key Laboratory of Solar Energy Science and Technology, School of Energy and Environment, Southeast University, Nanjing, China

Bin Zhu and Baoyuan Wang
Faculty of Physics and Electronic Technology, Hubei Collaborative Innovation Center for Advanced Organic Materials, Hubei University, Wuhan, China
Department of Energy Technology, Royal Institute of Technology KTH, Stockholm, Sweden

Yixiao Cai
Ångström Laboratory, Department of Engineering Sciences, Uppsala University, Uppsala, Sweden

Jung-Sik Kim
Department of Aeronautical and Automotive Engineering, Loughborough University, Loughborough, UK

Junjiao Li
Nanjing Yunna Nano Technology Co., Ltd., Nanjing, China

Thomas Schröder and Wilhelm Kuckshinrichs
Institute of Energy and Climate Research – Systems Analysis and Technology Evaluation (IEK-STE), Forschungszentrum Jülich GmbH, Jülich, Germany

Akitoshi Hayashi and Masahiro Tatsumisago
Department of Applied Chemistry, Graduate School of Engineering, Osaka Prefecture University, Sakai, Osaka, Japan

Atsushi Sakuda
Department of Applied Chemistry, Graduate School of Engineering, Osaka Prefecture University, Sakai, Osaka, Japan
Department of Energy and Environment, Research Institute of Electrochemical Energy, National Institute of Advanced Industrial Science and Technology (AIST), Ikeda, Osaka, Japan

Emanuele Facchinetti and Sabine Sulzer
Lucerne Competence Center for Energy Research, Lucerne University of Applied Science and Arts, Horw, Switzerland

I. A. Grant Wilson
Environmental and Energy Engineering Group, Department of Chemical and Biological Engineering, The University of Sheffield, Sheffield, UK

Herek L. Clack
Department of Civil and Environmental Engineering, University of Michigan, Ann Arbor, MI, USA

Nicola Gasparini and Tayebeh Ameri
Institute of Materials for Electronics and Energy Technology (I-MEET), Friedrich-Alexander-University Erlangen-Nuremberg, Erlangen, Germany

Amaranda García-Rodríguez, Şebnem Bayseç, Alex Palma-Cando, Sybille Allard and Ulrich Scherf
Macromolecular Chemistry Group (buwmakro), Institute for Polymer Technology, BergischeUniversität Wuppertal, Wuppertal, Germany

Mario Prosa
Istituto per lo Studio dei Materiali Nanostrutturati (ISMN), Consiglio Nazionale delle Ricerche (CNR), Bologna, Italy

Athanasios Katsouras and Apostolos Avgeropoulos
Department of Materials Science Engineering, University of Ioannina, Ioannina, Greece

Georgia Pagona and Vasilis G. Gregoriou
Advent Technologies SA, Patras Science Park, Patra, Greece
National Hellenic Research Foundation (NHRF), Athens, Greece

Christos L. Chochos
Department of Materials Science Engineering, University of Ioannina, Ioannina, Greece, Advent Technologies SA, Patras Science Park, Patra, Greece

Christoph J. Brabec
Institute of Materials for Electronics and Energy Technology (I-MEET), Friedrich-Alexander-University Erlangen-Nuremberg, Erlangen, Germany
Bavarian Center for Applied Energy Research (ZAE Bayern), Erlangen, Germany

Wilhelm Kuckshinrichs, Thomas Ketelaer and Jan Christian Koj
Forschungszentrum Juelich, Institute for Energy and Climate Research – Systems Analysis and Technology Evaluation (IEK-STE), Juelich, Germany

Yuheng Tian, Rose Amal and Da-Wei Wang
School of Chemical Engineering, The University of New South Wales (UNSW), Sydney, NSW, Australia

Index